SCOTT W. STARRATT

Treatise on
Invertebrate Paleontology

Prepared under Sponsorship of
The Geological Society of America, Inc.

The Paleontological Society   The Society of Economic Paleontologists and Mineralogists
The Palaeontographical Society   The Palaeontological Association

Directed and Edited by
Raymond C. Moore

Part H
BRACHIOPODA

By Alwyn Williams, A. J. Rowell, H. M. Muir-Wood, Charles W. Pitrat, Herta Schmidt, F. G. Stehli, D. V. Ager, A. D. Wright, G. F. Elliott, T. W. Amsden, M. J. S. Rudwick, Kotora Hatai, Gertruda Biernat, D. J. McLaren, A. J. Boucot, J. G. Johnson, R. D. Staton, R. E. Grant, and H. M. Jope

VOLUME 2

The Geological Society of America, Inc.
and
The University of Kansas Press

1965

Second Printing 1980
Third Printing 1985*

Library of Congress Catalogue Card Number 53-12913
ISBN 0-8137-3008-2

* Distributed by the Geological Society of America, Inc., P.O. Box 9140, Boulder, Colorado 80301, from which current price lists of Parts in print may be obtained and to which all orders and related correspondence should be directed. Editorial office for the *Treatise*: Paleontological Institute, 121 Lindley Hall, The University of Kansas, Lawrence, Kansas 66045.

PART H

BRACHIOPODA

By Alwyn Williams, A. J. Rowell, H. M. Muir-Wood, Charles W. Pitrat, Herta Schmidt, F. G. Stehli, D. V. Ager, A. D. Wright, G. F. Elliott, T. W. Amsden, M. J. S. Rudwick, Kotora Hatai, Gertruda Biernat, D. J. McLaren, A. J. Boucot, J. G. Johnson, R. D. Staton, R. E. Grant, and H. M. Jope

VOLUME 2

SCOTT W. STARRATT

PENTAMERIDA

By Thomas W. Amsden and Gertruda Biernat

[Oklahoma Geological Survey and Polska Akademia Nauk, Warszawa, Polska]

Order PENTAMERIDA
Schuchert & Cooper, 1931

[*nom. transl.* Moore in Moore, Lalicker, & Fischer, 1952, p. 220 (*ex* suborder Pentameroidea Schuchert & Cooper, 1931, p. 247)] [Diagnosis prepared by T. W. Amsden]

Generally biconvex shells with pedicle spondylium; delthyrium may be unmodified or at least partly closed by deltidium. Brachial valve with brachial processes, commonly braced at their posterior end by supporting plates and terminating blindly except in Enantiosphenidae (Pentameracea) where they end in a loop. Shell impunctate. *M.Cam.-U.Dev.*

SYNTROPHIIDINA

[Materials for this suborder prepared by Gertruda Biernat]

The Syntrophiidina are a very interesting but uncommon group of brachiopods. They are difficult to find and not well known. They include forms placed for many years in the genus *Triplesia*. In 1932 Schuchert & Cooper described them as a superfamily of the Pentameroidea, suggesting relationships with the Orthacea and Pentameracea. In 1936 they were separated by Ulrich & Cooper as the suborder Syntrophioidea. As presently known, they start in the Middle

Cambrian with *Cambrotrophia* and range into the Devonian. The last-known representative is *Anastrophia* from the Lower Devonian.

MORPHOLOGY

The syntrophiid shell is subcircular, elliptical to subelliptical in outline, widest at the hinge or near mid-valve. In profile it is invariably biconvex, with convexity of the valves moderately to strongly unequal, the brachial valve usually being more convex. Some shells are nearly globular (e.g., *Camerella, Porambonites*). Both umbones are incurved, the brachial commonly more so. In a few genera (e.g., *Parastrophina, Anastrophia*) the brachial umbo is strongly swollen and incurved, covering the delthyrium.

The **fold** and **sulcus** are significant features, present in all specimens but developed to different degrees from very distinct to obscure. These variations appear in representatives of all families. Usually the sulcus is wide and shallow, the fold wide and low, both originating generally a little behind the mid-length. Toward the anterior margin they widen rapidly and occupy one-third to slightly more than half of the shell width. In some forms (e.g., *Cambrotrophia*) the sulcus is narrow and deep, with rounded bottom, the fold being prominent and in some forms angular (e.g., *Syntrophina*). The sulcus and fold may be weakly developed, nearly obscure, conspicuous at the front margin only or expressed as a central depression on the pedicle valve (e.g., *Tetralobula*).

The degree of folding of the **anterior commissure** is related to variations in appearance of the sulcus and fold. The anterior commissure is invariably uniplicate, very distinct in some Camerellidae but obscure in other forms (e.g., *Syntrophia, Idiostrophia*).

Interareas are usually present on both valves, commonly reduced and scarcely visible. In *Diaphelasma* they are very short and narrow. In some shells the pedicle interarea is weakly developed and that of the brachial valve is absent (e.g., *Neostrophia*).

Only a few types of ornament are distinguished in the syntrophiids. Some are smooth, with delicate concentric growth lines only; others exhibit concentric lamellae or they are finely costellate to distinctly costate. Ornamentation of the earlier forms usually is costellate but in some late forms (e.g., *Anastrophia*) costate. The shells of some genera (e.g., *Punctolira*) are ornamented by both radial costellae and distinct concentric lines, giving a fenestrate surface. In a few forms (e.g., *Syntrophopsis, Xenelasma, Syntrophia*) strong concentric lines or concentric lamellae only are present and the surface is lamellose or imbricate (e.g., *Imbricatia*). Smooth forms with delicate concentric growth lines are rare (e.g., *Cambrotrophia*). Some families (e.g., Camerellidae) have both smooth and costate representatives. Ornament is variable and has generic and specific value.

A **spondylium** is usually present. In a few genera only (e.g., *Xenelasma*) is it absent, having subparallel, discrete dental plates uniting with the valve floor. Some genera of the Huenellidae are provided with a transverse platform in the delthyrial cavity; this platform, usually called **pseudospondylium**, bears the muscles. The spondylium is common in syntrophiids and few types may be differentiated. In earlier forms

it is sessile for nearly all its length (e.g., *Syntrophopsis*); for its posterior half or third it is elevated anteriorly on a median septum which usually is low (e.g., *Tetralobula, Syntrophinella*). The **median septum** or ridge supporting the spondylium reaches nearly to mid-valve. Its length, thickness, and height are variable. In some shells the septum is so short that the spondylium appears to be completely free. This structure was observed in *Syntrophia torynifera* ULRICH & COOPER, 1938, and recognized as a free spondylium. In some genera (e.g., *Syntrophinella, Clarkella, Yangtzeella*) short, obscure, accessory septa in addition to the median septum are observed, and these are found on each side of the spondylium. A true **spondylium simplex** elevated on a prominent median septum is observed in some Clarkellidae. Later forms (e.g., Camerellidae, Parastrophinidae) have a **spondylium duplex.** The presence or absence of a spondylium has taxonomic significance for distinction of families and subfamilies and the degree of its differentiation has generic value.

The **cardinal process** in known syntrophiids is rare. When present it is a simple, low ridge with variable height (e.g., *Glyptotrophia, Alimbella*). In some earlier forms (e.g., *Cambrotrophia, Palaeostrophia, Plectotrophia*) the cardinal process is absent. Later genera (e.g., *Huenella, Huenellina, Tetralobula, Imbricatia*) have representatives lacking a cardinal process but provided with a rudimentary cardinal process or ridge. In the posterior part of the notothyrial cavity of *Diaphelasma* and *Syntrophinella* a horizontal plate for muscle attachment occurs. Some species of *Syntrophinella* have a well-developed orthoid cardinal process.

A **cruralium** is absent in the earlier syntrophiids and in later forms it is rare. Usually **brachiophores** are present, closely united to supporting plates. **Brachiophore plates** extend dorsomedially, attached to the valve floor or to the median ridge or septum. They may be very close, nearly united (e.g., Huenellidae) or separated widely (e.g., Clarkellidae). The last-mentioned feature, as also length and thickness of the brachiophore plates, is variable and seems to have some value for generic classification. In some forms the brachiophore plates are poorly developed. They are slender in *Palaeostrophia* but low and stout in *Mesonomia, Glyptotrophia,* and *Diaphelasma.* In some species (e.g., *Plectotrophia*) convergent brachiophore plates unite with a low septum to form a short, simple cruralium. This structure is well developed in *Syntrophia rotundata* WALCOTT and some Clarkellidae with a high, thin median septum. Species in which the cruralium is obscured may be observed. In some Camerellidae and Parastrophinidae the cruralium is well developed and may be of simplex or duplex type, with the median septum low.

STRATIGRAPHICAL DISTRIBUTION AND PHYLOGENY

The Syntrophiidina include 12 families with 40 genera. The families are small, each represented by one or a few genera. The largest families are the Clarkellidae and Camerellidae, each with seven genera. The Syntrophiidina range from Middle Cambrian to Lower Devonian, but each separate family is somewhat restricted in time.

The oldest known syntrophiid, *Cambrotrophia cambria,* confined to the Middle Cambrian, is the only representative of the Eostrophiidae. The specimens are poorly preserved. The interior, as far as known, is primitive, without spondylium and with rudimentary brachiophores. The shape and lateral profile of *C. cambria* resemble that of species of *Syntrophia* and suggest its relationship with syntrophioids.

The Huenellidae have a very short range, existing only through the Late Cambrian and Early Ordovician. They are divided into two subfamilies, chiefly on the basis of degree of development of the cardinal process. One of them, the Huenellinae, includes genera lacking a cardinal process and having a deep, usually sessile spondylium. The Mesonomiinae are characterized by a rudimentary cardinal process and recumbent brachiophore plates. The systematic position of *Mesonomia* is not yet clear. The presence of the brachial fold and pedicle sulcus in the umbonal part of the shell and its fascicostellate ornamentation may suggest some relationship with *Billing-*

sella; the recumbent character of the brachiophore plates relates this form to *Finkelnburgia.* The shape of *Mesonomia,* the presence of a low brachial fold, and the flattened pedicle sulcus beginning near the middle are very syntrophioid and suggest affinities with Syntrophiidina. Probably this genus is intermediate between Billingsellidae and Syntrophiidae.

The family Tetralobulidae has some connection with the Huenellidae. ULRICH & COOPER (1938) suggest the possibility of evolving the former from the latter by separation of the brachiophore plates. The Tetralobulidae include four genera which are confined to the Lower Ordovician. *Tetralobula* externally and in some internal features, mainly in the pattern of the pallial markings, is very close to *Syntrophia* and even *Syntrophinella. Punctolira,* in the development of the distant brachiophore plates meeting on the floor of the valve, seems to be closely related to *Tetralobula,* rather than to *Porambonites,* which it resembles in ornamentation.

The Alimbellidae, lately described from the Lower Ordovician of USSR, are represented by two genera, *Alimbella* and *Medesia.* They show some external and internal similarity to *Huenellina, Xenelasma,* and *Plectotrophia.*

The Clarkellidae, one of the largest families, existed from Late Cambrian to Early Ordovician. Usually its members have a spondylium simplex, four or more discrete and divergent brachial septa, and an apical horizontal plate in the brachial valve for attachment of the diductors. The Clarkellidae include both smooth and ornamented forms. Externally they are very much like *Syntrophia.* Internally, especially in the pattern of pallial markings, they resemble *Billingsella.*

All representatives of the Syntrophopsidae are characterized by external or internal similarity. *Hesperotrophia* seems to be close to *Tetralobula,* differing in a few internal details.

The Porambonitidae are restricted to Lower, Middle, and Upper Ordovician. Their systematic position is questionable. Recently they were referred by COOPER (1956) to the Syntrophiidina but earlier were believed by SCHUCHERT & COOPER

(1932) to be aberrant orthoids. This family includes three genera. SPJELDNAES (1956) doubted the validity of two of them—*Isorhynchus* and especially *Noetlingia*. SPJELDNAES recognized only one genus, *Porambonites* PANDER (1830) with "illegal" type-species *P. reticulata* PANDER, selected by TEICHERT (1930). DALL (1877) selected *P. intermedia* PANDER (1830) as the type-species, which was accepted by HALL & CLARKE (1894). The genus *Isorhynchus* KING (1850), characterized by subglobular outline and diverging dental and brachiophore plates, is proposed by him as a subgenus, with *Terebratulites aequirostris* SCHLOTHEIM as type-species. *Noetlingia* HALL & CLARKE (1893) is accepted provisionally as a subgenus, with type-species *Spirifer tscheffkini* DE VERNEUIL (1840). This species in its internal structure is very similar to *P. reticulata*. SPJELDNAES suggested the existence of a gradual transition between *Noetlingia* and *Porambonites; P. reticulata* should be one of the intermediate species. *Poramborthis* HAVLÍČEK (1949) from the Tremadoc of central Bohemia differs in many features from genera of the Porambonitidae and Syntrophiidina. The specimens are biconvex, finely costellate, without sulcus and fold, internally with convergent dental plates but never united, brachiophore plates short, divergent. The external appearance and internal character of *Poramborthis* suggest more similarity to the Orthidina than to Syntrophiidina.

The Syntrophiidae are a small group, differentiated into two subfamilies, the Xenelasminae, without a spondylium and with subparallel dental plates, and the Syntrophiinae, with a spondylium simplex. *Xenelasma* has a brachial structure with narrow and short septalium, features that are further developed in *Syntrophia*.

The Brevicameridae are represented by a single genus, *Brevicamera*. Their interior, especially of the brachial valve, is very interesting. In appearance of the spondylium they resemble *Camerella*, but the median septum is short and does not extend anteriorly to the spondylium. Brachiophores are short but somewhat bulbous, opposite to the socket. A small process

is present, serving as an accessory articulating nub. The sessile septalium is unlike the brachial structure of any known genus.

The Camerellidae earlier were included in the Pentameracea by SCHUCHERT & COOPER (1932), with suggestion that they could have been derived from the Syntrophiidae. Their relation with the Syntrophiidina was indicated by their inclusion doubtfully in this suborder by ULRICH & COOPER (1936). The Camerellidae conditionally include *Branconia* SCHUCHERT & COOPER (1932), described by GAGEL (1890) from erratic boulders of the Estonian Ordovician. The specimens of *Branconia,* illustrated by GAGEL (pl. 4, fig. 12-a-d) were very poorly preserved, which makes their classification difficult. GAGEL suggested some similarity of *Branconia* to rhynchonellids. The shape of specimens, with fold and sulcus bearing semiplications and showing a septum in each valve, resemble features of *Gypidula* or even *Sieberella*.

The Parastrophinidae range from Upper Ordovician to Lower Devonian and terminate the stock of Syntrophiidina. This family includes both biconvex and plano-convex forms that usually are provided with a spondylium duplex, septalium, and cardinal process. They show similarity to the Pentameridae. According to ULRICH & COOPER (1938), *Anastrophia* may separate syntrophioids from pentameroids.

Suborder SYNTROPHIIDINA
Ulrich & Cooper, 1936

[*nom. correct.* BIERNAT, herein (*pro* suborder Syntrophioidea ULRICH & COOPER, 1936, p. 627)]

Variable in size, usually unequally biconvex; brachial fold and pedicle sulcus present; exterior smooth, costate or costellate; delthyrium open; spondylium simplex, in some cases duplex, or lacking; brachiophores invariably united by supporting plates of variable length; septalium present in some families; cardinal process absent or rudimentary; brachial muscles not enclosed by lamellae. *M.Cam.-L.Dev.*

Superfamily PORAMBONITACEA
Davidson, 1853

[*nom. transl.* BIERNAT, herein (*ex* Porambonitidae DAVIDSON, 1853, p. 99)]

Characters of suborder. *M.Cam.-L.Dev.*

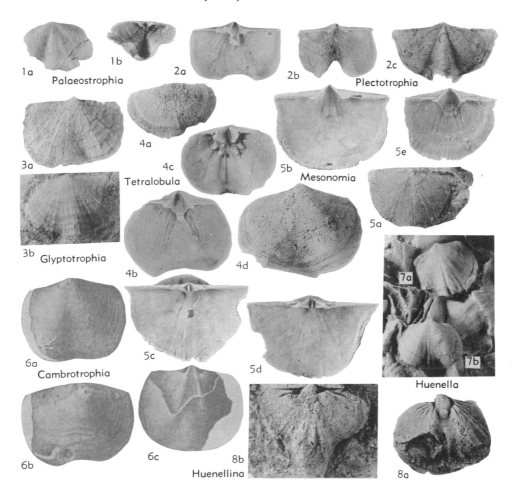

FIG. 398. Eostrophiidae *(6)*; Huenellidae (Huenellinae) *(1-2, 7-8)*; (Mesonomiinae) *(3, 5)*; Tetralobulidae *(4)* (p. H527-H530).

Family EOSTROPHIIDAE
Ulrich & Cooper, 1936

[Eostrophiidae ULRICH & COOPER, 1936, p. 627]

Spondylium absent; brachiophore plates poorly developed. *M.Cam.*

Cambrotrophia ULRICH & COOPER, 1937, p. 78 [*pro Eostrophia* ULRICH & COOPER, 1936, p. 627 (*non* DALL, 1890)] [*Syntrophia cambria* WALCOTT, 1908, p. 800; OD]. Small, wider than long; biconvex; smooth; weak fold and sulcus beginning on anterior half; spondylium absent; brachiophore plates rudimentary. *M.Cam.*, N.Am.-Austral.-USSR.——FIG. 398,6. *C. cambria* (WALCOTT), USA(Utah); *6a-c*, ped.v., brach.v., ped.v. views, *ca.* ×3 (848).

Family HUENELLIDAE
Schuchert & Cooper, 1931

[Huenellidae SCHUCHERT & COOPER, 1931, p. 247]

Medium size; exterior smooth, with concentric lines only or costellate to costate; pseudospondylium or sessile spondylium present; brachiophore plates usually weakly developed; cardinal process absent, rudimentary or simple, rodlike. *U.Cam.-L.Ord.*

Subfamily HUENELLINAE Schuchert & Cooper, 1931

[*nom. transl.* BIERNAT, herein (*ex* Huenellidae SCHUCHERT & COOPER, 1931, p. 247)] [=Palaeostrophiinae ULRICH & COOPER, 1936, p. 627]

Smooth, costellate or costate; with pseudo-

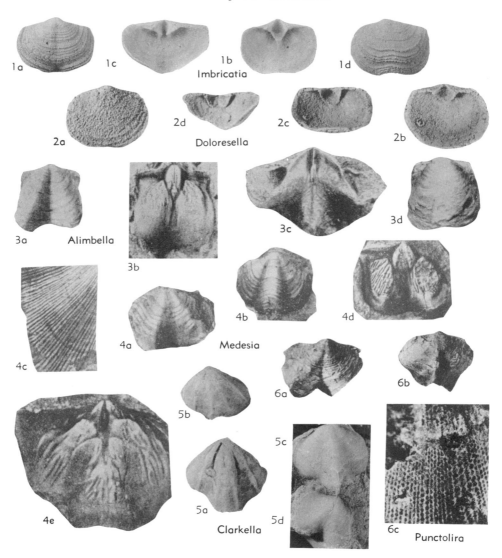

FIG. 399. Tetralobulidae *(1-2, 6)*; Alimbellidae *(3-4)*; Clarkellidae *(5)* (p. H530).

spondylium or sessile spondylium; brachiophore and supporting plates usually developed; cardinal process absent or scarcely visible. *U.Cam.-L.Ord.*

Huenella WALCOTT, 1908, p. 109 [*Syntrophia texana* WALCOTT, 1905, p. 294; OD]. Like *Plectotrophia* in outline but costate; fold marked by few costae; pseudospondylium or sessile spondylium present; cardinal process absent or scarcely visible. *U.Cam.*, N.Am.(USA.)-USSR.(N.Zem.). ——FIG. 398,7. *H. texana* (WALCOTT), USA (Wyo.); *7a,b*, brach.v. ext., ped.v. ext., ×2 (825).

Huenellina SCHUCHERT & COOPER, 1931, p. 247

[*Huenella triplicata* WALCOTT, 1924, p. 526; OD]. Differs from *Huenella* in having lateral septa at low angle to hinge; sulcus and fold strong, with 2 or 3 distinct costae on their medial part. *U.Cam.*, USSR(N.Zem.).——FIG. 398,8. *H. triplicata* (WALCOTT), USA(Mo.); *8a,b*, ped.v. and brach.v. int. molds, ×2 (729).

Palaeostrophia ULRICH & COOPER, 1936, p. 627 [*Syntrophia orthia* WALCOTT, 1905, p. 11; OD]. Like *Plectotrophia* but with subtriangular outline and smooth surface, with concentric lines only; short hinge; spondylium short, low, sessile; brachiophore plates delicate. *U.Cam.-L.Ord.*, China-N. Am.-USSR.——FIG. 398,1. *P. orthia* (WALCOTT).

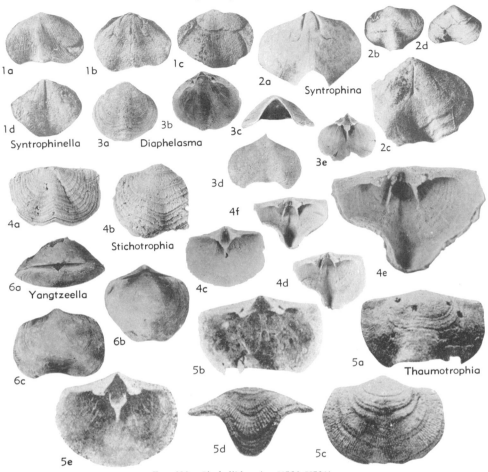

FIG. 400. Clarkellidae (p. *H530-H531*).

U.Cam.(L.Chaumit.), China; *1a,b*, brach.v. ext., int., ×2 (825).

Plectotrophia ULRICH & COOPER, 1936, p. 627 [**P. bridgei*; OD]. Moderate size; costellate; sulcus extending into distinct tongue; hinge wide; sessile spondylium short, supported by low, short septum; brachiophore plates convergent, in some species united in V-shaped structure; cardinal process absent. *U.Cam.*, N.Am.(USA-Can.).——FIG. 398, 2. **P. bridgei*, Wilberns F., USA(Tex.); *2a*, ped.v. int., *2b,c*, brach.v. int., ext.; all ×2 (825).

Subfamily MESONOMIINAE Ulrich & Cooper, 1936

[Mesnomiinae ULRICH & COOPER, 1936, p. 627]

Shells with crowded, distinct costellae; pseudospondylium present; brachiophores short, with recumbent supporting plates; cardinal process rudimentary or rodlike. *U. Cam.-L.Ord.*

Mesonomia ULRICH & COOPER, 1936, p. 627 [**Eoorthis iophon* WALCOTT, 1924, p. 507; OD]. Subquadrate, wider than long; finely costellate; broad fold and sulcus; brachiophores with short supporting plates; rudimentary or rodlike cardinal process usually present. *U.Cam.-L.Ord.*, N.Am. (Can.-USA)-USSR.——FIG. 398,5a. **M. iophon* (WALCOTT), L.Ord.(Mons F.), Can.; brach.v. ext., ×2 (825).——FIG. 398,5b-e. *M. magna* COOPER, U.Cam.(Ft. Sill F.), Okla.; *5b-e*, ped.v. int. (holotype), brach.v. int., brach.v. int., ped.v. int, ×2 (181).

Glyptotrophia ULRICH & COOPER, 1936, p. 627 [**G. imbricata*; OD]. Small, wider than long; hinge wide; internally like *Mesonomia* and *Huenella* but with external shape of *Syntrophina*; costellate, with very distinct concentric lamellae; cardinal process low and simple. *L.Ord.*, N.Am.-Eu. (USSR).——FIG. 398,3. **G. imbricata*, L.Ord. (Mons F.), Can.; *3a,b*, ped.v. ext., brach.v. ext., ×3 (825).

Family TETRALOBULIDAE
Ulrich & Cooper, 1936

[Tetralobulidae ULRICH & COOPER, 1936, p. 627]

Shells moderate in size, with fine costellae, in some genera with distinct concentric lamellae; brachiophore plates strong; spondylium supported by stout septum anteriorly; accessory septa thickened on inner sides. *L.Ord.*

Tetralobula ULRICH & COOPER, 1936, p. 628 [**T. delicatula*; OD]. Finely costellate; spondylium low; 2 lateral septa; brachiophores blunt; muscular platform quadrilobate, elevated; cardinal process absent or rudimentary. *L.Ord.*, N.Am.(USA)-Eu.(USSR).——FIG. 398,4. **T. delicatula*, Chepultepec F., USA(Va.); *4a*, brach.v. ext., ×2; *4b-d*, ped.v. int., brach.v. int., ped.v. ext., ×3 (825).

Doloresella SANDO, 1957, p. 122 [**D. concentrica*; OD]. Syntrophioid in outline, moderate size; surface with delicate, radial costellae and concentric lamellae like *Imbricatia* and *Stichotrophia*; with pseudospondylium; brachiophores short, supporting plates convergent. *L.Ord.(Beekmantown.)*, N.Am.——FIG. 399,2. **D. concentrica*, USA (Md.); *2a-c*, ped.v. ext., int., int., ×2; *2d*, brach.v. int., ×2 (706).

Imbricatia COOPER, 1952, p. 21 [**I. lamellata*; OD]. Like *Strichotrophia* in ornamentation; cardinal process absent; muscle scars well developed without strong callosities as in *Tetralobula*. *L.Ord.*, N.Am. (USA)-Eu. (USSR).——FIG. 399,1. **I. lamellata*, Cool Creek F., USA(Okla.); *1a,b*, ped. v. ext., int., ×2; *1c,d*, brach.v. int., ext., ×2 (181).

Punctolira ULRICH & COOPER, 1936, p. 628 [**P. punctolira*; OD]. Small, with syntrophioid outline; surface distinctly fenestrate; spondylium sessile posteriorly, supported anteriorly by low septum; brachiophore plates distant like those of *Tetralobula*. *L.Ord.*, N.Am.(USA).——FIG. 399, 6. **P. punctolira*, USA(Mo.); *6a*, ped.v. ext., ×1.5; *6b*, brach.v. ext., ×2; *6c*, brach.v. ext. ornament, ×10 (825).

Family ALIMBELLIDAE Andreeva, 1960

[Alimbellidae ANDREEVA, 1960, p. 293]

Shells moderate in size; smooth or finely costellate; sulcus and fold present or absent; pseudodeltidium rudimentary or lacking; cardinal process rodlike. *L.Ord.*

Alimbella ANDREEVA, 1960, p. 293 [**A. armata*; OD]. Smooth; like *Huenellina* and *Xenelasma* in outline and character of muscles. *L.Ord.(Tremadoc.)*, USSR(Urals).——FIG. 399,3. **A. armata*; *3a,b*, ped.v. ext., int. mold, ×1; *3c*, brach.v. int., ×3; *3d*, brach.v. ext., ×1 (37).

Medesia ANDREEVA, 1960, p. 295 [**M. uralica*; OD]. Subquadrate or trapezoidal in outline; sulcus and fold present; finely costellate; internally like *Alimbella* and *Plectotrophia*. *L.Ord.(Tremadoc.)*, USSR(Urals).——FIG. 399,4. **M. uralica*; *4a,b*, ped.v. ext., brach.v. ext., ×1; *4c*, ped.v. ext., ornament, ×4; *4d*, ped.v. int. mold, ×1.5; *4e*, brach.v. int., ×3 (37).

Family CLARKELLIDAE
Schuchert & Cooper, 1931

[Clarkellidae SCHUCHERT & COOPER, 1931, p. 247]

Shells small or moderate in size; smooth or costellate; with spondylium simplex; septa of brachial valve discrete, divergent. *U. Cam.-U.Ord.*

Clarkella WALCOTT, 1908, p. 110 [**Polytoechia? montanensis* WALCOTT, 1905, p. 295; OD]. Externally and internally like *Syntrophina* but brachial interior with 4 lateral septa. *L.Ord.*, E.Asia (Korea)-N. Am.-Eu. (USSR). —— FIG. 399,5. *C. mcgerriglei* ULRICH & COOPER, Hastings Creek F., Can.; *5a,b*, brach.v. ext. partly exfoliated, ×2; *5c,d*, brach.v. ext., ped.v. ext., ×2 (825).

Calliglypha CLOUD, 1948, p. 468 [**C. miseri*; OD]. Wide-hinged, syntrophioid; internally like *Diaphelasma*, externally like *Glyptotrophia*; with radial and concentric ornament giving cancellate and nodose appearance; brachiophore plates short, distantly divergent; collar-like callosity for diductors. *L.Ord.*, N.Am.-Eu.(USSR).——FIG. 401,1. **C. miseri*, USA; *1a,b*, ped.v. ext., int., ×2; *1c,d*, brach.v. ext., ped.v. int., ×3 (168).

Diaphelasma ULRICH & COOPER, 1936, p. 629 [**D. pennsylvanicum*; OD]. Surface with concentric lamellae only; brachial plates weak, widely separated, with callosities as in *Syntrophina*. *L.Ord.*, N.Am.-Eu.(USSR).——FIG. 400,3. **D. pennsylvanicum*, Longview F., USA(Pa.); *3a-c*, brach.v. ext., int., ant. (holotype); *3d,e*, ped.v. ext., int., all ×1.5 (825).

Stichotrophia COOPER, 1948, p. 473 [**S. lamellata*; OD]. Surface concentrically lamellose; lamellae strong, covered by distant costellae; internally like *Diaphelasma* and *Syntrophina*. *L.Ord.*, N.Am.-Eu. (USSR).——FIG. 400,4. **S. lamellata*, Longview Ls., USA(Va.); *4a*, brach.v. ext.; *4b,c*, ped.v. ext., int.; all ×2 (914a); *4d,f*, brach.v. int., brach.v. int., ×2; *4e*, brach.v. int., ×4 (913).

Syntrophina ULRICH in WELLER & ST. CLAIR, 1928, p. 74 [**Syntrophia campbelli* WALCOTT, 1912, p. 801; OD]. Externally and internally like *Clarkella* but with 2 long, slightly divergent septa in brachial valve. *L.Ord.*, E.Asia(China)-N.Am.-Eu. (USSR)-N.Afr.——FIG. 400,2. **S. campbelli* (WALCOTT), Chepultepec F., USA(Tenn.); *2a,c*, brach.v. and ped.v. int. molds, ×3; *2b,d*, brach.v. ext., ped.v. ext., ×2 (825).

Syntrophinella ULRICH & COOPER, 1934, p. 164 [**S. typica*; OD]. Like *Diaphelasma* but costel-

FIG. 401. Clarkellidae *(1)*; Syntrophopsidae *(2-3, 5)*; Lycophoriidae *(4)* (p. *H530-H532*).

late; spondylium low, sessile or simplex; small, lateral septa may be present; brachiophores short; supporting plates thin, strongly divergent. *L.Ord.,* E.Asia-N.Am.-Eu.(USSR).——FIG. 400,*1.* **S. typica*, Longview F., USA(Va.); *1a,b,* brach.v. ext., int. mold; *1c,d,* ped.v. ext., int. mold; all ×2 (825).

Thaumotrophia WANG, 1955, p. 342 [**T. sinensis*; OD]. Ornament as in *Tetralobula*; internally resembling *Diaphelasma* but brachiophore plates stout, converging toward floor of valve. *L.Ord.,* E.Asia-Eu.(USSR).——FIG. 400,*5.* **T. sinensis*, Liangchiashan Ser., China(Liaoning Prov.); *5a,b,* brach.v. ext., int.; *5c-e,* ped.v. ext., ant., int.; all ×2.5 (852).

Yangtzeella KOLAROVA, 1925, p. 219 [*Triplecia poloi MARTELLI, 1901, p. 302; OD]. Smooth; fold weak; sulcus shallow; spondylium simplex supported by 2 or 4 lateral septa. *M.Ord.-U.Ord.,* E.Asia-?Eu.(USSR). —— FIG. 400,*6.* **Y. poloi* (MARTELLI), Foppé & Neichiashan, China; *6a-c,* post., brach.v., ped.v. views, ×1 (729).

Family SYNTROPHOPSIDAE
Ulrich & Cooper, 1936

[Syntrophopsidae ULRICH & COOPER, 1936, p. 630]

Smooth or radially costellate; spondylium posteriorly sessile, anteriorly supported by low septal ridge; brachiophores convergent, united with floor of valve. *L.Ord.*

Syntrophopsis ULRICH & COOPER, 1936, p. 630 [**S. magna*; OD]. Brachial interior as in *Syntrophina* but different in having short, converging brachiophore plates; spondylium short, sessile. *L.Ord.,* N. Am.-Eu. (USSR)-Tasm. —— FIG. 401,*5.* **S. magna*, Black Rock F., USA(Ark.); *5a,b,* brach.v. ext., int. (holotype); *5c,d,* ped.v. ext., int. (holotype); *5e,* ant. (holotype); all ×1.5 (825).

Hesperotrophia ULRICH & COOPER, 1936, p. 630 [**H. obscura*; OD]. Exterior finely costellate; internally like *Syntrophopsis*; septum supporting anterior part of spondylium very low and short. *L.Ord.,* N.Am.-Eu.(USSR).——FIG. 401,*2.* **H. obscura*, Sarbach F., Can.(Alta.); *2a,b,* brach.v.

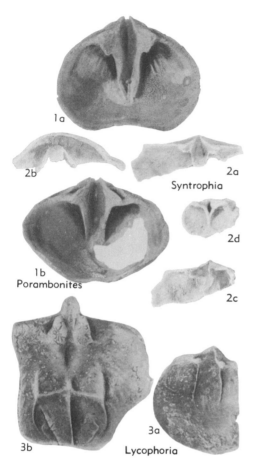

FIG. 402. Lycophoriidae *(3)*; Porambonitidae *(1)*; Syntrophiidae (Syntrophiinae) *(2)* (p. *H532-H534*).

int. mold, ped.v. ext., ✕2; *2c,* brach.v. ext., ✕6 (825).

Rhysostrophia ULRICH & COOPER, 1936, p. 630 [**R. nevadensis*; OD]. Exterior with distinct, radial costellae or costae; internally like *Syntrophopsis*. *L.Ord.,* N.Am.-Eu.(USSR).——FIG. 401,*3*. **R. nevadensis,* U. Pogonip F., USA(Nev.); *3a-c,* ped.v. ext. (holotype), ant., brach.v. ext., ✕1.5 (825).

Family LYCOPHORIIDAE
Schuchert & Cooper, 1931

[Lycophoriidae SCHUCHERT & COOPER, 1931, p. 245]

Shell biconvex; dental plates well developed; cardinal process simple, rodlike, united with brachiophore plates. *L.Ord.-M.Ord.*

Lycophoria LAHUSEN, 1886, p. 221 [**Atrypa nucella* DALMAN, 1828, p. 130; OD]. Shell globular, costellate; teeth large, parallel to hinge; dental plates subparallel; cardinal process tripartite, formed by fusing brachiophore plates and cardinal process. *L.Ord.-M.Ord.,* Eu.(Baltoscandia). ——FIG. 401,*4;* 402,*3.* **L. nucella* (DALMAN), L.Ord.(Chazy), Popovka; 401,*4a-d,* post., ant., lat., brach.v. views, ✕1.5 (729); 402,*3a,* ped.v. int., ✕1.5; 402,*3b,* brach.v. int., ✕3 (729).

Family PORAMBONITIDAE
Davidson, 1853

[Porambonitidae DAVIDSON, 1853, p. 99]

Shells median to large, biconvex; radially ornamented with rows of pits in furrows; dental plates nearly parallel or divergent; brachiophore plates strong; dental and brachiophore plates in old individuals simulating spondylium and septalium. *L.Ord.-L.Sil.*

Porambonites PANDER, 1830, p. 95 [**P. intermedia*; SD HALL & CLARKE, 1895, p. 226]. Large, biconvex or convexoplane; teeth stout; dental plates coalesced, brachiophore plates fusing anteriorly with median septum. *L.Ord.-L.Sil.,* Australia-Eu. (Baltoscandia) - Asia (Himalayas) - N. Am. - Eu. (USSR).

[Although application of zoological rules relating to designation of the type-species of *Porambonites* seems to be straightforward and clear, authors have disagreed about it, variously citing *Terebratulites aequirostris* VON SCHLOTHEIM (1820), *Porambonites intermedia* [*recte intermedius*] PANDER (1830), and *P. reticulatus* PANDER (1830). The first-mentioned of these was explicitly chosen by DAVIDSON (1853, p. 99), since PANDER made no original designation of type among the 30 species assigned by him to the genus; however, DAVIDSON's subsequent designation, antedating others is invalid because *P. aequirostris* is not one of PANDER's original included group. The fact that KING (1850, p. 112) designated this species as the type of a new genus named *Isorhynchus* has no bearing on the type-species of *Porambonites*. Next, DALL (1877, p. 57) mentioned *P. intermedia* as PANDER's first-described species but did not explicitly "select" it as the type-species. HALL & CLARKE (1895, p. 226), however, definitely named *P. intermedia* as the type-species of *Porambonites*. TEICHERT (1930, p. 182), giving no consideration at all to *P. intermedia*, concluded that *P. reticulatus* is best qualified on the basis of known morphological characters to represent *Porambonites* and therefore named it as the type-species. SCHUCHERT & LEVENE (1929, p. 100) cited *P. intermedia* as type-species, but SCHUCHERT & COOPER (1932, p. 104) rejected this designation in favor of *P. reticulatus,* mainly on the ground of asserted unrecognizability of *P. intermedia,* especially in view of the loss of its type-specimens. Despite all this, unless changed by ICZN under its plenary powers, *P. intermedia* is the legally established type-species and if unrecognizable, the genus may be construed likewise to be unrecognizable or, alternatively, interpreted in agreement with a century or more of usage as having morphological characters displayed by *P. reticulatus* and other accepted species. This latter course does not conflict with the Zoological Code and here is adopted.]

P. (Porambonites). Subcircular in outline with sulcus beginning nearly at mid-length of pedicle valve; hinge line rather prominent. *L.Ord. (Skiddav.),* Glaukonit Kalk, Eu.(NW.USSR). ——FIG. 403,*8. P. (P.) reticulatus,* M.Ord. (Chazy.), Iswos on Walchow R.; *8a-c,* brach.v., lat., ant. views, ✕1 (729).——FIG. 402,*1. P.*

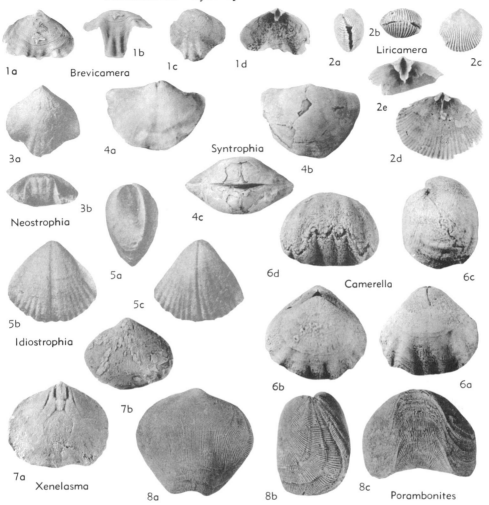

FIG. 403. Porambonitidae *(8)*; Syntrophiidae (Syntrophiinae) *(4)*, (Xenelasminae) *(7)*; Brevicameridae *(1)*; Camerellidae (Camerellinae) *(2, 3, 5, 6)* (p. *H*532-535).

(P.) schmidti NOETLING, Mohawk., Est.; *1a,b*, ped.v. int., brach.v. int., ✕1 (729).

P. (Equirostra) COOPER & MUIR-WOOD, 1951, p. 195 [*pro Isorhynchus* KING, 1850, p. 112 (*non* SCHOENHERR, 1833)] [**Terebratulites aequirostris* VON SCHLOTHEIM, 1820; OD]. Subglobular in profile, subtriangular in outline, compressed in front; long, slightly diverging dental and brachiophore plates. *U.Ord.(Caradoc.)*, Echinosphaerit Ls., Eu.

[KING's qualification of the type-species of *Isorhynchus* designated by his reading "*Terebratulites aequirostris* Schlotheim, as represented by DE VERNEUIL in *Geologie de la Russie d'Europe*, v. 2, pl. 3, fig. 1, 1845" does not affect validity of accepting the species named by VON SCHLOTHEIM as type-species of *Isorhynchus* (and hence of *Equirostra*), since Art. 70 of the Zoological Code (1961) stipulates that "it is to be assumed that an author correctly identifies the nominal species that he . . . designates as the type-species of a new or of an established genus."]

P. (Noetlingia) HALL & CLARKE, 1894, p. 229 [**Spirifer tscheffkini* DEVERNEUIL, 1845; OD (M)]. Hinge line wide and linear; internally similar to *P. (Porambonites)*. *L.Sil.*, Eu.(USSR).

Family SYNTROPHIIDAE Schuchert, 1896

[Syntrophiidae SCHUCHERT, 1896, p. 320]

Small to medium in size, with subparallel and discrete plates or spondylium simplex; septalium present. *L.Ord.*

Subfamily SYNTROPHIINAE Schuchert, 1896

[*nom. transl.* ULRICH & COOPER, 1936, p. 631 (*ex* Syntrophiidae SCHUCHERT, 1896, p. 320)]

Shells with spondylium simplex. *L.Ord.*

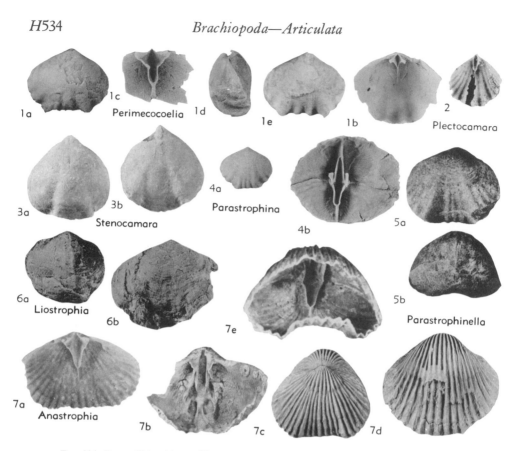

FIG. 404. Camerellidae (Camerellinae) *(1-2)*, (Stenocamerinae) *(3)*; Parastrophinidae *(4-7)* (p. *H535-H536*).

Syntrophia HALL & CLARKE, 1893, p. 270 [*Triplecia lateralis* WHITFIELD, 1886, p. 303; OD] [=*Syntrophia* HALL & CLARKE, 1894, p. 216 (syn. jr. hom.)]. Small to medium in size; exterior with concentric lines; long brachial septum supporting the beak short and shallow septalium. *L.Ord.*, N.Am.——FIG. 403,4. *S. lateralis* (WHITFIELD), Cassin F., USA(Vt.); *4a*, ped.v. ext., ×1; *4b*, brach.v. ext., ×1.5; *4c*, post. view, ×2 (825).——FIG. 402,2. *S. torynifera* ULRICH & COOPER, Smithville F. *(2a-c)*, Black Rock F. *(2d)*, USA(Ark.); *2a-c*, post. ped.v. int. views, ×2; *2d*, brach.v. int., ×2 (825).

Subfamily XENELASMATINAE
Ulrich & Cooper, 1936

[*nom. correct.* BIERNAT, herein (*pro* Xenelasminae ULRICH & COOPER, 1936, p. 631)]

With discrete, subparallel dental plates; septalium small, formed by brachiophore plates united with low septum. *L.Ord.*

Xenelasma ULRICH & COOPER, 1936, p. 631 [*X. syntrophioides*; OD]. Small, externally like *Syntrophia*; brachiophores short, supporting plates united with low septum to form septalium, as

in *Syntrophia*. *L.Ord.*, N.Am.——FIG. 403,7. *X. syntrophioides*, Longview F., USA(Va.); *7a*, ped. v. int. mold, ×3; *7b*, ped.v. ext., ×4 (825).

Family BREVICAMERIDAE Cooper, 1956

[Brevicameridae COOPER, 1956, p. 560]

Shell small, with paucicostate surface; short spondylium and sessile septalium. *Ord.*

Brevicamera COOPER, 1956, p. 560 [*B. camerata*; OD]. Both interareas reduced; semicostate; teeth small; spondylium short; brachiophore plates subparallel, with median callosity forming weak, sessile septalium. *Ord.*, N.Am.——FIG. 403,1. *B. camerata*, Pratt Ferry F., USA(Ala.); *1a-c*, ped.v. ext., ant., int., ×2; *1d*, brach.v. ext., ×2 (189).

Family CAMERELLIDAE
Hall & Clarke, 1894

[Camerellidae HALL & CLARKE, 1894, p. 355]

Shell usually biconvex; spondylium duplex or simplex; septalium; interarea weakly developed or obsolete. *L.Ord.-Sil.*

Subfamily CAMERELLINAE
Hall & Clarke, 1894

[*nom. transl.* BIERNAT, herein (*ex* Camerellidae HALL & CLARKE, 1894, p. 355)]

Spondylium simplex with short septum, which may serve as support for it. *L.Ord.-Sil.*

Camerella BILLINGS, 1859, p. 301 [**C. volborthi*; SD HALL & CLARKE, 1894, p. 219] [*=Rhyncho-camara* SCHUCHERT & COOPER, 1931, p. 248 (type, *R. plicata*)]. Shell biconvex, anteriorly costate; teeth strong; brachiophore plates short, supported by elongate septal ridges forming a septalium. *M. Ord.-Sil.,* N. Am.-Eu.(USSR).——FIG. 403,6. **C. volborthi,* Rockland F., Can.(Que.); *6a-d,* ped.v., brach.v., lat., ant. views, ×2 (189).

Idiostrophia ULRICH & COOPER, 1936, p. 631 [**I. perfecta*; OD]. Medium in size, with compressed outline; anterior commissure rectimarginate; interarea obsolete; internally like *Camerella. Ord.,* N.Am.-Eu.(USSR).——FIG. 403,5. **I. perfecta,* Mystic Congl., Can.(Que.); *5a-c,* lat., ped.v., brach.v. views, holotype, ×1.5 (189).

Liricamera COOPER, 1956, p. 592 [**L. nevadensis*; OD]. Moderate size; nearly circular; anterior commissure rectimarginate; multicostellate; spondylium deep, supported by thin, high septum; cruralium small, supported by long septum. *L.Ord.,* N.Am.——FIG. 403,2. **L. nevadensis,* Pogonip F., USA(Nev.); *2a-c,* lat., ant., brach.v. views (holotype), ×1; *2d,e,* brach.v. int., ped.v. int., ×2 (189).

Neostrophia ULRICH & COOPER, 1936, p. 631 [**N. subcostata*; OD]. Subpentagonal in outline; anterior commissure uniplicate and semicostate; sulcus shallow; fold low; interior like *Camerella. Ord.,* N.Am.-Eu.(USSR).——FIG. 403,3. **N. subcostata,* Mystic Congl., Can.(Que.); *3a,b,* brach.v., ant. views, ×2 (825).

Perimecocoelia COOPER, 1956, p. 593 [**P. semicostata*; OD]. In outline and ornamentation like *Parastrophina*; in brachial interior differs from *Parastrophina* in lacking alate plates. *Ord.,* N.Am.——FIG. 404,1. **P. semicostata,* Pratt Ferry F., Effna F., USA(Ala.-Va.); *1a,b,* brach.v. ext., int., ×2; *1c,* ped.v. int., ×2; *1d,* lat. view, ×2; *1e,* ped.v. ext., ×4 (189).

Plectocamara COOPER, 1956, p. 596 [**P. costata*; OD]. Rhynchonellid in outline, spondylium narrow; brachiophore supports united with floor of valve to form narrow apical cavity; brachial median septum like small ridge or absent. *M.Ord.,* N.Am.——FIG. 404,2; 405,2. **P. costata,* Ward Cove F. (404,2), Lincolnshire F. (405,2), USA (Tenn.); *404,2,* ped.v. int., ×3 (189); *405,2a-f,* ant., lat., ped.v., brach.v., post.v., brach.v. int. views, ×3 (189).

Subfamily STENOCAMARINAE Cooper, 1956

[Stenocamarinae COOPER, 1956, p. 602]

Shell smooth without spondylium but with septalium. *L.Ord.*

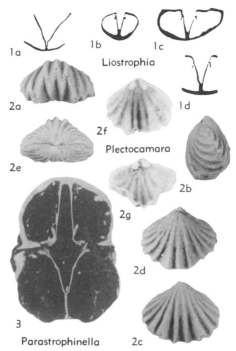

FIG. 405. Camerellidae (Camerellinae) *(2)*; Parastrophinidae *(1, 3)* (p. H535-H536).

Stenocamara COOPER, 1956, p. 602 [**S. perplexa*; OD]. Shape like *Camerella*; anterior margin uniplicate; smooth; dental plates subparallel; septalium short, supported by long septum. *L.Ord.,* N.Am.——FIG. 404,3. **S. perplexa,* Ellett F., USA(Va.-Tenn.); *3a,b,* brach.v. ext., ped.v. ext., ×2 (189).

Family PARASTROPHINIDAE
Ulrich & Cooper, 1938

[Parastrophinidae ULRICH & COOPER, 1938, p. 194]

Shell unequally biconvex; surface partially to strongly costate; with spondylium duplex; septalium sessile or elevated with subparallel and or converging brachial plates and alate plates. *M.Ord.-L.Dev.*

Parastrophina SCHUCHERT & LEVENE, 1929, p. 94 [*pro Parastrophia* HALL & CLARKE, 1893, p. 221 (*non* FOLIN, 1875)] [**Atrypa hemiplicata* HALL, 1847, p. 144; OD]. Moderate size; costate in anterior half; brachial interior differing from *Camerella* in having alate plates. *M.Ord.,* N.Am.-Eu.(USSR). —— FIG. 404,4a. **P. hemiplicata* (HALL), Martinsburg F., USA(W.Va.); brach.v. ext., ×1 (189).——FIG. 404,4b. *P. rotundiformis* (WILLARD), Prosser F., USA(Iowa); int. (ped. below), ×2 (189).

Anastrophia HALL, 1867, p. 163 [*pro Brachymerus* SHALER, 1865, p. 69 (*non* DEJEAN, 1834)] [**Pen-*

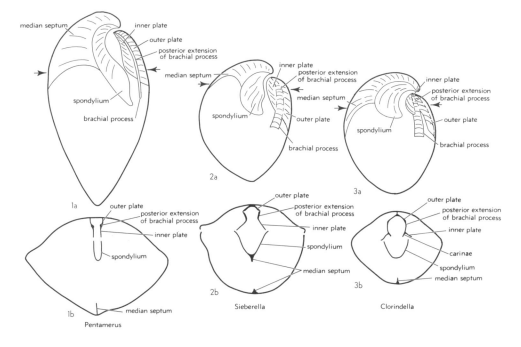

Fig. 406. Longitudinal sections *(1a, 2a, 3a)* (pedicle valve at left) and transverse sections *(1b, 2b, 3b)* (pedicle valve below) showing internal structure of three subfamilies of Pentameridae; heavy, unlettered arrows on longitudinal sections show position of transverse sections. *1.* Pentamerinae, *Pentamerus* sp. cf. *P. oblongus* Sowerby, Sil.(Reynales F.), Rochester, N.Y.——*2.* Gypidulinae, *Sieberella roemeri* Hall & Clarke, Sil.(Henryhouse F.), Pontotoc Co., Okla.——*3.* Clorindinae, *Clorindella areyi* (Hall & Clarke), Sil.(Irondequoit F.), Rochester, N.Y. [Explanation: *b,* brachial process; *bb,* posterior extension of brachial process; *c,* carinae; *i,* inner plate; *ms,* median septum; *o,* outer plate; *sp,* spondylium.]

tamerus verneuilli Hall, 1857, p. 104; OD]. Subtriangular in outline; costate; teeth stout; spondylium narrow, supported anteriorly by low, duplex septum; brachiophore plates parallel or nearly parallel, rarely united to form septalium; alate plates as in *Parastrophina.* Sil.-L.Dev., N.Am.-E. Eu.(USSR)-W.Eu.——Fig. 404,7. **A. verneuili* (Hall), L.Dev. (Helderberg.), *7a,b,* ped.v., brach.v. int. views, *7c,d,* brach.v. ext., ped.v. ext.; *7e,* int. (ped.v. below); all ×1.5 (729).

Liostrophia Cooper & Kindle, 1936, p. 355 [**L. glabra;* OD]. Like smooth *Parastrophina;* externally like *Syntrophia;* long septalium and alate processes just anterior to brachial supports, as in *Anastrophia.* U.Ord., N.Am.——Fig. 404,6; 405,1. **L. glabra,* Can.(Que.); *6a,b,* brach.v. ext., ×1 (194); *405,1a,* sec. through beak of ped.v., showing spondylium duplex, ×3; *405,1b-d,* secs. of brach.v., 1.25, 2.25, 3 mm. from beak, ×3 (194).

Parastrophinella Schuchert & Cooper, 1931, p. 248 [**Pentamerus reversus* Billings, 1857, p. 295; OD]. Biconvex, costate; spondylium with tendency to be sessile; septalium sessile, with subparallel brachial processes. U.Ord.-Sil., N.Am.-Eu.

(USSR).——Fig. 404,5. *P. ops* (Billings), Sil. (Chicotte F.), Anticosti; *5a,b,* ped.v., post. views, ×1 (729).——Fig. 405,3. **P. reversa* (Billings), Sil.(White Cliff), Anticosti; sec. showing spondylium and cardinalia, ×3 (729).

PENTAMERIDINA

[Materials for this suborder prepared by Thomas W. Amsden]

The Pentameridina are a suborder of middle Paleozoic brachiopods which comprise the superfamily Pentameracea, containing 43 genera and subgenera. They range from ?Middle Ordovician (Champlainian) to Upper Devonian (Senecan), but are most common in Lower Silurian to Lower Devonian strata. The Pentameridina tend to be larger than most middle Paleozoic brachiopods and include several species with very large shells. Some species of *Conchidium,* such as *C. alaskense* Kirk & Amsden and *C. vogulicum* (de Verneuil), are among the largest known brachiopods.

MORPHOLOGY
EXTERNAL FEATURES

Most pentameroids are moderately to strongly biconvex with swollen pedicle umbo and beak of the pedicle valve arched over the brachial valve. This development reaches an extreme in such forms as *Gypidula* of the Gypidulinae and *Conchidium* of the Pentamerinae. A few genera exhibit reversed convexity of the valves, however, the most conspicuous being *Capelliniella,* in which convexity of the brachial umbo exceeds that of the pedicle, and *Brooksina,* in which the pedicle valve is much flattened or even concave.

The fold and sulcus are absent or obscure in most Pentameracea, with exception of the Gypidulinae and Clorindinae, where this structure is generally present. The development of interareas is variable within this group; they are well developed in the Stricklandiidae and Gypidulinae, but poorly developed or absent in the Virgianidae and Pentamerinae.

Ornamentation ranges from costellate to costate to smooth; in a few genera it is granulose (e.g., *Devonogypa, Gypidulella*), or pitted (e.g., *Wyella*). If ribbing is present, it is commonly in the form of costae or plications, rather than costellae. The Pentameridae, Virgianidae, and Stricklandiidae have both smooth and ribbed representatives, whereas the Parallelelasmatidae include only paucicostate genera; the Enantiosphenidae contain only a single, noncostate genus.

INTERNAL FEATURES

The major structure inside of the pedicle valve is the **spondylium,** commonly called a **spondylium duplex,** which served as the seat of attachment for all muscles in this valve. In most pentameroids the spondylium is supported on a well-developed median septum (Fig. 406-409), but in some genera (e.g., *Harpidium*) the septum is abbreviated, or it may be completely lost (e.g., *Cymbidium, Holorhynchus*). The posterior end of the delthyrium may be closed by a deltidium which is generally concave downward (Fig. 410,2,3).

The brachial apparatus is tripartite, consisting of brachial processes which are braced at their posterior end by outer and inner plates (Fig. 406-410). The outer

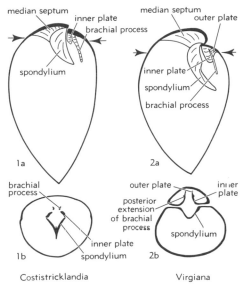

Fig. 407. Longitudinal sections *(1a, 2a)* and transverse sections *(1b,2b),* oriented as in Figure 406, showing internal structure of Stricklandiidae and Virgianidae; heavy, unlettered arrows on longitudinal sections show position of transverse sections.———1. Stricklandiidae, *Stricklandia (Costistricklandia) gaspeensis* (BILLINGS), Sil.(LaVielle F.), Chaleur Bay, Quebec.———2. Virgianidae, *Virgiana barrandei* (BILLINGS), Sil.(Becsie River F.), Anticosti Island, Quebec. [Explanation: *b,* brachial process; *i,* inner plate; *ms,* median septum; *o,* outer plate; *sp,* spondylium.]

plates may be discrete or they may unite to form a **cruralium.** In most pentameraceans these plates are long, extending far enough forward to enclose the brachial muscle area, but in the Stricklandiidae and Virgianidae the brachial apparatus is much shortened and the muscle area lies outside of the plates. Near the posterior end of the shell the upper edges of the inner plates curve outward to meet lateral walls of the valve; the sockets are located here and thus the inner and outer plates serve to brace the articulating mechanism and lophophore supports. The brachial processes extend forward beyond the inner and outer plates as free rodlike or bladelike structures which served to support at least a portion of the lophophore. In the Parallelelasmatidae, Pentameridae, Stricklandiidae, and Virgianidae, these processes terminate blindly, but in the Enantiosphenidae they end in a loop (Fig. 409,8).

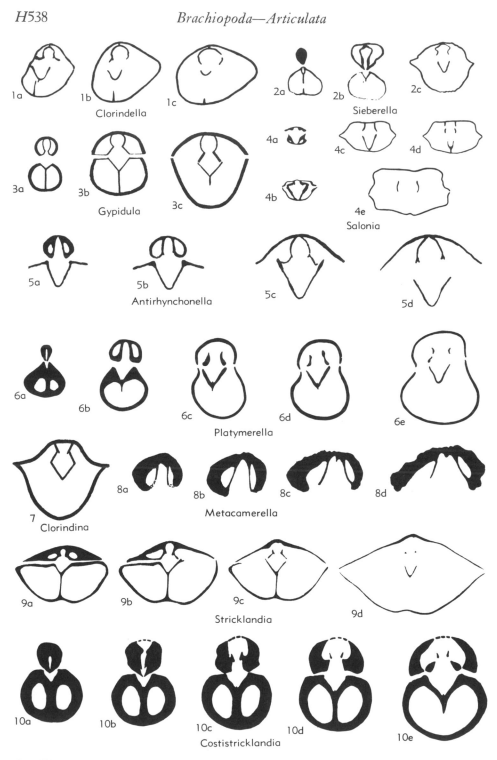

FIG. 408. Serial transverse sections (oriented as in Figure 406) of Parallelelasmatidae *(4,8)*, Stricklandiidae *(9,10)*, Virgianidae *(6)*, and Pentameridae *(1-3,5,7)*.

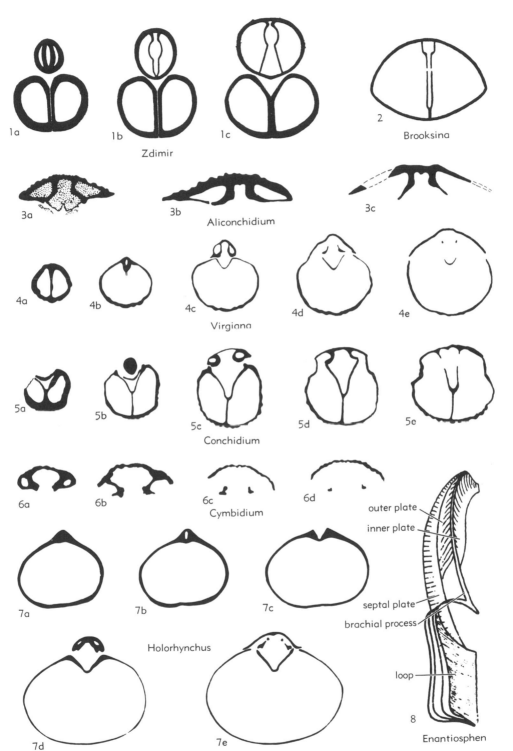

FIG. 409. Serial transverse sections (oriented as in Figure 406) of Virgianidae *(4,7)* and Pentameridae *(1-3,5,6)* and longitudinal drawing of Enantiosphenidae, *Enantiosphen (8)*.

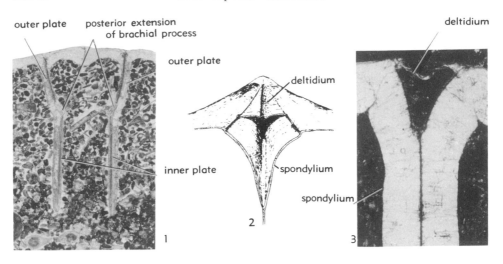

Fig. 410. Internal structures of *Pentamerus* sp. cf. *P. oblongus* Sowerby, Sil.(Reynales F.), N.Y.——*1.* Photomicrograph of brachial apparatus (×5) showing outer plates, posterior extension of brachial processes, and inner plates.——*2.* Drawing of silicified specimens (×3) showing deltidium and part of spondylium.——*3.* Photomicrograph of part of spondylium showing deltidium and spondylium, ×10.

CLASSIFICATION AND GEOLOGIC HISTORY

The Pentameridina first appeared in the ?Middle Ordovician (see remarks on Parallelelasmatidae below) and they range into the Upper Devonian (Fig. 411). They are rare in the Ordovician, four genera being represented in the Middle Ordovician and one in the highest Upper Ordovician. The Pentameracea first became fairly numerous in the Early Silurian and were common to abundant in the shelly faunas of the Late Silurian. They are moderately common in the Early Devonian (7 genera) and Middle Devonian (9 genera), but became rare in the Late Devonian (3 genera).

Division of the Pentameracea into families and subfamilies is based largely upon interior structures of the brachial valve, supplemented by such shell features as presence or absence of interareas, relative convexity of the valves, and development of the fold and sulcus. Five families are recognized: Parallelelasmatidae, Stricklandiidae, Virgianidae, Enantiosphenidae, and Pentameridae, the latter divided into three subfamilies (Pentamerinae, Gypidulinae, Clorindinae).

The oldest brachiopods now placed in the Pentameracea are the Parallelelasmatidae which comprise a small group (4 genera) confined to the Middle Ordovician (Fig. 411). They have a pauciplicate shell of moderate convexity, a spondylium supported on a median septum, and well-developed plates supporting the brachial processes. Some question exists concerning inclusion of this family in the Pentameracea, and it has been suggested (872, p. 232) that they should be referred to the Porambonitacea. The Parallelelasmatidae are separated from other representatives of the Pentameracea by almost the entire Upper Ordovician.

The Stricklandiidae comprise five genera and subgenera of Early and Middle Silurian brachiopods (Fig. 411). This family is characterized by an abbreviated brachial apparatus and well-developed interareas. The Virgianidae are similar to the Stricklandiidae in having short brachial plates, differing in their poorly developed interareas (Fig. 407). This family comprises three genera which range from Upper Ordovician (Ashgillian) through the Lower Silurian (Fig. 411). With exception of a single species of Pentamerinae (*Conchi-*

dium munsteri St. Joseph, 1937), all Late Ordovician and early and middle Llandoverian Pentameracea are referred to the Stricklandiidae and Virgianidae, characterized by their abbreviated brachial apparatus. It is not until fairly late in Llandoverian time that the Pentameridae with well-developed brachial plates became common.

The Pentameridae (31 genera) is the largest family of the Pentameracea. They are characterized by well-developed brachial plates which extend forward far enough to enclose the muscle field. These plates are tripartite, consisting of inner plates, brachial processes, and outer plates (Fig. 406). Except for a single species from the Late Ordovician of Norway (*Conchidium munsteri* St. Joseph, 1938), this family made first appearance in the late Llandoverian and ranged into the Late Devonian (Fig. 411). The youngest Pentameridae and youngest of the Pentameracea are found in the lower part of the Upper Devonian; they are species of *Gypidula* and *Pentamerella*.

The Enantiosphenidae is represented by a single genus *(Enantiosphen),* which is known only from the Middle Devonian of Great Britain and continental Europe (Fig. 411). *Enantiosphen* is the only pentameracean with a loop and is believed to be an offshoot of some stock of Pentameridae in which the distal ends of the brachial processes developed a connecting cross platform (Fig. 409,8).

Suborder PENTAMERIDINA
Schuchert & Cooper, 1931

[*nom. correct.* Amsden, herein (*pro* Pentameroidea Schuchert & Cooper, 1931, p. 247)]

Shells variable in size but tending to be large; commonly strongly biconvex; exterior smooth, costellate, costate, rarely pitted or granulose. Pedicle interior with well-developed spondylium, usually supported on septum, but free in a few genera. Lophophore supports consisting of rodlike or bladelike brachial processes which are unmodified except in Enantiosphenidae, where they terminate in a loop; at posterior end brachial processes are supported on plates which usually extend forward sufficiently to enclose brachial muscle field but which may be shortened to exclude muscle field. *?M.Ord., U.Ord.-U.Dev.*

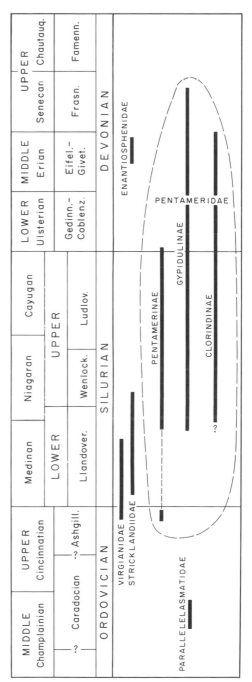

Fig. 411. Range of families and subfamilies of Pentameracea.

la lb lc
ld le
Didymelasma

2c

2d

2a   2b
        Parallelelasma

3a
Metacamerella

3b

FIG. 412. Parallelelasmatidae (p. *H*542).

Superfamily PENTAMERACEA
M'Coy, 1844

[*nom. transl.* SCHUCHERT, 1896, p. 320 (*ex* Pentameridae M'COY, 1844, p. 103)]

Characters of suborder. *?M.Ord., U.Ord.-U.Dev.*

?Family PARALLELELASMATIDAE
Cooper, 1956

[Parallelelasmatidae COOPER, 1956, p. 611]

Small to medium, subequally biconvex shells of moderate convexity. Surface generally smooth at posterior end, becoming paucicostate to pauciplicate anteriorly. Spondylium supported on median septum. Brachial plates well developed, discrete;

brachial processes relatively long, curved. [Taxonomic position doubtful, may belong in Porambonitacea (872).] *M.Ord.*

Parallelelasma COOPER, 1956, p. 611 [**P. pentagonum*; OD]. Subequally biconvex, paucicostate, with low brachial fold; brachial plates discrete, well developed. [May be synonym of *Metacamerella* (729, p. 231).] *M.Ord., SE.N.Am.*——FIG. 412,*2*. **P. pentagonum*, Pratt Ferry F., USA(Ala.); *2a,b*, brach.v. and ped.v. views, ×1, ×2; *2c*, ped.v. int. showing spondylium and teeth, ×6; *2d*, brach.v. int. showing brachiophores, ×8 (189).

Didymelasma COOPER, 1956, p. 615 [**D. longicrurum*; OD]. Small, weakly biconvex, pedicle sulcus and brachial fold; paucicostate; brachial plates relatively long, discrete. *M.Ord.(Wilderness)*, N.Am.——FIG. 412,*1*. **D. longicrurum*, Lebanon F., USA(Tenn.); *1a-e*, ped.v., brach.v., lat., ant., post. views, ×3 (189).

Metacamerella REED, 1917, p. 934 (*emend.* WILLIAMS, 1962, p. 231) [**Stricklandinia? balcletchiensis* DAVIDSON, 1883, p. 166; OD]. Subequally biconvex, shells subpentagonal in outline; paucicostate, with low brachial fold; brachial plates discrete, well developed. *M.Ord., Eu.*——FIG. 412,*3*; *408,8*. **M. balcletchiensis* (DAVIDSON), Balclatchie beds, Scot.; *412,3a,b*, brach.v. int., ext., ×2 (872); *408,8a-d*, brach.v. serial transv. secs., ×2 (703).

[SCHUCHERT & LEVENE (1929, p. 83) correctly cited the name *Stricklandinia? balcletchiensis* as published by DAVIDSON (1883, p. 166), this spelling of the specific name being evidently intentional, as shown by DAVIDSON'S use of it for species of other genera (Davidson, 1883, p. 160, 176, 210) and by mention of Balcletchie as source locality in several places. WILLIAMS (1962, p. 109, 228, 232) inaccurately recorded the spelling *balclatchiensis* as the form published by DAVIDSON. Seemingly, *balcletchiensis* is not validly emendable to *balclatchiensis*, even though present usage in Scotland (Ayrshire) recognizes the locality Balclatchie Bridge (SW. of Girvan) and Balclatchie beds.]

Salonia COOPER & WHITCOMB, 1933, p. 500 [**S. magnaplicata*; OD]. Trilobate, pauciplicate, with pedicle sulcus and brachial fold; brachial plates discrete, processes relatively long, curved. *M.Ord. (Trenton.)*, N.Am.——FIG. 408,*4*; *415,5*. **S. magnaplicata*, Salona F., USA(Pa.); *408,4a-e*, transv. secs. at 0.47, 0.61, 1.04, 1.33, 2.07 mm. from tip of ped.v. beak, all ×4 (197); *415,5a,b*, ped.v. and brach.v. views, ×2 (197).

Family STRICKLANDIIDAE
Schuchert & Cooper, 1931

[Stricklandiidae SCHUCHERT & COOPER, 1931, p. 248]
[=Stricklandiniidae HALL & CLARKE, 1894, p. 355; Stricklandidae AMSDEN, 1953, p. 146]

Smooth to costate, with well-developed interareas and generally elongate shells of moderate convexity. Pedicle spondylium relatively small, supported by short septum. Brachial apparatus much abbreviated; outer plates vestigial or absent, inner plates small; muscle area located in front of brachial plates. *L.Sil.-U.Sil.(Wenlock.).*

Stricklandia BILLINGS, 1859, p. 132 [**Atrypa lens* J. SOWERBY in MURCHISON, 1839, p. 637; SD OEHLERT, 1887, p. 1310] [=*Stricklandinia* BILLINGS, 1863, p. 370 (obj.)]. Large, smooth to weakly plicate, subcircular to elongate in outline, convexity moderate; spondylium supported on short septum; brachial apparatus relatively large, with small inner plates and in early species small outer plates which tend to be lost in later forms. *L.Sil.(Llandover.),* N.Am.(Appalachians-Ont.)-Eu.(Eng.-Norway).——FIG. 408,9; 414,8. **S. (S.) lens* (SOWERBY), Zone 6c, S. Norway; 408,*9a-d,* serial transv. secs., ×1 (702); 414, *8a,b,* lat. and brach.v. views, ×1 (Amsden, n).

[In the mistaken belief that *Stricklandia* BILLINGS, 1859, constituted a junior homonym of *Stricklandia* BUCKMAN, 1845 (p. 94), applied to a fossil plant, BILLINGS (1863) published the name *Stricklandinia* as replacement for his genus of Silurian brachiopods. HALL & CLARKE (1894, p. 250) selected "*Stricklandinia Gaspensis*" BILLINGS, 1859 (=*Stricklandia gaspeensis*) as the type-species of *Stricklandinia* (and hence of *Stricklandia*), but in view of OEHLERT's earlier designation of *Atrypa lens* SOWERBY (one of the eligible six originally included species of *Stricklandia*) as type-species, the cited publication by HALL & CLARKE lacks force, being nomenclaturally null and void. Hence, AMSDEN's (1953, p. 143) choice of *S. gaspeensis* as the type-species of *Costistricklandia* is admissible and entirely legal.]

Costistricklandia AMSDEN, 1953, p. 143 [**Stricklandia gaspeensis* BILLINGS, 1859, p. 134; OD]. Large, costate, short pedicle beak; spondylium and supporting septum short; brachial apparatus relatively large for this family; outer plates vestigial; inner plates of moderate size, adductor scars elongate, deeply impressed. *L.Sil.(U.Llandover.)-U.Sil.(Wenlock.),* N.Am.(N.Y.-Ont.-Que.-Anticosti)-Eu.(Eng.)-USSR(Novaya Zemlya).——FIG. 107,*1,* 100,*10,* 114,*4.* **C. gaspeensis* (BILLINGS), L.Landover. or Wenlock. (LaVielle F.), Que.; 407, *1a,b,* long., and transv. secs., ×1 (Amsden, n); 408,*10a-e,* transv. secs. at 3.1, 3.6, 3.8, 4.5, 5.7 mm. from tip of ped.v. beak, all ×1 (Amsden, n); 414,*4,* brach.v. view, ×1 (Amsden, n).

Kulumbella NIKIFOROVA, 1960, p. 61 [**K. kulumbensis*; OD]. Shells large, plano-convex to biconvex, with long hinge line; pedicle sulcus and brachial fold; surface marked with 2 sets of diagonal rugae intersecting to produce reticulate pattern; spondylium supported on short median septum; brachial processes short, supporting apparatus short. *L.Sil.(M.Llandover.),* USSR(Sib.)-?N.Am.——FIG. 413,*2.* **K. kulumbensis,* Sib.; *2a-c,* brach.v., lat., and ped.v. views, ×1 (600).

Microcardinalia BOUCOT & EHLERS, 1963, p. 51 [**Stricklandinia triplesiana* FOERSTE, 1890, p. 323; OD]. [In his original description FOERSTE (1885, p. 89) did not assign his new species *triplesiana* to any genus but later he referred it to *Stricklandinia* (=*Stricklandia*).] Small, subpentagonal in outline, smooth to sparsely costate; spondylium small, supporting septum short; brachial apparatus abbreviated, outer plates present in early species, absent in later ones; brachial adductor impres-

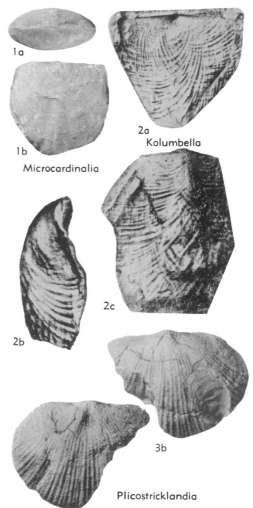

1a

1b
Microcardinalia

2a
Kolumbella

2c

2b

3b

3a
Plicostricklandia

FIG. 413. Stricklandiidae (p. *H543*).

sions separate, elongate. *L.Sil.(Llandover.),* USA (Ky.-Ill.-Mich.-Ohio)-G.Brit.(Scot.).——FIG. 413, *1.* **M. triplesiana* (FOERSTE), Brassfield F., Ohio; *1a,b,* post. and ped.v. views, ×1 (Amsden, n; 110).

Plicostricklandia BOUCOT & EHLERS, 1963, p. 55 [**Stricklandinia multilirata* WHITFIELD, 1878, p. 81; OD]. Similar to *Microcardinalia* but with costellate shell; outer plates vestigial. *L.Sil.(U. Llandover.) - U. Sil.(Wenlock.),* USA (Iowa-Wis.-Tex.)-Can.(Ont.)-?G.Brit.——FIG. 413,*3.* **P. multilirata* (WHITFIELD), Wenlock. (Hopkinton Dol.), Iowa; *3a,b,* ped.v., brach.v. views, ×1 (109).

Family VIRGIANIDAE Boucot & Amsden, 1963

[Virgianidae Boucot & Amsden, 1963, p. 296]

Smooth to costate, with interareas lacking or poorly developed. Spondylium moderate in size to small, supporting septum short or

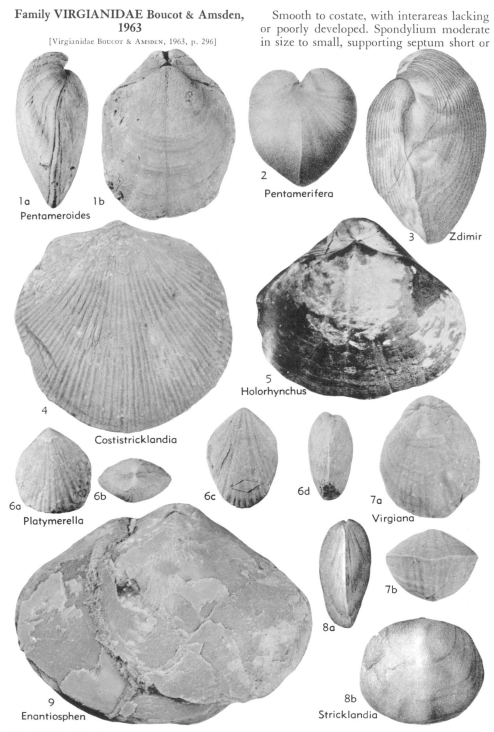

FIG. 414. Stricklandiidae *(4, 8)*; Virgianidae *(5-7)*; Pentameridae (Pentamerinae) *(1, 3)*, (Gypidulinae) *(2)*; Enantiosphenidae *(9)* (p. *H543, H547-H548, H551-H552*).

absent; spondylium and septum relatively thick-walled. Brachial apparatus much abbreviated; outer plates abbreviated or absent, inner plates small; muscle area located in front of brachial apparatus. *U.Ord.-L.Sil.*

Virgiana TWENHOFEL, 1914, p. 27 [**Pentamerus*

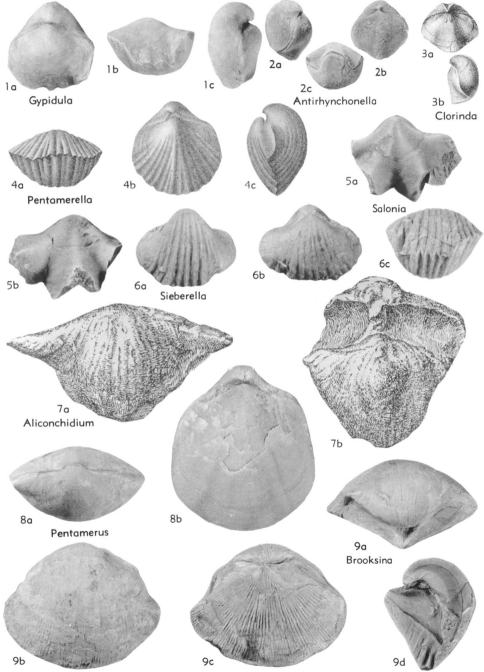

FIG. 415. Parallelelasmatidae *(5)*; Pentameridae (Pentamerinae) *(7-9)*, (Gypidulinae) *(1, 4, 6)* (Clorindinae) *(2, 3)* (p. *H542, H547-H548, H551*).

barrandi BILLINGS, 1857, p. 296; OD]. Strongly biconvex, costate shells with prominent pedicle beak and umbo which arch over brachial valve; pedicle valve commonly with low fold; spondylium of moderate size, supporting septum short; very short outer plates supporting long, rodlike

FIG. 416. Pentameridae (Pentamerinae) (p. *H547-H548*).

processes; small inner plates present. [The original spelling of *Pentamerus barrandi* is automatically correctable (Art. 32, c, Code, 1961) to *P. barrandei*.] *L.Sil.(M.Llandover.)*, N.Am.(USA-Can.-?Greenl.)-USSR(Sib.).——Fig. 407,2; 409,4; 414,7. *V. barrandei* (BILLINGS), Becsie River F., Que.(Anticosti); 407,2a,b, long. and transv. secs., ×1 (Amsden, n); 409,4a-e, transv. secs. at 2.2, 4.4, 5.9, 7.7, 11.5 mm. from tip of ped.v. beak, all ×1 (729); 414,7a,b, brach.v. and ant. views, ×1 (Amsden, n).

Holorhynchus KIAER, 1902, p. 68 [*H. giganteus*; OD]. Large, smooth, transversely elliptical in outline; spondylium free, no trace of supporting septum; outer plates absent or vestigial; small inner plates present. *U.Ord.(Ashgill.)*, Eu.(Norway-?Sweden).——Fig. 409,7; 414,5. *H. giganteus*, SE.Norway; 409,7a-e, serial transv. secs., ×1 (702); 414,5, brach.v. view, ×0.8 (702).

Platymerella FOERSTE, 1909, p. 70 [*P. manniensis*; OD]. Small, elongate elliptical in outline, moderately biconvex, multicostate; pedicle beak small. Spondylium supported on short septum; outer plates very short or absent, small inner plates present. *L.Sil.(M.Llandover.)*, USA(Tenn.-Ohio-Ill.).——Fig. 408,6; 414,6. *P. manniensis*, Tenn.; 408,6a-e, transv. secs. at 1.2, 1.9, 2.1, 2.6, 3.3 mm. from tip of ped.v. beak, all ×3 (Amsden, n); 414,6a-d, brach.v., post., ped.v., lat. views, ×3 (Amsden, n).

Family PENTAMERIDAE M'Coy, 1844

[Pentameridae M'Coy, 1844, p. 103]

Smooth, costate, costellate, granulose or pitted, with or without interareas, and generally having strongly biconvex shells. Pedicle spondylium well developed, commonly supported at least in part by septum. Brachial processes supported by well-developed plates which in some genera are parallel and discrete, and in others uniting to form cruralium; in both types brachial apparatus long, extending forward far enough to enclose muscle area; brachial processes rodlike or bladelike. *U.Ord.-U. Dev.*

Subfamily PENTAMERINAE M'Coy, 1844

[*nom. transl.* WAAGEN, 1883, p. 413 (*ex* Pentameridae M'Coy, 1844, p. 103)]

Moderate to large shells, with smooth, costate or costellate shells generally lacking well-developed interareas; fold and sulcus absent or weakly developed. Brachial processes long, rodlike; outer plates commonly discrete, but in few genera uniting to make cruralium. *U.Ord.-L.Dev.*

Pentamerus J. SOWERBY, 1813, p. 73 [*Pentamerus oblongus* J. DE C. SOWERBY, 1839; in MURCHISON,

1839, p. 641; by action of the ICZN]. [In 1954 the ICZN, Opinion 297, placed *Pentamerus* J. SOWERBY, 1813, on the Official List of Generic Names in Zoology; *Pentamerus oblongus* J. DE C. SOWERBY, 1839, was designated the type-species and added to the Official List of Specific Names in Zoology. The following names were placed on the Official Index of Rejected and Invalid Generic Names in Zoology: *Gypidia* DALMAN, 1828; *Trimurus* CALDWELL, 1934; *Miopentamerus* ALEXANDER (née CALDWELL), 1936; *Miopentamerus* WOODS, 1937. *Pentamerus laevis* J. SOWERBY, 1813, was placed on the Official Index of Rejected and Invalid Specific Names in Zoology.] Large, elongate, moderately biconvex, smooth surface; spondylium and supporting septum commonly extending forward less than half the length of pedicle valve; brachial plates discrete. *L.Sil.(U. Llandover.)-U. Sil. (Wenlock.)*, USA (N.Y.-Ohio-Ind.-Ky.-Ill.-Iowa-Wis.) - Can.(Ont. - Quebec-Anticosti)-Eng.-Sweden (Gotl.)-Est.-Asia(China)-USSR (Urals, Turkestan).——Fig. 406,1; 415,8. *P.* sp. cf. *P. oblongus* (SOWERBY), L.Sil.(U.Llandover.) (Reynales F.), N.Y.; 406,1a,b, long. and transv. secs., ×1 (Amsden, n); 415,8a,b, post. and brach. v. views, ×1 (Amsden, n).

?**Aliconchidium** ST. JOSEPH, 1942, p. 247 [*A. yassi*; OD]. Large, biconvex, costate; hinge line long, cardinal extremities commonly alate; pedicle palintrope prominent; spondylium and supporting septum well developed; brachial plates discrete. [This genus differs from most other Pentamerinae in having a prominent palintrope]. ?*U.Sil.*, Australia.——Fig. 409,3; 415,7. *A. yassi*, Hume Ser., New S. Wales; 409,3a-c, serial transv. secs. of brach.v., ×1.5 (703); 415, 7a,b, ped.v. and brach.v. views, ×1 (703).

Brooksina KIRK, 1922, p. 2 [*B. alaskensis*; OD]. Multicostellate, of moderate size; brachial valve strongly convex, pedicle valve gently convex, flat or concave; spondylium and supporting septum long, extending almost entire length of valve; brachial plates discrete. *U.Sil.(Ludlov.)*, Alaska-USSR(Ural Mts.-Turkestan).——Fig. 402,2; 415, 9. *B. alaskensis*, SE.Alaska(Kosciusco Is.); 409,2, transv. sec., ×1 (Amsden, n); 415, 9a-c, post., ped.v., brach.v. views, ×1; 415,9d, long. sec., ×1 (Amsden, n).

Callipentamerus (*see* p. H903).

Capelliniella STRAND, 1928, p. 38 [*pro Capellinia* HALL & CLARKE, 1894, p. 249 (*non* TRICHESE, 1874)] [*Capellinia mira* HALL & CLARKE, 1894, p. 249; OD]. Smooth shell differing from *Pentamerus* in having brachial valve deeper and more strongly convex than pedicle. *U.Sil.*, N.Am.-USSR (Turkestan).——Fig. 416,1. *C. mira* (HALL & CLARKE), Racine F., USA(Wis.); 1a,b, ped.v. and lat. views of int. mold (ped.v. at left), ×1 (Amsden, n).

Conchidium OEHLERT, 1887, p. 1311 [*Anomia bilocularis* HISINGER, 1799, p. 285; OD]. Rostrate, strongly biconvex, costate; spondylium partially or completely supported on median septum, extending forward more than half length of pedicle valve; brachial plates discrete. *U.Ord.(Ashgill.)-L.Dev.(Skala.),* cosmop.——Fig. 409,5; 416,4. **C. biloculare* (HISINGER), Sil., Sweden(Gotl.); 409,*5a-e,* transv. secs. at 4.2, 6.1, 8.1, 9.9, 12.1 mm. from tip of ped.v. beak, all ×1 (729); 416, *4a-b,* lat. and ped.v. views, ×1 (Amsden, n; 729).

[In 1954 the ICZN, Opinion 297, placed *Conchidium* OEHLERT, 1887, on the Official List of Generic Names in Zoology, type-species *Anomia bilocularis* HISINGER, 1799, by original designation. The following names were placed on the Official List of Specific Names in Zoology: *bilocularis* HISINGER, 1799, as published in the combination *Anomia bilocularis,* and *knighti* J. SOWERBY, 1813, as published in the combination *Pentamerus knighti* [=*Conchidium knighti*]. The following names were placed on the Official Index of Rejected and Invalid Generic Names in Zoology: *Conchidium* HISINGER, 1799; *Conchidium* BRONN, 1848; *Conchidium* WAHLENBERG, 1821.]

Cymbidium KIRK, 1926, p. 2 [**C. actum;* OD]. Multicostate; biconvex, brachial valve with greatest convexity; spondylium long, no median septum; brachial plates discrete, inner plates short. *U.Sil.(Ludlov.),* USA(Alaska-Nev.).——Fig. 409, 6; 416,*3.* **C. actum,* SE.Alaska(Kosciusko Is.); 409,*6a-d,* transv. secs. at 1.0, 3.0, 5.0, 6.0 mm. from tip of ped.v. beak, all ×1 (729); 416,*3a,b,* brach.v. and lat. views, ×1 (Amsden, n).

Harpidium KIRK, 1925, p. 1 [**H. insignis;* OD]. Smooth; biconvex; pedicle valve strongly convex, beaks of both valves arched, pedicle valve bent sharply over brachial; spondylium long, supporting septum short; brachial plates discrete. *U.Sil.,* Alaska-Greenl.-USSR(Ural Mts.).——Fig. 416,*5.* **H. insigne,* SE.Alaska (5a, Heceta Is.; 5b, Kosciusko Is.); *5a,b,* ped.v. view and long. sec., ×1 (729).

Jolvia SAPELNIKOV, 1960, p. 56 [**J. multiplexa;* OD]. Large, smooth to costate shells with well-developed spondylium and supporting septum; brachial apparatus with cardinal process. *U.Sil. (Wenlock.),* USSR(Ural Mts.).——Fig. 417,*8.* **J. multiplexa,* central Ural Mts., E. slope; brach.v. view, ×1 (707).

Lissocoelina SCHUCHERT & COOPER, 1931, p. 248 [**Pentamerus pergibbosus* HALL & WHITFIELD, 1875, p. 139; OD]. Smooth, strongly biconvex; pedicle valve rostrate, arched over brachial; spondylium supported on long median septum; brachial plates discrete. *L.Sil.(U.Llandover.)-U. Sil.(Wenlock.),* N.Am.(USA).——Fig. 416,*6.* **L. pergibbosa* (HALL & WHITFIELD), U.Sil.(Louisville F.), USA(Ky.); lat. view, ×1 (729).

?Pentamerifera KHODALEVICH, 1939, p. 22 [**Pentamerus taltiensis* CHERNYSHEV, 1893, p. 183; OD]. Smooth, biconvex shells with long spondylium and supporting septum; brachial apparatus similar to *Pentameroides (?).* [Internal characters poorly known.] *U.Sil.,* USSR(Ural Mts.).——FIG. 414,2. **P. taltiensis* (CHERNYSHEV), U.Sil., Urals E. slope; lat. view, ×1 (157).

Pentameroides SCHUCHERT & COOPER, 1931, p. 248 [**Pentamerus oblongus subrectus* HALL & CLARKE,

1894, p. 238; OD (M)]. Smooth, biconvex, external shape like *Pentamerus;* spondylium and supporting septum well developed; brachial plates uniting to make cruralium supported on median septum. *L.Sil.(up. Llandover.)-U.Sil(Wenlock.),* N. Am. (USA-Can.)-Eu. (Norway)-USSR (Ural Mts.).——FIG. 414,*1.* **P. subrectus* (HALL & CLARKE), U.Sil., USA(Iowa); *1a,b,* lat. and brach. v. views of int. mold, ×1 (729).

?Pleurodium WANG, 1955, p. 344 [**Conchidium tenuiplicatus* GRABAU, 1925, p. 80; OD]. Large, transversely elliptical, subequally biconvex; pedicle palintrope prominent, curved, apsacline; no fold or sulcus; strong, angular costae which do not bifurcate. Spondylium long, supporting septum short; brachial interior unknown. [This genus differs from most other Pentamerinae in having a well-marked palintrope.] *Sil.,* China(M. Yangtze Valley).——FIG. 417,5. **P. tenuiplicatum* (GRABAU); *5a,b,* ped.v. and ant. views, ×1 (852).

Rhipidium SCHUCHERT & COOPER, 1931, p. 249 [**Pentamerus knappi* HALL & WHITFIELD, 1872, p. 184; OD]. Costate, moderately to strongly biconvex; pedicle beak and umbo shorter and less prominent than in *Conchidium;* spondylium and supporting septum generally extending forward half or less than half of valve length; brachial plates discrete. *U.Sil.(Wenlock. or Ludlov.),* N. Am.(USA)-Eu.——FIG. 416,2. **R. knappi* (HALL & WHITFIELD), Louisville F., USA(Ky.); *2a,b,* brach.v. and lat. views, ×1 (729).

Subfamily GYPIDULINAE Schuchert & LeVene, 1929

[Gypidulinae SCHUCHERT & LEVENE, 1929, p. 15]

More or less galeatiform shells with interareas; fold and sulcus generally well developed, absent in few genera; exterior smooth, costate, pitted or granulose. Brachial apparatus commonly lyre-shaped in cross section, brachial processes bladelike; outer plates discrete or coalesced into cruralium. *L.Sil.-U.Dev.*

Gypidula HALL, 1867, p. 163 [**Gypidula typicalis* AMSDEN, 1953, p. 140 (pro *Pentamerus occidentalis* HALL, 1858, p. 514, *non* HALL, 1852); OD HALL, 1867, p. 380]. Elongate oval to subcircular in outline, pedicle valve swollen, beak arched over brachial; costate to multicostate; pedicle fold and brachial sulcus; brachial plates discrete. *L. Sil.(Llandover.)-U. Dev.,* N.Am.(USA-Can.)-Eu.-Asia-USSR-Afr.——FIG. 408,3; 415,1. **G. typicalis* AMSDEN, U.Dev.(Cedar Valley F.), USA(Iowa); 408,*3a-c,* serial transv. secs., ×2 (62); 415,*1a-c,* brach.v., ant., lat. views, ×1 (729).

[HALL (1867, p. 380) designated *Pentamerus occidentalis* HALL, 1858, as the type-species of his new genus *Gypidula. Pentamerus occidentalis* HALL, 1858, is a homonym of *Pentamerus occidentalis* HALL, 1852, the latter being a species of *Conchidium.* AMSDEN, 1953, p. 140, replaced *Pentamerus occidentalis* HALL, 1858 (not *Pentamerus occidentalis* HALL, 1852) with *Gypidula typicalis.*]

FIG. 417. Pentameridae (Pentamerinae) *(5, 8),* (Gypidulinae) *(1-4, 6-7)* (p. *H548, H550-H551).*

Barrandina BOOKER, 1926, p. 131 [*Pentamerus linguifera wilkinsoni* ETHERIDGE, 1892; OD]. Smooth, subgaleatiform shells with pedicle sulcus and brachial fold. Outer plates discrete (88). *U.Sil.,* New S. Wales.

?**Biseptum** KHODALEVICH & BREIVEL, 1959, p. 39 [*B. rectecostatum*; OD]. Shell large, costate, costae nonbifurcating; no fold nor sulcus. Outer plates uniting to make cruralium, supported on high septum. *M.Dev.,* USSR(Ural Mts.).——FIG. 418,4. *B. rectecostatum, 4a,b,* lat., transv. sec. post. part of ped.v., ×1, ×3 (468).

Devonogypa HAVLÍČEK, 1951, p. 3 [*D. spinulosa*; OD]. Large, subcircular to transversely elliptical, strongly biconvex; shallow brachial sulcus, low pedicle fold; surface smooth except for fine spines or granules arranged in irregular, horizontal to oblique rows. Pedicle and brachial interiors like *Gypidula. M.Dev.(Givet.),* Eu.(Czech.-Ger.).——

FIG. 417,7. *D. spinulosa,* Czech.; brach.v. view, ×1 (404).

Gypidulella KHODALEVICH & BREIVEL, 1959, p. 26 [*G. pennatula*; OD]. Hinge line straight, extended; pedicle fold and brachial sulcus; surface costate and tuberculate. Interior like *Sieberella. M. Dev.,* USSR (Ural Mts.).——FIG. 418,3. *G. pennatula, 3a,b,* ped.v., ant. views, ×1; *3c,* transv. sec., ×3 (468).

Gypidulina RZHONSNITSKAYA, 1956, p. 49 [*Pentamerus optatus* BARRANDE, 1847, p. 37; OD] [=*Sieberina* ANDRONOV, 1961 (obj.)]. Smooth to pauciplicate shells with sharply defined pedicle fold and brachial sulcus. Brachial plates uniting to form cruralium. *L.Dev.,* USSR(Ural Mts.-Novaya-Zemlya-Kuznetsk Basin)-Eu.——FIG. 417,3. *G. optata* (BARRANDE), Czech.; *3a,b,* ped.v. and ant. views, ×1 (468).

Ivdelinia ANDRONOV, 1961, p. 45 [*Gypidula ivdel-

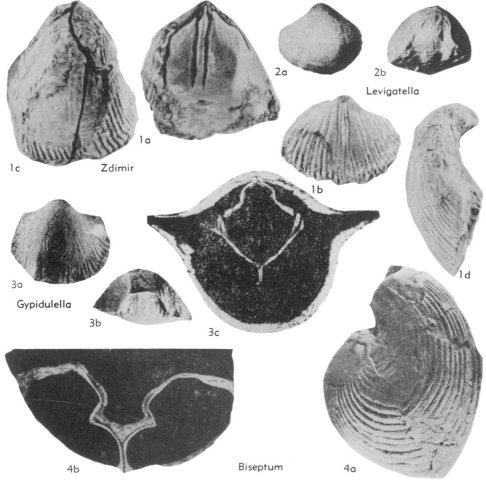

FIG. 418. Pentameridae (Gypidulinae) (p. *H*550-*H*551).

ensis KHODALEVICH, 1951, p. 22; OD]. Galeatiform, pedicle fold and brachial sulcus; costate and with fine concentric ridges crossing costae; costae split at anterior margin; brachial plates discrete or united into cruralium. *L.Dev.-M.Dev.,* USSR (Ural Mts.)-Eu.——FIG. 417,4. *I. acutolobata* (SANDBERGER), M.Dev., Kuznetsk Basin; *4a,b,* ped. v. view and ornament, ×1, ×5 (692).

Leviconchidiella RZHONSNITSKAYA, 1960, p. 47 [*Sieberella? vagranica* KHODALEVICH, 1951, p. 39; OD]. Costate shells without distinct fold and sulcus. Brachial plates discrete. Like *Gypidula* but lacking fold and sulcus. *M.Dev.,* USSR.—— FIG. 417,2. *L. vagranica* (KHODALEVICH), Kuzbas; *2a,b,* ped.v. and brach.v. views, ×1 (691).

Levigatella ANDRONOV, 1961, p. 38 [*Gypidula olga* KHODALEVICH, 1939, p. 15; OD]. Strongly biconvex, smooth, pedicle fold and brachial sulcus; brachial plates discrete as in *Gypidula. Sil.-U.Dev.,* USSR(Ural Mts.).——FIG. 418,2. *L. olga,* L.Dev.(Marginalis beds), Nadieja Reg.; *2a,b,* ped.v., ant. view, ×4 (466).

Pentamerella HALL, 1867, p. 163 [*Atrypa arata* CONRAD, 1841, p. 55; SD OEHLERT, 1887; p. 1312]. Costate to multicostate; pedicle sulcus and brachial fold; brachial plates united to form cruralium. *L. Dev.-U. Dev.,* N. Am. (USA-Can.)-Asia-USSR(Ural Mts.-Turkestan).——FIG. 415,4. *P. arata* (CONRAD), L.Dev.(Schoharie F.), USA (N.Y.); *4a-c,* ant., brach.v., lat. views, ×1 (729).

?**Procerulina** ANDRONOV, 1961, p. 76 [*Pentamerus acutolobatus procerulus* BARRANDE, 1879, p. 60; OD]. Like *Ivdelinia.* [Possibly synonym of *Ivdelinia.*] *L.Dev.-U.Dev.,* Eu.(Czech.-Ger.)-USSR(Urals).——FIG. 417,6. *P. procerulus* (BARRANDE), L.Dev., Czech.; *6a,b,* brach.v. and ped.v. views, ×1 (38).

Sieberella OEHLERT in FISCHER, 1887, p. 1311 [*Pentamerus sieberi* VON BUCH in BARRANDE, 1847, p. 103; OD]. Costate, pedicle fold and brachial sulcus; brachial plates uniting to form cruralium. *U.Sil.(Wenlock.)-L.Dev.,* N.Am.(USA-Can. - Greenl.)-Eu. (Ger.)-N. Afr. (Morocco)-Asia (Turkestan) - USSR (Ural Mts. - Kuznetsk Basin). ——FIG. 406,2. *S. roemeri* HALL & CLARKE, Sil. (Henryhouse F.), USA(Okla.); *2a,b,* long. and transv. secs., ×2 (Amsden, n).——FIG. 408,2; 415,6. *S. sieberi* (VON BUCH), L.Dev. (Konjeprus), Czech.; 408,*2a-c,* transv. secs. at 2.8, 4.0, 7.9 mm. from tip of ped.v. beak, all ×2 (729); 415,*6a-c,* ped.v., brach.v., ant. views, ×1 (Amsden, n).

Wyella KHODALEVICH, 1939, p. 21 [*Eichwaldia uralica* CHERNYSHEV, 1893, p. 179; OD]. Plicate to smooth shells with pitted exterior; pedicle fold and brachial sulcus; brachial plates discrete. *U. Sil.(Ludlov.),* USSR (Ural Mts.).——FIG. 417,1. *W. uralica* (CHERNYSHEV), Ural Mts.; *1a,b,* ped. v. and lat. views, ×1 (466).

Zdimir BARRANDE, 1881, p. 171 [*Zdimir solus* BARRANDE, 1881 (=*Porambonites ?robustus*

BARRANDE, 1879, p. 97); OD] [=*Conchidiella* KHODALEVICH, 1938, p. 32 (type, *Pentamerus pseudobaschkiricus* CHERNYSHEV, 1885, p. 55)]. Large shells with radial costellae which increase by bifurcation; fold and sulcus generally absent; brachial plates discrete. *M.Dev.,* Eu.(Czech.)-USSR.——FIG. 418,1. *Z. robustus,* Eifel. (Trebotov Ls.), Czech.; *1a-d,* brach.v. (exfol.), brach.v., ped.v., lat. view ped.v., all ×1 (115).——FIG. 409,1; 414,3. *Z. pseudobaschkiricus* (CHERNYSHEV), M.Dev., Ural Mts.; 409,*1a-c,* serial transv. secs., ×1 (466); 414,3, lat. view, slightly reduced (155, 466).

Subfamily CLORINDINAE Rzhonsnitskaya, 1956

[Clorindinae RZHONSNITSKAYA, 1956, p. 49]

Small to medium, biconvex, smooth to costate shells with more or less galeatiform profile. Brachial apparatus well developed, outer plates discrete or uniting to form cruralium; brachial processes bladelike, ventral edge of processes extending inside of inner plates as small carinae. *Sil.-M.Dev.*

Clorinda BARRANDE, 1879, p. 109 [*C. armata;* OD]. Smooth, pedicle sulcus and brachial fold. Brachial plates discrete. *Sil.-M.Dev.,* N.Am.(Anticosti-Greenl.)-Eu.-Asia(Turkestan).——FIG. 415, 3. *C. armata,* L.Dev. (E), Czech.; *3a,b,* brach.v. and lat. views, ×1 (53).

Antirhynchonella OEHLERT in FISCHER, 1887, p. 1311 [*Atrypa linguifera* J. DE C. SOWERBY in MURCHISON, 1839, p. 629; OD] [=*Barrandella* HALL & CLARKE, 1894, p. 241 (obj.)]. [In 1955, ICZN (Opinion 374) placed *Antirhynchonella* (type-species *Atrypa linguifera* J. DE C. SOWERBY, 1839) on The Official List of Generic Names; *Antirhynchonella* QUENSTEDT, 1871 (*nom. nud.*) and *Barrandella* HALL & CLARKE, 1894, were added to The Index of Rejected and Invalid Generic Names in Zoology]. Smooth, strongly biconvex; pedicle sulcus and brachial fold; brachial plates unite to form cruralium. *Sil.,* N.Am.(USA-Can.)-Eu.(Norway-G.Brit.)-Asia(Turkestan). —— FIG. 408,5; 415,2. *A. linguifera* (SOWERBY), U. Sil.(Wenlock.), Eng.; 408,*5a-d,* serial transv. secs., ×2 (729); 415,*2a-c,* lat., brach.v., ant. views, ×1 (729).

Clorindella AMSDEN, 1964, p. 236 [*Barrandella areyi* HALL & CLARKE, 1894, p. 368; OD]. Costate shells with pedicle-valve sulcus and brachial-valve fold; brachial plates uniting to form cruralium (34). *L.Sil.-U.Sil.,* N.Am.——FIG. 406,3; 408,1. *C. areyi* (HALL & CLARKE), L.Sil. (Clinton.=U.Llandover.), USA(N.Y.); 406,*3a,b,* long. and transv. secs., ×2; 408, *1a-c,* transv. secs. at 2.5, 3.1, 4.0 mm. from tip of ped.v. beak, all ×3 (Amsden, n).

Clorindina KHODALEVICH, 1939, p. 11 [*C. uralica;* OD]. Costate, pedicle valve deeper than brachial; pedicle sulcus and brachial fold; brachial plates discrete. *L.Dev.*, USSR.——Fɪɢ. 408,7. **C. uralica*, Urals; transv. sec., ×2 (466).

Family ENANTIOSPHENIDAE Torley, 1934

[Enantiosphenidae Tᴏʀʟᴇʏ, 1934, p. 93]

Specialized forms with brachial processes terminating in loop; supporting plates consisting of inner plates, brachial processes, outer plates; outer plates unite to form median septum. *M.Dev.*

Enantiosphen Wʜɪᴅʙᴏʀɴᴇ, 1893, p. 97 [**Megan-*

teris? vicaryi Dᴀᴠɪᴅsᴏɴ, 1882, p. 20; SD Hᴏʟᴢ-ᴀᴘꜰᴇʟ, 1912, p. 123]. Smooth, biconvex, transversely elliptical; spondylium supported on high, median septum; brachial processes extending forward and expanding to form broad plates connected to one another by transverse, subhorizontal plate, this transverse plate supported in center by median septum. *M.Dev.*, Eu.(G.Brit.-Ger.).——Fɪɢ. 409,8; 414,9. **E. vicaryi* (Dᴀᴠɪᴅ-sᴏɴ), Ger. (Bilveringen); 409,8, lat. view of brach. process and connecting septal plate, ant. extremity downward, ×2 (from 504); 414,9, brach.v., ×1 (815, 879).

RHYNCHONELLIDA

By D. V. Aɢᴇʀ, Rɪᴄʜᴀʀᴅ E. Gʀᴀɴᴛ, D. J. McLᴀʀᴇɴ, and Hᴇʀᴛᴀ Sᴄʜᴍɪᴅᴛ

[Imperial College of Science and Technology, London; United States Geological Survey, Washington, D.C.; Geological Survey of Canada, Ottawa; and Senckenbergische Naturforschende Gesellschaft, Frankfurt]

Order RHYNCHONELLIDA Kuhn, 1949

[*nom. correct.* Mᴏᴏʀᴇ in Mᴏᴏʀᴇ, Lᴀʟɪᴄᴋᴇʀ, & Fɪsᴄʜᴇʀ, 1952, p. 221 (*pro* order Rhynchonellacea Kᴜʜɴ, 1949, p. 104)]
[Diagnosis prepared by D. V. Aɢᴇʀ]

Articulate brachiopods, usually with rostrate shell, functional pedicle developed, delthyrium partially closed by deltidial plates. Mantle canals much branched, with one pair of main trunks in each mantle. Median septum commonly supporting septalium or hinge plates in brachial valve; dental plates usually present; spondylia normally absent. Recent representatives mostly with 2 pairs of metanephridia, lophophore spirolophous, with ventrodorsally directed cones supported by crura. Shell substance normally impunctate, rarely with inner fibrous layer punctate (583). *M.Ord.-Rec.*

Superfamily RHYNCHONELLACEA Gray, 1848

[*nom. transl.* Sᴄʜᴜᴄʜᴇʀᴛ, 1896, p. 323 (*ex* Rhynchonellidae Gʀᴀʏ, 1848, p. 438)] [Materials for this superfamily prepared by D. V. Aɢᴇʀ, D. J. McLᴀʀᴇɴ, and Hᴇʀᴛᴀ Sᴄʜᴍɪᴅᴛ as indicated by families]

Shell impunctate, commonly lacking spondylia. *M.Ord.-Rec.*

PALEOZOIC RHYNCHONELLACEA

By Hᴇʀᴛᴀ Sᴄʜᴍɪᴅᴛ and D. J. McLᴀʀᴇɴ

Separation of Paleozoic from Mesozoic and Tertiary rhynchonellaceans is an arbitrary and artificial arrangement which for

the present may be justified by the fact that the two groups have been studied from different points of view and seldom by the same workers. An additional factor is the scarcity of Lower and Middle Triassic rhynchonelloids. Paleozoic rhynchonellaceans currently are being subjected to a proliferation of genera, and there is no reason to suppose that this has ended. Recognition of the importance of detailed study of internal structures by means of a variety of techniques has resulted in the realization of the great complexity and abundance of forms in the superfamily, and this has not yet been fully exploited taxonomically.

Our present state of knowledge makes classification extremely difficult. Of 134 Paleozoic rhynchonellaceans here recognized (excluding homonyms and synonyms), 87 are definitely placed in 19 families or subfamilies and 47 are classed questionably in these groups or segregated as "Family Uncertain." Some of the family-group taxa are significant assemblages of related genera, whereas others merely represent a convenient, and presumably temporary, pigeonholing of morphologically similar forms. Difficulties in classification may arise from the methods used to examine interiors. Thus, while the study of internal structure by means of serial grinding techniques gives accurate information on interiors, it may nevertheless be difficult to interpret in terms of a shell interior examined in a different manner. Silicified specimens, internal molds, and prepared in-

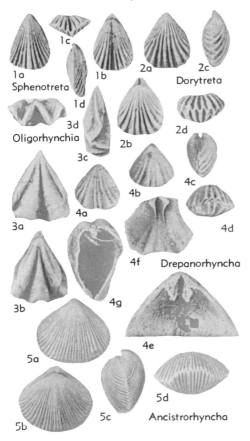

Fig. 419. Ancistrorhynchidae *(4-5)*; Oligorhynchiidae *(1-3)* (p. *H553-H554*).

teriors differ in appearance and are not always easy to interpret in mutually intelligible terms.

There is still little agreement on the morphological features of most value in defining genera and higher taxa. Different features are stressed by different workers and synonyms will certainly be discovered when some existing genera are more fully known. Nevertheless, ultimate recognition of accurately defined genera holds promise of rigorous stratigraphic refinement. Features of external morphology used in grouping genera into families include: degree and type of ornament, shape of shell, beak and beak ridges, interarea, shape of fold and sulcus, form of commissure, and presence of marginal spines. Muscle impressions appear significant features in the interior of

the pedicle valve and, in the brachial valve, all details of the cardinalia, shape and degree of development of hinge plates, septalium, median septum, and cardinal process are important characters.

In contrast to Mesozoic and Tertiary Rhynchonellacea, the form of the crura as yet has had little influence on classification of Paleozoic genera. In many genera the crura are unknown, although serial grinding often allows accurate observation of extremely delicate features. As far as now known, it seems that morphology of the crura is not as valuable in classification in Paleozoic rhynchonellaceans as in later forms.

Family ANCISTRORHYNCHIDAE Cooper, 1956

[*nom. transl.* H. SCHMIDT, herein (*ex* Ancistrorhynchinae COOPER, 1956, p. 618)] [Materials for this family prepared by HERTA SCHMIDT]

Small interarea in pedicle valve or both valves, ventral sulcus and dorsal fold present, delthyrium open, rounded costae extending from apex to anterior margin; commissure finely denticulate to undulate. Dental plates well developed; hinge plate divided; septa, cardinal process, and septalium wanting. *M.Ord.*

Ancistrorhyncha ULRICH & COOPER, 1942, p. 624 [**A. costata*; OD]. Small; sulcus and fold weakly developed, tongue short; pedicle valve with vestigial interarea and foramen. Costae numerous, fine. Crura long, slender, ending in hooklike expansions that point anterolaterally; socket walls strongly curved medially. *M.Ord.*, N.Am.——FIG. 419,5; 420,1. **A. costata*, USA(Okla.); 419,5a-d, ped.v., brach.v., lat., ant. views, ×2 (189); 420, 1a-e, ser. secs., ×5 (Schmidt, n).

Drepanorhyncha COOPER, 1956, p. 627 [**Porambonites ottawaensis* BILLINGS, 1862; OD]. Small to medium-sized; sulcus and fold well developed;

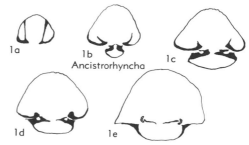

Fig. 420. Ancistrorhynchidae (p. *H553*).

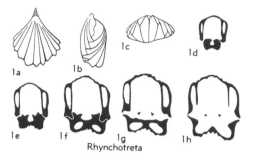

FIG. 421. Oligorhynchiidae (p. *H554*).

both valves with narrow interareas; costae not very numerous, medium fine. Pedicle valve with large teeth and elongate dental plates spaced closely together, making deep, narrow delthyrial cavity. Hinge plates concave; crural bases enlarged to concave plates; crura slender, very long. *M.Ord.,* N.Am.-Eu.——FIG. 419,4. **D. ottawaensis* (BILLINGS), Can.(Ont.); *4a-d,* pedv.v., brach.v., lat., ant. views, ×1; *4e,* brach.v. int., ×3; *4f,g,* ped.v. int., long. sec. showing crura, ×2 (189).

Family OLIGORHYNCHIIDAE
Cooper, 1956

[Oligorhynchiidae COOPER, 1956, p. 658] [Materials for this family prepared by HERTA SCHMIDT]

Small, triangular shells longer than wide, with erect beak; folding inverted at least in posterior part of shell (fold in pedicle and sulcus in brachial valve). Strongly plicated; teeth and dental plates well developed; hinge plate divided, supported by inner ridge or by plates including notothyrial cavity, or by both; median septum lacking, but inner dorsal swelling or ridge present, corresponding to outer sulcus; cardinal process wanting. *M.Ord., ?Sil.*

Oligorhynchia COOPER, 1935, p. 48 [**O. subplana;* OD]. Pedicle valve with strong median furrow between costae but without sulcus; posterior part of brachial valve with sulcus reverting anteriorly to become fold; few (3 or 4) strong, rounded angular costae. Dental plates strong, divergent. Hinge plate divided, attached to inner swelling produced by outer sulcus of valve, further supported by delicate converging plates; crura long, slender, nearly straight and extending almost directly anteriorly or obliquely toward pedicle valve. *M.Ord.,* N.Am.-Eu.——FIG. 419,3. **O. subplana,* USA(Tenn.); *3a-d,* ped.v., brach.v., lat., ant. views, ×4 (189).
Dorytreta COOPER, 1956, p. 666 [**D. bella;* OD]. Externally resembling *Sphenotreta,* but with sulcus of brachial valve reverting to fold anteriorly; foramen with thickened margin. Dental plates short. Crura shorter and stouter than those of

Sphenotreta, abruptly bent toward pedicle valve. *M.Ord.,* N.Am.——FIG. 419,2. **D. bella,* USA (Okla.); *2a-d,* ped.v., brach.v., lat., ant. views of holotype, ×4 (189).
?Rhynchotreta HALL, 1879, p. 166 [**Terebratula cuneata* DALMAN, 1828; OD]. Acutely triangular, with apical foramen; low median fold in pedicle valve beginning nearly at apex anteriorly becoming reversed to shallow sulcus; brachial valve with narrow depression at umbo, developing anteriorly into fold; sides of both valves in posterior parts abruptly bent, commissure thus lying nearly in plane; costae strong, subangular or rounded. Dental plates nearly parallel (in cross section); umbonal cavities rather narrow; teeth strong. Hinge plate divided, halves resting on thickened shell wall in posterior part of valve; hinge plates separated anteriorly from shell wall by cavities on either side of median ridge corresponding to outer sulcus; crural bases prominent; crura slender, nearly straight, extending almost anteriorly. *Sil.,* Eu.(Scand.-G.Brit.-Czech.)-N.Am.——FIG. 421,1. **R. cuneata* (DALMAN), Gotl.; *1a-c,* brach.v., lat., ant. views, ×1; *1d-h,* ser. secs. at 0.5, 0.8, 1.4, 1.55, and 1.7 mm. from post. extremity, ×4 (all from 702).
Sphenotreta COOPER, 1956, p. 663 [**S. cuneata;* OD]. Triangular to oval, with strong folding in pedicle valve and deep sulcation in brachial valve extending from posterior to anterior margin; costae numerous, rounded-angular. Dental plates short. Hinge plates small, triangular; crura long and slender, directed obliquely to anterior margin and slightly toward pedicle valve. *M.Ord.,* N.Am. ——FIG. 419,1. **S. cuneata,* USA(Tenn.); *1a-d,* ped.v., brach.v., ant., lat. views, ×4 (189).

Family RHYNCHOTREMATIDAE
Schuchert, 1913

[*nom. correct.* SCHUCHERT & LEVENE, 1929, p. 18, *et transl.* COOPER, 1956, p. 628 (*ex* Rhynchotreminae SCHUCHERT, 1913, p. 396)] [Materials for this family prepared by HERTA SCHMIDT]

Sulcus and fold well developed; costae strong, angular to rounded angular, beginning at apex, in most genera simple, crossed by concentric lamellae or striae; commissure denticulate. Hinge plates concave, separated by notothyrial cavity containing septiform cardinal process. *M.Ord.-M.Dev.*

Subfamily RHYNCHOTREMATINAE Schuchert, 1913

[*nom. correct.* SCHUCHERT & LEVENE, 1929, p. 18 (*pro* Rhynchotreminae SCHUCHERT, 1913, p. 396)] [=Lepidocyclidae COOPER, 1956, p. 657]

Dental plates and umbonal cavities in most genera distinct; notothyrial cavity formed by welding of hinge plates with median septum or ridge or callosity. [The relations of nominal genera assigned to this subfamily are not yet sufficiently cleared up; some of them may prove to be synonymous.] *M.Ord.-L.Dev.*

Rhynchotrema HALL, 1860, p. 68 [*Atrypa increbescens* HALL, 1847, p. 146; OD]. Small, rostrate, rounded triangular to transversely elliptical in outline; delthyrium narrow, only partially closed by narrow, elongate deltidial plates. Dental plates short; umbonal cavities small; teeth with large fossettes; muscle field triangular, adjustor scar large; dorsal median septum extending to middle of valve; notothyrial cavity small; cardinal process slender to thick. *M.Ord.-U.Ord.*, N.Am.-?Eu. ——FIG. 422,3. **R. increbescens* (HALL), M.Ord. (Trenton.), USA; *3a*, ped.v., ×1; *3b,c*, brach.v. and post. views, ×2 (189).

Ferganella NIKIFOROVA, 1937, p. 39 [**F. turkestanica*; OD]. Medium-sized to large, rounded to subpentagonal in outline; ventral beak subercct; deltidial plates obsolete. Teeth stout; dental plates and umbonal cavities present; notothyrial cavity large, oval, supported by thick median septum; cardinal process thin. *Sil.-L.Dev.*, Asia(Fergana)-Eu. (G. Brit.-Baltic). —— FIG. 422,4. **F. turkestanica*, Downton., Fergana; *4a-d*, ped.v., brach.v., lat., ant. views, ×1; *4e*, sec. in articulation zone, ×3 (599).

Hypsiptycha WANG, 1949, p. 17 [**H. hybrida*; OD]. Small, elongate rounded in outline, sulcus and fold strongly pronounced; ventral beak suberect; foramen large; deltidial plates well developed, moderately convex, uniting in mid-line; surface lamellae pronounced. Teeth strong; dental plates high, bounding narrow umbonal cavities; muscle field subcordate, with prominent adjustor scar; median ridge of brachial valve short. *U.Ord.*, N.Am.——FIG. 422,6. **H. hybrida*, USA(Iowa); *6a-d*, ped.v., brach.v., lat., ant. views, ×2; *6e*, ped.v. int., ×2; *6f*, brach.v. int. (beak portion), ×4 (851).

Lepidocyclus WANG, 1949, p. 12 [**L. laddi*; OD]. Medium-sized to large, old specimens globose; both beaks usually curved; delthyrium wide; deltidial plates large, conjunct along median line; surface lamellae strong. Teeth very stout, supported by strong shell thickening that encloses tubular delthyrial cavtiy; ventral muscle field large, flabelliform, deeply impressed; adjustor scar small; hinge plates strong; crura long, ending in hooklike expansions; dorsal median septum extending approximately to middle of shell. *U.Ord.*, N.Am.——FIG. 422,7. **L. laddi*, Maquoketa Sh., USA(Iowa); *7a-d*, ped.v., brach.v., lat., ant. views, ×1; *7e,f*, ped.v. int., brach.v. int., ×1 (851).——FIG. 423,3. *L. capax* (CONRAD), Cincinnat., USA(Ohio); *3a-c*, transv. secs. 14.2, 13.0, 12.8 mm. from ant. margin, ×3; *3d*, transv. sec. 13.2 mm. from ant. margin, ×4.5 (Schmidt, n).

?Pleurocornu HAVLÍČEK, 1961, p. 46 [**Rhyn-*

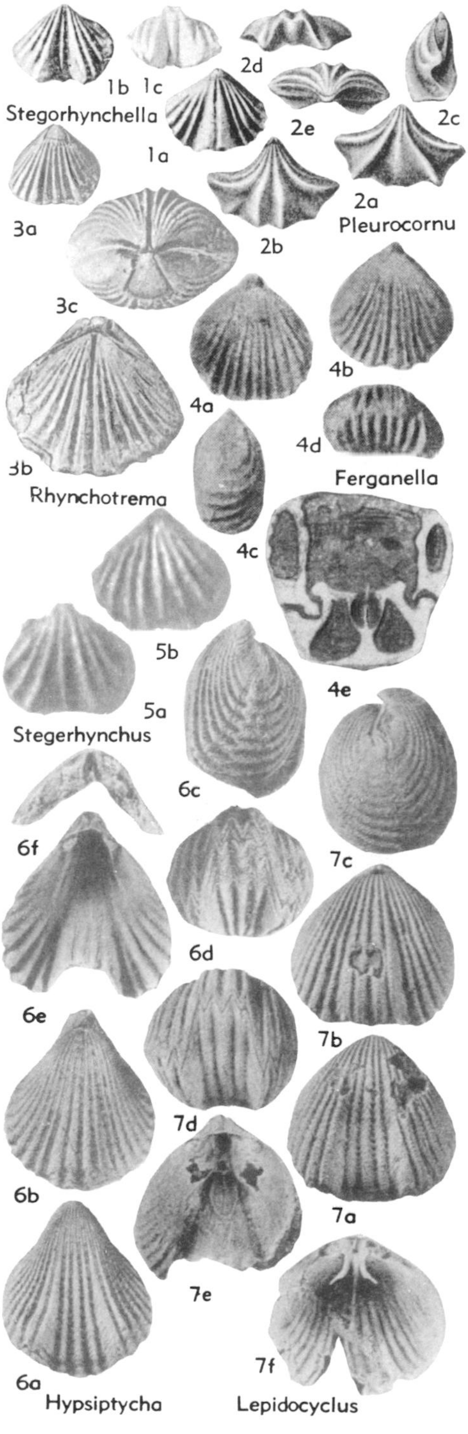

FIG. 422. Rhynchotrematidae (Rhynchotrematinae) (p. *H555-H556*).

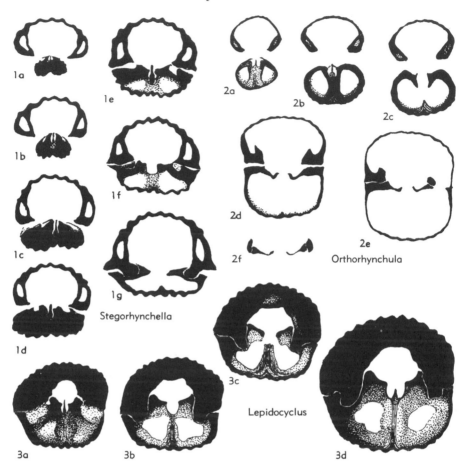

FIG. 423. Rhynchotrematidae (Rhynchotrematinae) *(1, 3)*, Orthorhynchulinae *(2)* (p. *H555-H557*).

chonella amissa BARRANDE, 1879; OD]. Medium-sized, rounded trigonal in outline, with narrow, protracted beak; ventral sulcus deep, defined by high costae; dorsal fold scarcely indicated; costae not numerous, strong, becoming very high anteriorly; commissure strongly denticulate. Teeth situated on thickened wall of valve; hinge plate entire, with small median cavity only in its upper-most part; cardinal process thin, septiform; median septum stout. *Sil.,* Eu.(Boh.).——FIG. 422,2; 424,*1. *P. amissum* (BARRANDE), Wenlock, Boh. (Lodenice); 422,*2a-e,* ped.v., brach.v., lat., ant., post. views, ×1 (53); 424,*1a-e,* ser. secs., ×3 (411a).

Stegerhynchus FOERSTE, 1909, p. 98 [**Rhynchonella* *(St.) whitii-praecursor* (*=S. praecursor*); OD]. Small, transversely elliptical in outline. Dental plates and umbonal cavities present. Interior of brachial valve with median longitudinal elevation posteriorly broadening and strengthened by shell thickening, leaving only narrow notothyrial

cavity; cardinal process very narrow. *M.Sil.,* N.Am. ——FIG. 422,*5. *S. praecursor* (FOERSTE), Clinton., USA(Tenn.); *5a,b,* ped.v. and brach.v. views, ×2 (305).

Stegorhynchella RZHONSNITSKAYA, 1959, p. 27 [**Stegerhynchus decemplicatus angaciensis* CHERNYSHEV, 1937, p. 29; OD]. Probably synonymous with *Stegerhynchus. U.Sil.,* Asia (Mongol.).——FIG. 422,*1;* 423,*1. *S. decempli-catus angaciensis* (CHERNYSHEV), Tuva; 422,*1a-c,* ped.v., brach.v., ant. views, ×1 (910); 423,*1a-g,* transv. secs., 7.45, 7.4, 7.3, 7.2, 7.0, 6.9, 6.8 mm. from ant. margin, ×6 (Schmidt, n).

Subfamily ORTHORHYNCHULINAE Cooper, 1956

[*nom. transl.* SCHMIDT, herein (*ex* Orthorhynchulidae COOPER, 1956, p. 669]

Dental plates reduced, scarcely visible because of thickened shell wall. Pair of crural plates starting from inner edges of hinge

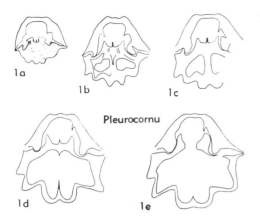

FIG. 424. Rhynchotrematidae (Rhynchotrematinae)
(p. *II556*).

plates and extending dorsally, may be obscured by callosity; dorsal septum present or reduced. *M.Ord.-M.Dev.*

Orthorhynchula HALL & CLARKE, 1893, p. 181 [*Orthis? linneyi* JAMES, 1881; OD]. Medium-sized, broadly elliptical in outline, with short, straight hinge line and interareas in both valves; delthyrium without deltidial plates; costae strong. Teeth blunt; dental plates visible in young specimens, umbonal cavities filled by adventitious testaceous matter in older shells; ventral muscle field short, subquadrate. Hinge plates strongly concave; crura very long; crural plates in older specimens united by callosity imbedding dorsal septum. *M.Ord.*, N.Am.——FIG. 423,2; 425, 3. *O. linneyi* (JAMES), USA(Ky.); 423,2a-f, transv. secs. 13.2, 13.0, 12.8, 12.5, 12.3, 12.0 mm. from ant. margin, ×3.75 (Schmidt, n); 425,3a,b, lat., post. views, ×1; 3c, brach.v. int. (tilted), ×2 (189).

Callipleura COOPER, 1942 [*Rhynchospira nobilis* HALL, 1860, p. 83; OD] [=*Cyclorhina* HALL & CLARKE, 1893, p. 206 (*non* PETERS, 1871)]. Medium-sized to large, broadly elliptical to pentagonal in outline; hinge line straight, short, laterally with winglike expansions; interareas in both valves; ventral beak truncated by large, round foramen; delthyrium very broad, only partially covered by small deltidial plates; costae crossed by fine concentric striae, crests of costae formed by row of knots, each knot corresponding to several striae. Teeth broad, attached to wall of valve and supported by thick converging dental plates forming pedicle cavity; crural plates resting on shell wall; cardinal process very delicate, conjoint with median ridge, both commonly imbedded in shell substance; crura ending in spoon-shaped processes. *M.Dev.(Hamilton.)*, N.Am.——FIG. 425,4. *C. nobilis* (HALL), USA(N.Y.); 4a, brach.v. view, ×1; 4b, ped.v. int., ×1 (914).

Latonotoechia HAVLÍČEK, 1960, p. 244 [*Terebratula latona* BARRANDE, 1847; OD]. Medium-sized to large, sulcus and fold commonly asymmetrical; foramen hypothyridid, anteriorly bound-

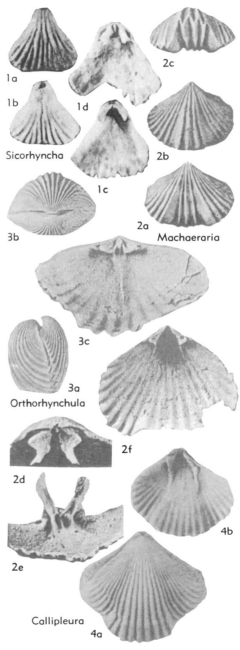

FIG. 425. Rhynchotrematidae (Orthorhynchulinae)
(p. *H557-H558*).

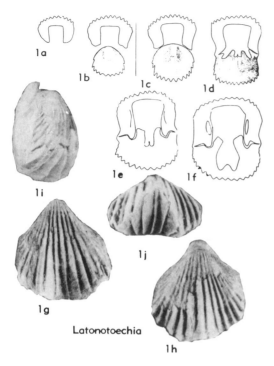

FIG. 426. Rhynchotrematidae (Orthorhynchulinae)
(p. *H558*).

ed by conjunct deltidial plates; dental plates scarce-ly discernible because of thickened shell wall. Hinge plates resting on callosity which fills out posterior part of valve, leaving free notothyrial cavity; crural plates not discernible. *L.Dev.,* Eu. (Boh.).——FIG. 426,*1*. **L. latona* (BARRANDE), Koněprusy; *1a-f,* ser. secs. at 17.7, 17.0, 16.3, 16.0, 15.3, and 14.8 mm. from ant. margin, ×2; *1g-j,* ped.v., brach.v., lat., ant. views, ×1 (411a).

Machaeraria COOPER, 1955, p. 55 [**Rhynchonella formosa* HALL, 1857, p. 76; OD]. Medium-sized; delthyrium partially closed by disjunct deltidial plates; foramen small. Teeth slender, with small fossettes; dental plates short, thin, bounding nar-row umbonal cavities; diductor scars elongate-flabellate; adductor scars small, elongate-oval. Socket ridges terminating in small teeth which articulate with fossettes of pedicle valve; crura curved, crescentic in section, with free ends blunt-ly pointed; crural plates meeting floor of valve to form narrow notothyrial cavity; cardinal proc-ess consisting of long, thin shaft and narrow crinkled myophore. *L.Dev.(Helderberg.),* N.Am. ——FIG. 425,*2*. **M. formosa* (HALL), USA(N.Y.); *2a-c,* ped.v., brach.v., ant. views, ×1; *2d,* brach.v. int. (post. portion), ×4; *2e,* brach.v. int. (tilted), ×4; *2f,* ped.v. int., ×3 (185).

Orthorhynchuloides WILLIAMS, 1962, p. 240

[**Hemithyris nasuta* M'COY, 1852, p. 203; OD]. Like *Orthorhynchula* but with costae dying out anteriorly and without massive callosity medianly between convergent crural plates. *M.Ord.,* Eu. (Scot.). [WILLIAMS.]

Sicorhyncha HAVLÍČEK, 1961, p. 28 [**Stegerhyn-chus trinacrius* HAVLÍČEK, 1956, p. 571; OD]. Small to medium-sized, trigonal to pentagonal in outline; ventral sulcus normally developed; dorsal fold may be absent; foramen permesothyridid to epithyridid, scarcely touching top of delthyrium; deltidial plates conjunct in their upper parts; costae high, angular, some of them bifurcating. Interior resembling that of *Latonotoechia*. *L.Dev.,* Eu.(Boh.-Fr.).——FIG. 425,*1*. **S. trinacria trinacria* (HAVLÍČEK), Boh.(Hlubocepy); *1a,b,* ped.v., brach.v. views, ×2.5; *1c,d,* ped.v. int., brach.v. int., ×2.5, ×2.8 (411a).

Zlichorhynchus HAVLÍČEK, 1963, p. 403 [**Z. hiatus*; OD]. Medium-sized; greatest width toward front; no fold or sulcus; anterior commissure uni-

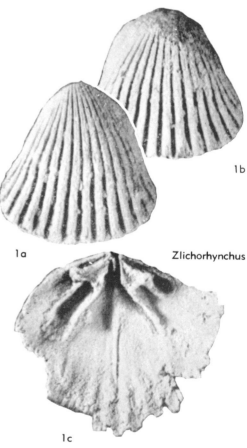

FIG. 427. Rhynchotrematidae (Orthorhynchulinae)
(p. *H558-H559*).

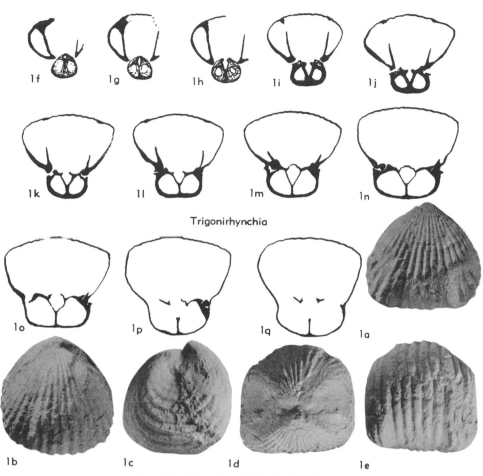

FIG. 428. Trigonirhynchiidae (p. H559).

plicate in adults; delthyrium large; deltidial plates minute; strong, rounded costae over whole shell. Dental plates thin and short; hinge plates triangular, concave, divided; small lamellar cardinal process; inner edges of hinge plates supported by crural plates that diverge anteriorly. *L.Dev.,* Eu. (Czech.).——FIG. 427,*1.* **Z. hiatus; 1a,b,* ped.v., brach.v. views, ×4; *1c,* brach.v., int., ×4 (413a). [MCLAREN.]

Family TRIGONIRHYNCHIIDAE
McLaren, n. fam.

[Materials for this family prepared by D. J. MCLAREN except as indicated otherwise]

Small to medium-sized; costae strong, simple, angular or subangular, extending from beak, rarely bifurcate; uniplicate, commissure serrate; fold and sulcus commonly developed. Dental plates present; median septum supporting well-formed septalium

that may be open or wholly or partly covered by plate uniting outer hinge plates; no cardinal process. *M.Ord.-L.Carb.(Miss.).*

Trigonirhynchia COOPER, 1942, p. 228 [**Uncinulina fallaciosa* BAYLE, 1878; OD] [*=Uncinulina* BAYLE, 1878 (*non* TERQUEM, 1862)]. [*non Trigonirhynchia* DAGIS, 1961, p. 94]. Medium-sized to large; rounded-triangular in outline; pedicle valve much less vaulted than brachial valve; front and flanks usually steeply sloping or truncated; sulcus and fold moderately deep but well defined; costae strong, generally simple, angulated, beginning in beaks; commissure denticulate. Dental plates convergent dorsally; brachial valve with large septalium covered in its anterior part by convex or flat plate. *L.Dev.-M.Dev.,* Eu.—— FIG. 428,*1.* **T. fallaciosa* (BAYLE), L.Dev., Fr. (Néhou); *1a-e,* ped.v., brach.v., lat., post., ant. views, ×1.2; *1f-q,* ser. transv. secs., ×2.8 (931d). [SCHMIDT.]

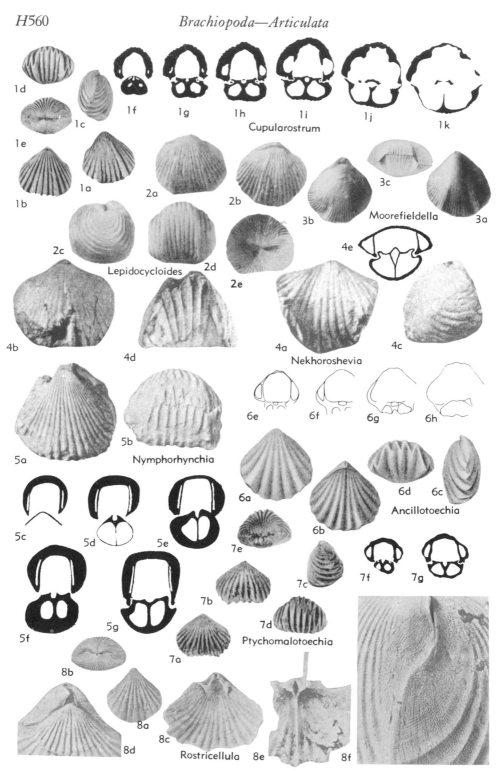

1d
1c
1e
1a
1b
1f
1g
1h
1i
1j
1k
Cupularostrum

2a
2b
3c
3b
Moorefieldella
3a

2c
2d
Lepidocycloides
2e
4e

4b
4d
4a
Nekhoroshevia
4c

5a
5b
Nymphorhynchia
6e
6f
6g
6h

5c
5d
5e
6a
6b
6d
6c
Ancillotoechia

5f
5g
7e
7b
7c
7f
7g

7d
Ptychomalotoechia
7a

8b
8a
8c
Rostricellula
8e
8f
8d

FIG. 429. Trigonirhynchiidae (p. *H561-H562*).

Ancillotoechia Havlíček, 1959, p. 78 [*Rhyn-chonella ancillans* Barrande, 1879; OD]. Similar to *Cupularostrum* but narrower, with smaller apical angle; fold high anteriorly. Septalium broad, entirely covered, supported by median ridge or strong septum, posteriorly broader than septalial cavity. *M.Sil.-U.Sil.*, Eu.——Fig. 429,6. *A. ancillans* (Barrande), U.Sil., Boh.; *6a-d*, ped.v., brach. v., lat., ant. views, ×2 (53); *6e-h*, ser. transv. secs. 7.55, 7.50, 7.45, 7.40 mm. from ant. margin, ×5 (411a).

Bathyrhyncha Fuchs, 1923, p. 854 [*B. sinuosa*; OD]. Medium-sized to large, inflated; pedicle beak incurved; sulcus extending from beak, trough-shaped; costae rounded, with narrow interspaces. Interior imperfectly known; dental plates short; muscle field deeply impressed, elongate; septalium apparently uncovered, septum strong, may extend to mid-length. *L.Dev.(Gedin.)*, Ger.-Belg.——Fig. 430,3. *B. sinuosa*; *3a*, ped.v. int. mold, ×1 (212b); *3b-d*, lat., post., ant. views of int. molds, ×2 (907a).

Cupularostrum Sartenaer, 1961 [*C. recticostatum*; OD]. Small to medium-sized; fold and sulcus developed anteriorly only, sulcus shallow; pedicle beak prominent; pedicle valve inflated; crest of fold falls to anterior margins; shell thick. Interior structures strong; dental plates short; U-shaped septalium, open posteriorly, covered with strong, arched, plate anteriorly, with median longitudinal ridge on ventral surface; cover persisting forward of articulation; septum stout, persists up to half shell length. [This genus includes many species formerly assigned to *Camarotoechia*.] *M.Dev.*, N.Am.(N.Y.-Yukon).——Fig. 429,1. *C. recticostatum*, M.Dev., USA(N.Y.); *1a-e*, ped.v., brach.v., lat., ant., post. views, ×1; *1f-k*, ser. transv. secs. at 1.3, 1.7, 2.0, 2.2, 2.4, 2.7 mm. from apex, ×3 (930c).

Hemiplethorhynchus von Peetz, 1898, p. 178 [*H. fallax*; OD] [=*Greenockia* Brown, 1952, p. 91]. Medium-sized, subpentagonal in outline, uniplicate; pedicle valve flattened, with sulcus developed from mid-length; beak small, incurved; brachial valve convex; fold well marked; costate, with abundant subangular costae. Dental plates small; hinge plates united anteriorly, divided posteriorly, forming triangular opening into septalium at apex. *L.Carb.*, USSR(Altay)-Can.(Alta.).—— Fig. 430,1a-d. *H. fallax*, Tournais., Altay; *1a-c*, ped.v., lat., ant. views, ×1; *1d*, brach.v. int., ×1 (711a).——Fig. 430,1e-g. *H. snaringensis* (Brown), L.Miss.(Banff), Can.(Alta.) [Type-species of *Greenockia*]; *1e*, ped.v., young spec., ×1; *1f*, brach.v., ×1; *1g*, post. view, int. mold, ×1 (907b).

?**Lepidocycloides** Nikiforova, 1961, p. 212 [*L. baikiticus*; OD]. Similar to *Lepidocyclus* but without deltidial plates. Internally with weak dental plates, strongly impressed muscle impressions and deep pedicle cavity; stout divided hinge

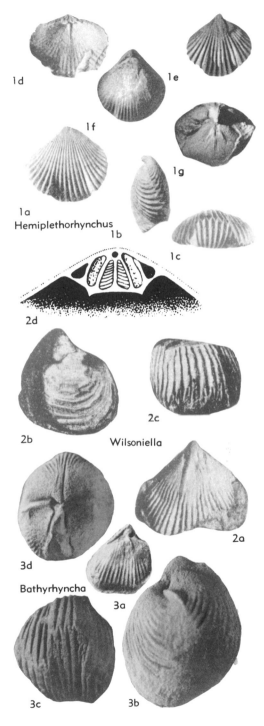

Fig. 430. Trigonirhynchiidae (p. *H561-H562*).

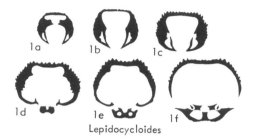

Lepidocycloides

Fig. 431. Trigonirhynchiidae (p. *H561-H562*).

plates; open septalium; without cardinal process. *U.Ord.*, USSR(Sib.).——Fig. 429,2; 431,*1*. *L. baikiticus*; *429,2a-e*, ped.v., brach.v., lat., ant., post. views, ×1; *431,1a-f*, ser. transv. secs. 14.1, 13.8, 13.5, 12.8, 12.1, 11.6 mm. from ant. margin, ×1.5 (602).

Moorefieldella GIRTY, 1911, p. 62 [*Rhynchonella Eurekensis* WALCOTT, 1884, p. 223; OD]. Small to medium-sized, ovate to subpentagonal, equivalve, uniplicate; low fold and sulcus developed anteriorly only; beak prominent, suberect; interarea wide; costellate, with even, rounded costellae; commissure smooth. Dental plates present; divided hinge plate; high septum. *U.Miss.(Meramec.)*, USA(Ark.-Okla.-Nev.).——Fig. 429,3. *M. eurekensis* (WALCOTT); *3a-c*, ped.v., brach.v., ant. views, ×1 (346a).

?Nekhoroshevia BUBLICHENKO, 1956 [*N. altaica*; OD]. High, cuboidal; septalium deep, entirely covered with convex plate and supported by high, slender median septum. *U.Dev.(Frasn.)*, USSR (Rudny Altay).——Fig. 429,4. *N. altaica*; *4a-d*, ped.v., brach.v., lat., ant. views, ×1; *4e*, transv. sec. near apex, ×2 (690).

Nymphorhynchia RZHONSNITSKAYA, 1956, p. 53 [*N. bischofioides*; OD]. Subtrigonal to subpentagonal; strongly marked, concave interareas; delthyrium open; fold and sulcus broad, poorly developed; costae strong, asymmetrical, flattened near front, with longitudinal grooves; no marginal spines; fine, closely set transverse notches on costae over entire shell. Dental plates close to lateral margins of shell; hinge plate massive; septalium open. *M.Dev.(Eifel.)*, Eu.(Boh.)-USSR (Kuznetsk).——Fig. 429,5. *N. bischofioides*, Kuznetsk; *5a,b*, brach.v., ant. views, ×1.5 (690); *5c-g*, ser. transv. secs., ×3 (689a).

Ptychomaletoechia SARTENAER, 1961, p. 7 [*Rhynchonella Omaliusi* GOSSELET, 1877; OD]. Similar to *Cupularostrum*, but with deeper, wider sulcus. Crural bases stronger; hinge plates more developed anteriorly; septalium uncovered. *U.Dev.(Famenn.)*, Eu.-N.Am.-Asia.——Fig. 429,7. *P. omaliusi* (GOSSELET), W.Eu.; *7a-e*, ped.v., brach.v., lat., ant., post. views, ×1; *7f,g*, ser. transv. secs. at 0.85, 1.15 mm. from apex, ×3 (930c).

?Rostricellula ULRICH & COOPER, 1942, p. 625 [*R. rostrata*; OD]. Subtriangular to subpentagonal, deltidial plates rudimentary; interarea narrow. Strong dental plates; teeth small, curved; septalium short; septum strong, extending to middle of valve or beyond; crura long, slender. *M.Ord.-U.Ord.*, *?Sil.*, N.Am.-Eu.-Asia.——Fig. 429,8. *R. rostrata*, M.Ord., N.Am.; *8a,b*, ped.v., post. views, ×1; *8c*, ped.v. int., tilted, ×2; *8d,e*, ped.v. beak, brach.v. int., ×4; *8f*, lat. view showing fine ornament, ×6 (188).

Sinotectirostrum SARTENAER, 1961a, p. 3 [*S. medicinale*; OD]. Similar to *Cupularostrum* but larger; fold and sulcus develop earlier; pedicle valve less inflated. Dental plates more persistent; septalium deep and narrow, covered anteriorly but weakly; covering commonly not preserved. *U.Dev. (Famenn.)*, W.Can.——Fig. 432,1. *S. medicinale*; *1a,b*, lat., post. views, ×1; *1c-f*, ser. transv. secs. at 2.0, 2.25, 3.0, 4.0 mm. from apex, ×3 (709a).

Wilsoniella KHALFIN, 1939, p. 83 [*W. prima*; OD] [=*Ussovia* KHALFIN, 1955, p. 239 (obj.)] [*non Wilsonella* NIKIFOROVA, 1937]. Large, subcuboidal, with inflated brachial valve; fold and sulcus developed only anteriorly; coarse, rounded costae with narrow interspaces. Dental plates rudimentary, teeth large; septalium opening into small foramen posteriorly; anteriorly, hinge plates united by double, ventrally ridged, triangular plate. *?L.Dev.*, USSR(Altay).——Fig. 430,2. *W. prima*; *2a-c*, ped.v., lat., ant. views, ×1; *2d*, brach.v. int., enlarged (690).

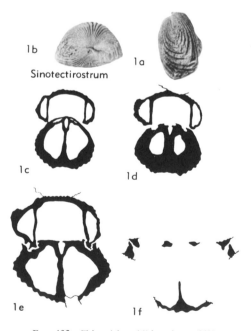

Sinotectirostrum

Fig. 432. Trigonirhynchiidae (p. *H562*).

Family UNCINULIDAE Rzonsnitskaya, 1956

[Uncinulidae RZHONSNITSKAYA, 1956, p. 125] [Materials for this family prepared by HERTA SCHMIDT except as indicated otherwise]

Shells occurring in two different shapes representing different growth stages: (1) high forms (most common), cubic to globose, with front and sides truncated by rectangular bending of valves, commissure commonly situated on truncated parts; (2) flat forms (rare), with valves meeting at acute angle. Sulcus and fold present or absent; tongue nearly always present; ventral beak curved to erect; foramen minute or wanting; costae in most species numerous, rounded or flattened, separated by linear furrows, each furrow of high forms projecting beyond margin of valve as slender spine which extends under costa of opposite valve (Fig. 433), but rarely spines are wanting; costae in truncated parts of shell much flattened and longitudinally grooved, crossed by undulating or zigzag-shaped transverse lines. Dental plates well developed, may be welded to wall of valve or umbonal cavities may be filled out by callus in older specimens; dorsal septum present; septalium in more primitive forms well developed and free, in more progressive forms partially or completely filled out or obsolescent. *Sil.-U.Dev., ?Perm.*

Subfamily UNCINULINAE Rzhonsnitskaya, 1956

[*nom. transl.* SCHMIDT, herein (*ex* Uncinulidae RZHONSNITSKAYA, 1956, p. 125)]

Cardinal process strongly developed, broad and low or narrower and projecting, with its basal part filling out posterior portion or whole cavity of septalium; myophore consisting of numerous vertical ridges; hinge plates conjunct also anteriorly from cardinal process. *L.Dev.-U.Dev., ?Perm.*

Uncinulus BAYLE, 1878, pl. 13, fig. 15 expl. [**Hemithiris subwilsoni* D'ORBIGNY, 1850, p. 92; SD OEHLERT, 1884, p. 423]. Roundish to pentagonal in outline; both valves convex; sulcus and fold moderately or weakly developed, in few species wanting; tongue usually well marked, but rarely absent; commissure even, appearing denticulate only if worn. Ventral muscle field divided by delicate septum or ridge; prominent oval diductors enclosing small round adductors; dorsal median septum rather high; septalium reduced, completely filled out; cardinal process broad, low, covering large part of hinge plate. *L.Dev.-U.Dev.,*

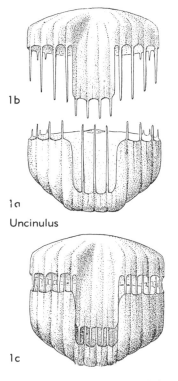

1b

1a
Uncinulus

1c

FIG. 433. Uncinulid marginal spines (schematic); *1a-c*, brach.v., ped.v., both valves connected, slightly opened (931a) (p. *H563*).

cosmop.——FIG. 434,6; 435,3. **U. subwilsoni* (D'ORBIGNY), L.Dev., Fr.; 434,*6a-e*, ped.v., brach. v., lat., ant. views, ped.v. int. mold, ×1; *6f*, brach.v. int., enl. (927); 435,*3a-l*, ser. secs., 13.1, 13.0, 12.8, 12.6, 12.3, 11.8, 11.7, 11.6, 11.5, 11.2, 10.2, 8.5 mm. from ant. margin, ×3 (Schmidt, n).

?**Fitzroyella** VEEVERS, 1959, p. 104 [**F. primula*; OD]. Small, subpentagonal in outline, rather flat, valves almost equally convex; hinge line short, in larger specimens nearly straight; beak suberect, with apical foramen; anterior and anterolateral parts of valves bent over perpendicularly; sulcus and fold developed, tongue short; rounded angular costae beginning at apex, increasing by intercalation and branching; commissure strongly denticulate. Dental plates strong but short, converging dorsally; brachial valve with short median ridge and rudimentary septalium; cardinal process not ascertained. *M.Dev.-U.Dev.*, Eu.-Australia.——FIG. 434,2. **F. primula*, U.Dev.(Frasn.), W.Australia (Fitzroy Basin); *2a-d*, ped.v., brach.v., ant., lat. views, ×3; *2e-k*, ser. secs. 0.65, 0.80, 0.90, 0.95, 1.05, 1.15 mm. from apex, ×7 (211).

Glossinulus SCHMIDT, 1942, p. 394 [**Rhynchonella adolphi mimica* BARRANDE, 1879, p. 178=*Glossinulus mimicus* (BARRANDE, 1879, p. 178); OD].

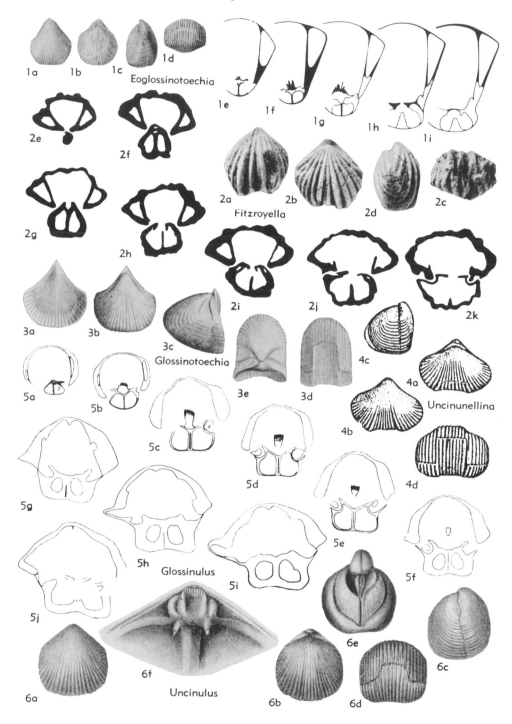

Fig. 434. Uncinulidae (Uncinulinac) (p. *H563, H565-H566*).

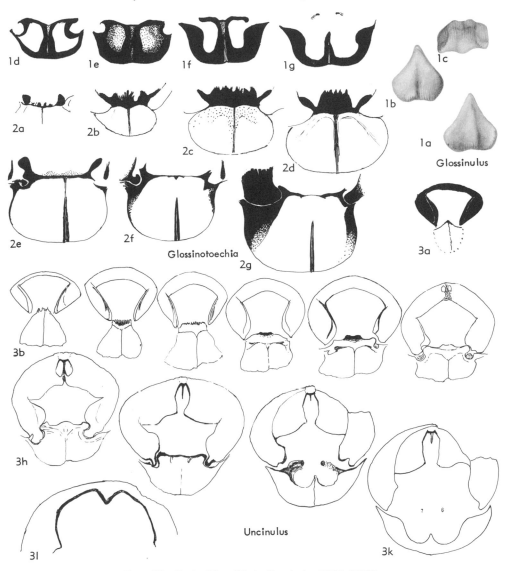

FIG. 435. Uncinulidae (Uncinulinae) (p. *H*563, *H*565).

Triangular to acute-oval in outline; pedicle valve flat to concave; beak nearly erect; tongue rectangular; zigzag lines on truncated parts of valves strongly pronounced. Umbonal cavities small; septalium reduced, completely filled; inner socket ridges prominent; cardinal process strongly developed. *L.Dev.-M.Dev.*

G. (Glossinulus). Cardinal process very long, tongue-shaped. *L.Dev.-M.Dev.,* Eu.——FIG. 434, 5; 435,*1.* **G. (G.) mimicus* (BARRANDE), L.Dev. (U.Ems.), Ger.(Eifel.); 435,*1a-c,* ped.v., brach. v., ant. view, ×1; 435,*1d-g,* ser. secs. of brach.v. ant. from cardinal process, 9.1, 8.9, 8.7, 8.1 mm.

from ant. margin, ×7 (Schmidt, n); 434,*5a-j,* sec. showing cardinal process, 6.4 mm. from ant. margin, ×4 (411a).

G. (Glossinotoechia) HAVLÍČEK, 1959, p. 81 [**Terebratula henrici* BARRANDE, 1847, p. 440; OD]. Cardinal process broader and shorter than in *Glossinulus. L.Dev.-M.Dev.,* Eu.-Afr.——FIG. 434,*3;* 435,*2.* **G. (G.) henrici* (BARRANDE), Boh.(Koněprusy); 434,*3a-e,* ped.v., brach.v., lat., ant., post. views, ×1 (53); 435,*2a-g,* ser. secs. of brach.v., 13.3, 13.0, 12.5, 12.0, 11.5, 11.3, 11.1 mm. from ant. margin, ×6 (Schmidt, n).

Eoglossinotoechia HAVLÍČEK, 1959, p. 81 [**E*

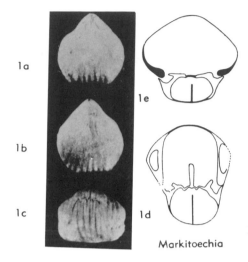

Markitoechia

Fig. 436. Uncinulidae (Uncinulinae) p. *H566*).

cacuminata; OD]. Resembling *Glossinotoechia*, but with pedicle valve more convex. Septalium small, its posterior part filled out by moderately large cardinal process; anterior part of septalial cavity free. *Sil.-L.Dev.,* Boh.——Fig. 434,*1.* **E. cacuminata* Havlíček, Sil., Dvorce; *1a-d,* ped.v., brach.v., lat., ant., views, ×1 (53); *1e-i,* ser. secs. 9.8, 9.6, 9.5, 9.35, 9.3 mm. from ant. margin, ×6 (411a).

Markitoechia Havlíček, 1959, p. 81 [**Uncinulus (Uncinulus) marki* Havlíček, 1956, p. 568; OD]. Exterior like *Uncinulus,* but with long, tongue-shaped cardinal process like that of *G. (Glossinulus).* *M.Dev.,* Eu.(Boh.).——Fig. 436,*1.* **M. marki* (Havlíček), Hlubočepy; *1a,b,* ped.v., brach.v., *1c,* ant. view, ×2.4; *1d,e,* secs. with cardinal process and anterior from cardinal process, ×7 (411a).

Plethorhyncha Hall & Clarke, 1893, p. 191 [**Rhynchonella speciosa* Hall, 1856, p. 81; SD Schuchert & LeVene, 1929, p. 99]. Large, longitudinally ovoid in outline; high forms higher than wide, with front and sides much flattened; sulcus and fold poorly developed or wanting; both beaks curved; cardinal margin of pedicle valve with considerable auriculate projections fitting into indentures of brachial valve margin; costae on truncated parts of shell with distinct longitudinal grooves and transverse zigzag lines; marginal spines present. Dorsal median septum well developed, in adult forms thickened; hinge plates in young specimens separated by small septalium, in old forms much thickened and coalesced, enclosing septalium. *L.Dev.,* N.Am.——Fig. 437,*1;* 438,*1.* **P. speciosa* (Hall), Oriskany, USA(Md.); *437,1a-d,* lat., ant., post. views, brach.v. int. (young shell), ×1 (396); 438,*1,* post. view (int. mold), ×1.5 (931c).

?**Uncinunellina** Grabau, 1932, p. 72 [**Uncinulus theobaldi* Waagen, 1884, p. 425; OD]. Broad, resembling *Uncinulus.* Interior insufficiently known. *Perm.,* Asia-Eu.(USSR).——Fig. 434,*4.* **U. theobaldi* (Waagen), India; *4a-d,* ped.v., brach.v., lat., ant. views, ×1 (845).

Subfamily HEBETOECHIINAE Havlíček, 1960

[*nom. transl.* Schmidt, herein (*ex* Hebetoechiidae Havlíček, 1960, p. 243)]

Primitive uncinulids, with high and flat forms developed as growth dimorphism; septalium pronounced, its cavity free or partially filled out by callus covering walls of septalium and, in some genera, projecting beyond hinge plates, forming incipient cardinal process. *Sil.-M.Dev.*

Hebetoechia Havlíček, 1959, p. 79 [*Terebratula hebe Barrande, 1847, p. 442; OD]. Sides and front of shell flattened, with longitudinally grooved costae; marginal spines present. Septalium filled out in its posterior part, with fill project-

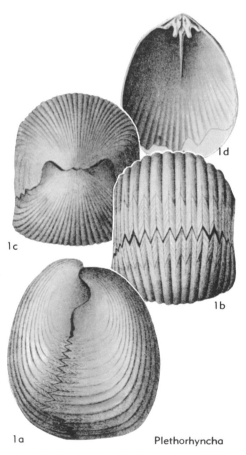

Plethorhyncha

Fig. 437. Uncinulidae (Uncinulinae) (p. *H566*).

ing above hinge plate as bilobate process; anterior part of septalium free. *U.Sil.(Ludlow)-L. Dev.*, Eu.(Boh.).——Fig. 439,*1*. **H. hebe* (Barrande), Sil., Boh.(Dlouhá Hora); *1a-d*, ped.v., brach.v., lat., ant. views, ×1 (53); *1e-h*, ser. secs. 9.9, 9.7, 9.5, 9.4 mm. from ant. margin, ×3 (411a).

?**Cassidirostrum** McLaren, 1961, p. 2 [**C. pedderi*; OD]. Medium-sized, gibbous, rounded to subpentagonal in outline; pedicle valve moderately vaulted, brachial valve inflated; beak erect to strongly incurved, defined by ridges; sulcus and fold inconspicuous; tongue moderately large, with rounded anterior margin; costae beginning at apex, angular, commonly simple; commissure denticulate. Dental plates convergent dorsally; umbonal cavities narrow; hinge plates not persisting throughout zone of articulation; septalium stretching farther forward than hinge plates, supported by strong median septum; septalial cavity filled out by callus extending over crural bases, forming bipartite process. *M.Dev.(L.Givet.)*, Can.(NW.Terr.).——Fig. 439,*2*; 440,*1*. **C. pedderi*, Can.(Anderson River); 439,*2a-e*, ped.v., brach.v., lat., ant., post. views, ×1; 440,*1a-d*, ser. secs. 2.8, 3.0, 3.3, 3.6 mm. from apex, ×2 (548a).

Estonirhynchia Schmidt, 1954, p. 236 [**Sphaerirhynchia (E.) estonica*; OD]. Almost perfectly globose; longitudinal furrows on costae and marginal spines wanting. Dental plates well developed, diverging dorsally, extending into zone of articulation; brachial interior like that of *Sphaerirhynchia*. *Sil.*, Eu.(Balt.)-?N.Am.——Fig. 441,*1*. **E. estonica* (Schmidt), Est.(Oesel); *1a-e*, ped.v., brach.v., ant., lat., post. views, ×2; *1f,g*, secs. in articulation zone of different specimens, ×4 (731b).

Lanceomyonia Havlíček, 1960, p. 243 [**Terebratula tarda* Barrande, 1847, p. 441; OD]. Resembling *Sphaerirhynchia* in shape; costae restricted to anterior halves of valves, broad, low, longitudinally grooved; marginal spines present. Ventral muscle field rather narrow, defined by ridge; dental plates distinct; septalium not covered; cardinal process wanting. *U.Sil.*, Eu.(Boh.).——Fig. 439,*3*. **L. tarda* (Barrande), Boh.(Dvorce); *3a,b*, brach.v., ant. view, ×1 (53); *3c-e*, ser. secs., 20.7, 20.0, 19.8 mm. from ant. margin, ×2 (411a).

?**Obturamentella** Amsden, 1958, p. 99 [**Wilsonia wadei* Dunbar, 1919, p. 52; OD]. Small with roundish to pentagonal outline; length, width, and breadth nearly equal; both valves gently convex and sharply deflected along margins; sulcus and fold shallow, narrow, restricted to anterior portion of shell; costae low and broad, rounded, beginning near apex; interspaces between costae narrow. Ventral muscle field deeply impressed; diductor scars elongate, enclosing small but deep adductor scars; muscle field

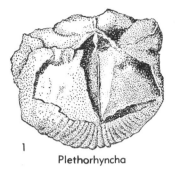

1 Plethorhyncha

Fig. 438. Uncinulidae (Uncinulinae) (p. *H566*).

divided in its whole length by low median septum; dental plates rudimentary; teeth scarcely defined from shell wall. Dorsal median septum low; septalium filled with shell material forming cardinal process of variable shape, either concave or flush with edges of septalium or projecting beyond hinge plate. *L.Dev.*, N.Am.(Tenn.-Okla.-Mo.).——Fig. 439,*7*. **O. wadei* (Dunbar); *7a-c*, ped.v., brach.v., lat. view, ×2; *7d,e*, ped.v. int., ×3, ×2; *7f,g*, different secs., ×18 (33).

?**Pectorhyncha** McLearn, 1918, p. 137 [**Atrypa obtusiplicata* Hall, 1852, p. 279; OD]. Small to medium-sized, gibbous to spheroidal; sulcus and fold weakly developed; costae beginning near apex, rounded, separated by narrow interspaces and crossed by concentric striae; median groove on anterior part of costae and marginal spines wanting. Dental plates at least partly cemented to shell wall; teeth longitudinally grooved, resting on shell wall; septalium not covered; dorsal median septum long, thickened posteriorly. *Sil.*, N.Am.(N.Y.).——Fig. 439,*4*. **P. obtusiplicata* (Hall), Niagaran, Lockport; *4a-c*, ped.v., brach.v., brach.v. int. views, ×1.5 (379).

Sphaerirhynchia Cooper & Muir-Wood, 1951, p. 195 [*pro Wilsonella* Nikiforova, 1937, p. 35 (*non* Carter, 1885)] [**Terebratula wilsoni* Sowerby, 1818, p. 38]. Globose-cubic, front and sides flattened; costae with longitudinal grooves in flattened parts of shell; marginal spines present. Dental plates reduced, restricted to apical parts of valve; hinge plates separated; septalium not covered; cardinal process wanting. *Sil.*, Eu.-?N.Am.——Fig. 439,*5*; 442,*3*. **S. wilsoni* (Sowerby), Wenlock., Eng.(Dudley); 439,*5a-d*, brach.v., lat., ant. views, ped.v., int., ×1 (229); 442,*3*, transv. sec. in articulation zone, ×5 (Schmidt, n).

Tadschikia Nikiforova, 1937, p. 35 [**Wilsonella (T.) wilsoniaformis*; OD]. Resembling *Sphaerirhynchia* in shape; costae on front and sides longitudinally grooved. Dental plates strong, subparallel; pedicle collar present in some shells; hinge plates united; septalial cavity reduced, filled out in its upper part by callus extending also over hinge plate and forming low process. *U.Sil.*, Asia

Fig. 439. Uncinulidae (Hebetocchiinae) (p. *H566-H567, H569*).

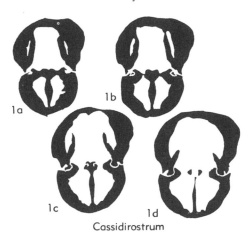

Cassidirostrum

Fig. 440. Uncinulidae (Hebetoechiinae) (p. H567).

(Turkestan).——Fig. 439,6. *T. wilsoniaformis* (NIKIFOROVA); *6a-e*, ped.v., brach.v., ant., lat., post. views, ×1; *6f*, sec. within articulation zone, ×2 (all from 599); *6g,h*, sections through septalium, ×2 (411a).

Subfamily HYPOTHYRIDININAE
Rzhonsnitskaya, 1956

[Hypothyridininae RZHONSNITSKAYA, 1956, p. 125]

Dorsal septum and septalium rudimentary or wanting; cardinal process and myophore resembling that of *Uncinulus*. *Sil.-U.Dev.*

Hypothyridina BUCKMAN, 1906, p. 323 [*pro Hypothyris* PHILLIPS, 1841, p. 55 (=KING, 1846, p. 28) (*non* HÜBNER, 1821)] [*Atrypa cuboides* SOWERBY, 1840, pl. 56, fig. 24; OD KING]. Medium-sized to large, high, cuboidal; sulcus and fold generally well defined, but low; tongue broad; costae numerous, low, rounded; commissure situated on angles of truncated front and sides or very near to angles. Dental plates distinct; dorsal median septum and septalium scarcely discernible; interior edges of hinge plates curved dorsally; hinge plates united by cardinal process. *M.Dev.-U.Dev.*, cosmop.——Fig. 443,2. *H. cuboides* (SOWERBY), probably U.Dev., Eng. (Plymouth); *2a,b*, ant. view, brach.v., ×1 (932). ——Fig. 442,1. *H. sp. cf. H. impleta* (SOWERBY), U.Dev., Ger.(Langenaubach); *1a,b*, transv. secs. of different young specimens, ×7 (Schmidt, n). ——Fig. 442,2. *H. procuboides* (KAYSER), M.Dev., Ger.(Eifel); *2a-d*, transv. ser. secs. 16.9, 16.7, 16.5, 16.4 mm. from ant. edge, ×5 (Schmidt, n).

Decoropugnax HAVLÍČEK, 1960, p. 244 [*Terebratula berenice* BARRANDE, 1847, p. 77; OD]. Small to medium-sized, rather flat, rounded trigonal to pentagonal in outline; sulcus and fold broad and shallow, tongue low, rectangular; costae numerous, fine, restricted to anterior halves of valves. Dental plates short, divergent; septalium

and cardinal process not observed. *U.Sil.*, Eu. (Boh.-Ural)-Asia(Fergana).——Fig. 443,1. *D. berenice* (BARRANDE), Boh.(Dlouhá Hora); *1a-d*, ped.v., brach.v., lat., ant. views, ×1 (53); *1e,f*, ser. secs., ×6 (411a).

Lorangerella CRICKMAY, 1963, p. 10 [*L. phaulomorpha*; OD]. Small-sized, subglobose; beak incurved, interarea not developed; fold and sulcus developed anteriorly; tongue vertical to recurved, broad, rounded; smooth posteriorly; rounded to subangular costae develop anteriorly, with narrow interspaces, grooved internally at margin. Dental plates slender, very short; hinge plates plane, inclined inwards; cardinal process supported, close to apex, by stout median ridge; crura cylindrical, hooked. *U.Dev.(L.Frasn.)*, Can.(Alta.).——Fig. 444,1. *L. phaulomorpha*; *1a-d*, ped.v., brach.v., ant., post. views, ×2.3 (915b); *1e-l*, ser. transv. secs. at 0.2, 0.6, 0.8, 0.9, 1.0, 1.2, 1.6, 1.9 mm. from apex, ×5 (McLaren, n). [McLaren.]

Subfamily HADRORHYNCHIINAE
McLaren, n.subfam.

[Materials for this subfamily prepared by D. J. McLaren]

Coarsely costate. Septalium very small, filled anteriorly; hinge plates wide, plane; no cardinal process; crura with small outside, lateral projections. *M.Dev.(Givet.).*

Hadrorhynchia McLAREN, 1961, p. 3 [*Pugnoides sandersoni* WARREN, 1944, p. 115; OD]. Medium-sized to large, subpentagonal to transversely elliptical; coarsely costate anteriorly, with rounded costae, umbones smooth, costae and interspaces grooved on interior of shell near commissure.

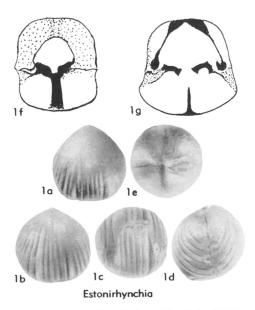

Estonirhynchia

Fig. 441. Uncinulidae (Hebetoechiinae) (p. H567).

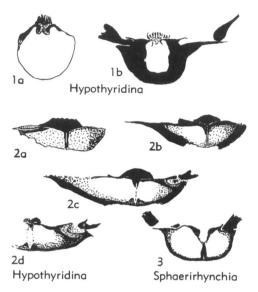

FIG. 442. Uncinulidae (Hebetoechiinae) *(3)*, (Hypothyridininae) *(1-2)* (p. *H567, H569*).

Dental plates widely spaced; hinge plates plane, divided by small septalium posteriorly, united anteriorly by 2 broad longitudinal ridges divided by median groove; septum long and slender. *M. Dev.,* W.Can.(Mackenzie Distr.-B.C.)-USA(Nev.)-USSR (Nov. Zemlya-NE.Sib.).——FIG. 445,*1.* *H. sandersoni* (WARREN), Givet., W.Can.(Mackenzie Distr.); *1a-c,* brach.v., lat., ant. views, ×1; *1d,* lat. view, detail, ×3; *1e-i,* ser. transv. secs. at 1.8, 2.05, 2.2, 2.65, 3.9 mm. from apex, ×3 (548a).

Family EATONIIDAE, Schmidt, n.fam.

[Materials for this family prepared by HERTA SCHMIDT]

Pedicle valve slightly convex, flanks tending to bend angularly toward opposite valve, beak slightly incurved, tongue marked; brachial valve moderately convex; costae beginning at apex, generally crossed by fine concentric striae. Dental plates cemented to shell wall (at least in old specimens); ventral muscle field well defined, in some genera deeply impressed and divided by small septum; hinge plates united by prominent cardinal process and welded with dorsal septum or ridge; cardinal process bilobate in most genera (at least in young shells) but may appear quadrilobate by dividing of each lobe, or may have trilobate form or may consist of simple process possibly developed by coalescing of bilobate form. *Sil.-L.Dev.*

Eatonia HALL, 1857, p. 90 [*Atrypa medialis* VANUXEM, 1852, p. 120; SD HALL & CLARKE, 1893, p. 205] [*non Eatonia* SMITH, 1875; *nec* CAMBRIDGE, 1898; *nec* MEUNIER, 1905] [=*Pareatonia* MCLEARN, 1918, p. 137 (obj.)]. Small to large, rounded in outline; deltidial plates large; ventral beak with ridges; lateral marginal parts of pedicle valve abruptly bending dorsally and partly covered by brachial valve; costae broad, most of them simple, but in addition fine costellae or radial striae and fine concentric striae may be present; commissure undulate to denticulate. Teeth small; ventral muscle field pentagonal, deeply impressed and bounded by prominent ridge; posterior parts of adductors enclosed in chambers on either side of small septum which continues posteriorly and anteriorly as low ridge dividing whole muscle field; cardinal process consisting of stout stem and 2 strong lobes with their elongate ends (myophores) excavated, stretching far into pedicle valve; crura stout at their bases, rounded in cross section, narrow and flattened distally, with tips expanded into transverse plates. *L.Dev.,* N.Am.——FIG. 446,*1;* 447, *3a-c.* *E. medialis* (VANUXEM), L.Helderberg, USA(Md.) (446,*1*), USA(N.Y.) (447,*3a-c*); 446, *1,* ped.v. int., ×1.3 (Schmidt, n); 447,*3a-c,* ped. v., brach.v., lat. views, ×1 (396).——FIG. 447, *3d. E. sinuata* HALL, L.Helderberg., USA(N.Y.); cardinal process and crural bases from below, ×2 (396).

Clarkeia KOZLOWSKI, 1923, p. 26 [*Terebratula antisiensis* D'ORBIGNY, 1847, p. 36; OD]. Medium-sized, rounded in outline, moderately convex; sulcus and fold distinct; ventral beak slightly incurved, with apical foramen; delthyrium wide, completely filled out by dorsal beak; costae strong, beginning at apex, those of fold bifurcating. Teeth thick; dental plates cemented to shell wall with

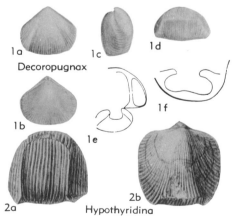

FIG. 443. Uncinulidae (Hypothyridininae) (p. *H569*).

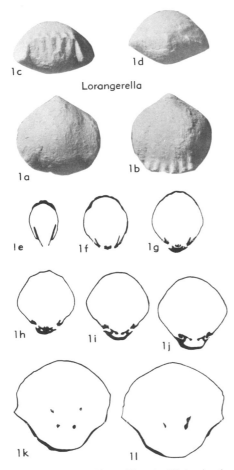

Fig. 444. Uncinulidae (Hypothyridininae) (p. H569).

anterior margins free; ventral muscle field flabelli-form, deeply impressed, bounded by ridge start-ing from dental plates; adductor scars very small, surrounded by large diductor scars; cardinal mar-gin of brachial valve much thickened in older specimens; muscular impressions marked, bounded by ridges; cardinal process prominent, bilobate in young shells, each lobe being divided into 2 sec-ondary lobes, but in older shells becoming tri-lobate by coalescence of median secondary lobes; dorsal septum thick, extending approximately 0.7 of length of valve; crura very strong at bases. *Sil.,* S.Am.(Bol.-Arg.).——Fig. 447,5. **C. anti-siensis* (D'Orbigny), Bol.; *5a,b,* ped.v., brach.v. views, ×1; *5c-e,* interiors of brach. valves of different ages, ×2; *5f,* ped.v. int., ×2 (922).

Costellirostra Cooper, 1942, p. 231 [**Atrypa pe-culiaris* Conrad, 1841, p. 56; OD]. Rounded to triangular in outline; posterolateral parts of ped-icle valve abruptly bent dorsally and covered by

brachial valve, as in *Eatonia*; costellae numerous, bifurcating; coarser costae may be indicated in marginal region; commissure denticulate, denticles being coarser and less numerous than costellae; lateral commissure in type-species shifted near edge of shell into plane of pedicle valve; frontal commissure shifted dorsally so that it crosses fold. Interior resembling that of *Eatonia*, but lobes of cardinal process more divergent anteriorly and myophores more compressed laterally. *L.Dev.,* N. Am.——Fig. 447,2. **C. peculiaris* (Conrad), USA(N.Y.); *2a,b,* brach.v., ant. views, ×1; *2c,* ped.v. int. mold, ×1 (396); *2d,* brach.v. int. showing cardinal process, crura, and muscular impressions, ×1 (384).

?**Eatonioides** McLearn, 1918, p. 45 [**E. lamellor-natus;* OD]. Exterior resembling *Eatonia,* but with concentric lamellae; radiating striae absent. Dor-sal septum present; septalium not filled out; car-dinal process wanting. *Sil.,* N.Am.——Fig. 447,1. **E. lamellornatus,* Arisaig, N.Scotia; *1a-d,* ped.v., brach.v., lat., ant. views, ×1 (550).

Eucharitina Schmidt, 1955, p. 121 [**Terebratula eucharis* Barrande, 1847, p. 424; OD]. Medium-sized to large, oval in outline; pedicle valve flat to concave, with large tongue strongly curved

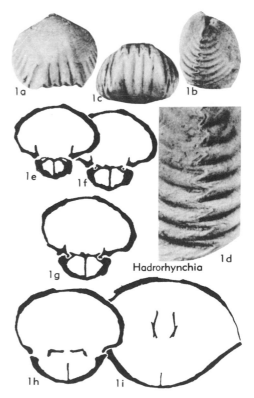

Fig. 445. Uncinulidae (Hadrorhynchiinae) (p. H569-H570).

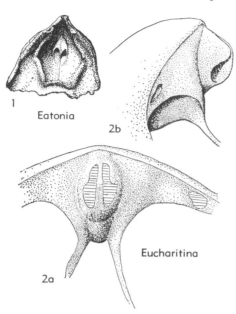

1

Eatonia

2b

Eucharitina

2a

FIG. 446. Eatoniidae (p. *H570-H572*).

dorsally; brachial valve convex but without prominent fold: flanks of pedicle valve and front of brachial valve abruptly bent toward opposite valve; costae flattened, with longitudinal grooves on truncated parts of shell; commissure denticulate, denticles prolonged and acuminated to marginal spines that taper uniformly from base to point, extending under costae of opposite valve. Ventral muscle field longitudinal oval, bounded by marked rim and divided by small septum; cardinal process large, its basal part filling out small septalium and uniting hinge plates; 2 lobes of cardinal process deeply excavated, process thus appearing quadripartite or, by coalescence of median parts, tripartite; crura long, slender. *L. Dev.*, Eu.-?Afr.——FIG. 446,2; 447,6. **E. eucharis* (BARRANDE), Boh.(Koněprusy); 446,2*a,b*, brach.v., cardinalia from vent. side and lat. view, ×4.7 (931c); 447,6*a-d*, ped.v., brach.v., lat., ant. views, ×1 (53); 447,6*e-i*, serial secs. at 29.1, 29.0, 28.9, 28.7, 28.6 mm. from ant. margin, ×? (411a).

Pegmarhynchia COOPER, 1955, p. 58 [**P. zimmi*; OD]. Small, rounded triangular to subpentagonal in outline; pedicle valve nearly plane in profile, with narrow, short sulcus near anterior margin, forming small tongue; deltidial plates small or lacking; foramen triangular; brachial valve gently convex, with small marginal fold; costae simple, rounded; commissure densely denticulate. Dental plates wanting; teeth ponderous, corrugated; muscle field very large, subtriangular, without prominent bounding ridge; hinge plates united to form broad, thick plate that commonly is elevated

medially into rounded boss, deep pit under beak; sockets transversely corrugated; crura short, broad, crescentic in section, with concave face directed inward; median ridge nearly obsolete to slightly elevated. *L.Dev.*, N.Am.——FIG. 447,4. **P. zimmi*, USA(N.Y.); *4a-d*, ped.v., brach.v., post., ant. views, ×2; *4e,f*, ped.v. int., brach.v. int., ×2 (185).

Tanerhynchia ALLAN, 1947, p. 442 [**Eatonia parki* ALLAN, 1935, p. 22; OD]. Medium-sized, transversely oval in outline; coarsely multicostate; commissure denticulate. Teeth strong, supported by plates not reaching floor of valve; muscle field restricted to posterior half of valve, impressed and bounded by low carinae; adductor scars divided by faint median ridge; dorsal median septum short; sockets large, corrugated; crura short and bluntly pointed; cardinal process with erect shaft and flat rugose myophore. *L.Dev.*, N.Z.——FIG. 447,7. **T. parki* (ALLAN); *7a,b*, ped.v. int., brach. v. int., ×1.5 (27).

Family PUGNACIDAE
Rzhonsnitskaya, 1956

[*nom. correct. et transl.* SCHMIDT, herein (*ex* Pugnaxinae RZHONSNITSKAYA, 1956, p. 125)] [Materials for this family prepared by HERTA SCHMIDT]

Small to large; sulcus, fold, and tongue generally well developed; beak commonly with ridges; costae not numerous, generally simple, coarse, angular or rounded-angular, in most genera restricted to anterior parts of valves, in some species wanting; commissure more or less denticulate, depending on strength or absence of costae; fine radial striae observed in many genera. Dental plates rarely wanting; hinge plates separated in their whole height; cardinal process absent; crural plates or ridges stretching dorsally from crural bases, connected with floor of valve at least in their upper parts, or uniting with median septum or ridge, forming very shallow septalium not reaching articulation zone; crural plates, generally narrowing in articulation zone, may be continued on crura, crural shape developing calcarifer type; median septum not strong, in some genera reduced to low ridge or absent. *L.Dev.-L.Carb.*

Pugnax HALL & CLARKE, 1893, p. 202 [**Terebratula acuminata* SOWERBY, 1822, p. 23; SD ICZN Opinion 420, 1956]. Small to large, commonly tetrahedral; sulcus and fold beginning in posterior halves of valves, broad but not strictly defined; brachial valve in some species elevated nearly to keel shape; pedicle valve rather flat or concave in posterior part; tongue narrowing anteriorly, with rounded or ogival anterior margin; costae strong

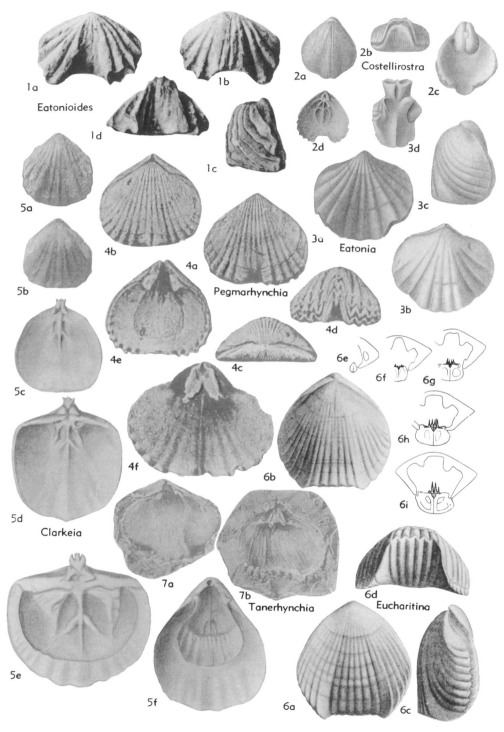

Fig. 447. Eatoniidae (p. *H*570-*H*572).

to weak or wanting; crural plates generally not uniting or forming septalium. *M.Dev.-Carb.*——Fig. 448,*1*. **P. acuminatus* (Sowerby), Eng. (Derbyshire) *(1a,b)*, Eire(Dublin) *(1c-q)*; *1a,b*, post., ant. views, ×1 (after 581); *1c-q*, ser. secs., ×2.8 (931d).

Corvinopugnax Havlíček, 1961, p. 36 [**Rhynchonella corvina* Barrande, 1847, p. 426; OD].

Subcuboidal with flanks and tongue steeply sloping; sulcus and fold well defined; sulcus shallow with flattened floor; fold not prominent; tongue broad, its anterior margin almost straight; anterior part of fold abruptly bent toward tongue, anterior commissure lying in plane; costae numerous, flattened and longitudinally grooved on steeply sloping parts of shell. Brachial valve

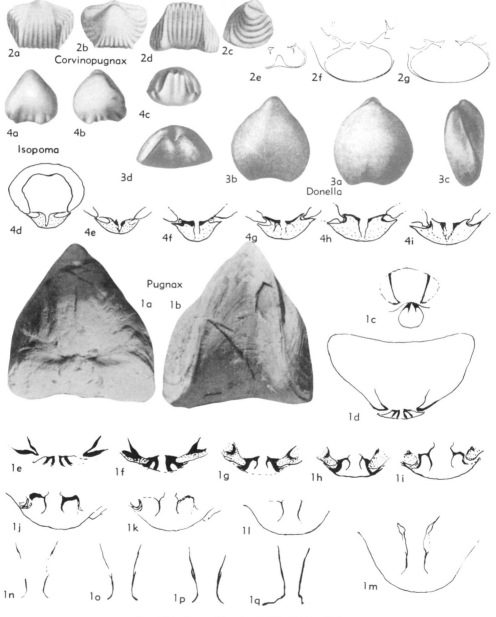

Fig. 448. Pugnacidae (p. *H572, H574-H575*).

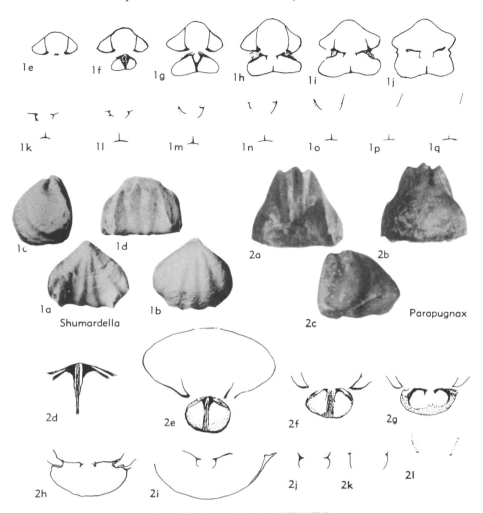

FIG. 449. Pugnacidae (p. *H575-H576*).

with short dorsal median ridge. *L.Dev.-M.Dev.,* Eu.(Boh.-Ger.).——FIG. 448,2. **C. corvinus* (BARRANDE), M.Dev., Boh.(Koněprusy); *2a-d,* ped.v., brach.v., lat., ant. views, ×1 (53); *2e-g,* ser. secs., ×3 (411a).

Donella ROTAY, 1931, p. 21 [**D. minima;* OD]. Small, resembling *Pugnax* in shape, with sulcus and fold not distinctly defined; surface usually smooth, rarely with 1 or 2 costae restricted to marginal parts of sulcus and fold and causing denticulation in commissure. Interior not sufficiently known; dental plates wanting; dorsal septum extending forward beyond zone of articulation. *L.Carb.,* USSR(Donetz Basin).——FIG. 448, *3. *D. minima; 3a-d,* brach.v., ped.v., lat., ant. views, ×2 (627a).

Isopoma TORLEY, 1934, p. 81 [**Terebratula brachyptycta* SCHNUR, 1853, p. 178; OD]. Small to medium-sized, rounded to pentagonal in outline; both valves nearly equally convex; ventral beak blunt; sulcus and fold weakly developed; tongue short; costae short and coarse; radial striae not observed. Interior of pedicle valve with low median ridge in posterior part; dental plates not discernible; crural plates delicate, not uniting. *M.Dev.,* Eu.——FIG. 448,4. **I. brachyptyctum* (SCHNUR), Couvin., Ger.(Eifel); *4a-c,* ped.v., brach.v., ant. views, ×1.5 (718b); *4d-i,* ser. secs., ×5 (931d).

Parapugnax SCHMIDT, 1964 [**P. brecciae* (=*Pugnax pugnus brecciae* SCHMIDT, 1941, p. 278); OD]. Pedicle valve flat or slightly convex, but not concave; sulcus and fold distinctly defined; anterior margin of tongue nearly straight or but faintly rounded. Median ridge or moderately high septum present in dorsal interior; crural plates

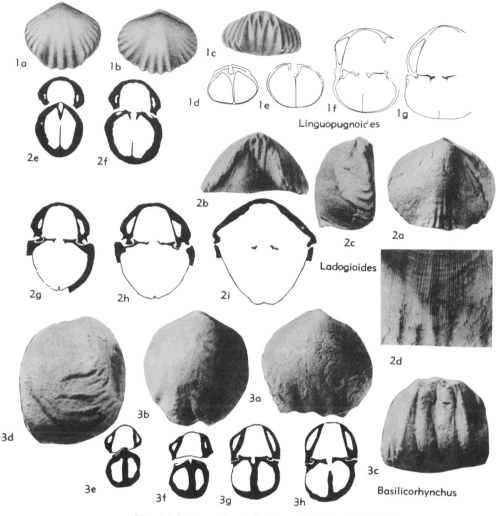

FIG. 450. Family Uncertain (?aff. Pugnacidae) (p. *H576-H577*).

approaching or uniting with median septum or ridge, forming flat septalium. *M.Dev.-U.Dev.*, Eu. ——FIG. 449,2. **P. brecciae* (SCHMIDT), Ger.(Dill distr.); *2a-c,* ped.v., brach.v., lat., ×1 (718c); *2d,* sec. near and parallel to surface, ×5; *2e-l,* ser. secs., ×3.5 (931d).

Shumardella WELLER, 1910, p. 512 [**Rhynchonella missouriensis* SHUMARD, 1855, p. 204; OD]. Medium-sized to large, triangular in outline; beak little incurved; sulcus and fold well marked, few rounded costae beginning in posterior halves of valves. Dental plates present; umbonal chambers rather large; dorsal median septum low, reaching about half length of valve; crural plates uniting to form very shallow septalium. *L.Carb.(Miss.),* N.Am.(Pa.-Iowa-Mo.).——FIG. 449,*1.* **S. missouriensis* (SHUMARD), USA(Mo.); *1a-d,* ped.v.,

brach.v., lat., ant. views, ×1 (178); *1e-q,* ser. transv. secs., ×4 (931d).

Family UNCERTAIN (?aff. PUGNACIDAE)

The following genera, though possibly related to Pugnacidae, are not included in this family on account of some differences.

Basilicorhynchus CRICKMAY, 1952, p. 1 [**Leiorhynchus basilicum* CRICKMAY, 1952, p. 600; OD]. Medium-sized, subglobular, tumid; sulcus and fold very short; flanks and tongue steeply sloping; costae confined to anterior region; fine radial striae present, smooth posteriorly; commissure denticulate. Dental plates strong, diverging dorsally; septum rather strong and high; septalium shallow,

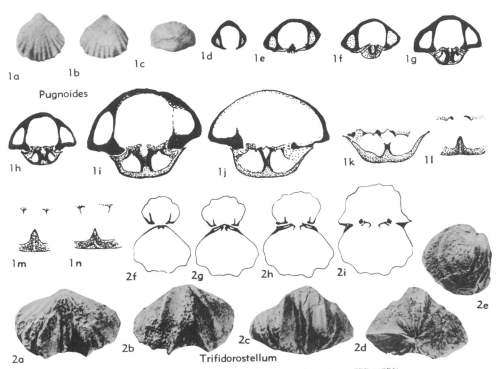

FIG. 451. Family Uncertain (?aff. Pugnacidae) (p. *H577-H578*).

persisting into articulation zone, uncovered. *U. Dev.(Famenn.)*, N. Am.(Can.)-W. Eu.-?Armenia. ——FIG. 450,*3*. **B. basilicum* (CRICKMAY), Can. (N.W.Terr.) *(3a-d)*, Belg. *(3e-h)*; *3a-d*, ped.v., brach.v., ant., lat. views, ×1.5 (203); *3e-h*, ser. secs., 0.5, 0.6, 0.9, 1.1 mm. from apex; ×2 (930b). [SCHMIDT.]

Ladogioides MCLAREN, 1961 [**L. pax*; OD] [=*Athabaschia* CRICKMAY, 1963, p. 10]. Exteriorly resembling *Pugnax*; medium-sized to large, with inflated brachial valve and shallow pedicle valve; sulcus rather deep but not sharply defined; fold gradually turning into general curvature of shell; tongue high, with rounded or ogival anterior margin; costae strong, angular, restricted to anterior parts of shell, or may be wanting; costellae or striae covering whole shell; commissure denticulate. Dental plates diverging dorsally; umbonal cavities narrow; septalium very shallow, not persisting to articulation zone; dorsal median septum very short. *U.Dev.(Frasn.)*, N. Am.-?Pol.——FIG. 450,*2*. **L. pax*, Can.; *2a-c*, brach., ant., lat. views, ×1; *2d*, surface detail, ×3; *2e-i*, ser. secs. at 1.0, 1.4, 1.7, 2.1, 2.6 mm. from apex, ×3 (548a). [SCHMIDT, MCLAREN.]

Linguopugnoides HAVLÍČEK, 1960 [**Rhynchonella (nympha) carens* BARRANDE, 1879 (=*Linguopugnoides carens* BARRANDE, 1879); OD]. Small to medium-sized, with broad fold and sulcus and

high tongue; deltidial plates wanting; costae strong, angular, dental plates developed, diverging dorsally; dorsal median septum moderately high. *Sil.-L.Dev.*, Eu.(Boh.-USSR)-M.Asia.——FIG. 450, *1*. **L. carens* (BARRANDE), Sil., Boh.; *1a-c*, ped.v., brach.v., ant. view, ×1 (53); *1d-g*, ser. secs., distance from front: 12.35, 12.15, 11.8, 11.6 mm.; ×2 (411a). [SCHMIDT.]

Pugnoides WELLER, 1910, p. 512 [**Rhynchonella ottumwa* WHITE, 1862; OD]. Small to medium-sized; strong costae beginning anteriorly from umbo but in posterior halves of valves; commissure strongly denticulate. Dental plates well developed; delthyrial cavity and umbonal cavities large; crural plates nearly approaching each other or uniting, forming septalium which reaches articulation zone; opening of septalium covered in its anterior part by angular plate. *?Dev., L.Carb., ?Perm.*, N.Am.-Eu.-?Asia.——FIG. 451,*1*. **P. ottumwa* (WHITE), Miss., USA(Iowa); *1a-c*, ped.v., brach.v., post. view; ×1 (178); *1d-n*, ser. secs., ×4 *(1d-h)*, ×6.5 *(1i-n)* (931d). [SCHMIDT.]

Trifidorostellum SARTENAER, 1961, p. 5 [**Leiorhynchus dunbarense* HAYNES, 1916, p. 38; OD]. [=*Pseudoleiorhynchus* ROZMAN, 1962, p. 122 (type, *Leiorhynchus uralicus* NALIVKIN, 1947, p. 90)]. Small to large, broad-ovate in outline, brachial valve rather inflated; ventral sulcus well defined, deep, rounded, beginning at short distance from beak, persisting on tongue; fold prominent, defined on either side by marked furrows; few strong rounded costae present, generally simple, median costae beginning at or not far from beak, lateral ones more distant from beak; commissure undulated to denticulated. Dental plates

slender, short; hinge plates separated; short crural plates present, not united; septum and septalium wanting. *U.Dev.(Famenn.),* N.Am.; *L.Carb.,* Asia(Urals-Kazakh.-Kuznetsk basin).——Fig. 451, *2.* *T. dunbarense* (Haynes), USA(Mont.); *2a-e,* ped.v., brach.v., ant., post., lat. views, ×1; *2f-i,* ser. secs., 0.8, 1.0, 1.1, 1.3 mm. from apex, ×2.9 (709a). [Schmidt.]

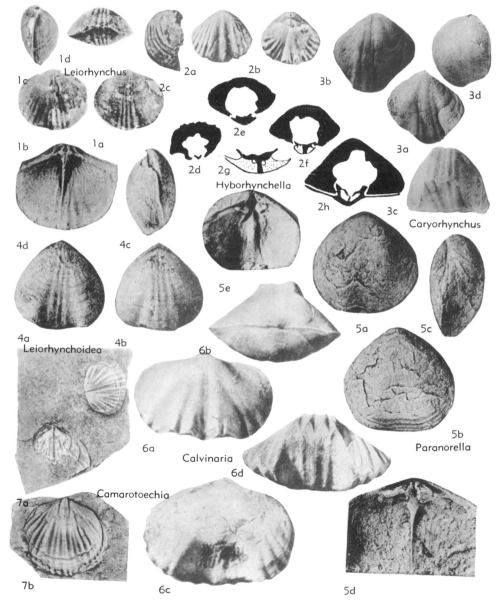

Fig. 452. Camarotoechiidae (Camarotoechiinae) (p. *H580-H582*).

Family CAMAROTOECHIIDAE
Schuchert & LeVene, 1929

[Camarotoechiidae SCHUCHERT & LEVENE, 1929, p. 18] [Materials for this family prepared by HERTA SCHMIDT except as indicated otherwise]

Medium-sized to large, round or elliptical in outline, sides and front never truncated; costae generally rounded. Brachial valve with high median septum or ridge; hinge plate entire, or divided only in its most anterior portion, or divided by small shallow open septalium; cardinal process in most genera wanting. *?L.Sil., U.Sil.-Perm.*

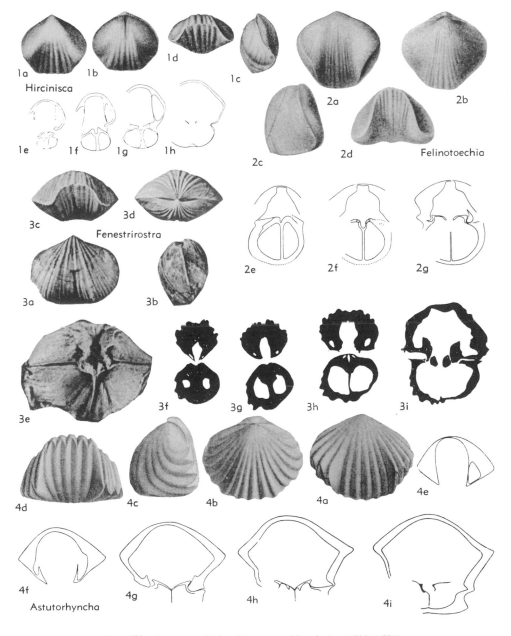

FIG. 453. Camarotoechiidae (Camarotoechiinae) (p. *H*580-*H*581).

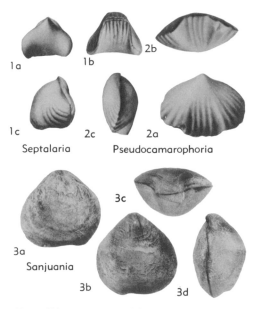

1a   1b   2b

1c   2c   2a

Septalaria   Pseudocamarophoria

3c

3a

Sanjuania

3b   3d

FIG. 454. Camarotoechiidae (Camarotoechiinae) *(3)*, (Septalariinae) *(1-2)* (p. H582-H584).

Subfamily CAMAROTOECHIINAE
Schuchert & LeVene, 1929

[Camarotoechiinae SCHUCHERT & LEVENE, 1929, p. 18] [=Leiorhynchinae STAINBROOK, 1945, p. 43; Nudirostrinae ROGER, 1952, p. 88] [Materials for this subfamily prepared by HERTA SCHMIDT and D. J. MCLAREN]

Costae with tendency to obliterate, especially on flanks. Crural bases slightly projecting ventrally, enclosing small median groove or shallow septalium; cardinal process absent. *?L.Sil., U.Sil.-Perm.*

Camarotoechia HALL & CLARKE, 1893, p. 189 [*Atrypa congregata* CONRAD, 1841, p. 55; OD]. Moderately high to flat, slightly inflated only in umbonal region; ventral sulcus and dorsal fold well defined, beginning at short distance from apex; costae low, unequal, starting near apex, those on sulcus and fold bifurcating, crossed by strong concentric striae; commissure undulated. Dental plates short, strong; dorsal muscular field long, narrow; dorsal septum long. *M.Dev., N.Am.* ——FIG. 452,7. *C. congregata* (CONRAD), USA (N.Y.); *7a,* two brach.v. int. molds, ×1.5; *7b,* ped.v. int. mold, ×1 (930b).

?Astutorhyncha HAVLÍČEK, 1961, p. 105 [*Rhynchonella Proserpina* BARRANDE, 1847, p. 420; OD]. Medium-sized to large, transverse, inflated; brachial valve high; fold and sulcus prominent; costae strong, rounded, simple; commissure serrate; umbones smooth. Dental plates short, divergent; septum high, slender; septalium if developed, shallow; crural bases project ventrally. [This genus probably is better placed in "Family Un-

certain."] *U.Sil.-L.Dev.,* Eu.(Czech.)-C.Asia.—— FIG. 453,4. *A. proserpina* (BARRANDE), M.Dev. (Eifel,), Boh.; *4a-d,* ped.v., brach.v., lat., ant. views, ×1 (53); *4e-i,* ser. secs. 19.4, 19.3, 18.8, 18.7, 18.5 mm. from ant. margin, ×2.5 (411a). [MCLAREN.]

Calvinaria STAINBROOK, 1945, p. 43 [*Rhynchonella ambigua* CALVIN, 1878, p. 727; OD]. Large, transversely elliptical in outline; ventral sulcus, dorsal fold, and tongue strongly developed; costae rounded to subangular, beginning anteriorly from umbo; commissure undulated to denticulated. Dental plates commonly divergent dorsally; umbonal cavities may be filled out by callus; hinge plate divided in its anterior part and detached from median septum within or posterior to zone of articulation. *U.Dev.,* cosmop.——FIG. 452,6; 455,1. *C. ambigua* (CALVIN), USA (Iowa); 452, *6a,b,* brach.v., post. view; 452,*6c,* ped.v.; 452,*6d,* ant. view, ×1 (768); 455,*1a-g,* ser. secs., 2.2, 2.6, 2.7, 2.95, 3.3, 3.5, 3.8 mm. from apex, ×2 (930a).

Caryorhynchus CRICKMAY, 1952 [*Leiorhynchus carya* CRICKMAY, 1952, p. 599; OD]. Very near to *Leiorhynchus,* perhaps synonymous; ventral sulcus deep; brachial valve inflated; costae confined to fold and sulcus. Interior with abundant callus; dental plates not distinguishable from thickened shell wall. *M.Dev.-U.Dev.,* N.Am.—— FIG. 452,3; 455,3. *C. carya* (CRICKMAY), U.Dev. (Frasn.), Can.(Alta.); 452,*3a-d,* ped.v., brach.v., ant., lat. views, ×1; 455,*3a-d,* ser. secs., ×5 (203).

?Felinotoechia HAVLÍČEK, 1961, p. 73 [*Atrypa astuta* BARRANDE var. *felina* BARRANDE, 1879, pl. 18; OD]. Medium-sized, transverse, inflated, with high brachial valve; well-developed fold and sulcus with very high, broad tongue; low, rounded costae confined to anterior part of fold and sulcus, less commonly on flanks. Pedicle valve thick-walled, without dental plates; muscle impressions strongly impressed; median septum high, slender, supporting deep narrow septalium. *U.Sil.-L.Dev.,* Eu.(Czech.).——FIG. 453,2. *F. felina* (BAR-RANDE), U.Sil.(Budňany); *2a-d,* ped.v., brach.v., lat., ant. views, ×1 (53); *2e-g,* ser. secs. at 0.1 mm. intervals, ×2 (411a). [MCLAREN.]

?Fenestrirostra COOPER, 1955, p. 56 [*Rhynchonella glacialis* BILLINGS, 1862, p. 143; OD]. Medium-sized to large, transverse, subequally biconvex; pedicle beak strongly incurved, foramen minute; thick deltidial plates; surface costellate; costellae rounded, uneven, increase by bifurcation and intercalation, and tending to be obliterated forward; entire surface finely capillate. Shell thick; dental plates short, stout; narrow, deep septalium, supported by stout median septum; hinge plates formed of thick socket ridges that pass into crural bases. [This genus probably is better placed in "Family Uncertain."] *L.Sil.,* Can.(Que.).——FIG. 453,3. *F. glacialis* (BILLINGS), Becsie, Anticosti; *3a-d,* ped.v., lat., ant., post. views, ×1; *3e,* rubber replica post. int., ×2; *3f-i,* ser. secs. 3.0, 3.6, 3.7, 4.2 mm. from apex, ×2 (185). [Mc-Laren.]

?**Hircinisca** Havlíček, 1960, p. 241 [**Atrypa hircina* Barrande var. de *Sapho* Barrande, 1879, pl. 90; OD]. Medium-sized, quadrate or transverse; smooth or weakly costate anteriorly on fold and sulcus, flanks always smooth. Long, thin dental plates, slightly divergent anteriorly; stout septum, open septalium. [This genus probably is better placed in "Family Uncertain."] *M.Sil.-U.Sil.,* Eu. (Czech.).——Fig. 453,*1a-d.* **H. hircina* (Barrande), U.Sil.(Wenlock.), Boh.; *1a-d,* ped.v., brach.v., lat., ant. views, ×1 (53).——Fig. 453, *1e-h. H. hebes* Havlíček, U.Sil.(Budňany), Boh.; *1e-h,* ser. secs. 9.0, 8.8, 8.5, 8.3 mm. from ant. margin, ×2 (411a). [McLaren.]

?**Hyborhynchella** Cooper, 1955, p. 59 [**H. bransoni;* OD]. Small, thick, rounded in outline; pedicle valve strongly swollen; brachial valve nearly plane to concave in profile; ventral sulcus and dorsal fold restricted to anterior halves of valves; tongue narrow; beaks curved, with short, obscure ridges; costae rounded, broadest on fold and sulcus; commissure strongly undulate to denticulate. Dental plates not discernible because of thickening of shell; teeth not defined from shell wall; hinge plate entire, slightly concave, with small median groove, supported by low septum which strongly thickens posteriorly; inner socket ridges prominent. *U.Dev.,* N.Am.——Fig. 452,*2.* **H. bransoni,* USA(N.Mex.); *2a,b,* ped.v., brach. v., ×2; *2c,* lat. view, ×2; *2d-h,* ser. secs., 0.5, 0.65, 0.8, 1.0, 1.15 mm. from apex, ×3, except *2g,* ×6 (185).

Leiorhynchoidea Cloud, 1944, p. 57 [**L. schucherti;* OD]. Round to broad elliptical, commonly rather flat, ventral sulcus and dorsal fold shallow, tongue short; costae weak to obsolescent. Ventral muscle field long-ovate, narrow adductors divided by low median ridge, laterally and anteriorly bounded by broader diductors. Dorsal muscle field narrow; hinge plate resting on thickened shell wall, its lateral parts concave, defined laterally by projecting socket plates, medially by prominent crural bases; median depression between crural bases with small ridge; sockets crenulated; dorsal median ridge strongly thickening before uniting with hinge plate. [Girty's (1911) *Leiorhynchus carboniferum* from the Moorefield Shale of Arkansas is judged to belong to *Leiorhynchoidea.*] *?U.Miss., U.Perm.,* N.Am.——Fig. 452,*4. *L. schucherti,* Mex.(Coahuila); *4a-c,* ped.v., brach.v., lat. view, ×1; *4d,* brach. int., young specimen, ×2 (912).

Leiorhynchus Hall, 1860, p. 75 [**Orthis quadracostata* Vanuxem, 1842, p. 168; SD Oehlert, 1887, p. 1308] [=*Liorhynchus* Oehlert, 1887 (obj.) (*non* Rudolphi, 1801); *Nudirostra* Cooper & Muir-Wood, 1951 (obj.)]. Rather near to *Camarotoechia;* globose, sulcus and fold weakly developed, beginning in or at short distance be-

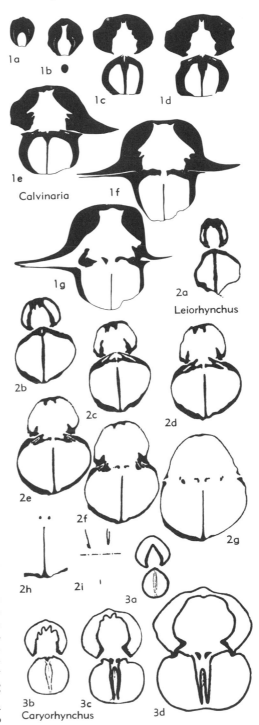

Fig. 455. Camarotoechiidae (Camarotoechiinae) (p. H580-H582).

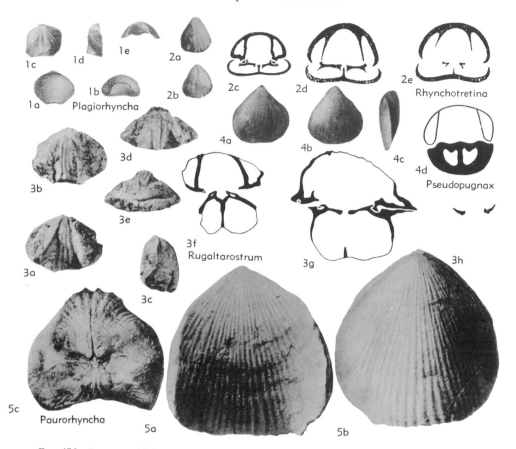

FIG. 456. Camarotoechiidae (Camarotoechiinae) *(1-3, 5),* (Septalariinae) *(4)* (p. *H582-H584).*

hind mid-length of valves; beak curved; ventral interarea limited by ridges; costae on flanks weak or obsolescent, on sulcus and fold stronger, low, rounded, of variable width, in type-species beginning near beaks, some of them bifurcating; commissure undulated. Dental plates well developed, nearly approaching each other ventrally, in some shells united to form spondylium duplex; dorsal septum high and long, detached from hinge plate in beginning of or posteriorly to zone of articulation; crura with trough-shaped ends. *M.Dev.-U. Dev.,* cosmop.——FIG. 452,1; 455,2. *L. quadracostatus* (VANUXEM), U.Dev., USA(N.Y.); 452, *1a-d,* ped.v., brach.v., lat., ant. views, ×1; 455, *2a-i,* ser. secs., 0.45, 0.8, 0.9, 0.95, 1.0, 1.1, 1.5, 1.8, 3.05 mm. from apex, ×3.2 (930b).

Paranorella CLOUD, 1944, p. 59 [*P. imperialis*; OD]. Biconvex, with pedicle valve the deeper, subcircular in outline; broad shallow ventral fold and dorsal sulcus restricted to anterior parts of valves; tongue very short, trapezoidal, or scarcely marked; costae wanting. Ventral muscle field broad, bounded by low ridge. Hinge plate re-

sembling that of *Leiorhynchoidea. U.Perm.,* N.Am.——FIG. 452,5. *P. imperialis,* Mex.(Coahuila); *5a-c,* ped.v., brach.v., lat. view, ×1; *5d,* brach.v. int., ×2; *5e,* ped.v. int., ×1 (912).

Sanjuania AMOS, 1958, p. 841 [*S. dorsisulcata*; OD]. Very near to *Paranorella,* perhaps synonymous. *L.Carb.,* Arg.——FIG. 454,3. *S. dorsisulcata* AMOS; *3a-d,* ped.v., brach.v., ant., lat. views, ×1 (905b).

?**Paurorhyncha** COOPER, 1942, p. 231 [*Rhynchonella Endlichi* MEEK, 1875, p. 46; OD]. Large, subtriangular; sulcus deep and very wide, extending from beak; brachial valve high; finely costate to costellate with rounded costellae, which may branch, extending from umbones, weak or absent on flanks. Small dental plates and teeth; foramen minute, deltidial plates vestigial; long median septum supporting V-shaped, open septalium. [This genus probably is better placed in "Family Uncertain."] *U.Dev.(Famenn.),* W.USA-?USSR [reported occurrence in M.Dev. of Kazakhstan probably incorrect].——FIG. 456,*5a,b.* *P. endlichi* (MEEK), Ouray, Colo.; *5a,b,* ped.v., brach.

v. views, ×1 (469).——FIG. 456,*5c*. *P. cooperi* STAINBROOK, Percha, N.Mex., post. view int. mold, ×1 (769). [McLAREN.]

?**Plagiorhyncha** McLEARN, 1918, p. 138 [**Rhynchonella Glassii* DAVIDSON, 1883, p. 155 (=*Atrypa depressa* J. DE C. SOWERBY, 1839, p. 629); OD]. Small to medium-sized; subequally biconvex; outline subcircular; lateral margins of brachial valve vertical; fold and sulcus developed anteriorly; tongue subrectangular or rounded; weakly costate or costellate or smooth; costae may be confined to fold and sulcus; may be finely capillate. Dental plates not developed; teeth small; muscle impressions strongly impressed; hinge plates divided; septum low; septalium open. *L.Sil.-U.Sil.(Wenlock.)*, Eu.-Can.(N.S.).——FIG. 456,*1*. **P. glassii* (DAVIDSON), Wenlock, Eng., L.Sil., Can.(N.S.); *1a,b*, brach.v., ant. views, ×1 (229); *1c-e*, brach. v., lat., ant. views, ×1 (550). [McLAREN.]

?**Rhynchotretina** KHALFIN, 1939, p. 175 [**R. aequivalvis*; OD]. Small, subtriangular, with acute apical angle and rounded anterior; almost equivalve, inflated, no fold and sulcus, rectimarginate; pedicle beak short, straight; costate anteriorly, smooth posteriorly. Interior imperfectly known; strong dental plates; no septalium; median septum protruding between divided hinge plates. [This genus probably is better placed in "Family Uncertain."] *L.Dev.*, USSR(Altay).——FIG. 456, *2*. **R. aequivalvis*; *2a,b*, ped.v., brach.v., views, ×1 (690); *2c-e*, ser. secs., ×10 (464). [McLAREN.]

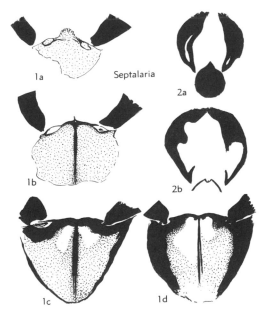

FIG. 457. Camarotoechiidae (Septalariinae) (p. H583-H584).

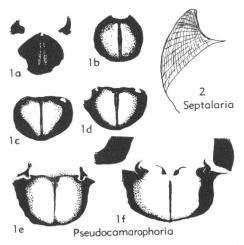

FIG. 458. Camarotoechiidae (Septalariinae) (p. H583-H584).

Rugaltarostrum SARTENAER, 1961, p. 6 [**Leiorhynchus madisonense* HAYNES, 1916, p. 39; OD]. Small to large, transverse; sulcus developed from near beak, wide; maximum height of shell at or near anterior margin; few, rounded, bifurcating costae on fold and sulcus, flanks smooth; may be finely capillate. Dental plates slender; dorsal septum supports wide cup-shaped septalium; hinge plates plane, inclined inward. *U.Dev.(Famenn.)*, N.Am.(Mont.-Idaho-Alta.-N.W.T.).——FIG. 456, *3*. **R. madisonense* (HAYNES), USA(Mont.); *3a-e*, ped.v., brach.v., lat., ant., post. views, ×1; *3f-h*, ser. secs. at 0.65, 0.95, 1.45 mm. from apex, ×9 (709a). [McLAREN.]

Subfamily SEPTALARIINAE Havlíček, 1960

[*nom. transl.* SCHMIDT, herein (*ex* Septalariidae HAVLÍČEK, 1960, p. 241)] [=?Pseudopugnaxinae LIKHAREV in RZHONSNITSKAYA, 1958, p. 114]

Hinge plate without median groove, its halves united by small plate slightly depressed between crural bases; latter developed as narrow ridges projecting dorsally. *L.Dev.-M.Dev., ?U.Perm.*

Septalaria LEIDHOLD, 1928, p. 41 [**Terebratula ascendens* STEININGER, 1853, p. 61; SD TORLEY, 1934, p. 74 (=*T. subtetragona* SCHNUR, 1851, p. 3)]. Medium-sized, commonly with sulcus and fold; prominent tongue always present, strongly bent dorsally in adult specimens; ventral beak but slightly incurved; costae beginning at or near apex, broader than furrows between them, low, rounded; marginal spines present in type-species. Dental plates short, bounding narrow umbonal cavities; upper part of hinge plate with small protruding rounded cardinal process which may be provided wtih parallel ridges; dorsal septum very high, triangular, with acute angle directed ventrally. *M.Dev.*, Eu.-Australia.——FIG. 454,*1*; 457,

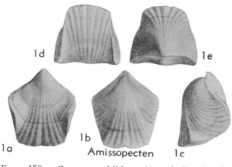

Fɪɢ. 459. Camarotoechiidae (Septalariinae) (p. H584).

1,2; 458,*2*. **S. subtetragona* (Schnur), Nohn., Ger.(Eifel); 454,*1a-c,* ped.v., ant., lat. views, ✕1 (718b); 457,*1,2,* transv. secs. of 2 specimens at *(1a-d)* 15.2, 15.0, 14.9, 14.8 mm. and *(2a,b)* 11.2, 11.1 mm. from ant. margin, ✕5 (Schmidt, n); 458,*2,* lat. view of dorsal septum, ✕3.3 (Schmidt, n).

Amissopecten Havlíček, 1960, p. 243 [**Terebratula velox* Barrande, 1847, p. 430; OD]. Resembling *Septalaria.* Dental plates wanting or obsolescent; cardinal process absent. *M.Dev.,* Eu.(Boh.).——Fɪɢ. 459,*1;* 460,*1.* **A. velox* (Barrande), Koněprusy Ls.; 459,*1a-e,* ped.v., brach.v., lat., ant., post. views, ✕1 (53); 460,*1a-d,* ser. secs., 16.3, 16.1, 15.9, 15.7 mm. from ant. margin, ✕5 (411a).

Pseudocamarophoria Wedekind, 1925, p. 197 [**Terebratula microrhyncha* C. F. Roemer, 1844, p. 65; OD]. Medium-sized to large, round to transverse elliptical; sulcus and fold strong, tongue large; beaks incurved; costae beginning near apex, irregular in strength, rounded, interspaces about as large as costae; commissure undulated. Umbonal cavities commonly filled out by callus; brachial interior resembling that of *Septalaria,* but without cardinal process. *L.Dev.-M.Dev.,* Eu.——Fɪɢ. 454,*2;* 458,*1.* **P. microrhyncha* (C. F. Roe-

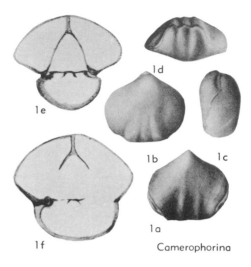

Fɪɢ. 461. Camerophorinidae (p. H584).

mer), *M.*Dev.(Couvin.), Ger.(Eifel); 454,*2a-c,* ped.v., ant., lat. views, ✕1 (718b); 458,*1a-f,* transv. secs. at 15.3, 15.0, 14.8, 14.6, 14.4, 14.0 mm. from ant. margin, ✕4 (Schmidt, n).

?Pseudopugnax Likharev, 1956, p. 56 [**P. planissimus*]. Small to medium-sized, triangular in outline, flat, with broad, shallow sulcus in anterior part of pedicle valve, without marked fold in brachial valve; sculpture wanting; commissure even. Dental plates slightly diverging dorsally; dorsal septum well developed; hinge plates connected by median plate extending anteriorly farther than lateral parts of hinge plate, detached from dorsal septum anteriorly from articulation. *U.Perm.,* N.Caucasus.——Fɪɢ. 456,*4.* **P. planissimus*; *4a-c,* ped.v., brach.v., lat. views, ✕1.5; *4d,* ser. sec., ✕3 (517).

Family CAMEROPHORINIDAE
Rzhonsnitskaya, 1958

[Camerophorinidae Rzhonsnitskaya, 1958, p. 115] [=Camarophorinidae Rzhonsnitskaya, 1956, p. 126 *(nom. vet.)*]
[Materials for this family prepared by D. J. McLaren]

Smooth posteriorly, weakly ribbed anteriorly; spondylium supported by low septum in pedicle valve; hinge plate entire, no septum or septalium in brachial valve. *M.Dev.*

Camerophorina Schmidt, 1941, p. 43 [**Terebratula pachyderma* Quenstedt, 1871, p. 200; OD]. Medium-sized, broad; commissure uniplicate; fold and sulcus developed anteriorly; most of shell smooth, few weak, rounded costae at front margin. Outer hinge plates joined by slightly convex single plate. *M.Dev.,* Ger.-Czech.——Fɪɢ. 461,*1.* **C. pachyderma* (Quenstedt), Eifel.; *1a-d,* ped. v., brach.v., lat., ant. views, ✕1; *1e,f,* ser. secs., ✕2.5 (718b).

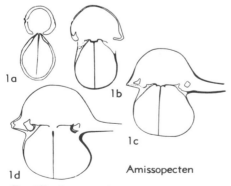

Fɪɢ. 460. Camarotoechiidae (Septalariinae) (p. H584).

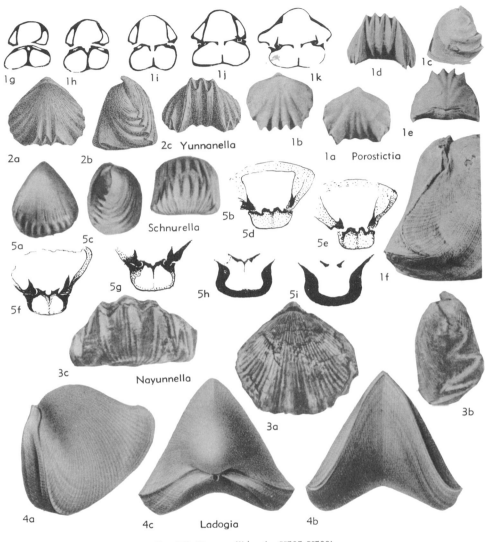

FIG. 462. Yunnanellidae (p. *H585-H588*).

Family YUNNANELLIDAE
Rzhonsnitskaya, 1959

[*nom. transl.* McLAREN, herein (*ex* Yunnanellinae RZHONSNIT-SKAYA, 1959, p. 28)] [=Iunnanellidae RZHONSNITSKAYA, 1956, p. 125; Junnanellidae RZHONSNITSKAYA, 1958, p. 112] [Materials for this family prepared by D. J. McLAREN except as indicated otherwise]

Smooth or paucicostate, with subangular to angular costae on anterior part of shell or, axially, from beak; entire shell costellate or capillate; uniplicate, commissure coarsely serrate. Dental plates present; strong median septum supporting deep open septalium. ?M.Dev., U.Dev., ?L. Miss.

Yunnanella GRABAU, 1923, p. 195 [**Rhynchonella Hanburii* DAVIDSON, 1853, p. 356; OD] [=*Yunnanellina* GRABAU, 1931 (obj.); *Junnanella, Junnanellina* RZHONSNITSKAYA, 1958, p. 113 (*nom. van.*)]. Small to medium; brachial fold high; anterior part of shell paucicostate on fold and flanks; whole shell finely capillate. Interior poorly known; wide septalium; strong septum. U.Dev. (Famenn.), China-USSR.——FIG. 462,2; 463,1. **Y. hanburii* (DAVIDSON), China(Yunnan); 462, *2a-c*, brach.v., lat., ant. views, ×1 (230a); 463,*1*, sec. near apex, enlarged (361a).

?Eoparaphorhynchus SARTENAER, 1961a, p. 2 [**E. maclareni*; OD]. Closely resembles *Paraphorhyn-*

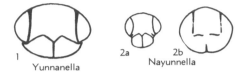

FIG. 463. Yunnanellidae (p. *H585-H586*).

chus; costellae reduced or absent. Septum strong; septalium deep, narrow. *U.Dev.(Famenn.)*, N.Am. (Can.:Alta., N.W.T.-USA:Nev.)-Eu.-Asia.——FIG. 464,*1.* **E. maclareni*, Can.(N.W.T.); *1a-e*, ped.v., brach.v., post., ant., lat. views, ✕1; *1f-j*, ser. sec. at 1,65. 2.2, 3.6, 5.2, 5.8 mm. from apex, ✕3.25 (709a).

?**Ladogia** NALIVKIN, 1941, p. 165 [**Terebratula Meyendorfii* DE VERNEUIL, 1845, p. 74; OD]. Medium-sized to large; brachial valve high, acuminate; pedicle valve flattened, with high, pointed tongue; prominent beak ridges; lateral commissure raised; noncostate; strong, flattened, bifurcating costellae. Dental plates bent inwards; hinge plates concave; septalium broad, U-shaped; septum

strong, persists forward of articulation. [This genus may be placed better in "Family Uncertain."] *M. Dev.(Givet.)-U.Dev.(Frasn.)*, E.Eu.——FIG. 462, *4*; 465,*1.* **L. meyendorfii* (VERNEUIL), U.Dev. (Frasn.); *462,4a-c*, lat., ant., post. views, ✕1 (841); 465,*1a-e*, ser. secs. at 1.8, 1.9, 2.6, 3.1, 4.1 mm. from apex, ✕2 (548b). [McLAREN.]

Nayunnella SARTENAER, 1961, p. 2 [*pro Yunnanella* GRABAU, 1931 (*non* GRABAU, 1923)] [**Yunnanella synplicata* GRABAU, 1931, p. 141; OD]. Similar to *Yunnanella*; shell flatter and more coarsely costellate; costellae develop into smooth costae anteriorly by fusion or widening. *U.Dev. (Famenn.)*, China.-USSR.——FIG. 462,*3*; 463,*2.* **N. synplicata* (GRABAU), China(Yunnan); 462, *3a-c*, brach.v., lat., ant. views, ✕2 (358); 463, *2a,b*, ser. sec., enlarged (361a).

?**Paraphorhynchus** WELLER, 1905, p. 260 [**P. elongatum*; SD SCHUCHERT & LEVENE, 1929, p. 93] [=*Paryphorhynchus* WELLER, 1914, p. 187 (*nom. van.*)]. Medium-sized or large, triangular to elongate triangular; fold and sulcus well developed; beak ridges prominent, rounded; shell entirely capillate. Dental plates approximate ventrally;

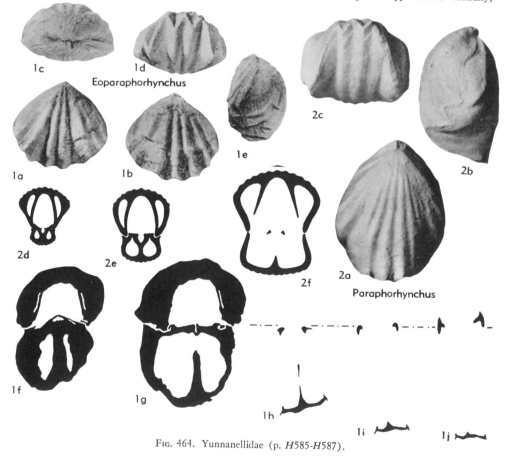

FIG. 464. Yunnanellidae (p. *H585-H587*).

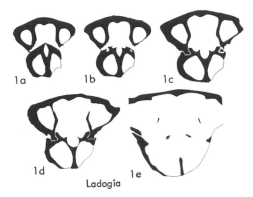

FIG. 465. Yunnanellidae (p. H586).

hinge plates plane. *L.Miss.(Kinderhook.)*, N.Am.
——FIG. 464,2. *P. elongatum*; *2a-c*, brach.v., lat., ant. views, ×1; *2d-f*, ser. secs. ×2.5 (858).

Porostictia COOPER, 1955, p. 62 [**Paraphorhynchus perchaensis* STAINBROOK, 1947, p. 316; OD]. Brachial valve deep; strongly uniplicate; posterior part of shell without costae; entire shell covered by fine capillae separated by rows of fine pits. *U. Dev.(Famenn.)*, USA (N.Mex.-Ariz.).——FIG. 462,1. **P. perchaensis* (STAINBROOK); *1a-e*, ped.v., brach.v., lat., ant., post. views, ×1; *1f*, lat. view, ×2; *1g-k*, ser. sec. at 1.5, 2.2, 2.6, 2.8, 3.0 mm. from apex, ×2 (185).

Schnurella SCHMIDT, 1964 [**Terebratula schnuri* DE VERNEUIL, 1840, p. 261; OD]. Triangular to longitudinal-oval in outline; rather high in adult specimens; pedicle valve flat, brachial valve strongly convex; flanks and front truncated in adult shells; sulcus wanting or very shallow and taking up whole width of valve; fold not discernible in convexity of brachial valve; tongue nearly as broad as valve; costellae terminating at beginning of coarse costae; costae flattened on truncated parts of shell. *M.Dev., ?U.Dev.*, Eu.(Ger.).——

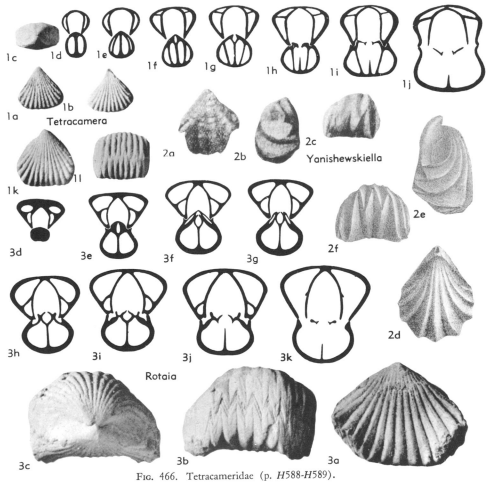

FIG. 466. Tetracameridae (p. H588-H589).

Yanishewskiella

Fig. 467. Tetracameridae (p. *H589*).

Fig. 462,5. *Schnurella schnuri,* Givet., Ger. (Eifel); *5a-c,* ped.v., ant., lat. views, ×1 (718b); *5d-i,* ser. secs., ×2 (931d). [Schmidt.]

Family TETRACAMERIDAE
Likharev in Rzhonsnitskaya, 1956

[*nom. transl.* Rzhonsnitskaya, 1958, p. 115 (*ex* Tetracamerinae Likharev in Rzhonsnitskaya, 1956, p. 126) [Materials for this family prepared by D. J. McLaren]

Subtriangular to wedge-shaped; lateral and anterior margins truncate, commissure plane or uniplicate, serrate; wholly costate with simple, subangular to rounded costae. Spondylium sessile, or supported on low septum in pedicle valve, with two lateral buttressing plates. Brachial valve with narrow outer hinge plates; septalium open or partially covered with inner hinge plates; strong median septum. *L.Carb.(Miss.).*

Tetracamera Weller, 1910, p. 503 [**Rhynchonella subcuneata* Hall, 1858, p. 11; OD]. Small or medium-sized; acute apical angle; commissure plane or weakly uniplicate; costae subangular; capillae may develop. Septalium open; lateral lamellae support hinge below dental sockets and rest on floor of brachial valve on either side of median septum. *Miss.,* N.Am.-?USSR.——Fig. 466,*1a-j.* **T. subcuneata* (Hall), U.Miss., USA (Ind.); *1a-c,* ped.v., brach.v., post. views, ×1; *1d-j,* ser. secs., ×2.5(858).——Fig. 466,*1k,l.* T. *arctirostrata* (Swallow), U.Miss., USA(Mo.); *1k,l,* ped.v., ant. views, ×1 (858).

Rotaia Rzhonsnitskaya, 1959, p. 30 [*pro Welleria* Rotay, 1941, p. 107 (*non* Ulrich & Bassler, 1923)] [**Rhynchonella subtrigona* Meek & Worthen, 1860, p. 451; OD]. Medium-sized or large; obtuse apical angle; commissure uniplicate,

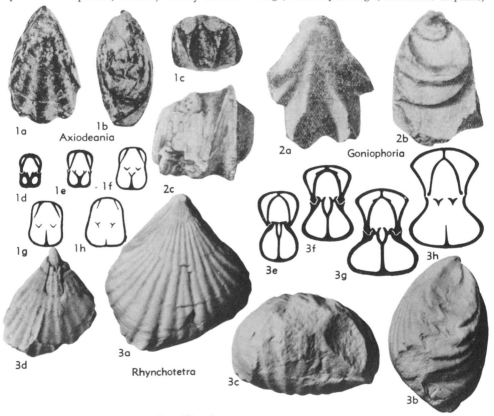

Fig. 468. Rhynchotetradidae (p. *H589*).

deeply serrate, downwardly deflected; costae rounded; low fold and shallow sulcus develop anteriorly. Septalium partly covered by discrete inner hinge plates. *L.Carb.(Miss.)*, N.Am.-Australia(New S. Wales)-?USSR.——Fɪɢ. 466,3. **R. subtrigona* (Mᴇᴇᴋ & Wᴏʀᴛʜᴇɴ), L.Miss., USA (Ill.); *3a-c,* ped.v., ant., post. views, ×1; *3d-k,* ser. secs., ×1.25 (858).

Yanishewskiella Lɪᴋʜᴀʀᴇᴠ, 1957, p. 139 [**Goniophoria angulata* Yᴀɴɪsʜᴇᴠsᴋɪʏ, 1910, p. 83 (=*?Anomia angulata* LɪɴɴÉ, 1767); OD]. Costae few, angular to subangular. Spondylium supported by septum and 2 lateral buttressing plates parallel to plane dividing valves. *L.Carb.*, USSR (S.Urals-Fergana)-?Eu.(Eng.).——Fɪɢ. 466,2*a-c;* 467,1. **Y. angulata* (Yᴀɴɪsʜᴇᴠsᴋɪʏ), Fergana; 466, *2a-c,* ped.v., lat., ant. views, ×1; 467,1, transv. sec. near beak, ×5 (517a).——Fɪɢ. 466,2*d-f.* **Y. angulata* (Yᴀɴɪsʜᴇᴠsᴋɪʏ)?, Eng.; *2d-f,* brach.v., lat., ant. views of LɪɴɴÉ's *Anomia angulata,* ×1 (229).

[Doubts relating to recognition of this nominal genus arise from (1) uncertainty that specimens from Fergana used by Lɪᴋʜᴀʀᴇᴠ in drawing up his diagnosis of *Yanishewskiella* are conspecific with *Goniophoria angulata* Yᴀɴɪsʜᴇᴠsᴋɪʏ from the southern Urals, including the holotype of this species, and (2) possibility that neither the Fergana nor Urals specimens are conspecific with LɪɴɴÉ's *Anomia angulata* from the British Isles. Lɪᴋʜᴀʀᴇᴠ's explicit original designation of the species *Goniophoria angulata* as the type-species of *Yanishewskiella* serves (under provisions of the Zoological Code) to tie this genus to the species named, defined by its holotype and associated specimens from the Urals region, regardless of the identity of shells studied by Lɪᴋʜᴀʀᴇᴠ in 1957 and his statement "if this Fergana species happens to differ from the holotype [of *Goniophoria angulata*] it should acquire a new species name."]

Family RHYNCHOTETRADIDAE
Likharev in Rzhonsnitskaya, 1956

[*nom. correct.* McLᴀʀᴇɴ, herein (*ex* Rhynchotetraidae Lɪᴋʜᴀʀᴇᴠ in Rᴢʜᴏɴsɴɪᴛsᴋᴀʏᴀ, 1956, p. 126)] [=Rhynchotetridae Lɪᴋʜᴀʀᴇᴠ in Rᴢʜᴏɴsɴɪᴛsᴋᴀʏᴀ, 1959, p. 30] [Materials for this family prepared by D. J. McLᴀʀᴇɴ]

Medium-sized to large, subovate, wedge-shaped posteriorly; apical angle acute; large concave interareas; uniplicate; fold and sulcus weakly developed; paucicostate to costate; angular to rounded costae that may branch, extend from beak; posterolateral margins may be smooth; commissure serrate; finely capillate. Dental plates converge ventrally or may join to form spondylium, sessile or supported by septum; hinge plate divided; deep, open septalium; strong septum; crural bases triangular. [May be closely related to Tetracameridae.] *L.Carb.-L. Perm.*

Rhynchotetra Wᴇʟʟᴇʀ, 1910, p. 506 [**Rhynchonella caput testudinis* Wʜɪᴛᴇ, 1862, p. 23; OD]. Large; costae rounded to subangular, few bifurcate. Dental plates converging ventrally to join floor of pedicle valve, or form spondylium supported by septum. [Type-species imperfectly

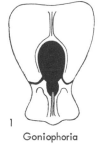

Goniophoria

Fɪɢ. 469. Rhynchotetradidae (p. *H589*).

known; Wᴇʟʟᴇʀ (1914, p. 205-7) would include forms with or without spondylium in this genus]. *Miss.,* N.Am.-?USSR.——Fɪɢ. 468,3*a-d.* **R. caput-testudinis* (Wʜɪᴛᴇ); Kinderhook., USA(Iowa); *3a-c,* brach.v., lat., ant. views, ×1; *3d,* int. cast ped.v., ×1 (858).——Fɪɢ. 468,3*e-h.* R. missouriensis Wᴇʟʟᴇʀ, Kinderhook., USA(Mo.); *3e-h,* ser. secs., ×2.5 (858).

Axiodeaneia Cʟᴀʀᴋ, 1917, p. 374 [**A. platypleura;* OD]. Narrowly triangular with rounded anterior; sides vertical; few strong, subangular costae extending from beak axially; posterolaterally smooth; commissure coarsely serrate; capillate. Strong, ventrally approximating dental plates. *L. Miss.,* USA(Mont.).——Fɪɢ. 468,1. **A. platypleura; 1a-c,* brach.v., lat., ant. views, ×1.3; *1d-h,* ser. secs., ×1.3 (161a).

Goniophoria Yᴀɴɪsʜᴇᴠsᴋɪʏ, 1910, p. 80 [**G. monstrosa;* SD Sᴄʜᴜᴄʜᴇʀᴛ & LᴇVᴇɴᴇ, 1929, p. 63]. High, laterally compressed near beak; few high, angular, commonly asymmetrical costae; strongly serrate commissure. Spondylium supported by strong median septum; septum in brachial valve protruding into cavity of septalium. [Lɪᴋʜᴀʀᴇᴠ (1957) suggested that species possessing a spondylium assigned to *Rhynchotetra* may belong here (e.g., *R. missouriensis* Wᴇʟʟᴇʀ).] *L.Carb.-L. Perm.,* USSR(Urals-Fergana)-Eu.-?N.Am.——Fɪɢ. 468,2. **G. monstrosa,* L.Carb., USSR(S.Urals); *2a-c,* brach.v., lat., ant. views, ×1 (517a).—— Fɪɢ. 469,1. G. carinata Yᴀɴɪsʜᴇᴠsᴋɪʏ, L.Carb., USSR(S.Urals), sec. near apex, ×2 (517a).

Family WELLERELLIDAE
Likharev in Rzhonsnitskaya, 1956

[*Wellerellidae* Lɪᴋʜᴀʀᴇᴠ in Rᴢʜᴏɴsɴɪᴛsᴋᴀʏᴀ, 1956, p. 125] [Materials for Paleozoic representatives of this family prepared by Hᴇʀᴛᴀ Sᴄʜᴍɪᴅᴛ (Mesozoic subfamilies by D. V. Aɢᴇʀ)]

Sulcus and fold moderately developed; costae commonly strong, angular to subangular; commissure denticulate. No septalium or cardinal process; hinge plate entire; dorsal septum and dental plates variously developed. *L.Carb.-U.Cret.*

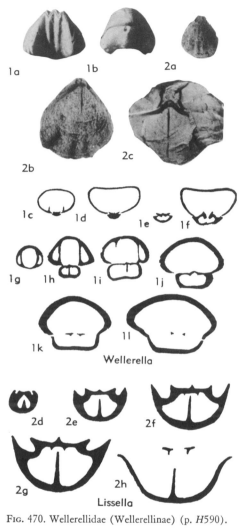

Wellerella

Lissella

FIG. 470. Wellerellidae (Wellerellinae) (p. H590).

Subfamily WELLERELLINAE Likharev in Rzhonsnitskaya, 1956

[*nom. transl.* LIKHAREV in RZHONSNITSKAYA, 1958, p. 114]
[Materials for this subfamily prepared by HERTA SCHMIDT]

Small to medium-sized; beak acute, little incurved. Dental plates commonly short or wanting; hinge plates united by flat or convex to keel-shaped plate; dorsal septum in most genera low or absent. *L.Carb.-U.Perm.*

Wellerella Group

Umbo smooth, costae beginning at moderate distance from apex. *U.Carb.-Perm.*

Wellerella DUNBAR & CONDRA, 1932, p. 286 [*W. tetrahedra*; OD]. Small, circular to subpentagonal in outline; fold and sulcus well developed; deltidial

plates leaving free oval foramen in front of beak; few simple subangular costae beginning away from apex. Dental plates short; hinge plates united by flat or keel-shaped plate; dorsal septum very short or wanting. *U.Carb.-Perm.*, N.Am.-S.Am.-Asia-Eu.(Ural region).——FIG. 470,*1a-f.* *W. tetrahedra*, M.Penn.(Marmaton), USA(Mo.); *1a,b*, ant., post. views, *ca.* ×2; *1c-f*, ser. secs., ?mag. (270).——FIG. 470,*1g-l.* *W. osagensis* (SWALLOW), U.Penn.(Shawnee Gr., Plattsmouth Ls.), USA (Neb.); ser. secs., ?mag. (270).

Lissella CAMPBELL, 1961, p. 452 [*L. booralensis*, p. 453; OD]. Small, oval, with sulcus and fold in anterior halves of valves; few rounded to subangular costae developed anteriorly. Dental plates very delicate; hinge plate robust with prominent median ridge; crural bases protruding dorsally; dorsal septum rather high and long, but not connected with hinge plate. *U.Carb.*, Australia.——FIG. 470,*2.* *L. booralensis*, Booral F., New S. Wales(Gloucester trough); *2a,b*, ped.v., ×1, brach.v., ×2; *2c*, posterodorsal view, ×2; *2d-h*, transv. secs. (from several int. molds), ×3 (143).

Pseudowellerella Group

Costae beginning at apex. *Permocarb.-U. Perm.*

Pseudowellerella LIKHAREV, 1956, p. 58 [*P. nikitchi*; OD]. Small, rounded, triangular in outline; sulcus and fold weakly developed; costae numerous, bifurcating. Dental plates wanting; hinge

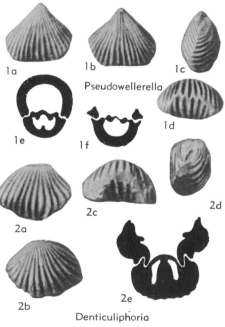

Pseudowellerella

Denticuliphoria

FIG. 471. Wellerellidae (Wellerellinae) (p. H590-H591).

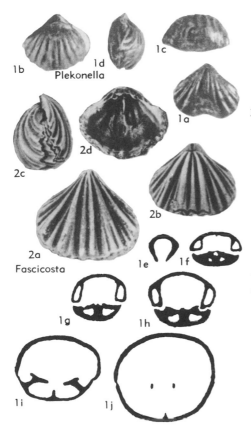

Fig. 472. Wellerellidae (Wellerellinae) (p. H591).

plates united by convex plate; dorsal median ridge short, low. *U.Perm.*, USSR(N.Caucasus).——Fig. 471,1. *P. nikitchi*; *1a-d*, ped.v., brach.v., lat., ant. views, ×1; *1e,f*, transv. secs, ×4 (517)

Denticuliphoria LIKHAREV, 1956, p 57 [*D. rara*; OD]. Resembling *Pseudowellerella* in shape, but with stronger and less numerous simple costae. Dental plates wanting; teeth very strong and long; accessory ventral denticles fitting into accessory sockets of brachial valve; hinge plates united by convex plate extending farther anteriorly than lateral parts of hinge plate; dorsal septum or ridge thick and short, becoming separated from hinge plate within zone of articulation. *U. Perm.*, USSR (N.Caucasus).——Fig. 471,2. *D. rara*; *2a-d*, ped.v., brach.v., ant., lat. views, ×1.5; *2e*, ser. sec., ×5 (517).

Fascicosta STEHLI, 1955, p. 71 [*Rhynchonella? longaeva* GIRTY, 1909, p. 322; OD]. Small, subpentagonal to rounded triangular in outline; sulcus and fold beginning near mid-length; tongue low; beak prominent, with subapical foramen and large conjunct deltidial plates; costae strong, increasing by bifurcation and implantation, crossed

by fine concentric lines; commissure denticulate to undulate. Dental plates small; teeth and sockets crenulate; hinge plate supported by low broad ridge, rarely by low septum. *Perm.(Word.-Capitan.)*, N.Am.(Tex.).——Fig. 472,2. *F. longaeva* (GIRTY), Cherry Canyon F., W.Tex.(Guadalupe Mts.; *2a-c*, ped.v., brach.v., lat. views, ×2; *2d*, brach.v. int., ×2 (774).

Plekonella CAMPBELL, 1953, p. 17 [*P. acuta*; OD]. Medium-sized, transversely oval; sulcus and fold developed in anterior parts of valves; costae simple. Dental plates well developed; hinge plate concave, divided only anteriorly from zone of articulation, with ridge directed ventrally; dorsal septum or ridge thick but short. *Permocarb.*, Australia (Queensl.).——Fig. 472,1. *P. acuta*, Inglesia Beds; *1a-d*, ped.v., brach.v., ant., lat. views, ×1; *1e-j*, transv. secs. (several specimens), ?mag. (139).

Allorhynchus Group

Genera having some external and internal features of Wellerellinae but lacking connecting plate between hinge-plates. [Loosely annexed to subfamily.] *L.Carb.-U.Perm.*

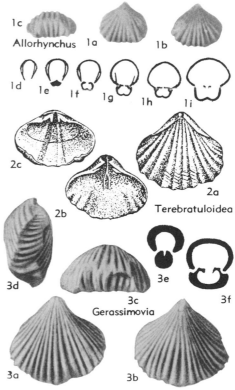

Fig. 473. Wellerellidae (Wellerellinae) (p. H592).

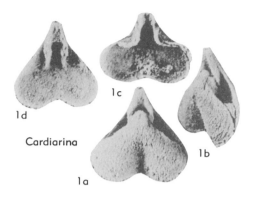

FIG. 474. Cardiarinidae (p. *H592*).

Allorhynchus WELLER, 1910, p. 509 [*Rhynchonella heteropsis* WINCHELL, 1865, p. 121; OD]. Small, transversely elliptical; ventral beak prominent; sulcus and fold distinct in anterior halves of valves; costae numerous, simple, angular, beginning at apex. Dental plates short, in typespecies closely approached to shell wall; dorsal septum absent but low ridge may be present. *L. Carb., ?Perm.,* N.Am.——FIG. 473,1. *A. heteropsis* (WINCHELL). L. Miss.(Kinderhook.), USA (Iowa); *1a-c,* ped.v., brach.v., ant. views, ×1 (178); *1d-i,* ser. secs., ×2.5 (856a).

Gerassimovia LIKHAREV, 1956, p. 59 [*G. gefoensis*; OD]. Exteriorly resembling *Pseudowellerella. U.Perm.,* USSR(N.Caucasus).——FIG. 473,3. *G. gefoensis*; *3a-d,* ped.v., brach.v., ant., lat. views, ×1; *3e,f,* ser. secs., ×4 (517).

Terebratuloidea WAAGEN, 1883, p. 410 [*T. davidsoni*]. Transversely elliptical; sulcus and fold well marked; ventral beak truncated by large foramen; costae simple, strong, beginning on beak, crossed by transverse lines. Dental plates and dorsal septum wanting. *Permocarb.,* Asia(India-China)-Eu. (USSR-Alps-Sicily).——FIG. 473,2. *T. davidsoni*; *2a-c,* brach.v., ped.v. int., brach.v. int., ×1 (396, after Waagen).

?Family CARDIARINIDAE Cooper, 1956

[Cardiarinidae COOPER, 1956, p. 527] [Materials for this family prepared by D. J. McLAREN]

Foramen apical, beak elongate, strong inner beak ridges, no crura, with elaborate parathyridium. *Penn.*

Cardiarina COOPER, 1956, p. 527 [*C. cordata*; OD]. Minute, heart-shaped; rectimarginate to sulcate; smooth. Pedicle valve with symphytium(?), excavated lateral margins; long dental plates that loop anteriorly round excavated area; teeth small, narrow, attached to inner side of dental plates. Brachial valve with narrow sockets, thin outer

socket ridges; high, stout, inner socket ridges; strong median carina, no septum. *Penn.,* USA(N. Mex.).——FIG. 474,1. *C. cordata*; *1a,b,* brach., side tilted views, ×15; *1c,d,* brach.v., ped.v. int., ×15 (188).

Family UNCERTAIN

Diabolirhynchia (see p. *H904*).

Dorsisinus SANDERS, 1958, p. 53 [*Centronella louisianensis* WELLER, 1914, p. 241; OD]. Very small, smooth-shelled; sulcate; delthyrium triangular, deltidial plates incipient; fold on pedicle, sulcus on brachial valve. Dental plates present; septum long, septalium open, crura long. [Interior suggests relationship with Trigonirhynchiidae, but external form may call for grouping with other smooth sulcate forms (e.g., *Paranorella, Sanjuania*).] *L.Miss.,* USA(Miss.Valley)-Mex.——FIG. 475,1. *D. louisianensis* (WELLER), Kinderhook., USA(Ill.) *(1a,b)*; L.Miss., Mex.(Sonora) *(1c,d)*; *1a,b,* ped.v., brach.v. views, ×4 (858); *1c,* brach. v. int., ×3; *1d,* oblique view int. both valves, ×4 (705a). [McLAREN.]

Katunia KULKOV, 1963, p. 54 [*K. subtrigonata*; OD]. Small to medium-sized, subtriangular to subpentagonal in outline; strongly inflated; uniplicate; fold and sulcus developed anteriorly; rounded to subangular costae (which may bifurcate) developed toward front; posterior smooth; commissure serrate. Thick-shelled; no dental plates; teeth supported by inturned shell margins; inner margins of massive divided hinge plate support rounded crura; no cardinal process or median septum; low ridge on floor of brachial valve. *L. Dev.,* USSR (Gorno - Altay) - ?Eu. (Karnic Alps).——FIG. 476,4. *K. subtrigonata,* Gorno-Altay; *4a-d,* ped.v., brach.v., lat., ant. views, ×1; *4e-i,* ser. secs., ×3 (493a). [McLAREN.]

Ladogifornix SCHMIDT, 1964 [*Terebratula fornicata* SCHNUR, 1853, p. 175; OD]. Medium-sized, subtetrahedral; pedicle valve flat to concave; brachial valve strongly convex; sulcus broad, deep, rounded; tongue high with curved margin; fold high, not distinctly defined; costae numerous, fine, more prominent near front, some of them branching; commissure finely denticulate. Pedicle valve without median ridge; dental plates nearly parallel in cross section; umbonal cavities large;

FIG. 475. Family Uncertain (p. *H592*).

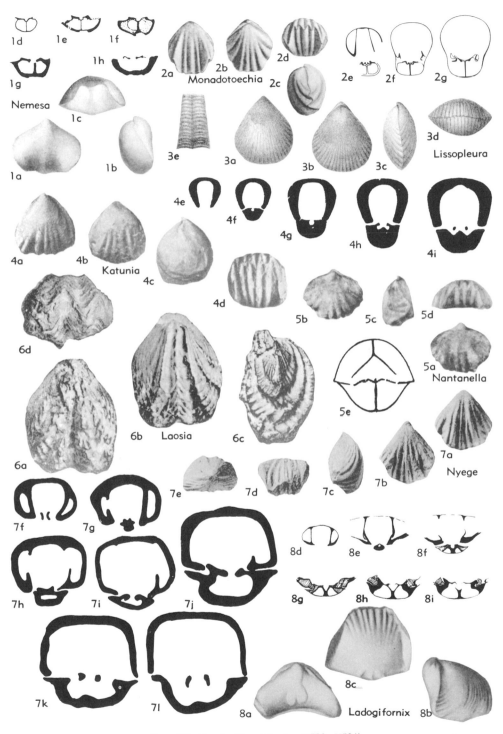

FIG. 476. Family Uncertain (p. *H*592, *H*594).

opening of septalium restrained by inner hinge plates; dorsal median septum extending to about half length of valve. *M.Dev.*, Eu.(Ger.).——Fig. 476,8. *L. fornicatus* (SCHNUR), Ger.(Eifel); *8a-c*, ped.v., lat., ant. views, ×1 (718b); *8d-i*, ser. transv. secs., ×2.5 (931d). [SCHMIDT.]

Laosia MANSUY, 1913, p. 83 [*L. Dussaulti*; OD]. Small, elongate subpentagonal in outline; apical angle acute; brachial valve more inflated than pedicle valve; pedicle beak erect; shell covered by few, coarse, angular, simple costae; median costa low on brachial valve, opposite deep, angular sulcus on pedicle valve; commissure coarsely serrate; strong concentric growth lines. Interior unknown. *Perm.*, Indo-China.——Fig. 476,6. *L. dussaulti*; *6a-d*, ped.v., brach.v., lat., ant. views, ×3 (532b). [McLAREN.]

Leiorhynchoides DOVGAL, 1953, p. 139 [*L. gratianovae*; OD]. Medium-sized, subpentagonal in outline; brachial valve more convex than pedicle; fold and sulcus present, margin uniplicate; rounded costae toward front of valves. Interior imperfectly known; impressed elongate diductor impressions and no dental plates in pedicle valve; median septum in brachial valve, no septalium, hinge plates reduced. *M.Dev.*, USSR(Altay). [McLAREN.]

Lissopleura WHITFIELD, 1896, p. 232 [*Rhynchonella aequivalvis* HALL, 1857, p. 66; OD]. Small, compressed, broadly ovate in outline; equivalve; beak small, erect to incurved; commissure plane to weakly uniplicate; without fold or sulcus; costate from beak; low rounded costae with narrow linear interspaces; fine concentric ornament. Dental plates present, and bilobed muscle impression in pedicle valve; strong median septum; hinge plate unknown. [This genus may be a compressed youthful form of rhynchonelloid.] *L.Dev.(Helderberg.)*, USA(N.Y.).——Fig. 476,3. *L. aequivalvis* (HALL); *3a-d*, ped.v., brach.v., lat., ant. views, ×1; *3e*, portion of costae showing concentric ornament, enlarged (384). [McLAREN.]

Monadotoechia HAVLÍČEK, 1960, p. 243 [*Terebratula Monas* BARRANDE, 1847, p. 444; OD]. Small, elongate, inflated; lateral margins almost vertical; uniplicate; shallow sulcus, low fold anteriorly only; few subangular costae developed toward front; smooth posteriorly; commissure serrate. Dental plates short, thin; hinge plate entire, concave, grooved posteriorly; supported by median septum. *L.Dev.*, Czech.-E.Eu.——Fig. 476,2a-d. *M. monas* (BARRANDE), Boh.; *2a-d*, ped.v., brach. v., lat., ant. views, ×2 (53).——Fig. 476,2e-g. *M. monadina* HAVLÍČEK, Boh.; *2e-g*, ser. transv. secs. 5.2, 5.15, 5.05 mm. from ant. margin, ×4 (411a). [McLAREN.]

Nantanella GRABAU, 1936, p. 70 [*N. mapingensis*; OD]. Small, transverse; uniplicate, with strong fold and sulcus in anterior part of shell; tongue broad, vertical; beak erect; interarea small; paucicostate anteriorly, costae subrounded to angular;

posterior smooth; commissure serrate. Interior incompletely known; spondylium supported by septum in pedicle valve; septum present in brachial valve; structure of hinge plate unknown. [GRABAU (1936) assigned the genus to the Camerophoriacea (=Stenoscismatacea).] *L.Perm.*, China.——Fig. 476,5. *N. mapingensis*; *5a-d*, ped.v., brach. v., lat., ant. views, ×1; *5e*, transv. sec. near apex, enlarged (362a). [McLAREN.]

Nemesa SCHMIDT, 1941, p. 41 [*N. nemesana*; OD]. Small, subpentagonal in outline; brachial valve more inflated than pedicle; beak incurved, interarea small; shallow sulcus with short tongue near front in pedicle valve; corresponding fold low or absent; uniplicate to sulciplicate; very few rounded shallow costae at front margin, commonly only 1 in sulcus, 2 on fold; remainder of shell smooth. Dental plates very close to lateral margins of shell, commonly lost in callus; divided hinge plates; open septalium supported by slender septum. *M.Dev.(Eifel.)*, Ger.——Fig. 476,1. *N. nemesana*; *1a-c*, ped.v., lat., ant. views, ×1.5; *1d-h*, ser. transv. secs. near apex, ×2.5 (718b). [McLAREN.]

Nyege VEEVERS, 1959, p. 113 [*N. scopimus*; OD]. Small to medium-sized, subpentagonal in outline; equivalve; intraplicate to sulcate; pedicle beak suberect; without interarea; high fold in pedicle and broad sulcus in brachial valves beginning near umbones; rounded costae with equal interspaces extending nearly from beak, may increase by bifurcation and intercalation. High, short dental plates; divided hinge plates, bifid cardinal process at apex; no septum; crura apparently straight. [VEEVERS (1959) included the genus in the Atrypoidea but failed to find spiralia. Externally it closely resembles species from the Famennian of the southern Urals assigned to *Plectorhynchella* by ROZMAN (1962) (e.g., *P. markovskii* ROZMAN), but she reported the presence of a septum supporting the hinge plate.] *U.Dev.(Famenn.)*, W. Australia.——Fig. 476,7. *N. scopimus*; *7a-e*, ped.v., brach.v., lat., ant., post. views, ×1.5; *7f-l*, ser. transv. secs. 0.5, 0.7, 0.9, 0.95, 1.05, 1.15, 1.35 mm. from apex, ×9 (838). [McLAREN.]

Payuella GRABAU, 1934, p. 150 [*P. obscura*; OD]. Medium-sized, transverse, flattened; sulcus on brachial valve; low fold on pedicle valve; ?sulcate; pedicle beak erect; weak, rounded costae extend nearly from beak; commissure serrate. Dental plates present, otherwise interior unknown. [GRABAU classified this genus with the Dielasmatidae, but described the shell as fibrous and made no mention of endopunctae.] *L.Perm.*, China.——Fig. 477,5. *P. obscura*; *5a-c*, brach.v., lat., post. views, ×2 (362). [McLAREN.]

Phoenicitoechia HAVLÍČEK, 1960, p. 242 [*Terebratula Phoenix* BARRANDE, 1847, p. 431; OD]. Small, high in anterior region; sulcus and fold beginning at distance from beak; sulcus shallow, tongue low; fold very low, in some shells not distinguishable from curvature of valve; costae not numerous, strong, rounded, restricted to marginal parts of valves; commissure strongly denticulate. Dental plates slender, thin; hinge plates divided, narrow; median septum supporting wide septalium. *L.Dev.,* Eu.(Boh.).——Fig. 477,3. *P.

phoenix (Barrande), Boh.(Kroněprusy); *3a-d,* ped.v., brach.v., lat., ant. views, ×1 (53); *3e-h,* ser. transv. secs., ×7.5 (Schmidt, n). [Schmidt.]

Plectorhynchella Cooper & Muir-Wood, 1951, p. 195 [*pro Monticola* Nalivkin, 1930, p. 86 (*non* Boie, 1822)] [**Athyris collinensis* Frech, 1902,

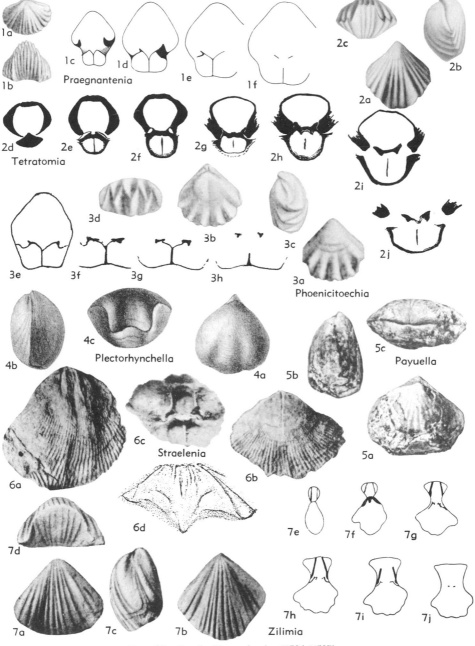

Fig. 477. Family Uncertain (p. *H594-H597*).

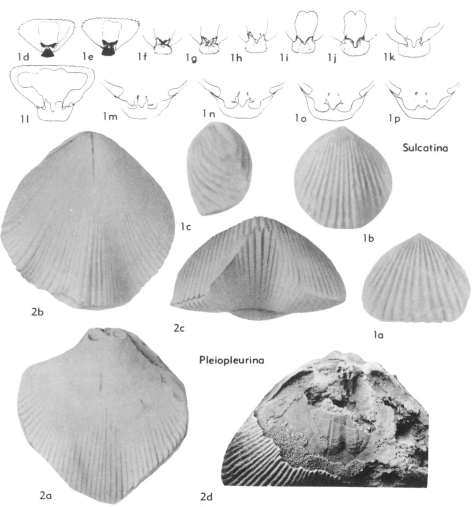

Sulcatina

Pleiopleurina

FIG. 477A. Family Uncertain (p. *H596-H597*).

p. 99; OD]. Small, subpentagonal, inflated; intra-plicate to sulcate; pedicle beak nearly straight to erect; pedicle valve fold and brachial valve sulcus confined to anterior part of shell; smooth posteriorly; may develop low, irregular, bifurcating costae anteriorly; strong concentric micro-ornament. Interior poorly known; dental plates present; dorsal median septum, and possibly divided hinge plates. [HAVLÍČEK (1961, p. 203) stated that Bohemian species assigned to this genus may belong to the Atrypoidea and not the Rhynchonelloidea. The affinities of the type-species are not clear, nor is it certain that all species assigned to the genus are, in fact, congeneric.] *?L.Dev., U. Dev., ?L.Carb.*, Eu., USSR.——FIG. 477,4. **P. collinensis* (FRECH), U.Dev., Karnic Alps; *4a-c*, brach.v., lat., and ant. views, ×2 (311b). [MC-LAREN.]

Pleiopleurina SCHMIDT, 1964 [**Atrypa pleiopleura* CONRAD, 1841, p. 55; OD]. Large, not high; pedicle valve faintly convex; brachial valve moderately convex; sulcus and tongue broad; fold prominent in anterior part; costae numerous, relatively fine, beginning on beaks, rounded-angular; commissure denticulate, situated on edges. Ventral muscle field far removed posteriorly, oval, longitudinally divided by low ridge; dorsal septum short; hinge plates united by stout cardinal process consisting of 2 tubes or funnels; crura thick. *L.Dev.*, N.Am.——FIG. 477A,2. **P. pleiopleura*; *2a-c*, ped.v., brach.v., ant. view, ×1 (384); *2d*, Oriskany, N.York, posterior part of brach.v. with cardinal process, ×1.5 (931d). [SCHMIDT.]

Praegnantenia HAVLÍČEK, 1961, p. 99 [**Terebratula praegnans* BARRANDE, 1847, p. 428; OD]. Similar to *Phoenicitoechia*; more inflated, with high fold and tongue; anterior commissure depressed below crest of brachial valve; numerous angular costae increasing by bifurcation, developed anteriorly only; umbones smooth; commissure strongly serrate. *L.Dev.,* Eu.(Czech.).——Fɪɢ. 477,*1.* *P. praegnans* (Bᴀʀʀᴀɴᴅᴇ); *1a,b,* brach.v., ant. views, ×1 (53); *1c-f,* ser. transv. secs. 7.2, 7.0, 6.85, 6.75 mm. from ant. margin, ×4 (411a). [Mᴄ-Lᴀʀᴇɴ.]

Protorhyncha Hᴀʟʟ & Cʟᴀʀᴋᴇ, 1893, p. 180 [*Atrypa dubia* Hᴀʟʟ, 1847, p. 21; OD]. Types of type-species lost; interior unknown. [Unrecognizable, not certainly a rhynchonellacean.] *M.Ord.,* N.Am.

Straelenia Mᴀɪʟʟɪᴇᴜx, 1935, p. 10 [*Rhynchonella Dunensis* Dʀᴇᴠᴇʀᴍᴀɴɴ, 1902, p. 108=*Rhynchonella Dannenbergi* Kᴀʏsᴇʀ mut. nov. *minor* Dʀᴇᴠᴇʀᴍᴀɴɴ, 1902, p. 107; OD] [=*Dinapophysia* Mᴀɪʟʟɪᴇᴜx, 1935, p. 5]. Large, transverse to subquadrate, with greatest width about mid-length; inequivalve; fold and sulcus weakly developed; commissure shallowly uniplicate; numerous, rounded, simple costae, which may extend from umbones. Dental plates present; strong median septum supporting undivided hinge plate with surface bearing longitudinal ridges and furrows posteriorly and rounded median elevation anteriorly. *L.Dev.,* W.Eu.-?N.Afr.——Fɪɢ. 477,*6.* *S. dunensis* (Dʀᴇᴠᴇʀᴍᴀɴɴ), Belg.-Ger.; *6a,* ped. v., int. mold, ×1; *6b,* brach.v., int. mold, ×1; *6c,* cast of cardinalia of both valves, ×2 (529a); *6d,* hinge plate, ×6 (931c). [MᴄLᴀʀᴇɴ.]

Sulcatina Sᴄʜᴍɪᴅᴛ, 1964 [*Trigonirhynchia sulcata* Cᴏᴏᴘᴇʀ, 1942, p. 234; OD]. Medium-sized to large; rounded trigonal in outline; pedicle valve flat, brachial valve strongly convex; sulcus and fold beginning at distance from apex, broad, not strictly defined; anterior margin of tongue curved; costae simple, strong, angular, beginning at apex; commissure denticulate, situated on edge. Interior of apical parts filled by callus; dental plates very deep; hinge plates separated by deep cavity; walls of cavity extending to valve wall; 2-winged process above and partly on hinge plates. *Sil.,* N.Am.-?Eu.——Fɪɢ. 477A,*1.* *S. sulcata; 1a-c,* ped.v., brach.v., lat. view, ×1 (178); *1d-p,* Indiana (Waldron), ser. secs., ×2 (931d). [Sᴄʜᴍɪᴅᴛ.]

Tetratomia Sᴄʜᴍɪᴅᴛ, 1941, p. 13 [*Terebratula tetratoma* Sᴄʜɴᴜʀ, 1851, p. 4; OD]. Small, subpentagonal in outline; equivalve, moderately convex; beak incurved; uniplicate, with strong fold and sulcus developed from near beak; tongue trapezoidal; costate nearly from beak, with simple, angular costae; commissure weakly serrate. Dental plates commonly fused with shell walls; hinge-plate entire; no septalium; median septum not joined to hinge plate, becoming detached from floor of valve and extending forward of articulation as unsupported plate, *M.Dev.(Eifel.),* Ger. ——Fɪɢ. 477,*2.* *T. tetratoma* (Sᴄʜɴᴜʀ); *2a-c,*

ped.v., lat. views, ×2 (718b); *2d-j,* ser. transv. secs., ×4 (719a). [MᴄLᴀʀᴇɴ.]

Zilimia Nᴀʟɪᴠᴋɪɴ, 1947, p. 93 [*Rhynchonella polonica* Gürɪᴄʜ, 1896, p. 291; OD] [=*Zilimia* Nᴀʟɪᴠᴋɪɴ, 1937, p. 107 *(nom. nud.)*]. Large, subtriangular to subpentagonal in outline; brachial valve strongly inflated, pedicle valve flattened to concave; beak long, straight; uniplicate, sulcus very broad, shallow, tongue high and rounded; fold high, developed only anteriorly; costae numerous, rounded, and increase by bifurcation over whole shell; commissure smooth. Dental plates strong; hinge plates divided; no septalium, cardinal process, or median septum. *U. Dev. (Famenn.),* Pol. - USSR (Urals). —— Fɪɢ. 477,*7a-d.* *Z. polonica* (Gürɪᴄʜ), Ural; *7a-d,* ped.v., brach.v., lat., ant. views, ×1 (690).—— Fɪɢ. 477,*7e-j.* *Z. mugodjarica* Rᴏᴢᴍᴀɴ, Ural; *7e-j,* ser. transv. secs. 1.5, 2.5, 5.0, 7.0, 7.5, 8.0 mm. from apex, ×1 (683a). [MᴄLᴀʀᴇɴ.]

MESOZOIC AND CENOZOIC RHYNCHONELLACEA
By D. V. Aɢᴇʀ

Mesozoic and Cenozoic rhynchonellaceans are segregated in the *Treatise,* partly because of separate authorship and partly because of the great taxonomic break between the Paleozoic and later forms. The separation is reflected not only in the classification, but also in the characters used in diagnosis and even in the morphological terminology. The separation is largely artificial, since the author has no doubt that most or all of the Mesozoic lineages can be traced back into the Paleozoic, but it cannot be resolved at present because the groups have been studied in different ways and the same information is not available about both. This arises particularly from the dependence placed by Mesozoic workers on serial section studies.

To a certain degree the same reservations apply to the apparent break in the rhynchonellacean lineages at the end of the Mesozoic, but at that time there does seem to have been a considerable reduction and restriction of the stocks, both geographically and ecologically.

Numerically, described rhynchonellacean genera from Mesozoic rocks are nearly five times greater than those reported from Cenozoic deposits, including Recent. The combined assemblage of Triassic, Jurassic, and Cretaceous rhynchonelloids contains 105 genera (not counting 17 nominal genera classed as synonyms), whereas the total number of known Cenozoic genera amounts only to 20 (likewise omitting nominal genera considered to be synonyms, 8 in number). A single genus *(Aetheia)* is included in the count of both Mesozoic and Cenozoic forms, which indicates a surprising degree of classificatory separateness.

The large number of genera erected for Mesozoic brachiopods, particularly those proposed for Jurassic rhynchonelloids and terebratuloids by S. S. BUCKMAN (1918) in a single publication, has been much criticized by specialists in other fields. The preponderance of named Jurassic forms over those from Triassic and Cretaceous strata actually does not reflect relative complexity of the groups occurring in the three systems. Certainly, the rhynchonelloids were quite as varied in late Triassic times as in the Jurassic, and those of the former period still constitute the greatest gap in our knowledge of the Brachiopoda.

The validity of the named genera is a very subjective matter, though almost any grouping of related forms among the thousands of species which have been called *"Rhynchonella"* is likely to be useful. Unfortunately, different criteria have been used by different authors, e.g., BUCKMAN (1918) placed great emphasis on muscle scars and internal plates, WISNIEWSKA (1932) assigned major significance to forms of the crura, and LEIDHOLD (1920) considered microscopic structure of the shell to have greatest weight. Almost every morphological character, considered apart from others, can be shown to be unsatisfactory as a basis for classification in one or more groups, and few genera have yet been fully described both internally and externally. Although further investigations will undoubtedly show that many of the nominal genera are most usefully regarded as synonyms, there seems to be little value in suppressing them now, since later studies of internal and microscopic structures may call for their resurrection. Also, it is probable that several new generic names will be needed, especially for designation of Triassic forms.

The arrangement of genera in subfamilies proposed here should be regarded as an attempt to bring together members of related but independently evolving stocks.

MORPHOLOGICAL FEATURES
CRURA

Eight different forms of crura have so far been distinguished among the Mesozoic and Cenozoic rhynchonelloids, and three more are added here (Fig. 478). These are not, however, of equal importance and distinctiveness. Undoubtedly the basic and most separate types are the three originally defined by ROTHPLETZ (1886), and named by him "radulifer," "falcifer," and "septifer."

Radulifer. The radulifer type of crura are unspecialized hook- or rodlike structures which project from the hinge plates toward the pedicle valve. This is the simplest form of crura and may well be the basic type. What may be regarded as the typical radulifer crus is shown in Fig. 478,*1,* but commonly the crura are much straighter than this and have various terminal processes. These processes, which are often overlooked by students of the group, were mentioned in ROTHPLETZ's original description, and in fact are implied by the name of the form. The essential character of radulifer crura, as seen in transverse sections, is that they arise on the ventral side of the hinge plates and project (albeit only slightly) into the pedicle valve. They are accompanied by a more or less well-developed dorsal median septum.

Falcifer. The falcifer type was originally defined by ROTHPLETZ as sickle-shaped, as the name indicates, but the most essential character of this type is that they arise on the dorsal side of the hinge plates and project into the brachial valve as broad bladelike processes (Fig. 478,*2*). They are characteristically accompanied by a very reduced dorsal median septum or none at all.

Septifer. The septifer type refers to crura which have the form of septa that descend directly from the dorsal side of the hinge plates to the floor of the brachial valve (Fig. 478,*3*). In his original description, ROTHPLETZ implied that septifer crura might arise from the falcifer type simply by the crura coming into contact with the brachial valve. The converse is more likely to be true, since, in the main, the septifer types precede the falcifer types in time. What is more, some genera such as *Sulcirostra* as revised by AGER (1959) and *Crurirhynchia*

Dagis (1961) appear to show combinations of the radulifer and septifer types.

Arcuifer. Wisniewska (1932) redefined the Rothpletz types with the aid of serially

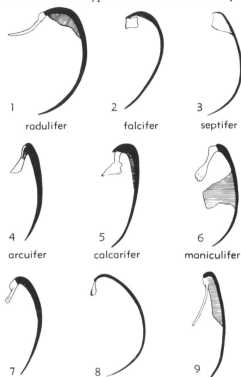

FIG. 478. Types of crura in Mesozoic and Cenozoic Rhynchonellida, illustrated by longitudinal sections of brachial valves through left crus with median septum (shaded), if present, not in same plane (*1,6,7* drawn from previous illustrations, others reconstructed from serial transverse sections) (Ager, n).——*1. Gibbirhynchia amalthei* (Quenstedt), L.Jur., Ger.; *ca.* ×5.——*2. Lacunosella visulica* (Oppel), U.Jur., Fr.; *ca.* ×1.75.——*3. Septocrurella deflexoides* (Uhlig), M.Jur., Rumania; *ca.* ×3.5.——*4. Monticlarella czenstochowiensis* (Roemer), U.Jur., Pol.; *ca.* ×5.——*5. Kallirhynchia platiloba* (Muir-Wood), M.Jur., Eng.; *ca.* ×3.75. ——*6. Mannia nysti* (Davidson), Mio., Belg.; *ca.* ×12.——*7. Grammetaria bartschi* (Dall), Rec., Philip.; *ca.* ×4.——*8. Cirpa kiragliae* (Ager), L. Jur., Turkey; *ca.* ×3.5.——*9. Peregrinella whitneyi* (Gabb), L.Cret., USA; *ca.* ×3.5. [Two other types of crura (canalifer, cilifer) are not suitable for representation in longitudinal sections, since their essential characters are only recognizable in transverse section, the canalifer type being concavoconvex and the cilifer type flattened in the plane of the commissure). Similarly the arcuifer type (Fig. 1,4) can be clearly distinguished only in transverse sections, which show the lateral origin of the crura and their inwardly facing concave faces.]

ground Polish material and added a fourth type named arcuifer. She recognized this only in the little-known genus *Monticlarella* (Fig. 478,4), however. The crura are hammer-shaped in cross section, with arcuate "heads" which are concave toward each other. Several members of the subfamily Norellinae are now known to have crura which approach this form, but it may also grade into the radulifer type, and is not really well enough known to be properly evaluated now.

Calcarifer. Muir-Wood (1934) added a fifth type, termed calcarifer, which she recognized particularly in the genera *Kallirhynchia* and *Rhynchonelloidella*. It was said to be characterized by a dorsally directed process at the distal end of each crus (Fig. 479,1,2), but when a reconstruction was made from the only complete set of transverse sections available (Fig. 478,5), it proved to be close to the falcifer shape. On general morphological grounds also,

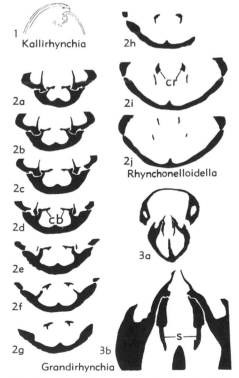

FIG. 479. Calcarifer type crura in Rhynchonellidae (Tetrarhynchiinae) (Ager, n).——*1, Kallirhynchia; 2, Rhynchonelloidella; 3, Grandirhynchia.*

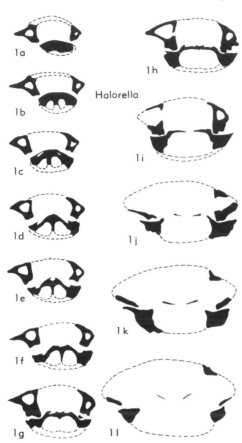

Halorella

Fig. 480. Serial transverse sections of *Halorella amphitoma* (Bronn), Trias., Austria, showing cilifer type of crura (figures indicate distance from pedicle-valve beak in mm.), *ca.* ×2.5 (810).

the genera with calcarifer-type crura appear to belong to the dominantly falcifer subfamily (i.e., Lacunosellinae).

Maniculifer. Cooper (1959) described two further types designated as maniculifer and spinulifer, in his work on Cenozoic rhynchonelloids, but these are certainly derived from the radulifer form or merely variants of it. The maniculifer type has curious handlike processes at the end of straight, ventrally directed crura (Fig. 478, *6*). As has already been stated, various terminal processes are commonly developed on radulifer crura, though these commonly are lost in the process of fossilization and preparation for study. Cooper described the maniculifer type in the Cenozoic Crypto-

poridae, but it and other variants of the radulifer type certainly occur far back in the Mesozoic and possibly earlier.

Spinulifer. The spinulifer type also appears to be no more than a variant of the radulifer type, in this case with the crura laterally compressed (Fig. 478,7). Such bladelike supports are found from very early in rhynchonelloid history and do not appear to have any evolutionary significance.

Prefalcifer. One additional type has been named by the present author (Ager, 1962). It is termed prefalcifer. In this type the crura are straight (i.e., in the plane of the commissure) and are slightly compressed (Fig. 478,*8*). It is regarded as a variant of the falcifer type, but is worthy of distinction because it precedes the falcifer in time and particularly characterizes an important early Mesozoic group (Cirpinae). As with true falcifer forms, the dorsal median septum is characteristically reduced.

Three further types are named herein, all variants of the radulifer type, but they characterize particular families and subfamilies and are easily recognized in serial transverse sections.

Mergifer. The mergifer type is characterized by long crura, radulifer in form, but very close together and parallel, and arising directly from the swollen edge of a high dorsal median septum (Fig. 478,9). It is well seen in the Peregrinellinae and in several Paleozoic genera, for example, as illustrated recently by Havlíček (1961) in *Plagiorhyncha* from the Silurian of Czechoslovakia. In cross section, the crura and high septum have the form of a two-pronged pitchfork (Latin, *merga*).

Canalifer. In the canalifer type the ventrally directed radulifer crura are folded longitudinally in the form of a dorsally facing channel or gutter (Latin, *canalis*). In other words, the crura are V- or U-shaped in cross section. In some shells (e.g., *Curtirhynchia*) a further lateral flange occurs, giving the crura Z-shaped cross sections. This type is especially characteristic of the Cyclothyridinae and is one of the best criteria for recognizing that group.

Cilifer. The cilifer type also has crura of radulifer form, but flattened in the plane of the commissure between the valves, and they form direct prolongations of the horizontal hinge plates (Fig. 480), with or without a lateral flange. They are chisel-like in appearance (Latin, *cilio*), and characterize the Triassic Halorellinae, which may be a very ancient stock. DAGIS (1963) has now shown that these continue ventrally by turning suddenly through a right angle and forming parallel, slightly crescentic blades.

SHELL STRUCTURE

A character which is potentially of great value in the study of Mesozoic and Cenozoic rhynchonelloids is the form of the "shell-mosaic" or *Schuppenpanzerstruktur*. This is a scaly pattern produced by the grouping together of the calcite fibers which form the shell. It is only seen on the inner surface, in exceptionally well-preserved specimens, and usually takes the form of elongated polygons or ellipses. It has long been known in living species, but LEID-HOLD (1920) figured it in Upper Jurassic forms from Germany and the present author (AGER, 1957) figured it in the type-species of *Rhynchonella* from the Upper Jurassic (Volgian) of the Moscow region. The variability observed in the few species so far studied suggests that it may be an extremely useful classificatory tool; much further work is needed on such microscopic shell structures, especially since they have proved so successful in other groups.

DENTICULA

Structures named denticula appear only to have been described in Mesozoic forms. They are small toothlike projections developed in the pedicle valve externally of the main teeth and fitting into accessory sockets in the brachial valve. Presumably they reinforced the articulation. In one small family (Austrirhynchiidae) they occur instead in the brachial valve.

SEPTALIUM

The septalium is a structure characteristic of many Mesozoic rhynchonelloids, and there is every reason to believe that it is identical with the similar structure in Paleozoic rhynchonelloids (which is usually called a *cruralium*) and with the "small chamber" described by COOPER (1959) in Cenozoic forms.

LEIDHOLD (1928) introduced the term and described it as a chamber formed by splitting of the dorsal median septum ventrally. This is, in fact, the impression given by the majority of specimens, especially when the shell has been recrystallized and the

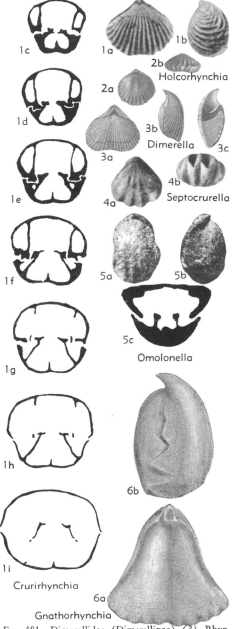

FIG. 481. Dimerellidae (Dimerellinae) *(3)*, Rhynchonellininae) *(1-2, 4-6)* (p. *H602-H603*).

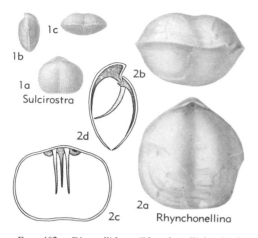

FIG. 482. Dimerellidae (Rhynchonellininae) (p. H602-H603).

finer sutures obliterated. However, when the material is sufficiently well preserved, it can be seen that the septalium is, in fact, formed by two plates (septalial plates) which descend from the inner edges of the hinge plates to meet the median septum (Fig. 479,*3b*). This was shown by WISNIEWSKA (1932, p. 6), MUIR-WOOD (1936, p. 50), COOPER (1959, p. 10) and the present author (AGER, 1956, p. 22), all dealing with different forms, and it is reasonable to suppose that this is the normal state of affairs.

Great stress has been placed by some workers on the presence or absence of the septalium as a criterion in classification, but the present author has often found it so little developed as to be misleading, and it is often overlooked (or overemphasized) because of the precise orientation of the transverse sections in a particular case.

Family DIMERELLIDAE Buckman, 1918

[Dimerellidae BUCKMAN, 1918, p. 72]

Usually sulcate and very small (though the Halorellinae and Peregrinellinae are exceptionally large), deltidial plates commonly reduced, crura very long; dorsal septum may be prominent (3, 136). ?*Dev., Trias.-L.Cret.*

Subfamily DIMERELLINAE Buckman, 1918

[*nom. transl.* AGER, 1959, p. 330 (*ex* Dimerellidae BUCKMAN, 1918)]

Dorsal median septum very strong. *Trias.*

Dimerella ZITTEL, 1870, p. 220 [**D. gümbeli;* OD]. Shell depressed, slightly sulcate, with wide, straight

hinge line, capillate; beak high, erect; wide, open delthyrium. Crura long, radulifer; dorsal median septum very high, rising anteriorly. *Trias.*, Eu.——FIG. 481,*3*. **D. guembeli,* Ger.; *3a,b,* brach. v., lat. views, ×2; *3c,* long. sec., ×2 (900).

Subfamily RHYNCHONELLININAE Ager, 1959

[Rhynchonellininae AGER, 1959, p. 330]

Shell with little or no dorsal median septum and extremely long septifer crura. *U. Trias.-U.Jur.*

Rhynchonellina GEMMELLARO, 1876?, p. 29 [**R. suessi;* OD] [=*Terebratulopsis* DEGREGORIO, 1930 (obj.)]. Medium-sized, biconvex, rectimarginate to sulcate, smooth; beak strong, with wide delthyrium and rudimentary deltidial plates. Crura very long, touching ventral valve. *U.Trias.-L.Jur.,* S. Eu.——FIG. 482,*2*. **R. suessi,* Sicily; *2a,b,* brach. v., ant. views, ×0.7 (329); *2c,d,* brach.v. int., half of shell in lat. view showing long crura and median septum of brach.v. (reconstr.), enlarged (3).

Capillirostra COOPER & MUIR-WOOD, 1951, p. 195 [*pro Rhynchonellopsis* BÖSE, 1894, p. 57 (*non* VINCENT, 1893)] [**Rhynchonellina? finkelsteini* BÖSE, 1894, p. 77; OD]. Like *Sulcirostra* but small, depressed, with grooves delimiting dorsal muscle scars and shorter crura. [Probably a juvenile form and only doubtfully included here.] *U.Jur.(Oxford.),* Eu.

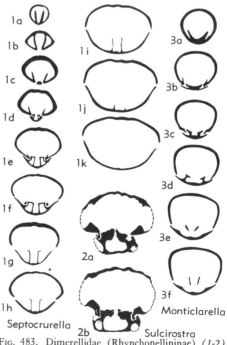

FIG. 483. Dimerellidae (Rhynchonellininae) *(1-2),* Norellinae *(3)* (p. H603-H604).

Carapezzia Tomlin, 1930, p. 24 [*pro Geyeria* Carapezza & Schopen, 1899, p. 248 (*non* Buchecker,
1876)] [**Rhynchonellina (Geyeria) globosa*
Carapezza & Schopen, 1899, p. 248; OD]. Like
Rhynchonellina but very massive, incurved beak
and fine capillae on well-preserved specimens;
crura long. *L.Jur.,* Eu.(Sicily).

Caucasella Moisseev, 1934, p. 187 [**Rhynchonella
trigonella* Rothpletz, 1886, p. 154; OD]. Triangular, flattened anteriorly, with many sharp
costae, no perceptible fold; beak small, incurved,
well-developed planareas. Crura septifer. *U.Jur.,*
Eu.(Alps-S.USSR).——Fig. 487,2. **C. trigonella*
(Rothpletz), Switz.; *2a,b,* brach.v., ant. views,
×1 (679).

Crurirhynchia Dagis, 1961, p. 96 [**C. kiparisovae*;
OD]. Medium-sized, transversely oval; multicostate throughout; beak low. Ridgelike median
septum; crura arising from oblique septa, more
or less fused with socket bases. *U.Trias.(Nor.-
?Rhaet.),* USSR(Caucasus)-?C.Eu.——Fig. 481,1.
*C. kiparosovae, USSR(Caucasus); *1a,b,* brach.v.,
lat. views, ×1; *1c-i,* ser. transv. secs., ×2.5
(211).

Gnathorhynchia Buckman, 1918, p. 29 [**Rhynchonella liostraca* Buckman, 1886, p. 217; OD].
Like *Holcorhynchia* but triangular in outline.
Dorsal septum strong; crura septifer. [Doubtfully
separable from *Holcorhynchia.*] *Jur.(Bajoc.-Cal-
lov.),* Eu.-USA(Calif.).——Fig. 481,6. **G. lio-
siraca* (Buckman), Eng.; *6a,b,* brach.v., lat. views,
×4 (229).

Holcorhynchia Buckman, 1918, p. 28 [**Rhynchonella standishensis* Buckman, 1901, p. 245;
OD]. Small, subcircular, depressed, posteriorly
sulcate, with many fine costae anteriorly after long
smooth stage; beak small, hypothyridid. *U.Trias.-
L.Jur.(Pliensbach.-Toarc.),* Eu.-Asia (Anatolia-
Japan).——Fig. 481,2. **H. standishensis* (Buckman), Eng.; *2a,b,* brach.v., ant. views, ×1 (138).

Omolonella Moisseev, 1936, p. 39 [**O. omolonen-
sis*; OD]. Medium-sized, smooth, with few faint
costae anteriorly; shell wall very thick. Ventral
median ridge; short dorsal septum and septalium;
crura arising on strong septa. *U.Trias.,* Sib.-Eu.
(Alps)-N.Am.(Alaska).——Fig. 481,5. **O. omo-
lonensis,* Sib.; *5a,b,* brach.v., lat. views, ×1 (567);
5c, transv. sec., ×2 (567).

Septocrurella Wisniewska, 1932, p. 63 [**Rhynchonella Sanctae Clarae* Roemer, 1870, p. 247;
OD]. Small, sulcate, with few rounded costae.
Beak small, upright. Crura short, septifer, supported by crural plates; dorsal septum a low
ridge. *Jur.(Callov.-Oxford.-?Tithon.),* Eu.——Fig.
481,4; 483,1. **S. sanctaeclarae* (Roemer), Pol.;
481,*4a,b,* brach.v., ant. views, ×1; 483,*1a-k,*
transv. secs. of beak region (ped.v. above), ×2
(893).

Sulcirostra Cooper & Muir-Wood, 1951, p. 195
[*pro Rhynchonellopsis* deGregorio, 1930, p. 5
(*non* Vincent, 1893; *nec* Böse, 1894)] [**Rhyn-

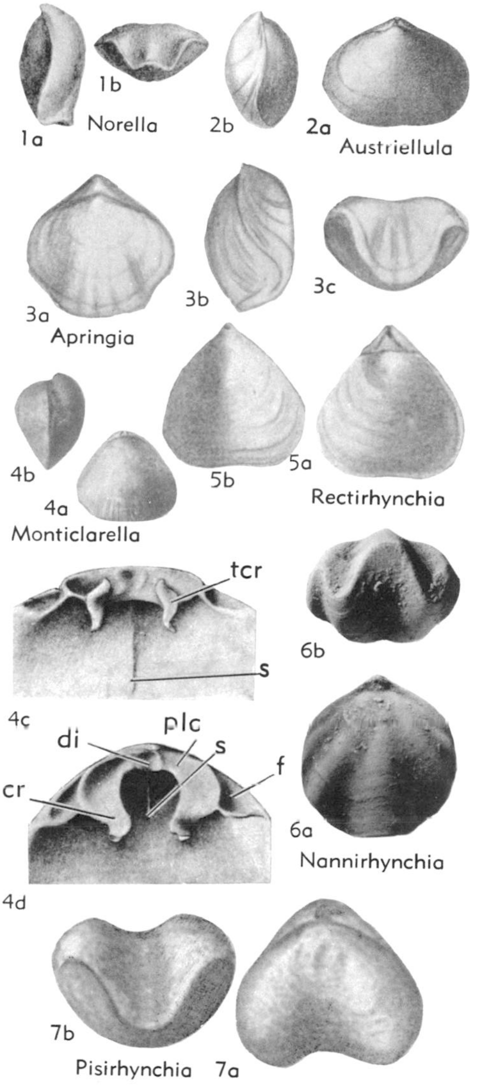

Fig. 484. Dimerellidae (Norellinae) (p. *H604*).

chonellina seguenzae Gemmellaro, 1876?, p.
34; OD]. Like *Rhynchonellina* but costate.
No septalium; very short median septum
and lateral septa supporting massive hinge plates.
U.Trias.-L.Jur., S.Eu.-?Asia(Anatolia).——Fig.
482,1; 483,2. **S. seguenzae* (Gemmellaro), Sicily;
482,*1a-c,* brach.v., lat., ant. views, ×1 (329);
483,*2a,b,* transv. secs. of beak region (ped.v.
above), ×2 (3).

Subfamily NORELLINAE Ager, 1959

[Norellinae Ager, 1959, p. 330]

Small, mostly smooth shells with small

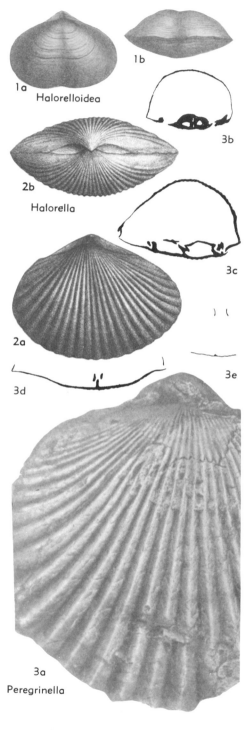

1a
1b
Halorelloidea

2b
Halorella

2a

3d
3b
3c
3e

3a
Peregrinella

Fig. 485. Dimerellidae (Hallorellinae) *(1-2)*,
(Peregrinellinae) *(3)* (p. H605).

delthyria. Crura arcuifer (where known).
M.Trias.-U.Jur., ?L.Cret.

Norella Bittner, 1890, p. 315 [*Rhynchonella re-fractifrons* Bittner, 1890, p. 39 (ICZN, 1961, Op. 633)]. Smooth, subcircular, some shells with slight plication, sulcate; beak small, incurved. Dental plates long; no dorsal median septum; crura arcuifer. *M.Trias.-U.Trias.*, Eu.(Alps).——Fig. 484,*1*. **N. refractifrons* (Bittner), Austria; *1a,b*, lat., ant. views, ×1 (76).

Apringia deGregorio, 1886, p. 22 [**A. giuppa*; OD]. Like *Pisirhynchia* but larger, with wider sulcation, which may be asymmetrical, and less obvious costae. *L.Jur.(Toarc.)*, Eu.——Fig. 484,*3*. **A. giuppa*, Sicily; *3a-c*, brach.v., lat., ant. views, ×1 (918).

Austriellula Strand, 1928, p. 37 [*pro Austriella* Bittner, 1890, p. 314 (*non* Tenison-Woods, 1881)] [**Rhynchonella dilatata* Suess, 1855, p. 29; OD] [=*Jacobella* Patte, 1926, p. 125 (*non* Jeannet, 1908); *Austriellina* Schuchert & Le-Vene, 1929, p. 119 (obj.)]. Smooth, triangular, rectimarginate or slightly sulcate; some species uniplicate. *M.Trias.-U.Trias.*, Eu.(Alps)-SE.Asia. ——Fig. 484,*2*. **A. dilatata* (Suess), Austria; *2a,b*, brach.v., lat. views, ×1 (792).

Monticlarella Wisniewska, 1932, p. 55 [**Rhyn-chonella czenstochowiensis* Roemer, 1870, p. 247; OD]. Small, posteriorly sulcate, capillate anteriorly after smooth stage. Dorsal septum faint; crura arcuifer. *U.Jur.(Oxford.-Kimmeridg.)*, *?L.Cret.*, Eu.——Fig. 483,*3*; 484,*4*. **M. czenstochowiensis* (Roemer), Pol.; 483,*3a-f*, transv. secs. of beak region (ped.v. above), ×2 (893); 484,*4a,b*, brach. v., lat. views, ×1; 484,*4c,d*, brach.v. int., ped.v. int., ×6 (893).

Nannirhynchia Buckman, 1918, p. 67 [**N. sub-pygmaea*; OD]. Minute, globose; sulcate, with median uniplication; fold well marked, finely capillate, with few rounded costae anteriorly, very fine spines; beak massive, incurved, foramen small. No median septum; crura arcuifer. *L.Jur.(Toarc.)-M.Jur.(Bajoc.)*, Eu.——Fig. 484,*6*. *N. pygmaea* (Morris), Eng.; *6a,b*, brach.v., ant. views, ×10 (Ager, n).

Pisirhynchia Buckman, 1918, p. 28 [**Rhynchonella pisoides* Zittel, 1869, p. 129; OD]. Small, globose, sulcate, ventral fold low, with few rounded costae after long smooth stage; no um-bonal callosities. *L.Jur.*, S.Eu.——Fig. 484,*7*. **P. pisoides* (Zittel), Italy; *7a,b*, brach.v., ant. views, ×3 (938).

Rectirhynchia Buckman, 1918, p. 74 [**Rhyn-chonella lopensis* Moore, 1855, p. 114; OD]. Minute, smooth, depressed; sulcate, with strong ventral fold; straight hinge line. Beak large, hypo-thyridid. *M.Jur.(Bajoc.)*, Eu.——Fig. 484,*5*. **R. lopensis* (Moore), Eng.; *5a,b*, brach.v., ped.v. views, ×4 (229).

Subfamily HALORELLINAE Ager, n. subfam.

Large, with wide, straight hinge line, rectimarginate or nearly so, commonly with opposite sulci, may be asymmetrical; high hypothyridid beak. Dental plates widely spaced, septum very small or absent; crura direct prolongations of hinge plates, flattened in plane of commissure (cilifer type). [This and the next subfamily extend back at least to the Devonian. Its members may have lived only in · a restricted geosynclinal environment, which would explain their infrequent appearance in the known stratigraphical record. The subfamilies may be the ancestral stock from which most, if not all, of the other Mesozoic rhynchonellids evolved.] *?Dev., Trias.*

Halorella BITTNER, 1884, p. 107 [**Rhynchonella amphitoma* BRONN, 1832, p. 162; SD HALL & CLARKE, 1894, p. 832] [=*Barzellinia* DEGREGORIO, 1930, p. 8 (type, *B. primogenita*; OD)]. Medium-sized to large, subcircular to laterally oval, with many sharp costae; rectimarginate to slightly uniplicate, commonly developing opposite sulci in each valve, sulci may be asymmetrical; beak sharp, suberect; aperture elongate. Crura may be unusually long; median septum short. [Devonian fossils from Morocco and Nevada referred to *Halorella* are doubtful representatives of the genus in the view of HERTA SCHMIDT.] *?Dev.,* Morocco - USA (Nev.); *Trias.,* Eu. (Alps)-S.Asia-USA(Ore.).——FIG. 480,*1*; 485,*2*. **H. amphitoma* (BRONN), Austria; 480,*1a-l,* serial transv. secs., ×1 (Ager, n); 485,*2a,b,* brach.v., post. views, ×1 (76).

Halorelloidea AGER, 1960, p. 159 [**Halorella rectifrons* BITTNER, 1884, p. 107; OD]. Like *Halorella* but usually smaller and smooth, or with only few irregularly developed costae; 2 sharp opposite sulci well developed or with sharp dorsal fold; commonly asymmetrical; no median septum. *U. Trias.,* Eu. (Alps-S. Asia).——FIG. 485,*1*. **H. rectifrons* (BITTNER), Austria; *1a,b,* brach.v., ant. views, ×1 (76).

Subfamily PEREGRINELLINAE Ager, n. subfam.

Large, coarsely costate dimerellids, pentameroid-like in appearance. Dental plates much reduced or absent; crura radulifer in form, but set very close together throughout their length and arising directly from swollen ventral edge of dorsal median septum (mergifer type). [Though at present including only one Cretaceous genus, there are undescribed forms both in the Cretaceous and the Jurassic, which indicate a connection with the Rhynchonellininae

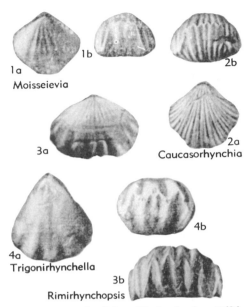

FIG. 486. Wellerellidae (Cirpinae) (p. *H*606-*H*608).

and with certain Paleozoic genera such as *Plagiorhyncha* of the Silurian.]. *?Dev., M. Jur.-L.Cret.*

Peregrinella OEHLERT, 1887, p. 1305 [**T. peregrina* VON BUCH, 1835, p. 73 (*non* SCHLOTHEIM, 1813, =**Terebratula multicarinata* LAMARCK, 1819, p. 253); OD]. Large, circular, biconvex, rectimarginate, with many strong costae; beak massive, incurved. Hinge plates wide, flat; dental plates oblique, short, teeth small; crura long, close together; septum long. *L.Cret.(Valangin.-Hauteriv.),* Eu.(Alps-Carpathians)-Calif.——FIG. 485, *3*. **P. multicarinata* (LAMARCK), Fr.; *3a,* brach.v. view, ×1 (907); *3b-e,* serial transv. secs., ×1 (934).

Family WELLERELLIDAE Likharev in Rzhonsnitskaya, 1956

[Wellerellidae LIKHAREV in RZHONSNITSKAYA, 1956, p. 125]

Uniplicate, no septalium or cardinal process, entire hinge plates; dorsal septum and dental plates variously developed (694). *L. Carb.-U.Cret.*

Subfamily WELLERELLINAE Likharev in Rzhonsnitskaya, 1958

[See p. *H*590, prepared by HERTA SCHMIDT]

Subfamily CIRPINAE Ager, n. subfam.

Multicostate, hinge plates fused; dorsal median septum usually very much reduced; beak generally small and incurved, with

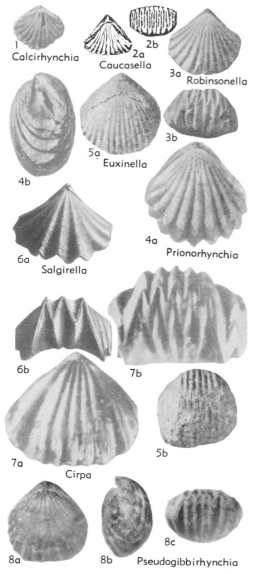

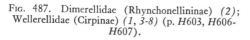

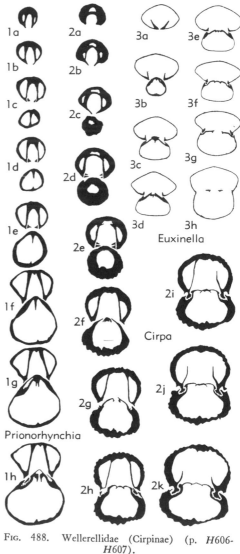

Fig. 487. Dimerellidae (Rhynchonellininae) *(2)*; Wellerellidae (Cirpinae) *(1, 3-8)* (p. *H603, H606-H607*).

well-developed planareas. Deltidial plates thick and distinctive; crura prefalcifer (where known). *Trias.-U.Jur.*

Cirpa DEGREGORIO, 1930, p. 40 [*Rhynchonella (C.) primitiva* (probably=*R. briseis* GEMMELLARO, 1874, p. 77), OD]. Outline subtriangular, rectangular and flattened anteriorly; with low fold, uniplication, and few strong, sharp costae; beak small, marked planareas; deltidial plates double.

Hinge plates flat, fused; median septum very short; crura prefalcifer. *L.Jur.(Pliensbach.)*, Eu.-Anatolia.——Fig. 487,7. *C. langi* AGER, Eng.; *7a,b*, brach.v., ant. views, ×2 (1).——Fig. 488,2. *C. briseis* (GEMMELLARO), Sicily; *2a-k*, transv. secs. at 0.6-2.6 mm. from tip of beak (ped.v. above), ×3.25 (1).

Calcirhynchia BUCKMAN, 1918, p. 30 [*C. calcaria; OD*]. Small; with wide uniplication, low fold and many sharp costae, no posterior smooth stage; beak small, incurved. Crura prefalcifer. *L.Jur.(Hettang.)*, Eu.——Fig. 487,1. *C. calcaria*, Eng.; brach.v. view, ×1 (138).

Caucasorhynchia DAGIS, 1963, p. 63 [*C. kuenensis;*

Fig. 488. Wellerellidae (Cirpinae) (p. *H606-H607*).

OD]. Medium-sized, subcircular to pentagonal in outline, biconvex, low uniplication and fold; many blunt costae throughout growth, branching anteriorly. Ridgelike median septum; hinge plates flat, almost fused; prefalcifer crura, lateral umbonal cavities very narrow. *U.Trias.(Nor.-Rhaet.)*, USSR(Caucasus).——Fig. 486,2. **C. kuenensis; 2a,b*, brach.v., ant. views, ×1 (212a).

Euxinella MOISSEEV, 1936, p. 41 [**E. iatirgvantaensis*; OD]. Globose, flattened anteriorly, with strong uniplication,.no distinct fold, multicostate; beak massive. Septalium absent; dorsal median septum hardly visible; crura ?prefalcifer. *Trias.*, Asia(Sib.)-N.Am.(W.Can.).——Fig. 487,5; 488, 3. **E. iatirgvantaensis; 487,5a,b*, brach.v., ant. views, ×1 (567); *488,3a-h*, transv. secs. of beak region (ped.v. above), enlarged (567).

Hagabirhynchia JEFFERIES, 1961, p. 5 [**H. arabica*; OD]. Small, sulcate to uniplicate, strong costae. Beak massive with weak ventral septum. Hinge plates separate but no true septalium; low persistent dorsal median septum; crura prefalcifer. *U.Trias.(Nor.)*, Arabia.——Fig. 489,1. **H. arabica*, Oman; *1a,b*, brach.v., ant. views, ×2.5; *1c-i*, transv. secs. beak region (ped.v. above), ×5 (437).

Moisseievia DAGIS, 1963, p. 46 [**M. moisseievi*; OD]. Subpentagonal to transversely elliptical, very small beak, strongly uniplicate; costae mainly confined to fold, lateral areas smooth. Well-developed pedicle collar and double deltidial plates; fused hinge plates, no dorsal median septum, prefalcifer crura. *U.Trias.(Nor.-Rhaet.)*, USSR(Caucasus).——Fig. 486,1. **M. moisseievi*; *1a,b*, brach.v., ant. views, ×1 (212a).

Prionorhynchia BUCKMAN, 1918, p. 62 [**Terebratula serrata* SOWERBY, 1825, p. 168; OD]. Medium-sized to large, without interarea but planareas well developed; rectimarginate or uniplicate; costae strong, sharp; beak very small, incurved; deltidial plates narrow, thick. Dorsal septum very short; crura prefalcifer. *L.Jur.*, Eu.-Anatolia-Indo-China-?Timor.——Fig. 487,4; 488,1. **P. serrata* (SOWERBY), Eng.; *487,4a,b*, brach.v., lat. views, ×1; *488,1a-h*, transv. secs. at 0.4-2.0 mm. from tip of beak (ped.v. above), ×2.5 (1).

Pseudogibbirhynchia AGER, 1962, p. 108 [**Rhynchonella Moorei* DAVIDSON, 1852, p. 82; OD]. Small, globose, brachial valve flattened posteriorly; low uniplication, multicostate. Very short median septum, strengthened deltidial plates, prefalcifer crura. *L.Jur.*, Eu.——Fig. 487,8. **P. moorei* (DAVIDSON), Eng.; *8a-c*, brach.v., lat., ant. views, ×2 (1).

Rimirhynchopsis DAGIS, 1963, p. 71 [**R. triadicus*; OD]. Medium-sized, laterally ovate, small beak, strong deltidial plates; strongly uniplicate, low fold; many fine capillae posteriorly, strong costae anteriorly. Shell thick, umbonal cavities small; low massive median septum and septalium, massive teeth and denticula, flat hinge plates, prefalci-

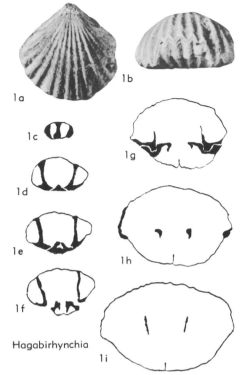

Hagabirhynchia

FIG. 489. Wellerellidae (Cirpinae) (p. H607).

fer crura. *U.Trias.(Nor.-Rhaet.)*, USSR(Caucasus). ——Fig. 486,3. **R. triadicus; 3a,b*, brach.v., ant. views, ×1 (212a).

Robinsonella MOISSEEV, 1936, p. 45 [**R. mastakanensis*; OD]. Triangular, depressed, with strong uniplication and many sharp costae, no smooth stage; beak small, incurved. Septalium absent; median ridge in ventral valve; dorsal septum massive. *Trias.*, Sib.——Fig. 487,3. **R. mastakanensis; 3a,b*, brach.v., ant. views, ×1 (567).

Salgirella MOISSEEV, 1936, p. 48 [**Rhynchonella albertii* OPPEL, 1861, p. 546; OD]. Medium-sized, uniplicate, with acute apical angle and very strong, sharp costae, no anterior flattening; beak small, incurved; deltidial plates double; median septum short. [Possibly a synonym of *Cirpa*, but it cannot be confirmed that the Siberian form, on which the genus was founded, is the same as OPPEL's.] *L.Jur.*, Eu.-Sib.——Fig. 487,6. *S. albertii* (OPPEL), Ger.; *6a,b*, brach.v., ant. views, Ger.; ×1 (928).

Squamirhynchia BUCKMAN, 1918, p. 63 [**Terebratula triplicata squamiplex* QUENSTEDT, 1871, p. 72; OD]. Depressed, brachial valve nearly flat, with low uniplication and fold, strong branching costae, and no smooth stage; beak strong, upright, with large foramen. Low persistent septum, shal-

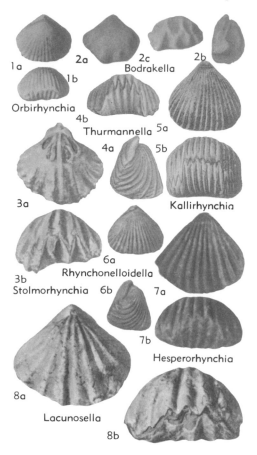

Fig. 490. Wellerellidae (Lacunosellinae) (p. *H608-H609*).

low septalium; crura prefalcifer, concave dorsally, at distal ends, double deltidial plates. *L.Jur. (Sinemur.-L.Pliensbach.),* Eu.——Fig. 500,6. **S. squamiplex* (Quenstedt), Ger.; brach.v. view, ×1 (651).

Trigonirhynchella Dagis, 1963, p. 41 [*nom. subst. pro Trigonirhynchia* Dagis, 1961, p. 94 (*non* Cooper, 1942)] [**Trigonirhynchia trigona* Dagis, 1961, p. 95; OD]. Triangular, small acute beak, no interarea; smooth posteriorly, with few sharp costae anteriorly. Divergent dental plates, no pedicle collar, ridgelike median septum, prefalcifer crura. *U.Trias.(Nor.),* USSR(Caucasus)-?Eu. (Alps).——Fig. 486,4. **T. trigona; 4a,b,* brach. v., ant. views, ×1 (212a).

Subfamily LACUNOSELLINAE Smirnova, 1963

[Lacunosellinae Smirnova, 1963, p. 15]

Usually small, multicostate, commonly asymmetrical, characterized mainly by absence or very slight development of dorsal

median septum and septalium, and presence of falcifer or calcarifer crura. *L.Jur.-U.Cret.*

Lacunosella Wisniewska, 1932, p. 30 [**Rhynchonella arolica* Oppel, 1866, p. 294; OD]. Medium-sized, subtriangular, with few strong costae commonly branching, in some shells asymmetrical, crura falcifer. No dorsal septum or septalium; teeth strong. *U.Jur.(Oxford.-Tithon.),* ?*L. Cret.,* Eu.——Fig. 490,8. **L. arolica* (Oppel), Pol.; *8a,b,* brach.v., ant. views, ×1 (893). [Whether this is the same species as first inadequately figured by Heer (1864) is doubtful.]

Bodrakella Moisseev, 1936, p. 47 [**Rhynchonella bodrakensis* Moisseev, 1934, p. 182; OD]. Small, uniplicate. Dorsal median septum reduced to ridge. *L.Jur.,* Sib.——Fig. 490,2. **B. bodrakensis* (Moisseev); *2a-c,* brach.v., lat., ant. views, ×1 (566). [May belong to the Cirpinae.]

Hesperorhynchia Warren, 1937, p. 2 [**H. superba;* OD]. Medium-sized, subtriangular, with moderate uniplication and low fold; costae strong and few, no smooth stage; beak small, incurved. No dorsal septum, dental plates short. *U.Cret.,* Can. ——Fig. 490,7. **H. superba,* Sask.; *7a,b,* brach.v., ant. views, ×1 (936).

Kallirhynchia Buckman, 1918, p. 31 [**Rhynchonella concinna* var. *yaxleyensis* Davidson, 1878, p. 206; OD]. Medium-sized, almost convexiplanate; well-developed uniplication, flat fold; multicostate after short posterior smooth stage; beak hypothyridid, suberect. Dorsal septum short, low; crura long, calcarifer to ?falcifer. *Jur. (Bathon.-?Callov.),* Eu.-India-?Japan-USA(Calif.). ——Fig. 479,1; 490,5. **K. yaxleyensis* (Davidson), Eng.; *479,1,* part of long. sec. through beak region, brach.v. on right, enlarged (576); *490, 5a,b,* brach.v., ant. views, ×1 (576).

Orbirhynchia Pettitt, 1954, p. 29 [**O. orbignyi;* OD]. Biconvex, with low arcuate uniplication and very slight dorsal fold, which may be asym-

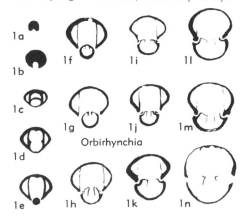

Fig. 491. Wellerellidae (Lacunosellinae) (p. *H608-H609*).

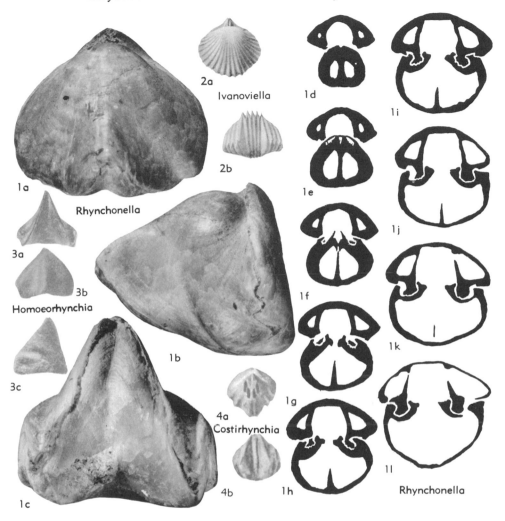

FIG. 492. Rhynchonellidae (Rhynchonellinae) (p. *H*610-*H*611).

metrical; with many rounded costae, smooth posteriorly; beak small, crura falcifer. Dorsal median septum and septalium absent. *Cret.(Alb.-Senon.)*, NW.Eu.——FIG. 490,*1*; 491,*1*. **O. orbignyi*, Eng.; 490,*1a,b*, brach.v., ant. views, ×1; 491,*1a-n*, transv. secs. of beak region (ped.v. above), ×3 (639).

Rhynchonelloidella MUIR-WOOD, 1936, p. 49 [**Rhynchonella varians* var. *smithi* DAVIDSON. 1878, p. 213; OD]. Medium-sized, uniplication strong, dorsal fold low, with many sharp costae; beak small, massive, incurved. Dental plates long; dorsal septum short; crura calcarifer. *M.Jur.-U. Jur.*, NW.Eu.——FIG. 479,*2*; 490,*6*. **R. smithi* (DAVIDSON), Eng.; 479,*2a-j*, transv. secs. of beak region (ped.v. above), *2d*, showing crural base

(*cb*) and *2i* showing crura (*cr*), ×2 (579); 490, *6a,b*, brach.v., lat. views, ×1 (579).

Stolmorhynchia BUCKMAN, 1918, p. 46 [**S. stolidota*; OD]. Very variable in size and shape, uniplicate, may be asymmetrical; few sharp costae developed anteriorly; beak small, suberect. Dorsal septum feeble or absent; muscle scars impressed; crura falcifer. [Genus probably requires subdivision and needs confirmation of occurrence.] *L. Jur., ?L.Cret.*, Eu.-N.Afr.-India.——FIG. 490,*3*. **S. stolidota*, Eng.; *3a,b*, brach.v. (mold) and ant views, ×1 (138).

Thurmannella LEIDHOLD, 1920, p. 357 [**Terebratula Thurmanni* VOLTZ, 1833, p. 172; OD]. Medium-sized, almost convexiplanate; with strong uniplication and slight fold, many costae, smooth

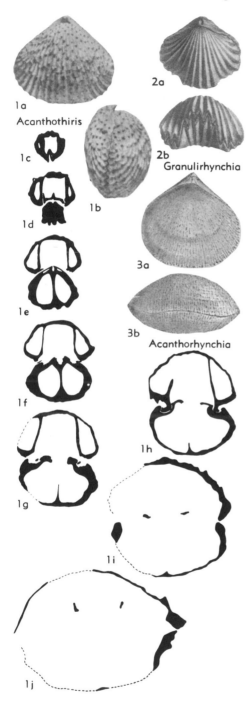

posteriorly. Crura falcifer, strongly curved ventrally, distal points directed toward each other and almost meeting. *U.Jur.(Oxford.)*, Eu.——Fig. 490,4. **T. thurmanni* (Voltz), Eng.; *4a,b,* lat., ant. views, ×1 (229).

Family RHYNCHONELLIDAE Gray, 1848

[Rhynchonellidae Gray, 1848, p. 438]

Shell without prominent median septum in brachial valve and none in pedicle valve; crura comparatively short, cardinal process absent; anterior margin of valves rectimarginate or uniplicate (3, 810). *Trias.-U.Cret.*

Subfamily RHYNCHONELLINAE Gray, 1848

[*nom. transl.* Gill, 1871, p. 25 (*ex* Rhynchonellidae Gray, 1848, p. 438)]

Shell form cynocephalous, with strong, sharp dorsal folds and uniplications; long smooth stages posteriorly, only few costae anteriorly; crura radulifer. *Trias.-U.Jur.*

It has often been suggested that cynocephalous rhynchonellids are polyphyletic, homeomorphic end forms. This is difficult to prove or disprove, but the group seems to constitute a continuous stock, with no significant differences between its members. It ranges from the Triassic to the type-genus of the family at the end of the Jurassic and does not depart far from the main stock (4).

Rhynchonella Fischer, 1809, p. 35 [**R. loxiae;* OD] [=*Eurhynchonella* Leidhold, 1920, p. 352 (obj.)]. Small to medium in size, triangular; dorsal fold high, ventral sulcus somewhat flattened; few sharp costae anteriorly; beak small, incurved. Dental plates strong, septalium shallow, dorsal septum short; crura radulifer. [The so-called capillae of this genus are probably just a matter of the preservation of the fibrous shell at the type locality of the type-species.] *U.Jur. (Volg./Portland.)*, Eu.——Fig. 492,1. **R. loxiae,* USSR(near Moscow); *1a-c,* brach.v., lat., ant. views, ×4; *1d-l,* transv. secs. at 1.5-2.5 mm. from tip of beak, ped.v. above, ×4.6 (2).

Costirhynchia Buckman, 1918, p. 39 [**C. costigera;* OD]. Small, globose, with high fold and few costae; small beak with slitlike foramen. Median septum long. *M.Jur.(Bajoc.)*, Eu.——Fig. 492,4. **C. costigera,* Eng.; *4a,b,* brach.v. (mold) and ant. views, ×1 (136).

Homoeorhynchia Buckman, 1918, p. 36 [**Terebratula acuta* J. Sowerby, 1816, p. 115 (*non* J. de C. Sowerby, 1825); OD]. Small to medium in size, with high, sharp dorsal fold and few sharp costae anteriorly; beak small, incurved. Dorsal septum short; crura fairly long, radulifer; dorsal muscle-scars anterior. *?Trias.,* Eu.(Alps); *L.Jur.-M.Jur.,* Eu.-W.N.Am.——Fig. 492,3. **H. acuta*

Fig. 493. Rhynchonellidae (Acanthothyridinae) (p. *H611*).

(J. Sowerby), Fr.; *3a-c*, ant., brach.v., lat. views, ×1 (1).

Ivanoviella Makridin, 1955, p. 83 [*Rhynchonella alemanica* Rollier, 1917, p. 151] [=*]vanoviella* Makridin, 1955, p. 83 *(nom. van.)*]. Like *Homoeorhynchia,* but with more costae which develop earlier. Massive spoon-shaped crura. *U.Jur. (Callov.-Oxford.),* Eu.-Asia.——Fig. 492,2. *I. alemanica* (Rollier), Fr.; *2a,b,* brach.v., ant. views, ×1 (377).

Subfamily ACANTHOTHYRIDINAE Schuchert, 1913

[*nom. correct.* Ager, herein *(pro* Acanthothyrinae Schuchert, 1913, p. 400) (name based on junior synonym of *Acanthothiris,* here retained in accordance with Zool. Code, 1961, Art. 40)]

Mesozoic rhynchonellids having only spinosity (may be incipient) in common. [A very doubtful grouping.] *M.Jur.-U.Jur.*

Acanthothiris d'Orbigny, 1850, p. 323 [*Anomia spinosa* Linné, 1788, p. 3346; SD Buckman & Walker, 1889, p. 50] [=*Acanthothyris* Paetel, 1875, p. 1 *(nom. van.)*]. Globose, uniplicate, with low dorsal fold; many bifurcating costae, not smooth posteriorly, spinose throughout; beak small, incurved. Dorsal septum ridgelike; crura radulifer. *M.Jur.(Bajoc.-Bathon.),* Eu.-Asia.—— Fig. 493,1. *A. spinosa* (Linné), Eng.; *1a,b,* brach. v., lat. views, ×1 (136); *1c-j,* serial transv. secs., ×4 (Ager, n).

Acanthorhynchia Buckman, 1918, p. 69 [*Acanthothyris panacanthina* Buckman & Walker, 1889, p. 53; OD]. Medium-sized, biconvex or very slightly uniplicate, without fold; capillate, spinose; beak sharp, suberect. Dorsal septum short, feeble; crura falcifer. *Jur.(Bajoc.-Portland.),* Eu.-N.Afr.-India.——Fig. 493,3. *A. panacanthina* (Buckman & Walker), Eng.; *3a,b,* brach.v., ant. views, ×1 (229).

Granulirhynchia Buckman, 1918, p. 64 [*Rhynchonella granulata* Upton, 1905, p. 83; OD]. Wide, depressed, with low, wide fold and many sharp costae covered with fine granules; beak fairly strong, suberect, foramen rimmed. Dorsal septum strong; dorsal muscle-scars broad. *M.Jur. (Bajoc.),* Eu.——Fig. 493,2. *G. granulata* (Upton), Eng.; *2a,b,* brach.v., ant. views, ×1 (935).

Subfamily TETRARHYNCHIINAE Ager, n. subfam.

Multicostate, some with short smooth stage posteriorly, uniplicate, with moderate dorsal fold; beak small, usually incurved; delthyrium usually small, foramen not rimmed. Crura radulifer, usually in form of simple hooks. [These are "ordinary-looking" rhynchonellids, as generally understood, and as such constitute the bulk and probably the main stock of the Mesozoic forms.] *U. Trias.-L.Cret.*

Tetrarhynchia Buckman, 1918, p. 41 [*Terebratula tetraëdra* Sowerby, 1812, p. 191 =*Tetrarhynchia tetrahedra* (Sowerby) *(nom. correct.,* Ager, 1956, p. 7); OD]. Medium-sized, laterally expanded, subtriangular, with many fairly sharp costae, short smooth stage posteriorly; beak small, incurved. Median septum short, septalium deep;

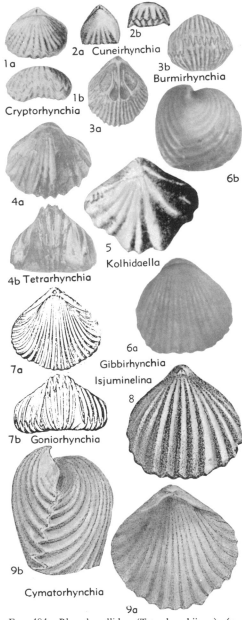

Fig. 494. Rhynchonellidae (Tetrarhynchiinae) (p. H611-H614).

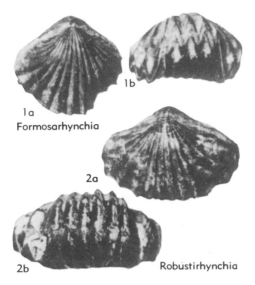

1a
Formosarhynchia

1b

2a

2b   Robustirhynchia

FIG. 495. Rhynchonellidae (Tetrarhynchiinae) (p. *H612, H614*).

crura radulifer. *Jur.(Sinemur.-Bajoc.)*, Eu.-N.Am. (NW.Can.).——FIG. 494,4; 496,1. **T. tetrahedra* (SOWERBY), Eng.; 494,4a,b, brach.v., ant. views, ×1; 496,1a-k, transv. secs. of beak region (ped.v. above), ×1.75 (1).

Burmirhynchia BUCKMAN, 1918, p. 49 [**B. gutta*; OD]. Medium-sized, globose; with slight uniplication, indistinct fold, and many rounded costae; beak massive, hypothyridid. Dental plates divergent; median septum strong. *M.Jur.(Bathon.-Callov.)*, Eu.-Somaliland - M.East - India - Burma-China - ?Japan - ?Australia. —— FIG. 494,3. **B. gutta*, Burma; *3a,b*, brach.v. mold and ant. views, ×1 (138).

Cryptorhynchia BUCKMAN, 1918, p. 66 [**Rhynchonella pulcherrima* KITCHIN, 1900, p. 52; OD]. Small, uniplicate, dorsal fold moderate; ornament reticulate; beak sharp, suberect. Teeth and sockets projecting well into brachial valve, having appearance of lateral septa; no septalium; crura radulifer, bladelike, converging ventrally. *M.Jur.(Bathon.)*, Eu.-India-N.Z. —— FIG. 494,1. **C. pulcherrima* (KITCHIN), Cutch; *1a,b*, brach.v., ant. views, ×1 (478).

Cuneirhynchia BUCKMAN, 1918, p. 35 [**Rhynchonella dalmasi* DUMORTIER, 1869, p. 331; OD]. Small, depressed, uniplicate, convexi-planate or convexi-concave; pronounced smooth stage with few blunt costae anteriorly; beak small, upright. Hinge plates massive, dorsal septum long and low. *?Trias.*, Eu.(Alps); *L.Jur.(Sinemur.-Pliensbach.)*, Eu.-Anatolia.——FIG. 494,2. **C. dalmasi* (DUMORTIER), Fr.; *2a,b*, brach.v., ant. views, ×1 (916).

Cymatorhynchia BUCKMAN, 1918, p. 53 [**Rhyn-*

chonella cymatophorina BUCKMAN, 1910, p. 105 (=*R. cymatophora* BUCKMAN, 1895, *non* ROTHPLETZ, 1886); OD]. Medium-sized to large, with strong dorsal fold and uniplication; many sharp costae, no smooth stage; beak small, hypothyridid. Dorsal septum strong; no septalium; crura radulifer. *M.Jur.(Bajoc.)*, Eu.——FIG. 494,9. **C. cymatophorina* (BUCKMAN), Eng.; *9a,b*, brach.v., lat. views, ×1 (229).

Formosarhynchia SEIFERT, 1963, p. 177 [**F. formosa*; OD]. Like *Cymatorhynchia* but dorsal valve less inflated posteriorly and definite septalium present. *M.Jur.(Bajoc.)*, Eu.——FIG. 495,1. **F. formosa*, Ger.; *1a,b*, brach.v., ant. views, ×1 (735a).

Gibbirhynchia BUCKMAN, 1918, p. 43 [**G. gibbosa*; OD]. Small, globose, with strong uniplication, multicostate; beak small, incurved, with 2 deep, narrow muscle impressions. Crura short, radulifer. *L.Jur.(Sinemur.-Toarc.)*, Eu.-Anatolia-Iran. ——FIG. 494,6. **G. gibbosa*, Eng.; *6a,b*, brach.v., lat. views, ×2 (905a).

Goniorhynchia BUCKMAN, 1918, p. 52 [**G. goniaea*; OD]. Medium-sized, wide; with strong uniplication and dorsal fold; many strong, sharp costae; not smooth posteriorly; beak small, suberect. Dorsal septum weak; crura radulifer; with much internal secondary thickening. *M.Jur.(Bathon.)*, Eu. ——FIG. 494,7. **G. goniaea*, Eng.; *7a,b*, brach.v., ant. views, ×1 (909).

Grandirhynchia BUCKMAN, 1918, p. 40 [**G. grandis*; OD]. Large, laterally expanded, uniplicate; with few strong blunt costae and pronounced smooth stage posteriorly; beak large, suberect, sharp beak ridges. Very deep septalium, long median septum; crura long, radulifer. *L.Jur.(Pliensbach.)*, G. Brit. - Greenl. —— FIG. 479,3; 497,11. **G. grandis*, Scot.; *479,3a*, transv. sec. 4.5 mm. from

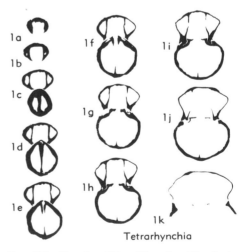

Tetrarhynchia

FIG. 496. Rhynchonellidae (Tetrarhynchiinae) (p. *H611-H612*).

tip of beak (ped.v. above), ×2; 479,*3b*, part of brach.v. in same sec. showing septalial plates *(s)*, ×7; 497,*11a,b*, lat., ant. views, ×1 (1).

Isjuminelina MAKRIDIN, 1960, p. 254 [*Rhyn-chonella pseudodecorata* ROLLIER, 1917, p. 139; OD] [=*Isjuminella* MAKRIDIN, 1955, p. 85

(nom. nud.) (type, "*Rhynchonella decorata* VON BUCH" does not exist)]. Large, globose, uniplicate, thick-shelled. Dorsal septum supporting septalium; crura short. [Insufficiently known; probably a synonym of one of the other genera listed herein.] *U.Jur.(Oxford.),* Eu.-S.USSR.——

FIG. 497. Rhynchonellidae (Tetrarhynchiinae) (p. *H612, H614*).

FIG. 494,8. **I. pseudodecorata* (ROLLIER), Ger.; brach.v. view, ×2 (626).

Kolhidaella MOISSEEV, 1939, p. 189 [**K. kolhidaensis*; OD] [=*Gagriella* MOISSEEV, 1939, p. 183 (type, *G. abhasiaensis*)]. Large, pentagonal, strongly folded and uniplicate, strong costae. [Insufficiently known; may be close to *Lacunosella*.] *L. Cret.*, USSR(Caucasus).——FIG. 494,5. **K. kolhidaensis*; oblique brach.v. view, ×1 (627).

Kutchirhynchia BUCKMAN, 1918, p. 54 [**Rhynchonella concinna* var. *kutchensis* KITCHIN, 1900, p. 48; OD]. Medium-sized, globose, with strong uniplication and dorsal fold; no smooth stage, many costae; beak short, suberect. Dental plates very long, dorsal septum strong but short. [Possibly a synonym of *Cymatorhynchia*.] *M.Jur.* (*Bathon.*), Eu.-India.——FIG. 497,8. **K. kutchensis* (KITCHIN), Cutch; *8a,b*, brach.v., ant. views, ×1 (478).

Microrhynchia MUIR-WOOD, 1952, p. 124 [**M. barnackensis*; OD]. Small, globose, with fine costae anteriorly; uniplicate (may be asymmetrical), fold ill-defined; beak slightly incurved. Well-developed dorsal septum, no septalium; crura calcarifer. *M.Jur.*, Eu.——FIG. 497,4. **M. barnackensis*, Eng.; *4a,b*, brach.v., ant. views, ×4.5 (582).

Mosquella MAKRIDIN, 1955, p. 6 [**Terebratula oxoptychia* FISCHER, 1843, p. 118; OD]. Like *Russirhynchia*, but with more costae, septum joined to hinge plates by callous thickening, crura thin and bladelike. *U.Jur.*(*Volg.*), USSR.——FIG. 497,10. **M. oxoptychia* (FISCHER); *10a,b*, ped.v., lat. views, ×1 (694).

Piarorhynchia BUCKMAN, 1918, p. 34 [**Rhynchonella lineata* var. *radstockiensis* DAVIDSON, 1878, p. 210; OD] [=*Tropiorhynchia* BUCKMAN, 1918]. Medium-sized, globose to depressed, equivalved, uniplicate, dorsal fold low; pronounced smooth stage posteriorly, rounded costae anteriorly; beak small, incurved. Dorsal septum massive; crura radulifer; with thick horizontal hinge plates. *U.Trias.*, Eu.(Alps)-W.Can.; *L.Jur.*, Eu.-N.Afr.——FIG. 497,2. **P. radstockiensis* (DAVIDSON), Eng.; *2a,b*, lat., ant. views, ×1 (229).

Quadratirhynchia BUCKMAN, 1918, p. 42 [**Q. quadrata*; OD]. Medium-sized to large; strong, wide uniplication, with many, very sharp costae; beak small, incurved. Pedicle collar present; median septum very short; crura radulifer. *L.Jur.* (*U.Pliensbach.*), W.Eu.——FIG. 497,6. **Q. quadrata*, Eng.; *6a,b*, brach.v., ant. views, ×1 (1).

Rhynchonelloidea BUCKMAN, 1918, p. 38 [**Rhynchonella ruthenensis* REYNÈS, 1868, p. 101; OD]. Medium-sized, with strong dorsal fold and uniplication; few fairly sharp costae after smooth stage; beak small but clear and erect. Dorsal septum strong, short; crura radulifer, distally concave. [Perhaps attributable to the Rhynchonellinae.] *L.Jur.-M.Jur.*, Eu.——FIG. 497,1. **R. ruthenensis* (REYNÈS), Fr.; *1a,b*, brach.v., ant. views, ×1 (136).

Robustirhynchia SEIFERT, 1963, p. 174 [**Terebratula Ehningensis* QUENSTEDT, 1857, p. 497; OD]. Like *Goniorhynchia* but very wide, with wide uniplication and thinner shell. *U.Jur.* (*Callov.*), Eu.——FIG. 495,2. **R. ehningensis* (QUENSTEDT), Ger.; *2a,b*, brach.v., ant. views, ×1 (735a).

Rudirhynchia BUCKMAN, 1918, p. 44 [**R. rudis*; OD]. Small, subtriangular, uniplicate, dorsal fold low, with few strong, fairly sharp costae, smooth posteriorly; beak strong, sharp, projecting, slightly incurved. Dorsal septum and septalium strong; crura radulifer. *L.Jur.*(*Pliensbach.*), Eu.——FIG. 497,5. **R. rudis*, Eng.; *5a,b*, brach.v., lat. views, ×2 (1).

Russirhynchia BUCKMAN, 1918, p. 52 [**Rhynchonella fischeri* ROUILLIER, 1847, p. 394; OD]. Medium-sized to large, globose, with strong uniplication, dorsal fold, and many very strong costae; beak short, suberect. Dorsal septum strong; crura radulifer; with much internal secondary thickening. *U.Jur.*(*Kimmeridg.-Volg.*), Eu.(USSR-W. Eu.).——FIG. 497,12. **R. fischeri* (ROUILLIER); *12a,b*, ant., post. views, ×1 (929).

Sakawairhynchia TOKUYAMA, 1957, p. 126 [**S. tokomboensis*; OD]. Small, subpentagonal, with strong uniplication and flattened fold, about 10 to 15 subangular costae; beak sharp, upright. Septalium shallow, with median projection; crura radulifer. *U.Trias.*(*Carn.*), Asia(Japan-Himalayas)-Eu.(Alps)-N.Am.(W.Can.). —— FIG. 497,7. **S. tokomboensis*, Japan; *7a,b*, brach.v. and lat. views of int. mold, ×1 (812).

Scalpellirhynchia MUIR-WOOD, 1936, p. 477 [**Terebratula scalpellum* QUENSTEDT, 1851, p. 453; OD]. Small, biconvex, flattened anteriorly, uniplication low, with costae anteriorly; beak short, erect. Dorsal septum long, supporting wide septalium; crura radulifer. *L.Jur.*, Eu.——FIG. 497,3. **S. scalpellum* (QUENSTEDT), Ger.; *3a,b*, brach.v., ant. views, ×2 (579).

Somalirhynchia WEIR, 1925, p. 79 [**S. africana*; OD]. Large, uniplicate, dorsal fold low, multicostate; beak strong, incurved, with small foramen, hypothyridid. Dorsal median septum long, strong; muscle scars well marked; crura radulifer, enlarged distally. *U.Jur.*(*Oxford.*), E.Afr.-M.East.——FIG. 497,9. **S. africana*, Somaliland; *9a,b*, brach.v., ant. views, ×1 (577).

Subfamily CYCLOTHYRIDINAE Makridin, 1955

[*nom. correct.* AGER, herein (*pro* Cyclothyrisinae MAKRIDIN, 1955, p. 82)] [Cyclothyridae proposed by PHILLIPS, 1841, p. 55, for *Epithyris* and *Hypothyris* is not an available family group name under Article 11e of the International Code]

Multicostate, rarely with posterior smooth area; beak massive, with hypothyridid, rimmed foramen (i.e., deltidial plates produced into short tube around pedicle). Dorsal median septum usually very much reduced or absent, septalium lacking; crura canalifer type. Characteristically strongly costate, some shells with fine capillae pos-

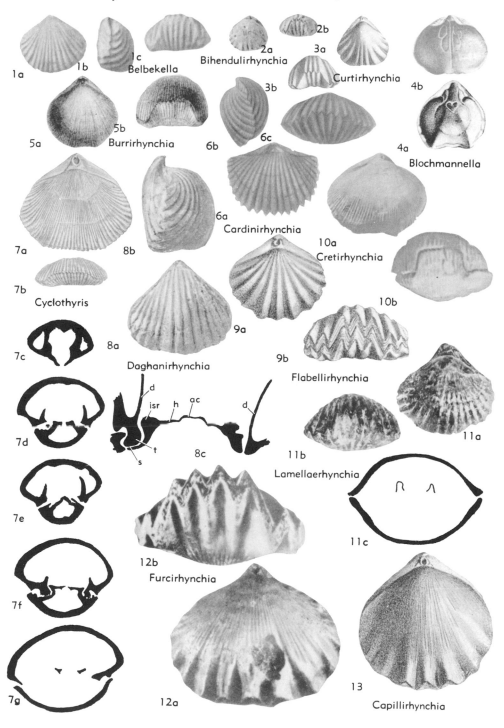

Fig. 498. Rhynchonellidae (Cyclothyridinae) (p. *H616-H617*).

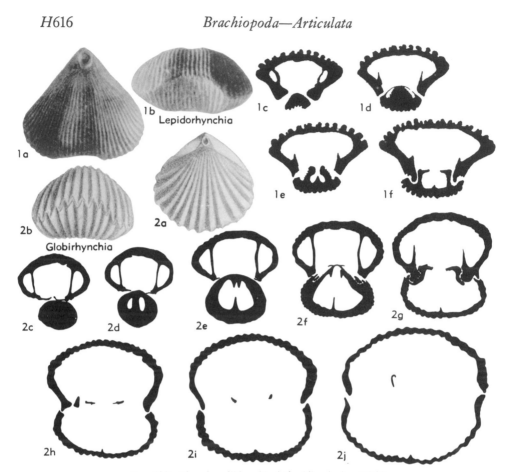

Fig. 499. Rhynchonellidae (Cyclothyridinae) (p. *H617*).

teriorly passing into costae anteriorly; may be asymmetrical. *L.Jur.-U.Cret.*

Cyclothyris M'Coy, 1844, p. 103 [**Terebratula latissima* J. DE C. SOWERBY, 1829, index (=**T. lata* J. DE C. SOWERBY, 1825, p. 165) (*non* J. SOWERBY, 1815), p. 227; OD]. Large, wide, depressed, with low arcuate uniplication, commonly asymmetrical with many fine costae, may be capillate posteriorly; beak erect. Dorsal septum very short or absent. *Cret.(Apt.-Cenoman.)*, Eu.-N.Am.——FIG. 498,7. **C. latissima* (J. DE C. SOWERBY), Eng.; *7a,b*, brach.v., ant. views, ×1, ×0.7 (229); *7c-g*, serial transv. secs., ×2 (629).

Belbekella MOISSEEV, 1939 [**B. airgulensis*; OD]. Globose, subtriangular, uniplicate, without distinct fold, multicostate, lacking smooth stage; beak massive, erect. Strong teeth and denticula; septum may be absent. *Cret.*, USSR(Crimea-Caucasus-C.Asia)-?W.Eu.——FIG. 498,1. **B. airgulensis*; *1a-c*, brach.v., lat., ant. views, ×1 (925).

Bihendulirhynchia MUIR-WOOD, 1935, p. 104 [**B. afra*; OD]. Small, smooth and sulcate posteriorly, uniplicate anteriorly, with low fold and about 10 costae; beak erect, hypothyridid. Pedicle collar supported by septum; dorsal septum short and low; no septalium; crura short, radulifer. *U.Jur. (L. Kimmeridg.)*, Somaliland.——FIG. 498,2. **B. afra*; *2a,b*, brach.v., ant. views, ×1 (577).

Blochmannella LEIDHOLD, 1920, p. 356 [**Rhynchonella Friereni* BRANCO, 1879, p. 128; OD]. Like *Septaliphoria* but with long, strong median septum and well-marked muscle scars. *M.Jur. (Bajoc.)*, Eu.——FIG. 498,4. **B. friereni* (BRANCO), Ger.; *4a,b*, ped.v. int., brach.v. int., ×2 (503).

Burrirhynchia OWEN, 1962, p. 58 [**Rhynchonella leightonensis* LAMPLUGH & WALKER, 1903, p. 261; OD]. Like *Cretirhynchia* but for disjunct deltidial plates, thinner hinge plates and weaker median septum. *L.Cret.(Apt.-Alb.)*, Eu.——FIG. 498, 5. **B. leightonensis* (LAMPLUGH & WALKER), Eng.; *5a,b*, brach.v., ant. views, ×1 (923).

Capillirhynchia BUCKMAN, 1918, p. 58 [**Rhynchonella wrighti* DAVIDSON, 1852, p. 69; OD]. Large, globose, uniplicate, with low fold; capillae all over shell, strong sharp costae anteriorly; beak strong, suberect. *Jur.(Bajoc.-Callov.)*, Eu.-N.Am.

(Calif.).——Fig. 498,*13*. **C. wrighti* (Davidson), Eng.; brach.v. view, ×1 (229).

Cardinirhynchia Buckman, 1918, p. 74 [**Terebratula acuticosta* Zieten, 1830, p. 58; OD]. Wide, hinge line nearly straight; multicostate; with incipient uniplication; broad beak with large foramen; deltidial plates narrow. *M.Jur.(Bajoc.-?Bathon.)*, Eu.——Fig. 498,*6*. **C. acuticosta* (Zieten), Ger.; *6a-c*, brach.v., lat., ant. views, ×1 (937).

Cretirhynchia Pettitt, 1950, p. 1 [**Terebratula plicatilis* J. Sowerby, 1816, p. 37; OD]. Biconvex, uniplicate, dorsal fold low; smooth or with many low, round costae; beak short. Large teeth; low median septum, no septalium. [Probably requires division.] *U.Cret.*, NW.Eu.——Fig. 498,*10*. **C. plicatilis* (Sowerby), Eng.; *10a,b*, brach.v., ant. views, ×1 (639).

Curtirhynchia Buckman, 1918, p. 36 [**Rhynchonella oolitica* Davidson, 1852, p. 81; OD]. Small, uniplicate, with low dorsal fold; blunt costae anteriorly, smooth posteriorly; beak sharp, suberect, hypothyridid. Dorsal septum, no septalium, pedicle collar supported by short septum; crura radulifer, distally concave. *M.Jur.(Bajoc.)*, NW. Eu.——Fig. 498,*3*. **C. oolitica* (Davidson), Eng.; *3a,b*, brach.v., ant. views, ×1 (229).

Daghanirhynchia Muir-Wood, 1935, p. 82 [**D. daghaniensis*; OD]. Medium in size, uniplicate, with distinct dorsal fold, costae few; beak acute, incurved. Dental plates strong, dorsal septum weak, divided hinge plates united by thin lamella anteriorly; crura nearly horizontal. *U.Jur.(Callov.)*, E.Afr.——Fig. 498,*8*. **D. daghaniensis*, Br. Somaliland; *8a,b*, brach.v., lat. views, ×1; *8c*, transv. sec. showing accessory plate *(ac)* uniting hinge plates *(h)*, inner socket ridge *(isr)*, tooth *(t)* in socket *(s)*, and dental plates *(d)*, enlarged (577).

Flabellirhynchia Buckman, 1918, p. 65 [**Rhynchonella lycettii* Davidson, 1852, p. 65; OD]. Wide, depressed, with low fold and many strong, sharp costae; anterior margin thickened; beak strong, with large foramen. Dorsal septum feeble. *M.Jur.(Bajoc.)*, Eu.-N.Am.(Calif.).——Fig. 498,*9*. **F. lycettii* (Davidson), Eng.; *9a,b*, brach.v., ant. views, ×1 (229).

Furcirhynchia Buckman, 1918, p. 59 [**F. furcata*; OD] [=*Lineirhynchia* Buckman, 1918 (type, *Lineirhynchia cotteswoldiae* (Upton), 1899, p. 129)]. Small to medium-sized, depressed to cynocephalous; .capillate posteriorly, with few strong, sharp costae anteriorly; beak strong, upright, with large oval foramen. Dorsal median septum long; crura radulifer. *U.Trias.-L.Jur.*, Eu.-W.Canada.——Fig. 498,*12*. **F. furcata*, Eng.; *12a,b*, brach.v., ant. views, ×2 (1).

Globirhynchia Buckman, 1918, p. 48 [**Rhynchonella subobsoleta* Davidson, 1852, p. 91; OD]. Small to medium-sized, globose; with arcuate uniplication and low dorsal fold; many sharp costae, no smooth stage; beak massive, suberect,

hypothyridid. Dorsal septum long and low; no septalium; dorsal muscle scars linear; crura radulifer, hooked dorsally. *M.Jur.(Bajoc.)*, NW.Eu.-USA(Calif.).——Fig. 499,*2*. **G. subobsoleta* (Davidson), Eng.; *2a,b*, brach.v., ant. views, ×1 (229); *2c-j*, serial transv. secs., ×4 (Ager, n).

Lamellaerhynchia Burri, 1953, p. 274 [**Terebratula rostriformis* Roemer, 1836, p. 40 (=*T. multiformis* Roemer, 1839, *partim*); OD]. Medium-sized, multicostate; uniplicate, rectimarginate, or asymmetrical; beak strong, projecting, suberect. Dorsal septum ridgelike; crura distally concave. [Possibly a synonym of *Belbekella*.] *L.Cret.(U. Valangin.-Barrem.)*, W.Eu.-N.Am.(Tex.).——Fig. 498,*11*. **L. rostriformis* (Roemer), Ger.; *11a,b*, brach.v., ant. views, ×1; *11c*, transv. sec. of beak region, ×2.4 (138).

Lepidorhynchia Burri, 1956, p. 689 [**L. dichotoma*; OD]. Medium-sized, biconvex, with many fine branching costae; rectimarginate or slightly uniplicate, sulcus in both valves; beak high, with large foramen. Dental plates poorly developed; median septum a low ridge. *L.Cret.(U.Hauteriv.)*, *?U.Cret.(Cenoman.)*, Eu.——Fig. 499,*1*. **L. dichotoma*, Switz.; *1a,b*, brach.v., ant. views, ×2; *1c-f*, transv. secs. of beak region, ×4 (138).

Malwirhynchia Chiplonker, 1938, p. 306 [**M. transversalis*; OD]. Small, low uniplication and ill-defined fold; many fine costae, bifurcating anteriorly. Beak short, suberect, with large foramen. Weak dorsal septum, crura ?calcarifer. *?L. Cret.(Alb.)*, *U. Cret.(Cenoman.)*, India.——Fig. 500,*1*. **M. transversalis*, India; brach.v. view, ×1 (911).

Maxillirhynchia Buckman, 1918, p. 55 [**M. implicata*; OD]. Small; low rectangular fold and uniplication after short early sulcate stage. Capillate throughout, strong costae anteriorly. Beak sharp, incurved, hypothyridid. [Doubtfully included here.] *L.Jur.(Pliensbach.-Toarc.)*, Eu.—— Fig. 500,*9*. **M. implicata*, Eng.; *9a-c*, brach.v., lat., ant. views, ×3 (Ager, n).

Parvirhynchia Buckman, 1918 [**Rhynchonella parvula* Eudes-Deslongchamps, 1862; p. 29; OD]. Small, depressed, with low fold and uniplication; few blunt costae, capillate throughout; beak strong, erect. Dorsal septum low; crura ?calcarifer. *Jur. (Bajoc.-Oxford.)*, Eu.-Japan.——Fig. 500,*4*. **P. parvula* (Eudes-Deslongchamps), Eng.; *4a,b*, brach.v., lat. views, ×2 (229).

Plicarostrum Burri, 1953, p. 281 [**P. hauteriviense*; OD]. Medium-sized, with many sharp costae; cynocephalous, nearly convexiplanate; beak projecting. Thick dental plates almost fused with shell wall; median septum a low ridge, arising late; crura distally concave. *L.Cret.(Hauteriv.)*, Eu.——Fig. 500,*2*. **P. hauteriviense*, Switz.; *2a,b*, brach.v., lat. views, ×1 (138).

Praecyclothyris Makridin, 1955, p. 84 [**Rhynchonella moeschi donetziana* Makridin, 1952; OD]. Like *Septaliphoria* but with a flared rim inside the foramen. *U.Jur.(Callov.-Kimmeridg.)*, USSR-

?W.Eu.——Fig. 500,*8*. **P. moeschi donetziana* (Makridin); *8a,b*, brach.v., ant. views, ×1 (694).
Ptilorhynchia Crickmay, 1933, p. 877 [**P. plumasensis*; OD]. Medium-sized, triangular, subglobose; uniplication strong, with blunt costae anteriorly after long smooth stage; beak small, upright, hypothyridid. Dental plates divergent. *U.Jur.(?Kimmeridg.)*, USA(Calif.).——Fig. 500,*3*. **P. plumasensis*; *3a,b*, brach.v., ant. view, ×1 (915a).

Ptyctorhynchia Buckman, 1918, p. 47 [**Rhynchonella pentaptycta* Buckman, 1910, p. 103; OD]. Small, globose, with low wide uniplication and fold; few very strong costae anteriorly; beak small, suberect; foramen hypothyridid, slightly rimmed. *M.Jur.(Bajoc.)*, Eu.——Fig. 500,*5*. **P. pentaptycta* (Buckman), Eng.; *5a,b*, brach.v., lat. views, ×1 (229).
Rhactorhynchia Buckman, 1918, p. 50 [**R. rhacta*;

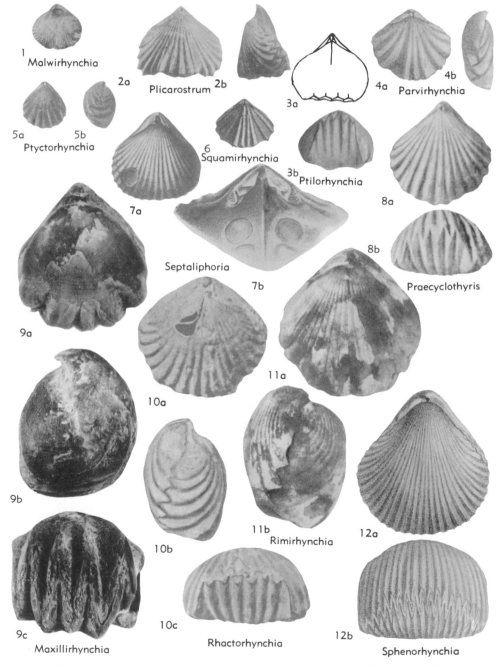

Fig. 500. Wellerellidae (Cirpinae) *(6)*; Rhynchonellidae (Cyclothyridinae) *(1-5, 7-12)* (p. *H607-H608, H617-H619*).

OD]. Medium-sized to large, globose, with feeble, commonly asymmetrical dorsal fold and many strong, sharp costae; beak strong, slightly incurved, hypothyridid. Dorsal septum strong; crura raduli-fer; muscle scars expanded. [An important, long-ranging genus that may need subdivision.] *?L.Jur.,* SE. Asia; *M. Jur. (Bajoc.), ?U. Jur. (Kimmeridg.),* Eu.-N. Afr.-?E. Afr.-India-N. Am.——FIG. 500,*10.* **R. rhacta,* Eng.; *10a-c,* brach.v., lat., ant. views, ×1 (138).

Rimirhynchia BUCKMAN, 1918, p. 60 [**R. rimosi-formis* BUCKMAN, 1918, p. 60=**Rhynchonella anglica* ROLLIER, 1917, p. 92; OD]. Like *Furci-rhynchia* (to which it is closely related) but very globose, with massive incurved beak. *L.Jur. (Sinemur.-L.Pliensbach.),* W.Eu.-W.Canada.——FIG. 500,*11.* **R. anglica* (ROLLIER), Eng.; *11a,b,* brach.v., lat. views, ×2 (1).

Septaliphoria LEIDHOLD, 1921, p. 354 [**Rhyncho-nella arduennensis* OPPEL, 1857, p. 608 (=*R. in-constans* D'ORBIGNY, 1850, p. 24, *non* SOWERBY, 1821); OD]. Medium-sized to large, uniplicate, commonly asymmetrical, costate; strong, high beak with large hypothyridid foramen. Septalium supported by short median septum; crura strong, concave dorsally. *U.Jur.(Oxford.)-?L.Cret.(Val-angin.),* Eu.-?E.Afr.——FIG. 500,*7.* **S. arduen-nensis* (OPPEL), Ger.; *7a,* brach.v. view, ×1; *7b,* ped.v. int., ×2 (503).

Sphenorhynchia BUCKMAN, 1918, p. 30 [**Tere-bratula plicatella* J. DE C. SOWERBY, 1825, p. 167; OD]. Medium-sized to large, globose, wedge-shaped; with arcuate uniplication, dorsal fold raised in some shells, with many sharp costae; beak small, massive, suberect. Dorsal septum strong. [Doubtfully included here.] *Jur.(Bajoc.-Callov.),* Eu.-?Asia(Afghan.).——FIG. 500,*12.* **S. plicatella* (J. DE C. SOWERBY), Eng.; *12a,b,* brach. v., ant. views, ×1 (229).

Stiirhynchia BUCKMAN, 1918, p. 68 [**Rhyncho-nella dorsetensis* DAVIDSON, 1884, p. 177; OD]. Small to medium-sized, depressed, uniplicate, with many fine dichotomizing capillae, no costae; beak small, sharp. Dental plates short; dorsal septum feeble. *M.Jur.(.Bajoc.), ?U.Jur.(Kimmeridg.),* Eu. ——FIG. 501,*1.* **S. dorsetensis* (DAVIDSON), Eng.; brach.v. view, ×2 (229).

Suiaella MOISSEEV, 1956, p. 20 [**S. weberi;* OD]. Like *Belbekella* but smaller, with brachial valve less convex than pedicle, and crura less curved. *L. Cret.(Barrem.),* USSR (Crimea-Caucasus).—— FIG. 501,*3.* **S. weberi; 3a-c,* brach.v., lat., ant. views, ×1.5 (925).

Sulcirhynchia BURRI, 1953, p. 271 [**Rhynchonella valangiensis* DELORIOL, 1864, p. 442; OD]. Me-dium-sized, with many sharp costae, slight sulcus in median fold; beak projecting. Median septum soon reduced to low ridge; crura radulifer, slight-ly concave distally. *L.Cret.(U.Valangin.-?L.Apt.),* Eu.——FIG. 501,*2.* **S. valangiensis* (DELORIOL), Switz.; *2a,b,* brach.v., lat. views, ×1; *2c,* diagram. ant. view (brach.v. above), ×1 (138).

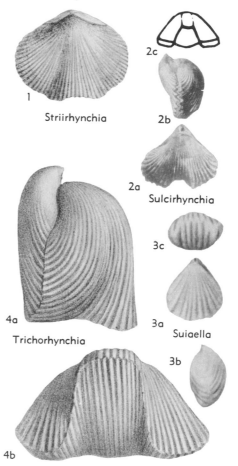

FIG. 501. Rhynchonellidae (Cyclothyridinae) (p. H619).

Trichorhynchia BUCKMAN, 1918, p. 58 [**Rhyn-chonella deslongchampsii* DAVIDSON, 1852, p. 253; OD]. Large, with wide uniplication, strong flat-tened dorsal fold, many fine costae; beak massive. [Possibly belongs to Tetrarhynchiinae.] *Jur. (Pliensbach.-Bajoc.),* NW.Eu.——FIG. 501,*4.* **T. deslongchampsii* (DAVIDSON), Fr.; *4a,b,* lat., ant. views, ×1 (229).

Family SEPTIRHYNCHIIDAE Muir-Wood & Cooper, 1951

[Septirhynchiidae MUIR-WOOD & COOPER, 1951, p. 5]

Unusually large, pentameroid in appear-ance, with cardinal process in dorsal valve and median septum in ventral valve (577, 926). *U.Jur.*

Septirhynchia MUIR-WOOD, 1935, p. 106 [**Rhyn-chonella azaisi* COTTREAU, 1924, p. 581; OD]. Thick-shelled, with many strong, subangular costae, fold low. Long ventral median septum

FIG. 502. Septirhynchiidae (p. *H619-H620*).

and strong dental plates; dorsal median septum low; small knoblike cardinal process; long radulifer crura. *Jur.(Callov.-Kimmeridg.)*, E.Afr.-W. Asia(Sinai).——FIG. 502,*1*. *S. pulchra* MUIR-WOOD & COOPER, Abyssinia; *1a,b*, brach.v. int., ped.v. int., ×1.3 (926).

Family AUSTRIRHYNCHIIDAE Ager, 1959

[Austrirhynchiidae AGER, 1959, p. 325]

Shell extremely expanded laterally, with cardinal process and dorsal denticula (3). [Perhaps should be classed as subfamily of Dimerellidae.] *Trias.*

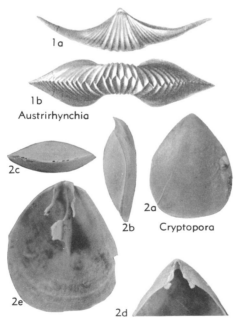

FIG. 503. Austrirhynchiidae *(1)*; Cryptoporidae *(2)* (p. *H620, H622*).

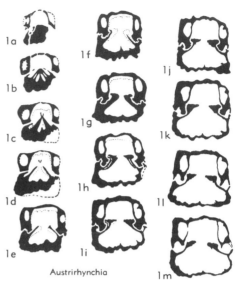

FIG. 504. Austrirhynchiidae (p. *H620*).

Austrirhynchia AGER, 1959, p. 325 [*Terebratula cornigera* SCHAFHÄUTL, 1851, p. 407; OD]. Small, triangular, anterolateral angles considerably extended; with short, wide uniplication, multicostate, well-developed planareas; beak incurved, with large hypothyridid foramen. Dorsal septum very short, with bilobed cardinal process, dorsal denticula present; crura radulifer. *U.Trias.(Carn.-Rhaet.)*, Eu.(Alps).——FIG. 503,*1*; 504,*1*. *A. cornigera* (SCHAFHÄUTL), Austria; 503,*1a,b*, brach. v., ant. views, ×1 (933); 504,*1a-m*, transv. secs. of beak region, 0.3-1.2 mm. from tip of beak (ped.v. above), ×3.7 (3).

Family CRYPTOPORIDAE Muir-Wood, 1955

[Cryptoporidae MUIR-WOOD, 1955, p. 76]

Large deltoid foramen slightly restricted by elongate, triangular, elevated deltidial plates; crura long, maniculifer, continuous with socket ridges; median septum elevated; cardinal process a lobate thickening between socket ridges; single pair of nephridia (193, 583). *Eoc.-Rec.*

Cryptopora JEFFREYS, 1869, p. 136 [*Atretia gnomon* JEFFREYS, 1876, p. 251; OD (M)] [=*Atretia* JEFFREYS, 1870, p. 421 (type, *A. gnomon* JEFFREYS, 1876, p. 251); *Mannia* DAVIDSON, 1874, p. 156 (type, *M. nysti*; OD, M);*Neatretia* FISCHER & OEHLERT, 1891, p. 122 (obj.)]. Subtrigonal, rectimarginate to broadly sulcate, smooth. Beak moderately long, nearly straight; deltidial plates rudimentary, disjunct. Thickened plate elevated above floor in apex of pedicle valve.

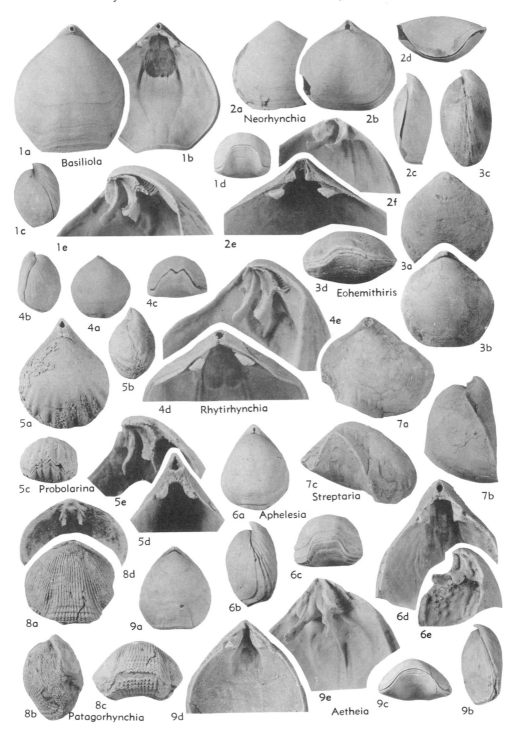

Fig. 505. Basiliolidae (Basiliolinae) *(1-5, 7)*, (Aphelesiinae) *(6)*, (Aetheiinae) *(8-9)* (p. *H622-H623*).

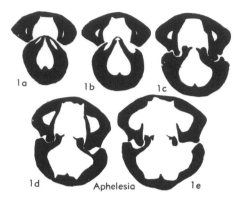

Fig. 506. Basiliolidae (Aphelesiinae) (p. *H622-H623*).

Small bilobed cardinal process; median septum rising anteriorly in brachial valve. *Eoc.-Rec.,* Eu.-N. Am. - Atl. - Afr.-Australia.——Fig. 503,2. **C. gnomon* (JEFFREYS), Rec., off E.USA(Mass.); *2a-c,* ped.v., lat., ant. views, ×6; *2d,* post. ped.v. int., ×8; *2e,* brach.v. int., slightly oblique, ×8 (193).

Family BASILIOLIDAE Cooper, 1959

[Basiliolidae COOPER, 1959, p. 25]

Smooth or semicostate; conjunct deltidial plates and small auriculate foramen; well-developed pedicle collar; broad falcifer crura supported by outer hinge plates or socket ridges; median septum in brachial valve reduced to ridge or absent (193). *Cret.-Rec.*

Subfamily BASILIOLINAE Cooper, 1959

[Basiliolinae COOPER, 1959, p. 25]

Brachial valve with crura attached to broad outer hinge plates, no median septum (193). *Eoc.-Rec.*

Basiliola DALL, 1908, p. 442 [**Hemithyris beecheri* DALL, 1895, p. 717; OD] [=*Basiola* THOMSON, 1915, p. 390 *(nom. null.)*; *Neohemithyris* YABE & HATAI, 1934, p. 587 (type, *Rhynchonella lucida* GOULD, 1861, p. 323; OD)]. Subpentagonal uniplicate, smooth. Inconspicuous fold in strongly convex brachial valve. Beak small, foramen submesothyridid. Uniplication in pedicle valve fitting into re-entrant in brachial valve. Strong complex pedicle collar; long crescentic falcifer crura. *Plio.-Rec.,* Japan-Indon.-Pac.——Fig. 505,1. **B. beecheri* (DALL), Rec., off Hawaii; *1a,b,* brach.v. view, ped.v. int., ×2; *1c,d,* lat., ant. views, ×1; *1e,* oblique view brach.v. int. showing crura, ×4 (193).

Eohemithiris HERTLEIN & GRANT, 1944, p. 55 [**E. alexi*; OD] [=*Eohemithyris* COOPER, 1959, p. 30 *(nom. van.)*]. Like *Basiliola b*ut more nearly equivalve and without elaborate pedicle collar. *Eoc.-*

Rec., N.Am.-Fiji-Australia.——Fig. 505,3. **E. alexi,* Eoc., USA(Calif.); *3a-d,* ped.v., brach.v., lat., ant. views, ×2 (193).

Neorhynchia THOMSON, 1915, p. 388 [**Hemithyris strebeli* DALL, 1908, p. 441; OD]. Pentagonal, deeply sulcate, smooth. Beak short, hypothyridid. Short crura. *Rec.,* Pac.——Fig. 505,2. **N. strebeli* (DALL); *2a-d,* ped.v., brach.v., lat., ant. views, ×2; *2e,* ped.v. int., ×4; *2f,* oblique view brach. v. int., ×4 (193).

Probolarina COOPER, 1959, p. 37 [**Rhynchonella holmesii* DALL, 1903, p. 1536; OD]. Subpentagonal to subtrigonal, uniplicate, costate anteriorly. Beak long, pointed, nearly straight; foramen hypothyridid to submesothyridid. Strong pedicle collar and dental plates. Crura scimitar-like. *Eoc.,* N.Am.——Fig. 505,5. **P. holmesii* (DALL), USA (N.Car.); *5a,* brach.v. view, ×3; *5b,c,* lat., ant. views, ×2; *5d,* ped.v. int., ×4; *5e,* oblique view brach.v. int., ×6 (193).

Rhytirhynchia COOPER, 1957, p. 8 [**Hemithyris sladeni* DALL, 1910, p. 440; OD]. Like *Basiliola* but costate anteriorly; dental plates much reduced. *Plio.-Rec.,* Pac.-Ind.O.——Fig. 505,4. **R. sladeni* (DALL), Rec., Ind.O.; *4a-c,* brach.v., lat., ant. views (lectotype), ×1; *4d,* ped.v. int. (lectotype), ×3; *4e,* oblique view brach.v. int., ×4 (193).

Streptaria COOPER, 1959, p. 38 [**Terebratula debuchii* MICHELOTTI, 1938, p. 4; OD]. Pentagonal, sharply uniplicate, asymmetrical; may be faintly costate anteriorly. Beak short, foramen rimmed, dental plates reduced; pedicle collar poorly developed. *Eoc.-Mio.,* Eu.-N.Afr.-Cuba.——Fig. 505, 7. **S. debuchii* (MICHELOTTI), M.Mio., Italy (Sicily); *7a-c,* brach.v., lat., ant. views, ×2 (193).

Subfamily APHELESIINAE Cooper, 1959

[Aphelesiinae COOPER, 1959, p. 41]

Crura attached directly to side of socket ridge; brachial valve with thick median ridge (193). *Eoc.-Plio.*

Aphelesia COOPER, 1959, p. 41 [**Anomia bipartita* BROCCHI, 1814, p. 469; OD]. Subtrigonal to subpentagonal, uniplicate, smooth, with incipient costae anteriorly; beak elongated, hypothyridid, foramen rimmed; no septalium. *Eoc.-Plio.,* Medit.——Fig. 505,6; 506,1. **A. bipartita* (BROCCHI), Plio., Italy(Sicily); *505,6a-c,* brach.v., lat., ant. views, ×1; *505,6d,e,* ped.v. int., brach.v. int.,

Aetheia

Fig. 507. Basiliolidae (Aetheiinae) (p. *H623*).

×2 (193); 506,*1a-e*, transv. secs. of beak region (ped.v. above), ×1.8 (193).

Subfamily AETHEIINAE Cooper, 1959
[Aetheiinae COOPER, 1959, p. 42]

With minute foramen, concave deltidial plates, dental plates reduced to obsolete, inner hinge plates thick (193). *Cret.-Mio.*

Aetheia THOMSON, 1915, p. 389 [*Waldheimia(?) sinuata* HUTTON, 1873, p. 36 (=?*Terebratula gualteri* MORRIS, 1850, p. 329); OD] [=*Thomsonica* COSSMANN, 1920, p. 137 (obj.)]. Elongate-oval to triangular, uniplicate, smooth; beak small, erect, submesothyridid. Hinge teeth thick, attached to valve wall; deltidial plates conjunct; no dental plates; median septum short, stout; inner hinge plates filling intercrural space; cardinal process small, crura very long. *U.Cret.-Mio.*, N.Z.——FIG. 505,9; 507,1. *A. gualteri* (MORRIS), Mio.; 505,*9a-c*, brach.v., lat., ant. views, ×1; 505, *9d*, ped.v. int., ×2; 505,*9e*, oblique view brach.v. int., ×4 (193); 507,1, brach.v. view, ×1 (810).

Patagorhynchia ALLAN, 1938, p. 199 [*Rhynchonella patagonica* VON IHERING, 1903, p. 334; OD]. Subcircular to subpentagonal, uniplicate; finely costate with strong growth lines anteriorly; beak small, nearly straight, submesothyridid. *Eoc.*, S. Am.——FIG. 505,8. *P. patagonica* (VON IHERING), Arg.; *8a-c*, brach.v., lat., ant. views, ×1; *8d*, brach.v. int., ×1 (193).

Family HEMITHYRIDIDAE Rzhonsnitskaya, 1956
[*num. transl. et correct.* AGER, herein (*ex* Hemithyrinae RZHONSNITSKAYA, 1956, p. 126; based on jr. obj. syn. of *Hemithiris* D'ORBIGNY, 1847)]

With strong, slender, curved radulifer crura attached to small outer hinge plates by their posterodorsal face or to thick socket ridges; crura radulifer, distally pointed and horizontally flattened (193). *Eoc.-Rec.*

Hemithiris D'ORBIGNY, 1847, p. 342 [*Anomia psittacea* GMELIN, 1790, p. 3348; SD D'ORBIGNY, 1847, p. 342] [=*Hemithyris* BRONN, 1848, p. 246 (obj.)]. Trigonal, uniplicate, finely costate with intermediate striae. Beak long, suberect, hypothyridid; ventral median ridge posteriorly, low dorsal median ridge. *Mio.-Rec.*, N.Hemis.——FIG. 508,*1a-c*. *H. psittacea* (GMELIN), Rec., off Alaska; *1a-c*, brach.v., lat., ant. views, ×1 (193). ——FIG. 508,*1d,e*. *H. woodwardi* (DAVIDSON), Rec., off Japan; *1d,e*, ped.v. int., brach.v. int., ×3, ×4 (193).

Notosaria COOPER, 1959, p. 48 [*Terebratula nigricans* SOWERBY, 1846, p. 91; OD]. Subpentagonal, uniplicate, dorsal fold low; finely costate, growth lines anteriorly; beak nearly straight to suberect; large hypothyridid foramen, deltidial plates disjunct; low dorsal median ridge. *Mio.-Rec.*, Eu.-N.Z.-S.Ind.O.(Kerguelen Is.).——FIG. 508,2. *N.

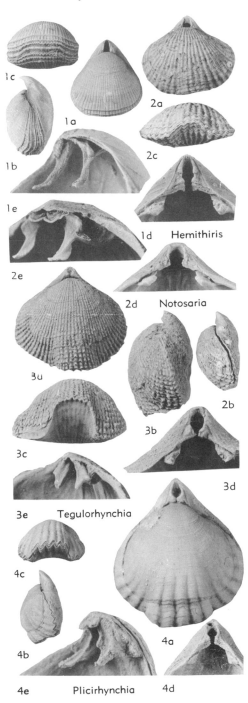

1c
1a
1b
1e
2a
2c
1d Hemithiris
2e
2d Notosaria
3a
3b
3c
2b
3d
3e Tegulorhynchia
4c
4b
4e Plicirhynchia
4a
4d

FIG. 508. Hemithyrididae (p. *H623-H624*).

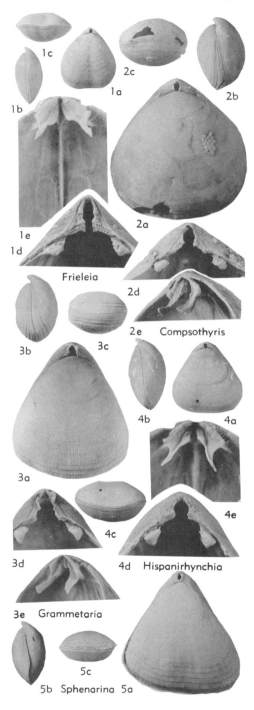

FIG. 509. Frieleiidae (p. H624-H625).

nigricans (SOWERBY), Rec., N.Z.; 2a-c, brach.v., lat., ant. views, ×1; 2d,e, ped.v. int., brach.v. int., ×2, ×4 (193).

Plicirhynchia ALLAN, 1947, p. 493 [*Rhynchonella plicigera VON IHERING, 1897, p. 270; OD]. Subtrigonal to subpentagonal, uniplicate; striate posteriorly, costate anteriorly; beak long, acute; large hypothyridid foramen; deltidial plates thick and conjunct; thick bilobed cardinal process; small dorsal median ridge. Eoc., S.Am.(Arg.)-?Antarctic. ——FIG. 508,4. *P. plicigera (VON IHERING), Arg.; 4a, brach.v. view, ×2; 4b,c, lat., ant. views, ×1; 4d,e, ped.v. int., brach.v. int., ×2, ×4 (193).

Tegulorhynchia CHAPMAN & CRESPIN, 1923, p. 175 [*Rhynchonella squamosa HUTTON, 1873, p. 37; OD]. Trigonal to subpentagonal, uniplicate, dorsal fold low; costate and lamellose, some shells with hollow spines; beak long, upright; foramen large, hypothyridid; deltidial plates usually conjunct; crura short; dorsal median septum short, low. Oligo.-Rec., Australia-Pac.O.——FIG. 508, 3a-c. *T. squamosa (HUTTON), Mio., N.Z.; 3a-c, brach.v., lat., ant. views, ×1 (193).——FIG. 508, 3d-e. T. doederleni (DAVIDSON), Mio. or Plio., Okinawa; 3d,e, ped.v. int., brach.v. int., ×4 (193).

Family FRIELEIIDAE Cooper, 1959

[Frieleiidae COOPER, 1959, p. 53] [=Hispanirhynchiidae COOPER, 1959, p. 59]

Usually capillate to costellate, triangular; dental plates strong, spinulifer crura short and straight, septalium small. (Cooper, 1959, emended.) [Includes the family Hispanirhynchiidae (193) since its specified characters correspond entirely with those of this family.] ?Eoc., ?Mio., Plio.-Rec.

Frieleia DALL, 1895, p. 713 [*F. halli, p. 714; OD]. Elongate, oval to subtrigonal; shell thin, rectimarginate to ligate, smooth to faintly costate; beak short, nearly straight to suberect, hypothyridid, deltidial plates thick, disjunct, with long divergent crura, long slender dorsal median septum supporting short septalium; inner hinge plates strongly developed. ?Mio., Rec., W.USA-N. Pac.——FIG. 509,1. *F. halli, Rec., N.Pac.; 1a-c, brach.v., lat., ant. views, ×1; 1d,e, ped.v. int., brach.v. int., ×4, ×6 (193).

Compsothyris JACKSON, 1918, p. 188 [*Rhynchonella racovitzae JOUBIN, 1901, p. 5; OD]. Trigonal, broad gentle uniplication, dorsal fold inconspicuous, with fine radial striae; hypothyridid; ridgelike dorsal septum supporting small septalium. Rec., Antarctic.——FIG. 509,2. *C. racovitzae (JOUBIN); 2a, brach.v. view, ×2; 2b,c, lat., ant. views, ×1; 2d,e, ped.v. int., brach.v. int., ×4 (193).

Grammetaria COOPER, 1959, p. 58 [*Hemithyris bartschi DALL, 1920, p. 289; OD]. Elongate trigonal, rectimarginate, capillate; beak small, suberect; hypothyridid; deltidial plates auriculate.

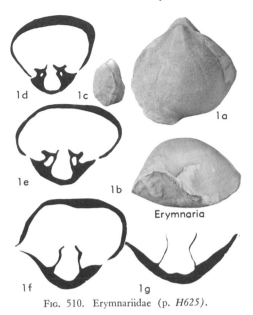

FIG. 510. Erymnariidae (p. *H625*).

conjunct; crura short; dorsal median ridge stout, supporting wide septalium. *Rec.,* Pac.——FIG. 509,*3*. *G. bartschi* (DALL), Phillip.; *3a*, brach.v. view, ×2; *3b,c*, lat., ant. views, ×1; *3d,e*, ped.v. int., brach.v. int., ×4 (193).

Hispanirhynchia THOMSON, 1927, p. 159 [*Rhynchonella cornea* FISCHER, 1887, p. 171; OD]. Elongate trigonal, rectimarginate to ligate to slightly uniplicate; fine capillae and with growth lines, beak short, suberect, with large hypothyridid foramen, deltidial plates disjunct; crura bladelike, dorsal median ridge low, thick. *?Eoc.,* Cuba; *Rec.,* N.Atl.-E.Pac.——FIG. 509,*4*. *H. cornea* (FISCHER), Rec., off Morocco; *4a-c*, brach.v., lat., ant. views, ×1; *4d,e*, ped.v. int., brach.v. int., ×4, ×6 (193).

Sphenarina COOPER, 1959, p. 62 [*Rhynchonella sicula* SEGUENZA, 1870, p. 461; OD]. Like *Hispanirhynchia* but lacking dorsal median ridge, deltidial plates conjunct, auriculate. *Plio.,* Medit. ——FIG. 509,*5*. *S. sicula* (SEGUENZA), Sicily; *5a*, brach.v. view, ×2; *5b,c*, lat., ant. views, ×1 (193).

Family ERYMNARIIDAE Cooper, 1959
[Erymnariidae COOPER, 1959, p. 64]

Shell with septifer crura (193). *Eoc.*

Erymnaria COOPER, 1959, p. 64 [*Terebratula polymorpha* MASSALONGO, 1850, p. 18; OD]. Trigonal to subpentagonal; uniplicate, some shells asymmetrical; smooth or irregularly costate anteriorly; beak short, foramen hypothyridid, deltidial plates conjunct. *Eoc.,* Medit.-Cuba.——FIG. 510,*1*. *E. polymorpha* (MASSALONGO), Italy; *1a,b*, brach.v., ant. views, ×2; *1c*, lat. view, ×1; *1d-g*, transv. secs. of beak region (ped.v. above), ×4 (193).

Superfamily STENOSCISMATACEA Oehlert, 1887 (1883)

[*nom. correct.* MUIR-WOOD, 1955, p. 69 (*pro* Stenoscismacea SHROCK & TWENHOFEL, 1953, p. 317; *nom. transl. et correct. ex* Stenoschismatinae OEHLERT, 1887, p. 1304)] [=Camerophoriacea WAAGEN, 1883 (*nom. transl.* GRABAU, 1936, p. 70, *ex* Camerophoriinae WAAGEN, 1883, p. 435)] [Materials for this superfamily prepared by RICHARD E. GRANT]

Camarophorium in brachial valve; also typically with spondylium in pedicle valve. *M.Dev.-U.Perm.*

Three morphological terms that pertain especially to stenoscismatacean shells need to be noted and explained. These are **camarophorium,** an elongate large trough-shaped structure located on a high median septum duplex in the brachial valve; **intercamarophorial plate,** a short, low median septum on the posterior mid-line of the camarophorium extending to the underside of the hinge plate but independent of the septum supporting the camarophorium; and **stolidium,** a thin marginal extension of one or both valves which forms a narrow to broad frill protruding at a distinct angle to the main contour of the shell.

Family ATRIBONIIDAE Grant, 1965
[Atriboniidae GRANT, 1965, p. 1]

Stolidium lacking, costae weak, fine or absent, spondylium typically sessile in apex, intercamarophorial plate present or absent. *M.Dev.-U.Perm.*

Subfamily ATRIBONIINAE Grant, 1965
[Atriboniinae GRANT, 1965, p. 29]

Intercamarophorial plate strong, extending anteriorly beyond undivided hinge plate. *M.Dev.-L.Perm.*

Atribonium GRANT, 1965, p. 37 [*A. simatum*; OD]. Small (average length of adults, 8-10 mm.); outline and profile subtrigonal; anterior surface flattened; commissure uniplicate; valve edges overlapping slightly along posterior slopes; costae low, rounded, beginning about mid-length on adults. Pedicle valve geniculate near anterior margin; beak nearly straight to suberect; delthyrium constricted by opposite beak and by pair of small conjunct or nearly conjunct deltidial plates; foramen slit-shaped; sulcus broad, shallow; interior with spondylium sessile near apex in most species, then elevated on low median septum duplex, extending from beak about 0.3 length of valve. Brachial valve more strongly convex than pedicle valve, anterior margin geniculate, fold commonly flat, standing above flanks only near anterior margin; interior with undivided hinge plate bearing low cardinal boss at apex; camarophorium short,

FIG. 511. Atriboniidae (Atriboniinae) (p. *H625-H627*).

relatively flat longitudinally and transversely, braced to underside of hinge plate by short inter-camarophorial plate. *M. Dev.(Onondag.)-L. Carb. (L.Miss.)*, USA-S.Can.-USSR(Urals).——FIG. 511, *3a-e*; 512,*2k-s*. **A. simatum*, M.Dev., USA(Mich.); 511,*3a-e*, brach.v., ped.v., lat., ant., brach.v. int. views, ×2 (365); 512,*2k-s*, ser. transv. secs., ×4 (orig. length, 11.5 mm., figures indicate distance from ped.v. beak in mm.; *c*, camarophorium; *cp*, cardinal process; *hp*, hinge plate; *icp*, inter-camarophorial plate; *sp*, spondylium); *2k* (0.6), *c* small swelling on septum, *icp* clearly duplex; *2l* (0.7), *c* wider; *2m* (0.8), *cp* higher; *2n* (1.0), *cp* at max. height; *2o* (1.2), *hp* detached; *2p* (1.5), *icp* thin, crural bases visible; *2q* (1.6), *icp* absent; *2r* (1.9), crura present; *2s* (2.5), crura absent (365).——FIG. 511,*3f-i*; 512,*2a-j*. *A. cooperorum* GRANT, M.Dev., USA(Mich.); 511, *3f-i*, brach. v., ped.v., lat., ant. views, ×2 (365); 512,*2a-j*, ser. transv. secs., ×4 (orig. length, 9.1 mm.; figures and abbrev. as above); *2a* (0.3), *sp* sessile; *2b* (0.5), *c* and *cp* visible, *sp* sessile; *2c* (0.8), *sp* elevated; *2d* (1.1), *hp* detached, *icp* separated; *2e* (1.4), dental plates detached; *2f* (1.8), *hp* small, *icp* absent; *2g* (2.1), *hp* absent; *2h* (2.5), *c* wide; *2i* (2.9), *c* wider; *2j* (4.2), septal remnant of *sp*, *c* high and narrow, reaching 4.5 mm. from beak (365).

Camerisma GRANT, 1965, p. 63 [**C. prava*; OD] [=*Laevicamera* GRABAU, 1936 *(nom. nud.)*] [*see Psilocamara*]. Length up to 20 mm.; outline oval or subpentagonal; strongly biconvex; shell walls thick; commissure strongly uniplicate; costae absent or weak, confined to anterior region; posterolateral valve edges strongly overlapping. Pedicle valve with beak thick, blunt, tightly curved against dorsal umbo, entirely closing delthyrium and foramen; sulcus shallow, with narrow median trough or slight flattening; interior with spondylium elevated on low median septum duplex. Brachial valve more strongly convex than pedicle valve; fold highly arched, crest bluntly or sharply ridged, symmetrical or skewed to one side; interior with large fimbriate cardinal boss at apex of hinge plate; strong camarophorium curving ventrally on high median septum duplex, braced to underside of hinge plate by thick duplex intercamarophorial plate. *Miss.*, USA(Alaska); *L.Perm.(Artinsk.)*, Yugosl.-USSR.——FIG. 511,2; 512,*3*. **C. prava*, Miss., Alaska; 511,*2a-d*, brach.v., ped.v., lat., ant. views, ×1 (365); 512,*3a-d*, ser. transv. secs., ×2.7 (orig. length, 15 mm., figures indicate distance from ped.v. beak in mm.; *c*, camarophorium; *cp*, cardinal process; *hp*, hinge plate; *icp*, inter-camarophorial plate; *sp*, spondylium); *3a* (2.0), *cp* low; *3b* (2.6); *3c* (2.7), *c*, *icp*, *sp* septa all duplex, with wedged insertion of septa into shell; *3d* (3.0), *hp* thin, crura visible, *c* slightly thickened at base of septum (365).

Sedenticellula COOPER, 1942, p. 231 [**Camarophoria hamburgensis* WELLER, 1910, p. 500; OD].

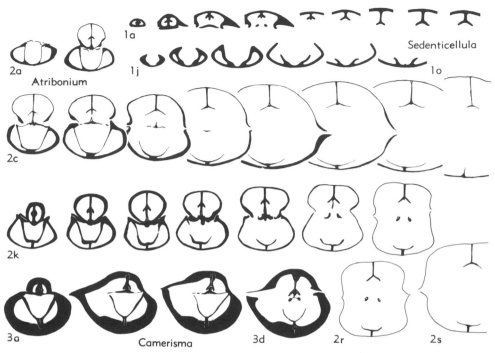

Fig. 512. Atriboniidae (Atriboniinae) (p. *H625-H627*).

Medium-sized (length of adult, approx. 12 mm.) flatly to rather strongly biconvex; outline elongate or transverse; commissure weakly uniplicate; costae low, narrow, beginning at beaks or in posterior third of shell, simple, bifurcating or intercalating; valve edges not overlapping. Pedicle valve with short beak, nearly straight to suberect; delthyrium small, constricted by dorsal beak; deltidial plates not observed; foramen small, open; sulcus shallow, broad; interior with spondylium sessile in posterior, elevated farther forward on low median septum. Brachial valve more convex than pedicle valve; fold low, crest flattened; interior with shallow camarophorium; intercamarophorial plate low, thick. *L.Miss.,* USA(Ill.-Tex.).——Fig. 511,*1a-c*; 512,*1.* **S. hamburgensis* (WELLER), Ill.; 511,*1a-c,* brach.v., ped.v., ant. views, ×2 (365); 512,*1a-o,* ser. transv. secs., *1a-i,* brach.v.; *1j-o,* ped.v. showing sessile spondylium; all ×2.5 (858).——Fig. 511,*1d-h. S. sacra* GRANT, Tex.; *1d-h,* brach.v., ped.v., lat., ant., post. views, ×2 (365).

Subfamily PSILOCAMARINAE Grant, 1965

[Psilocamarinae GRANT, 1965, p. 29]

Intercamarophorial plate absent or rudimentary, hinge plate divided or short. *U. Carb.-U.Perm.*

Psilocamara COOPER, 1956, p. 523 [**P. renfroarum*; OD] [=*Levicamera* GRABAU, 1934 *(nom. nud.)*]

[*see Camerisma*]. Small (average length of adults, 5-7 mm.), smooth to weakly costate; outline subpentagonal; commissure strongly uniplicate; posterolateral valve edges with little or no overlap. Pedicle valve with short beak, straight to slightly incurved; sulcus shallow, some shells with weak median grove; delythrium small, constricted by dorsal beak and pair of small disjunct deltidial plates; foramen slit-shaped; interior with deep spondylium on low median septum duplex. Brachial valve with high fold sloping smoothly to flanks, crest bluntly ridged; interior with short undivided hinge plate; camarophorium gently curved ventrally; intercamarophorial plate absent. *M.Penn.-L.Perm.,* USA(Tex.); ?*L.Perm.,* China (Nantan-Yunnan).——Fig. 513,*1*; 514,*1.* **P. renfroarum,* M.Penn., Tex.; 513,*1a-e,* brach.v., ped.v., lat., ant., post. views, ×2 (188); 514,*1a-g,* ser. transv. secs. (figures indicate distance from ped.v. beak in mm.), *1a* (?); *1b* (0.3); *1c* (0.65); *1d* (0.77); *1e* (1.0); *1f* (1.2); *1g,* (1.0); *1a, 1b-f, 1g,* different specimens, intercamarophorial plate absent in *1a, 1d, 1g; 1a,* ×3; *1b-f,* ×2.7; *1g,* ×3.3 (188, 365).

Camarophorina LIKHAREV, 1934, p. 211 [**Camarophoria antisella* BROILI, 1916, p. 58; OD]. Small (length, approx. 8 mm.) flatly to rather strongly biconvex; commissure strongly sulcate; costae absent or very weak; valve edges not overlapping. Pedicle valve inflated in umbonal region, profile

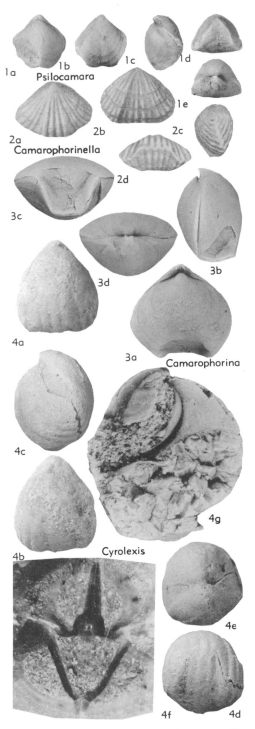

1a 1b 1c 1d
Psilocamara

2a 2b 2c
Camarophorinella

2d

3c 3b 3d

4a 3a
Camarophorina

4c

4g

4b **Cyrolexis**

4e

4f 4d

FIG. 513. Atriboniidae (Psilocamarinae) (p. *H627-H629*).

flat along crest of fold; beak short, sharp, suberect; delthyrium small, nearly filled by dorsal beak; deltidial plates unknown; fold standing above flanks about 0.7 of distance in front of beak; interior with spondylium elevated on high median septum duplex. Brachial valve with sulcus distinctly depressed only near anterior margin; interior with short undivided hinge plate bearing cardinal boss near apex; camarophorium thick near apex, edges meeting hinge plate, curving ventrally on high septum duplex; intercamarophorial plate absent. *L.Carb.*, Eng.; *U.Perm.*, Timor.
——FIG. 513,*3*; 514,*3*. *C. antisella* (BROILI), *U. Perm.*, Timor; 513,*3a-d*, brach.v., lat., ant., post. views, ×2 (365); 514,*3a-g*, ser. transv. secs., ×3.3 (orig. length, 11 mm., figures indicate distance from ped.v. beak in mm.; *c,* camarophorium; *cp,* cardinal process; *hp,* hinge plate; *icp,* intercamarophorial plate; *sp,* spondylium); *3a* (0.8), *c* not shown but visible through clear shell, *sp* elevated; *3b* (1.2), *c* thick, without *icp, cp* and *hp* visible; *3c* (1.6), *icp* absent; *3d* (1.8), *sp,* septum and plates very thin; *3e* (2.6), *sp* absent, *c* with duplex septum; *3f* (3.0), *c* trough deep; *3g* (3.4), *c* trough high, pushed onto septum, *c* ending 3.9 mm. from beak (365).

Camarophorinella LIKHAREV, 1936, p. 63 [*C. caucasica*; OD]. Transversely subpentagonal, holocostate, with costae simple, bifurcating or intercalating; commissure uniplicate; valve edges not overlapping. Pedicle valve with shallow sulcus distinct from flanks; beak short; delthyrium, deltidial plates and foramen unknown; interior with deep spondylium; median septum low and thick near apex, thinner and higher toward anterior margin. Brachial valve slightly more convex than pedicle valve; fold low, flat-crested; interior with hinge plate divided as far back as low cardinal boss at apex of valve; camarophorium high, curving ventrally, sides attached to underside of hinge plate; intercamarophorial plate absent. *U.Perm.*, USSR(N.Caucasus).——FIG. 513,*2*; 514,*2*. *C. caucasica*; 513,*2a-d*, brach.v., ped.v., lat., ant. views, ×1 (518); 514,*2a-e*, ser. transv. secs., ×2 (*c,* camarophorium; *cp,* cardinal process; *hp,* hinge plate; *sp,* spondylium); *2a, cp* visible; *2b, hp* divided, joined to *c* edges; *2c, 2d, c* and *sp* detached from shell proximally, *hp* reduced; *2e, c* and *sp* thin (518).

Cyrolexis GRANT, 1965, p. 88 [*C. haquei*; OD]. Globular (length, 12-14 mm.); commissure uniplicate; costae low, rounded, beginning about midlength; valve edges abutting at anterior margin, overlapping broadly on posterior slopes. Pedicle valve with inflated umbonal region; beak short, incurved against dorsal umbo, closing delthyrium; sulcus shallow, beginning far forward; interior with spondylium sessile in posterior part, anteriorly elevated on low median septum duplex. Brachial valve with fold low, crest flat, elevated only in anterior third of shell; interior with undivided hinge plate; camarophorium curved strongly ventrally, posterior edges touching hinge plate; intercamarophorial plate absent or rudimentary. *L. Perm. - U. Perm.,* USSR (Urals-E. Sib.)-Pak. (Salt Range.).——FIG. 513,4; 514,4. **C. haquei,* L. Perm., Pak.; 513,4a-e, brach.v., ped.v., lat., ant., post. views, ×2; 513,4f, transv. sec. 0.5 mm. from ped.v. beak, ×6; 513,4g, profile along mid-line, ×4; 514,4a-j, ser. transv. secs., ×2.7 (orig. length, 12.9 mm., figures indicate distance from ped.v. beak in mm.; *c,* camarophorium; *cp,* cardinal process; *hp,* hinge plate; *icp,* intercamarophorial plate); *4a* (0.9), *icp* absent; *4b* (1.1), *c* edges meeting *hp* with thick filling between, *cp* high; *4c* (1.3), *icp* marked by small dot of shell; *4d* (1.6), *c* separating from crura; *4e* (1.8), *c* separate from crura; *4f* (1.9), crura thinner, divergent; *4g* (2.2), crura near edges of *sp,* separated widely; *4h* (2.4), wide overlap of valves, crura absent; *4i* (2.8), *sp* nearly absent; *4j* (3.5), *c* septum separated from valve floor, wide overlap of valve edges, *c* disappearing at 4.5 mm. (all 365).

Family STENOSCISMATIDAE Oehlert, 1887 (1883)

[*nom. transl. et correct.* MUIR-WOOD, 1955, p. 91 (*ex* Stenoschismatinae OEHLERT, 1887, p. 1304)] [=Camerophoriidae WAAGEN, 1883 (*nom. transl.* GRABAU, 1936, p. 70) (*ex* Camerophoriinae WAAGEN, 1883, p. 435)]

Outline rhynchonelliform or uncinuliform, stolidium incipient, well developed, or degenerate, camarophorium and spondylium large, intercamarophorial plate present; early representatives small and weakly costate, late forms typically large, strongly or completely costate. *M.Dev.-U.Perm.*

Subfamily STENOSCISMATINAE Oehlert, 1887 (1883)

[*nom. correct.* MUIR-WOOD, 1955, p. 91 (*pro* Stenoschismatinae OEHLERT, 1887, p. 1304)] [=Camerophoriinae WAAGEN, 1883, p. 435]

Rhynchonelliform, with incipient or well-developed stolidium, costae beginning in front of beaks. *M.Dev.-U.Perm.*

Stenoscisma CONRAD, 1839, p. 59 [*non Stenocisma* HALL, 1847, 1867; *nec Stenoschisma* HALL & CLARKE, 1894; *Stenochisma* GRABAU & SHIMER, 1907] [**Terebratula schlotheimi* VON BUCH, 1835; OD] [=*Camerophoria* KING, 1844 (*nom. nud.*); *Camerophoria* KING, Aug., 1846, p. 89 (obj.); *Camarophoria* HERRMANNSEN, Dec., 1846, p. 161 (*nom. van.*); *Stenoschisma* OEHLERT, 1887, p. 1309 (*nom. van.*)]. Small to large (length to 35 mm.) subtrigonal or pentagonal; strongly uniplicate; costae on fold, flanks or both, beginning near beaks or far in front of them, rounded or sharp; broad stolidium around anterior margins of adults; posterolateral edges of pedicle valve flattened, strongly overlapped by edges of brachial valve.

Pedicle valve beak long for genus, nearly straight to tightly incurved; deltidial plates conjunct or disjunct; foramen oval, open or completely closed; sulcus distinctly depressed; interior with large

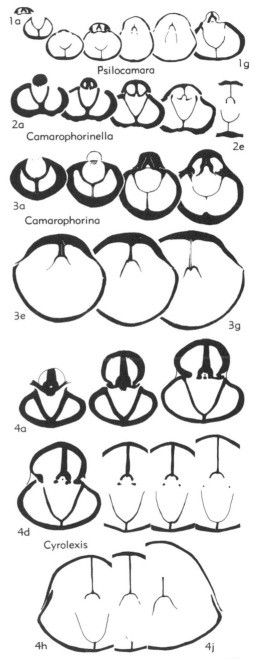

Psilocamara

Camarophorinella

Camarophorina

Cyrolexis

FIG. 514. Atriboniidae (Psilocamarinae) (p. *H627-H629*).

spondylium elevated on low median septum duplex; weak muscle marks in spondylium; adductors narrow, median; diductors large, surrounding adductors; adjustors undifferentiated in apical part of spondylium; gonocoel troughs deep, transverse, one on each side beginning at anterior edge of median septum; mantle canals beginning as mesial pair near origin of gonocoels, bifurcating toward margins, extending onto stolidium. Brachial valve with high, distinct fold; interior with broad undivided hinge plate; large, low cardinal boss at apex, finely fimbriate for diductor muscle

FIG. 515. Stenoscismatidae (Stenoscismatinae) *(2, 4)*, (Torynechinae) *(1, 3)* (p. *H629-H632*).

attachment; crura extending from hinge plate, bowed distally and ventrally; camarophorium long, deep, ventrally curved on high median septum duplex, intercamarophorial plate strong; muscle marks weak on concave surface of camarophorium; anterior adductors small, paired along mid-line; posterior adductors larger, lateral and posterior; mantle canals as in opposite valve. *L. Carb.,* Eu.; *U.Perm.,* cosmop.——FIG. 515,*4a-e*; 516,*1.* **S. schlotheimi* (VON BUCH), U.Perm., Ger.; 515,*4a-e*, brach.v., lat., ant., post., lat. int. views, ×2 (365); 516,*1a-i,* ser. transv. secs., ×2.5 (858).——FIG. 515,*4f-i. S. venustum* (GIRTY), L.Perm., USA(Tex.); *4f-h,* brach.v., ped.v., lat. int. showing relationships of camarophorium, spondylium, and crura, ×1.5; *4i,* brach.v., int., ×2 (all 365).

Coledium GRANT, 1965, p. 95 [**C. erugatum*; OD]. Small, rarely large (average length about 10 mm.); commissure uniplicate; costae few, rounded, weak or absent, beginning far forward; valve edges overlapping at posterior margin; stolidium narrow and sporadic or absent. Pedicle valve with beak slightly attenuate, suberect to incurved; deltidial plates small and disjunct or absent; foramen small, rarely closed by incurvature of beak; sulcus shallow, beginning far forward; interior with spondylium on low median septum duplex, rarely sessile in posterior region. Brachial valve with fold not sharply raised above flanks; interior with hinge plate and camarophorium as in *Stenoscisma;* intercamarophorial plate short. *M.Dev.-Penn.,* USA; *Perm.,* Timor.——FIG. 515,*2a-d*; 516,*2.* **C. erugatum,* U.Miss., Okla.; 515,*2a-d,* brach.v., ped.v., lat., ant. views, ×2; 516,*2a-k,* ser. transv. secs., ×4 (orig. length, 11.4 mm.; figures indicate distance from ped.v. beak in mm.; *c,* camarophorium; *cp,* cardinal process; *hp,* hinge plate; *icp,* intercamarophorial plate; *sp,* spondylium); *2a* (1.2), *cp* hardly discernible; *2b* (1.5), *cp* and *hp* present; *2c* (2.1), *cp* large, fimbriate; *2d* (2.6), no *cp*; *2e* (2.9); *2f* (3.0), *hp* reduced, *sp* low, narrow; *2g* (3.4), *icp* low, *sp* absent, septum low; *2h* (4.0), *icp* nearly absent; *2i* (4.2); *2j* (4.7), *c* high, wide, flexed; septum thin; *2k* (5.5), *c* high, strongly flexed transversely, septum detached from valve floor (all 365).——FIG. 515, *2e-i. C. evexum* GRANT, L.Miss., Tex.; *2e-i,* brach.v., ped.v., lat., ant., post. views, ×2 (365).

Subfamily TORYNECHINAE Grant, 1965

[Torynechinae GRANT, 1965, p. 31]

Uncinuliform, with stolidium greatly reduced (less commonly absent), costae beginning at beaks. ?*U.Carb., L.Perm.*

Torynechus COOPER & GRANT, 1962, p. 1128 [**T. caelatus*; OD]. Rounded subtrigonal in outline and profile, anterior surface flattened; length of adult about 18 mm.; commissure uniplicate; costae fine, sharp, numerous, beginning at beaks, intercalating and bifurcating; posterolateral valve edges over-

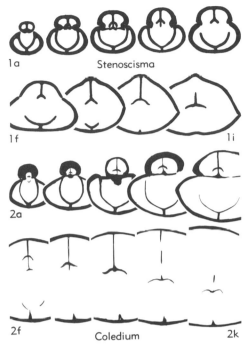

FIG. 516. Stenoscismatidae (Stenoscismatinae) (p. H629-H631).

lapping slightly; anterior margins protruding slightly, indicating incipient or decadent stolidium. Pedicle valve sharply geniculate near anterior margin; beak elongate, attenuate, suberect to erect; deltidial plates small; foramen small, oval; sulcus depressed below flanks only near front margin; interior with spondylium on low median septum duplex; muscle marks weak in spondylium; transverse gonocoel troughs shallow; pattern of mantle canals as in *Stenoscisma.* Brachial valve sharply geniculate in anterior region; fold low, broad, flatcrested; interior as in *Stenoscisma;* intercamarophorial plate short, thick, may be buried in callus; muscle marks and mantle canals as in *Stenocisma. L.Perm.(Leonard.);* USA(Tex.).——FIG. 515,*1.* **T. caelatus; 1a-e,* brach.v., lat., ant., lat. int., int. profile views, ×1 (365).

[The spondylium-bearing rhynchonellid brachiopods found in Leonard beds of the Glass Mountains, western Texas, which now are assigned to the genus *Torynechus,* were first described and figured by R. E. KING (1931, p. 112, pl. 35, fig. 6-7), who introduced for them the new genus *Uncinuloides* and designated as its type-species a form named *Rhynchonella Guadalupae* SHUMARD (1858, p. 295). Although the type of SHUMARD's species has been lost, other specimens collected in the Guadalupe Mountains from the Capitan Limestone, the source of *R. guadalupae,* possess *Wellerella*-like dental plates instead of a spondylium in the pedicle valve. Clearly, they are not the same as shells which KING named *Uncinuloides guadalupensis* (SHUMARD), one of which was selected by COOPER & GRANT as the holotype of *Torynechus caelatus.* The specific name *guadalupensis* is not the same as *guadalupae,* and thus argument might be advanced that the type-species of *Torynechus* actually is *Uncinuloides guadalupensis* KING (*non guada-*

FIG. 517. Rhynchoporidae (p. *H632*).

lupae SHUMARD) (=*Torynechus caelatus* COOPER & GRANT); this can be countered effectively by pointing out that KING's usage of *U. guadalupensis* makes evident that *guadalupensis* is a subsequent spelling with status of an "unjustified emendation" which ranks as a junior objective synonym of *guadalupae* (1961 Code, Art. 33,a,ii). *Uncinuloides* then is tied to a misidentified type-species (Art. 70) calling for adjudication by ICZN; instead, COOPER & GRANT have decided to settle the issue themselves by adopting the disposition given in Art. 70,a,iii, which is to accept the species named by the designator, regardless of the misidentification. Challenge is unlikely and accordingly *Torynechus*, with type-species *T. caelatus*, is here recognized as valid and *Uncinuloides* is left for needed further research.]

Septacamera STEPANOV, 1937, p. 146 [*Camarophoria kutorgae* CHERNYSHEV, 1902, p. 90] [=*Septocamera* LIKHAREV, 1960, p. 249 *(nom. null.)*]. Large (length about 25 mm.), profile rounded subtrigonal, with flattened anterior surface; commissure strongly uniplicate; costae strong, simple, beginning at beaks; valve edges apparently not overlapping. Pedicle valve geniculate near anterior margin; beak short, suberect; foramen open; sulcus shallow, distinct from flanks; interior with posterior part of spondylium sessile, elevated farther forward on relatively high median septum,

possibly also braced by pair of short lateral buttress plates. Brachial valve with distinct fold standing above flanks only near anterior edge; interior with large camarophorium on high septum, strongly curved ventrally; presence of intercamarophorial plate uncertain. *?U.Carb., L.Perm.(Sakmar.),* USSR (Urals - Timan); *L. Perm.(Word equiv.),* USA(Ore.)-Can.(Arctic).——FIG. 515,3. *S. kutorgae* (CHERNYSHEV), L.Perm., Urals; *3a-d,* brach. v., ped.v., lat., ant. views, ×1 (518).

Superfamily RHYNCHOPORACEA Muir-Wood, 1955

[Rhynchoporacea MUIR-WOOD, 1955, p. 91 (erroneously attributed to MOORE, 1952, who classed Rhynchoporacea as suborder)] [Materials for this superfamily prepared by D. J. McLAREN]

Shell punctate, lacking spondylium or camarophorium. *Miss.-Perm.*

Family RHYNCHOPORIDAE Muir-Wood, 1955

[Rhynchoporidae MUIR-WOOD, 1955, p. 91]

Characters of superfamily. *Miss.-Perm.*

Rhynchopora KING, 1865, p. 124 [*Terebratula Geinitziana* DE VERNEUIL, 1845, p. 83; OD] [=*Rhynchoporina* OEHLERT, 1887, p. 1305 (obj.)]. Subtriangular to subpentagonal; costate; uniplicate; fold on brachial and sulcus on pedicle valve; tongue high; shell flattened anteriorly. Dental plates present; hinge plate entire, supported posteriorly by septum and septalium; crura directed anteriorly. *L.Carb.-Perm.,* Eu.-Asia-N.Am.-S. Am.——FIG. 517,1a,b. *R. geinitziana* (DE VERNEUIL), U. Perm., USSR(Russ.platform); *1a,b,* brach.v., lat. views, ×1 (841).——FIG. 517,1c-f. *R. triznae* SOKOLSKAYA, L.Carb.(Tournais.), USSR (Kuznetsk basin); *1c-e,* ped.v., lat., ant. views, ×1; *1f,* shell microstructure, ×25 (711a).

SPIRIFERIDA

By A. J. BOUCOT,[1] J. G. JOHNSON,[1] CHARLES W. PITRAT,[2] and R. D. STATON[3]

[[1]California Institute of Technology, [2]University of Massachusetts, and [3]Museum of Comparative Zoölogy at Harvard College]

Order SPIRIFERIDA Waagen, 1883

[*nom. correct.* MOORE in MOORE, LALICKER, & FISCHER, 1952, p. 221 (*pro* order Spiriferacea KUHN, 1949, p. 104, *nom. transl. ex* suborder Spiriferacea WAAGEN, 1883, p. 447)] [*emend.* BOUCOT, JOHNSON, & PITRAT, herein] [Diagnosis prepared by A. J. BOUCOT, J. G. JOHNSON, & R. D. STATON]

Articulate brachiopods with spiral brachidium (except Leptocoeliidae); jugum present or absent. Shell punctate or impunctate, lacking pseudopunctae; mostly biconvex, rarely plano-convex, with relatively large body cavity; cicatrix of attachment uncom-

mon; delthyrium open or closed, circular foramen present or absent. *M.Ord.-Jur.*

Suborder ATRYPIDINA Moore, 1952

[*nom. correct.* BOUCOT, JOHNSON, & STATON, herein (*pro* suborder Atrypacea MOORE, 1952, p. 221)] [=suborder Atrypoidea MUIR-WOOD, 1955, p. 91] [Materials for this suborder prepared by A. J. BOUCOT, J. G. JOHNSON, and R. D. STATON]

Impunctate, mostly biconvex spire-bearing brachiopods, commonly with narrow cardinal margin. Interarea low, obsolescent, or lacking. Pedicle-valve beak may be trun-

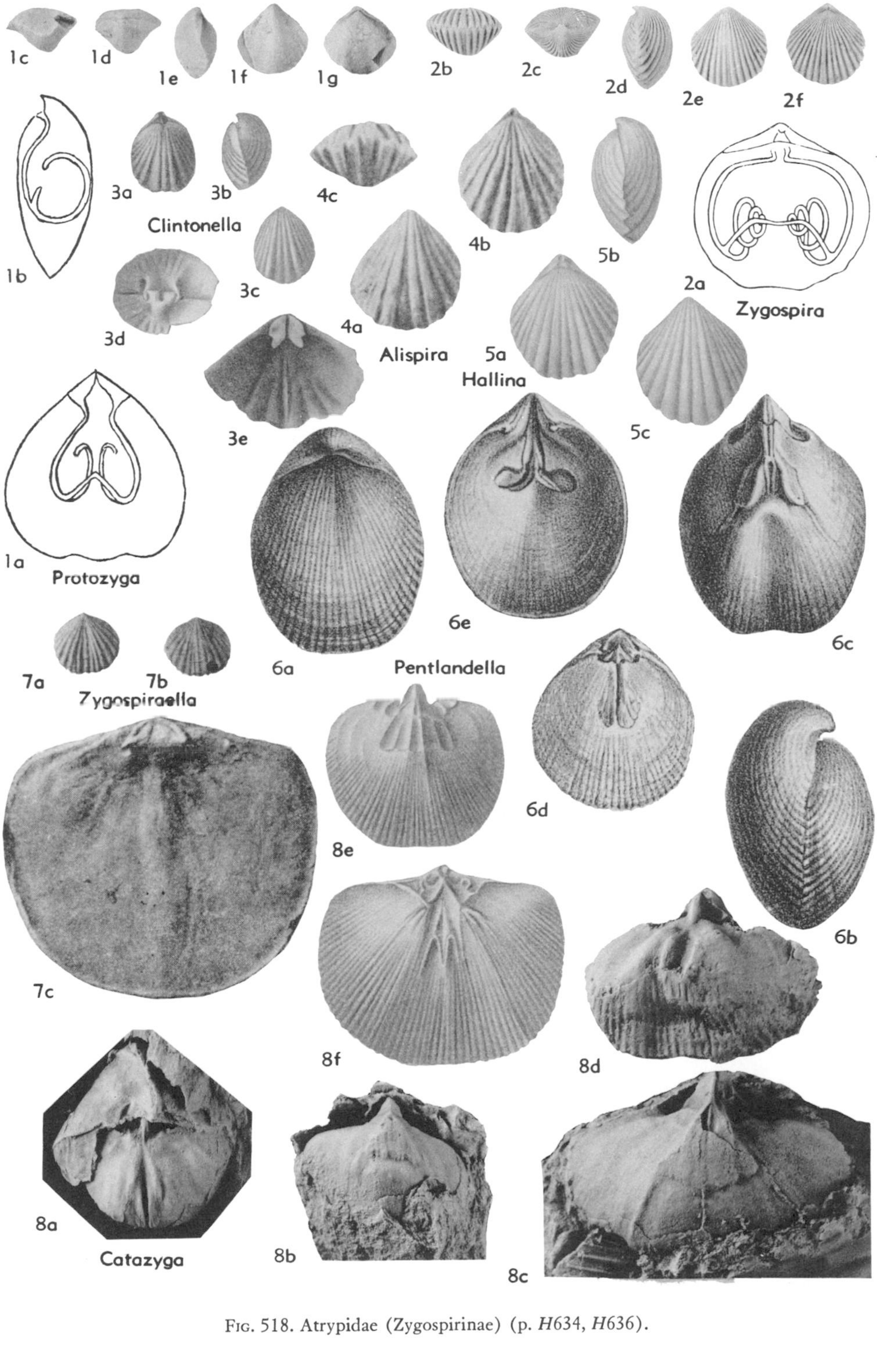

FIG. 518. Atrypidae (Zygospirinae) (p. *H634, H636*).

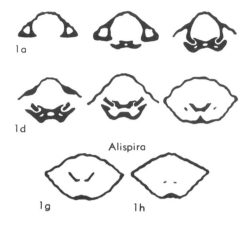

FIG. 519. Atrypidae (Zygospirinae) (p. H634).

cated by foramen or foramen may reside within delthyrium. Spiralia directed medially, dorsomedially, or laterally; crura projecting to join primary lamellae of spiralia at posteromedial position, or deflected laterally more or less parallel to plane of commissure to join primary lamellae in posterolateral position; simple jugum commonly present connecting primary lamellae. *M.Ord.-U. Dev.*

Superfamily ATRYPACEA Gill, 1871

[*nom. transl.* SCHUCHERT & LEVENE, 1929, p. 19 (*ex* Atrypidae GILL, 1871, p. 25)]

Spiralia directed medially or dorsomedially. *M.Ord.-U.Dev.*

Family ATRYPIDAE Gill, 1871

[Atrypidae GILL, 1871, p. 25]

Shell plicate or costate. *M.Ord.-U.Dev.*

Subfamily ZYGOSPIRINAE Waagen, 1883

[Zygospirinae WAAGEN, 1883, p. 449] [=Clintonellinae POULSEN, 1943, p. 40]

Primitive small, biconvex forms, costate or multiplicate, growth lines rarely prominent. Deltidial plates conjunct. Spiralia directed medially or dorsomedially; simple jugum present, situated posteriorly or anteriorly. *M.Ord.-L.Sil.*

Zygospira HALL, 1862, p. 154 [*Atrypa modesta* SAY in HALL, 1847, p. 141; OD] [=*Anazyga* DAVIDSON, 1882, p. 128 (type, *Atrypa recurvirostra* HALL, 1847, p. 140)]. Unequally biconvex, elongate or transverse shells, pedicle valve more convex, commonly with ventral fold and dorsal sul-

cus, simple plications; foramen mesothyridid, deltidial plates conjunct, beak ridges strong and well defined. Dental plates lacking; hinge plates disjunct, parallel medially, diverging ventrally, and supported by myophragm; spiralia directed submedially with dorsal inclination; jugum a simple band curving toward middle of valve, origin of jugum variable, anterior or posterior. *M.Ord.-U. Ord., ?L.Sil.,* Eu.(G.Brit.)-N.Am.——FIG. 518, *2a.* **Z. modesta* (SAY), U.Ord., USA(Ohio); brach.v. int. showing brachidium (diagram.), ×3 (396).——FIG. 518,*2b-f. Z. circularis* COOPER, M. Ord.(Carters F.), USA(Tenn.); *2b-f,* ant., post., lat., ped.v., brach.v. views, ×2 (189).

Alispira NIKIFOROVA, 1961, p. 243 [**A. gracilis;* OD]. Inequally biconvex, elongate, costate shells, pedicle valve more convex; costae increasing in number anteriorly by bifurcation and implantation, crossed by fine, closely spaced growth lines. Dental plates present; conjunct inner hinge plates supported by low crural plates; median septum lacking; spiralia dorsomedially directed; jugum situated posteriorly. *L.Sil.,* Asia.——FIG. 518,*4;* 519,*1.* **A. gracilis;* 518,*4a-c,* ped.v., brach.v., ant. views, ×2; 519,*1a-h,* serial secs., ×5 (602).

Catazyga HALL & CLARKE, 1893, p. 157 [*Athyris headi BILLINGS, 1862, p. 147; OD] [=*Orthonomaea* HALL, 1893, p. 159 (type, *Orthis? erratica* HALL, 1847, p. 288)]. Inequally biconvex shells, pedicle valve slightly more convex and valves slightly bisulcate; surface finely costellate. Dental plates obsolete; adductor platform developed in pedicle valve; hinge plates essentially as in *Zygospira;* myophragm present; spiralia directed medially; jugum U-shaped, arising well posterior. *M.Ord.-L.Sil.(low. Llandovery),* Eu.(G.Brit.)-N.Am.——FIG. 518,*8a-d. C.* sp., Ashgill, USA (Maine); *8a,b,* post. view, int. mold, ped. int. mold, ×2; *8c,* ped. int. mold, ×3; *8d,* brach. int. mold, ×4 (Boucot, Johnson, & Staton, n).——FIG. 518,*8e,f. C. erratica* (HALL), U.Ord., USA(N.Y.), type-sp. of *Orthonomaea; 8e,f,* ped.v. and brach.v. int. molds, ×2 (396).

Clintonella HALL & CLARKE, 1893, p. 159 [**C. vagabunda;* OD]. Subequally biconvex shells with simple plications and dorsal fold and ventral sulcus; beak slightly incurved; growth lines imbricate. Short dental plates present; ventral diductor muscle field flabellate, enclosing pair of cordate adductors; hinge plates disjunct, subparallel medially, divided into posterior and anterior lobes, and supported by stout myophragm; spiralia reported to be present (HALL & CLARKE, 1893, p. 160) but their disposition and nature of jugum unknown. *L.Sil.(Clinton),* N.Am.——FIG. 518,*3.* **C. vagabunda,* USA(N.Y.); *3a-c,* brach.v., lat., ped.v. views, ×1; *3d,e,* post. int. (both valves), brach.v. int. showing cardinalia, ×3 (396).

Hallina WINCHELL & SCHUCHERT, 1892, p. 291 [**H. saffordi;* OD]. Externally like *Zygospira,*

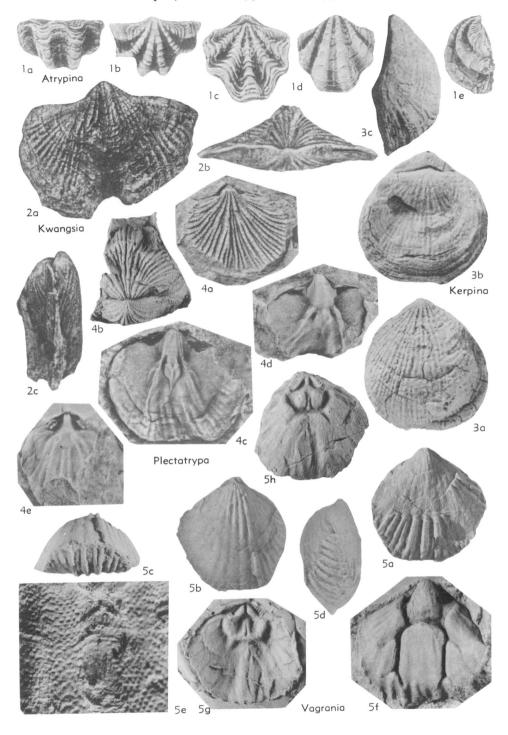

FIG. 520. Atrypidae (Atrypininae) *(1)*, (Carinatininae) *(2-5)* (p. *H. 636-H637*).

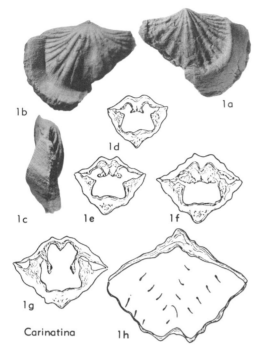

FIG. 521. Atrypidae (Carinatininae) (p. *H636-H637*).

internally like *Protozyga* but jugum U-shaped in *Hallina*. *M.Ord.,* N.Am.——FIG. 518,5. **H. saffordi,* Trenton., USA(Tenn.); *5a-c,* brach.v., lat., ped.v. views, ×5 (396).

Pentlandella BOUCOT, 1964, p. 104 [**Rhynchonella pentlandica* HASWELL, 1865, p. 31; OD]. Small suboval costate shells with pedicle valve more convex, costae bifurcated, growth lines subdued; anterior commissure rectimarginate; beak incurved. Dental plates lacking; pedicle valve diductor scars widely divergent anteriorly, impressed into thickened secondary shell material; brachial valve with median septum and septalium. *Sil.* *(late upper Llandover., ?Wenlock.),* Eu.(G.Brit.-Est.).——FIG. 518,6. **P. pentlandica* (HASWELL), Wales; *6a-e,* brach.v., lat., ped.v. int. mold, brach. v. int. mold, ped.v. int., ×3 (229).

Protozyga HALL & CLARKE, 1893, p. 151 [**Atrypa exigua* HALL, 1847, p. 141; OD]. Unequally biconvex shells, pedicle valve more convex, with beak slightly incurved; brachial valve sulcate; shells may be paucicostate marginally. Dental plates present; hinge plates disjunct, divergent; dorsal myophragm may be present; spiralia slightly submedially directed, making about one volution; jugum simple, short, anteriorly situated. *M. Ord.,* Eu.(G.Brit.)-N.Am.——FIG. 518,1. **P. exigua* (HALL), Trenton., USA(N.Y.); *1a,b,* brach. and lat. views of brachidium (diagram.), ×4

(396); *1c-g,* ant., post., lat., ped.v., brach.v. views, ×2 (189).

Zygospiraella NIKIFOROVA, 1961, p. 237 [**Terebratula duboisi* DE VERNEUIL in MURCHISON, 1845, p. 97; OD]. Unequally biconvex or plano-convex subcircular shells, pedicle valve more convex; anterior commissure rectimarginate or with faint ventral fold and dorsal sulcus; surface covered by bifurcating costae and more or less prominent growth lines. Short dental plates may be present; hinge plates discrete, bearing crural lobes; dorsal myophragm may be present; spiralia directed dorsomedially; jugum unknown. *L.Sil.,* N.Am.-Asia. ——FIG. 518,7. **Z. duboisi* (DE VERNEUIL), *7a-c,* brach.v., ped.v., ×1, brach.v. int. view, ×3 (602).

Subfamily ATRYPININAE McEwan, 1939

[*nom. transl.* BOUCOT, JOHNSON, & STATON, 1964, p. 808 (*ex* Atrypinidae McEWAN, 1939, p. 619]

Pauciplicate, plano-convex, with lamellose growth lines. *L.Sil.-L.Dev.*

Atrypina HALL & CLARKE, 1893, p. 161 [**Leptocoelia imbricata* HALL, 1857, p. 108; OD]. Inequally biconvex or plano-convex shells, pedicle valve more convex, beak slightly incurved; anterior commissure rectimarginate or deflected ventrally; pauciplicate, plications low and rounded, growth lines lamellose. Dental plates lacking; hinge plates forming bilobed cardinal process that rests on thick myophragm; spiralia directed dorsomedially; jugum posterior, V-shaped, pointing anteriorly. *L.Sil.(U.Llandovery)-L.Dev.,* N.Am.-S. Am.(Venez.)-Eu.(Eng.-Podolia-Boh.-Urals). —— FIG. 520,1. *A. hami* AMSDEN, L.Dev.(Haragan), USA(Okla.); *1a-e,* ant., post., brach.v., ped.v., lat. views, ×3 (33).

Subfamily CARINATININAE Rzhonsnitskaya, 1960

[Carinatininae RZHONSNITSKAYA, 1960, p. 261]

Costate biconvex or plano-convex, with conjunct deltidial plates. *U.Ord.-M.Dev.*

Carinatina NALIVKIN, 1930, p. 104 [**Orthis ari-*

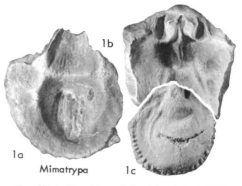

FIG. 521A. Atrypidae (Carinatininae) (p. *H637*).

maspus EICHWALD, 1840, p. 108; OD]. Subequally biconvex, plicate shells with straight hinge line, brachial valve commonly more convex, low ventral fold and dorsal sulcus present; plications commonly irregular and coarse, bifurcating somewhat rarely, but costellate frill may be present; ventral beak straight; deltidial plates conjunct. Dental plates present; hinge plates disjunct; brachidium essentially as in *Atrypa. L.Dev.-M.Dev.,* Eu.-Asia-N.Am.——FIG. 521,*1a-c.* *C. arimaspa* (EICHWALD), U.Ems.-Couvin.(Vagran F.), Ural Mts.; *1a-c,* ped.v., brach.v., side view, ×1 (8).——FIG. 521,*1d-h.* *C. minuta* SIEHL, M.Dev.(Greifensteiner Kalk), Ger.; *1d-h,* serial secs., ×4.5 (744).

Kerpina STRUVE, 1961, p. 333 [*K. vineta vineta*; OD]. Resembles *Atrypa,* but with reversed convexity; pedicle valve convex, with nearly straight beak; brachial valve slightly convex or plane; fold and sulcus lacking; broad conjunct deltidial plates fill area between beak ridges. *M.Dev.(U. Eifel.),* Eu.——FIG. 520,*3.* *K. vineta goniorhyncha* STRUVE, Ger.; *3a-c,* ped.v., brach.v., lat. views, ×2 (788).

Kwangsia GRABAU, 1931, p. 204 [*K. yohi*; OD] [=*Kwangsiella* GRABAU, 1932, p. 54 (obj.)]. Subequally biconvex costate shells with long hinge line; costae increasing anteriorly by bifurcation and crossed by prominent growth lines; well-developed dorsal fold and ventral sulcus present. Interior unknown. *M.Dev.,* China.——FIG. 520,*2.* *K. yohi*; *2a-c,* ped.v., post., lat. views, ×1 (358).

Minatrypa STRUVE, 1964, p. 436 [*Terebratula prisca* var. *flabellata* ROEMER, 1844, p. 66; OD]. Coarsely costate shells of variable shape, subequally biconvex to convexi-plane, concentric growth lines faint or lacking; pedicle valve beak straight or nearly straight with apical foramen posterior to well-developed, conjunct deltidial plates. Dental plates lacking on large specimens, ventral muscle impressions raised anteriorly on transverse platform; brachial valve with ponderous hinge plates and crural lobes flanked by deep sockets, corrugated laterally and commonly lacking longitudinal ridge; adductor scars deeply impressed or defined by muscle-bounding ridges; spiralia and jugum unknown. *M.Dev.,* Eu., N.Am. (Nev.).——FIG. 521A,*1.* *M. flabellata* (ROEMER), Ger.; *1a-c,* ped.v. int., brach.v. int., ped.v. int. mold, ×1.3 (*1a,b,* 932a; *1c,* Boucot, Johnson, & Staton, n).

Nalivkinia BUBLICHENKO, 1928, p. 982 [*Atrypa grünwaldtiaeformis* VON PEETZ, 1901, p. 147; OD]. Elongate, subequally, biconvex, costate shells, costae only rarely increasing in number anteriorly by bifurcation or implantation, fold and sulcus lacking, but anterior commissure commonly deflected dorsally. Dental plates present; hinge plates discrete; sockets noncrenulate; spiralia dorsomedially directed, jugum simple, short, situated posteriorly. *Sil., ?L.Dev.,* Eu.(USSR).——FIG. 524, *1.* *N. gruenwaldtiaeformis* (VON PEETZ); *1a-c,* ped. v., brach.v., ant. views, ×1 (125).

Plectatrypa SCHUCHERT & COOPER, 1930, p. 278 [*Terebratula imbricata* SOWERBY in MURCHISON, 1839, p. 624; OD]. Subequally biconvex, costate shells with dorsal fold and ventral sulcus; growth lines prominent or subdued, costae bifurcating anteriorly; beak incurved, deltidial plates conjunct. Short dental plates present or obsolescent; pedicle valve musculature confined, nonflabellate; hinge plates discrete; median septum lacking; spiralia directed dorsomedially; jugum short, simple, situated posteriorly. *U.Ord.-L.Dev.,* cosmop.——FIG. 520,*4.* *P. imbricata* (SOWERBY), L.Sil.(Llandovery), Can.(N.B.); *4a,b,* ped.v., post. ext.v., ×1.5; *4c-e,* ped. int. mold, ped. int. mold. post. view, brach. int. mold, ×2 (Boucot, Johnson, Staton, n).

Spirigerina D'ORBIGNY, 1849, p. 42 [*Terebratula marginalis* DALMAN, 1828, p. 143; SD ALEKSEEVA, 1960, p. 64] [=*Spirigerina* D'ORBIGNY, 1847, p. 268 *(nom. nud.)*]. Suboval to pentagonal, finely costate, biconvex shells with dorsal fold and ventral sulcus. Dental plates present; pedicle-valve diductor scars elongate, impressed, nonflabellate; hinge plates defining sockets medially and bearing crural lobes; spiralia directed dorsomedially. *Sil.(U.Llandover.-Ludlov.),* Eu.(Gotl.-G.Brit.)-N.Am.(N.Greenl.).

Vagrania ALEKSEEVA, 1959, p. 389 [*Atrypa kolymensis* NALIVKIN, 1936, p. 17; OD] [=*Dentatrypa* BREIVEL, 1959, p. 57 (type, *Atrypa kolymensis* NALIVKIN, 1936)]. Subequally biconvex, plicate shells, brachial valve commonly more convex; with or without dorsal fold and ventral sulcus; some plications bifurcating anteriorly; fine, nodose, growth lines present; ventral beak straight; conjunct deltidial plates present. Dental plates present, prolonged anteriorly as long ridges; brachial valve adductor scars confined, deeply impressed, non-elongate; cardinalia and brachidium as in *Atrypa. ?U. Sil.(?Ludlov.),* L.Dev.-M.Dev., Eu. (Ural Mts.)-Asia-N.Am.(Nev.-Yukon Terr.-Bathurst Is.).——FIG. 520,*5a-e.* *V. kolymensis* (NALIVKIN), U.Ems.-Eifel., Ural Mts.; *5a-d,* ped.v., brach.v., ant., lat. views, ×1; *5e,* view of fine surface ornament, ×7 (8).——FIG. 520,*5f-h.* *Vagrania* sp., Ems:(Stuart Bay F.), Can.(Bathurst Is.); *5f,* ped. int. mold, ×1.5; *5g,h,* cast of brach. int., brach.v. int. mold, ×2 (113).

Zejszneria SIEMIRADZKI, 1922, p. 172 [*Orthisina davyi* BARROIS, 1886, p. 194; OD]. Resembles *Carinatina* externally. Interarea pronounced. Conjunct deltidial plates present in front of a circular foramen. Cardinal process and dorsal myophragm present. Brachidium unknown. May equal *Carinatina. M.Dev.,* Eu.

Subfamily ATRYPINAE Gill, 1871

[*nom. transl.* WAAGEN, 1883, p. 448 (*ex* Atrypidae GILL, 1871, p. 25)] [=Punctatrypinae RZHONSNITSKAYA, 1960, p. 262]

Costate biconvex or convexi-plane Atrypidae, commonly with lamellose growth lines. Conjunct deltidial plates lacking. *L. Sil.-U.Dev.*

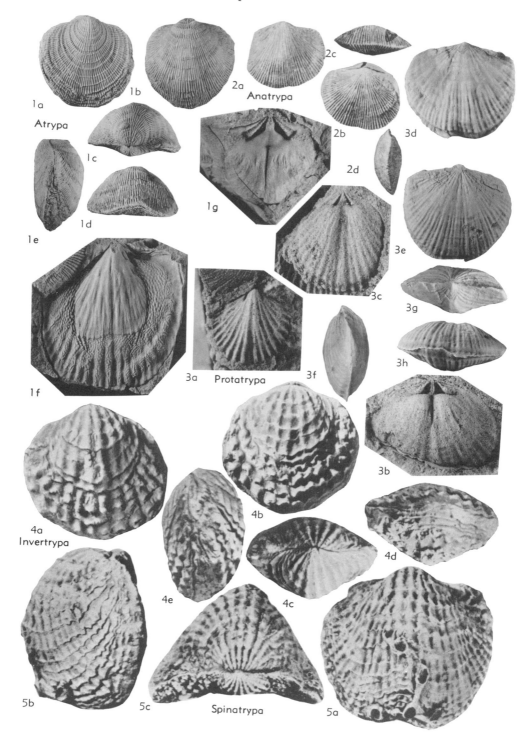

FIG. 522. Atrypidae (Atrypinae) (p. H639-H641).

la   lb

Desquamatia

lc

ld

FIG. 522A. Atrypidae (Atrypinae) (p. H639).

Atrypa DALMAN, 1828, p. 93 [*Anomia reticularis* LINNÉ, 1758, p. 702 (holotype, SD ALEXANDER, 1949); SD DAVIDSON, 1853, p. 90] [=*Cleiothyris* PHILLIPS, 1841, p. 55; *Mikrothyris* QUENSTEDT, 1868, p. 30]. Unequally biconvex or convexi-plane, costate shells with brachial valve more convex; dorsal fold and ventral sulcus present or absent; anterior commissure rectimarginate or deflected slightly toward brachial valve; costae increasing in number anteriorly by bifurcation and implantation, generally prominent growth lines may develop as frills; ventral beak incurved; interarea lacking; no conjunct deltidial plates in mature specimens. Dental plates short or obsolescent; ventral diductor muscle field flabellate; hinge plates discrete, diverging widely, sockets crenulate; crural lobes present, consisting of horizontal plates that connect outer hinge plates with crural bases; diductor area longitudinally striate; stout myophragm may be present; median septum absent; spiralia directed dorsomedially; jugum simple, short, disposed posteriorly, with jugal processes that may or may not be united posteromedially. *L. Sil.(U. Llandovery) - U. Dev.(Frasn.)*, cosmop. ——FIG. 522,*1a-e*. *A.* sp. cf. *A. reticularis* (LINNÉ), Sil.(Hemse Marl), Gotl.; *1a-e*, ped.v., brach.v., post., ant., lat. views, ×1 (113).—— FIG. 522,*1f,g*. *A.* sp., L.Dev.(Stonehouse F.), Nova Scotia; *1f,g*, ped.v. int. mold, brach.v. int. mold post. view, ×1.5 (Boucot, Johnson, & Staton, n).

Anatrypa NALIVKIN, 1941, p. 172 [*Orthis micans* VON BUCH, 1840, p. 56; OD]. Subcircular, biconvex shells, pedicle valve more convex, brachial valve tending to be sulcate and nearly flat; ventral beak straight; delthyrium broad, closed by deltidial plates in front of circular foramen. Dental plates present; brachial interior unknown. *M.Dev.*

(Givet.)-U.Dev.(Frasn.), Eu.——FIG. 522,2. *A. micans* (VON BUCH), Pskov beds; *2a-d*, ped.v., brach.v., ant., lat. views, ×1 (594).

Atrypinella KHODALEVICH, 1939, p. 45 [*A. biloba*; OD]. Subequally biconvex, transverse or circular shells with faintly developed ventral fold and dorsal sulcus. Rudimentary dental plates present or lacking; hinge plates discrete, supported by stout median septum forming septalium; spiralia dorsomedially directed, jugum unknown. *L.Dev. (Marginalis beds)*, Eu.(Ural Mts.).

Desquamatia ALEKSEEVA, 1960, p. 421 [*D. khavae*; OD]. Externally like *Atrypa*, but with finer costae and subdued growth lines; ventral beak only slightly incurved; conjunct deltidial plates present. Dental plates well developed; cardinalia and brachidium essentially as in *Atrypa. Dev.*, cosmop.——FIG. 522A,*1*. *D. khavae*, M.Dev., USSR(east slope, N.Urals); *1a-c*, ped.v., brach.v., lat. views, ×1; *1d*, post. part of brach.v. view, ×2 (8).

Invertrypa STRUVE, 1961, p. 334 [*Spinatrypa kelusiana* STRUVE, 1956, p. 385; OD]. Resembles *Spinatrypa*, but with reversed convexity; shells inequally biconvex, pedicle valve more convex, fold and sulcus lacking. Rudimentary dental plates may be present. Teeth crenulated; cardinalia essentially as in *Spinatrypa*; spiralia dorsomedially

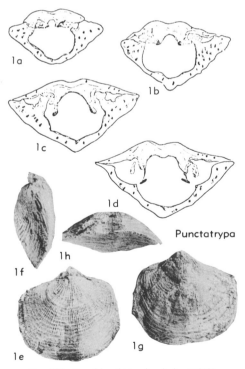

la

lb

lc

ld

Punctatrypa

lh

lf

le   lg

FIG. 523. Atrypidae (Atrypinae) (p. H640).

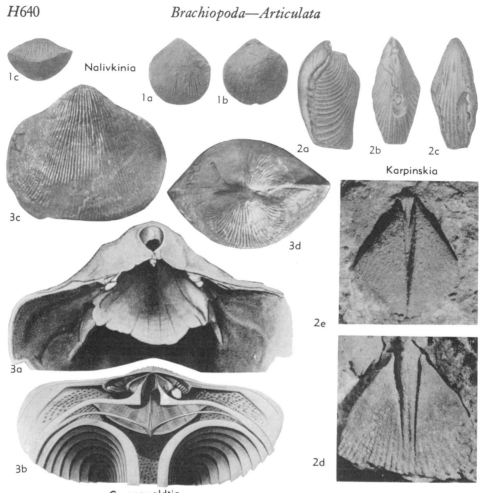

FIG. 524. Atrypidae (Carinatininae) *(1)*, (Karpinskiinae) *(2)*, (Palaferellinae) *(3)* (p. *H637, H641*).

directed. *M. Dev. (U. Eifel.),* Eu.-N. Am.——FIG. 522,*4. *I. kelusiana* (STRUVE), Ger.; *4a-e,* ped.v., brach.v., post., ant., lat. views, ×2 (787).

Protatrypa BOUCOT, JOHNSON, & STATON, 1964, p. 809 [**P. malmoeyensis*; OD]. Subequally biconvex, subcircular, transverse, or elongate shells without a well-developed fold and sulcus; brachial valve may bear a shallow furrow posteriorly; ornament as in *Atrypa.* Dental plates present or obsolescent; ventral muscle field nonflabellate; striated area for diductor attachment lacking in brachial valve; sockets rarely crenulated; stout myophragm may be present; brachidium as in *Atrypa. L. Sil.,* Eu.-Asia-N. Am.-S. Am.(Venez.). ——FIG. 522,*3a-c. P.* sp., Ede Qtzt., Sweden; *3a-c,* ped.v. int. mold, brach.v. int. mold, brach.v. int. mold, ×3 (Boucot, Johnson, & Staton, n).——FIG. 522,*3d-h. *P. malmoeyensis,* Malmøya, Norway; *3d-h,* ped.v., brach.v., lat., post., ant. views, ×1.5 (113).

Punctatrypa HAVLÍČEK, 1953, p. 8 [**P. nalivkini*; OD]. Biconvex, finely costate shells commonly lacking distinct fold and sulcus, or brachial valve may be slightly sulcate; fine, evenly spaced growth lines crossing costae make evenly reticulate pattern, hollow spine bases situated in concentric rows on growth lamellae at their intersections with costae; ventral beak suberect. Dental plates lacking; hinge plates disjunct; brachidium unknown. *L.Dev.-M.Dev.,* Eu.-Asia.——FIG. 523,*1. P.* sp. aff. *P. granulifera* (BARRANDE), M.Dev.(Greifensteiner Kalk), Ger.; *1a-d,* serial secs., ×4.5; *1e-h,* brach.v., lat., ped.v., ant. views, ×1.5 (744).

Spinatrypa STAINBROOK, 1951, p. 196 [*pro Hystricina* STAINBROOK, 1945, p. 49 (*non* MALLOCH, 1932)] [**"Atrypa hystrix* var. *occidentalis* HALL, 1858" (*errore pro A. aspera* var. *occidentalis* HALL, 1858, p. 515) (=*A. occidentalis, nom. transl.* STAINBROOK, 1938, p. 241); OD]. External conformation like *Atrypa,* but with few rounded plications crossed by lamellose, spinose growth lines. Internal structures essentially as in *Atrypa*. *L.Dev.(U.Ems.)-U.Dev.(Frasn.)*, cosmop.——FIG. 522,5. *S. coriacea* CRICKMAY, M.Dev., Can. (N.W.T.); *5a-c*, brach.v., lat., post. views, ×1 (549).

Subfamily KARPINSKIINAE Poulsen, 1943

[Karpinskiinae POULSEN, 1943, p. 40]

Strongly biconvex, elongate, with long dental plates. *Sil.-M.Dev.*

Karpinskia CHERNYSHEV, 1885, p. 48 [**K. conjugula*; OD] [=*Notoconchidium* GILL, 1950, p. 242 (type, *Pentamerus tasmaniensis* ETHERIDGE, 1883, pl. 2, fig. 1; SD BOUCOT, JOHNSON, & STATON, herein) (=*Notoconchidium* GILL, 1951, p. 187, type, *N. thomasi* GILL, 1951, p. 188; OD, syn. hom.)]. Unequally biconvex, elongate, costate shells of trapezoidal transverse cross section; brachial valve more convex; thickness commonly greater than width posteriorly; ventral lateral slopes abruptly angular. Dental plates long and thick; hinge plates discrete, forming bilobed cardinal process; dorsal median septum may be present; spiralia essentially as in *Atrypa*, jugum unknown. *L.Dev.-M.Dev.*, Eu.(Ural Mts.-Carnic Alps)-Asia-Australia,Tasmania-Victoria). ——FIG. 524,2a-c. **K. conjugula*, USSR(Urals); *2a-c*, lat., brach.v., ped.v. views, ×1 (396).——FIG. 524, *2d,e*. *K. thomasi* (GILL), L.Dev., Victoria; *2d,e*, ped.v. int. mold, brach.v. int. mold, ×2 (339).

Subfamily PALAFERELLINAE Spriestersbach, 1942

[nom. transl. STRUVE, 1955, p. 211 (ex Palaferellidae SPRIESTERSBACH, 1942, p. 187)]

Pedicle valve with raised muscle platform and chamber below. *L.Dev.-M.Dev.*

Gruenewaldtia CHERNYSHEV, 1885, p. 46 [**Terebratula latilinguis* SCHNUR, 1851, p. 7; OD] [=*Palaferella* SPRIESTERSBACH, 1942, p. 187]. Subequally biconvex, costate shells lacking fold and sulcus, but with anterior commissure commonly deflected dorsally; costae bifurcating anteriorly, growth lines subdued; ventral beak strongly incurved. Short dental plates present; muscle platform supported by 2 or more radially disposed septa present in pedicle valve; outer hinge plates crenulate in brachial valve; inner hinge plates present, forming small septalium, supported by short, thin myophragm; crural lobes and brachidium essentially as in *Atrypa*. *M.Dev.(Eifel.)*, Eu.——FIG. 524,3. **G. latilinguis* (SCHNUR), Rommersheimer Sch., Ger.; *3a,b*, ped.v. int., brach.v. int., ×3; *3c,d*, ped.v., post. views, ×1.5 (786).

Falsatrypa HAVLÍČEK, 1956, p. 584 [**F. admiranda*; OD]. Small, costate, irregularly subcircular shells lacking fold and sulcus; growth lines strongly lamellose. Dental plates lacking; ventral muscle platform present as in *Gruenewaldtia*; cardinalia

and brachidium unknown. *L.Dev.(U.Ems.)*, Eu. (Boh.).

Family LISSATRYPIDAE Twenhofel, 1914

[nom. transl. BOUCOT, JOHNSON, & STATON, 1964, p. 811 (ex Lissatrypinae TWENHOFEL, 1914, p. 31)]

Shell smooth. *M.Ord.-M.Dev.*

Subfamily LISSATRYPINAE Twenhofel, 1914

[Lissatrypinae TWENHOFEL, 1914, p. 31] [=Glassiinae SCHUCHERT & LEVENE, 1929, p. 20]

Inner hinge plates present. *L.Sil.-M.Dev.*

Lissatrypa TWENHOFEL, 1914, p. 31 [**L. atheroidea*; OD]. Subequally biconvex, suboval or subcircular shells, lenticular in profile, lacking well-developed fold and sulcus; ventral beak incurved. Teeth large, attached to wall of valve; dental plates lacking; hinge plates triangular, ponderous, disjunct, with parallel inner edges, or they may be conjunct and supported by stout myophragm in brachial valve posterior; jugum originating posteriorly. *L.Sil.(U.Llandovery)-L.Dev.(Borszczów)*, N. Am.-Eu.-Australia(Vict.).——FIG. 525,5a. *L. leprosa*, L.Dev.(Borszczów), Podolia; brach.v. int. (cardinalia and brachidium), ×6 (487).——FIG. 525,5b,c. **L. atheroidea*, Jupiter Cliffs E3, Can. (Anticosti Is.); *5b,c*, ped.v. int. mold, brach.v. int. mold, ×3 (476).

Australina CLARKE, 1913, p. 348 [**A. jachalensis*; OD]. Unequally biconvex or plano-convex subcircular shells with shallow dorsal sulcus; ventral beak suberect, foramen submesothyridid. Dental plates lacking; teeth attached directly to wall of valve; pedicle valve with short myophragm that bifurcates and splays laterally; inner hinge plates disjunct, somewhat ponderous, as in *Lissatrypa*, supported by stout myophragm in brachial posterior; spiralia directed dorsomedially; jugum unknown. *?U.Sil.(Wenlock.)*, S.Am.(Arg.).——FIG. 525,3. **A. jachalensis*; *3a-h*, serial secs., ×3; *3i,j*, brach.v., ped.v. views, ×1; *3k,l*, brach.v. int., ped. v. int. mold, ×1.5 (147).

Glassia DAVIDSON, 1881, p. 11 [**Atrypa obovata* SOWERBY in MURCHISON, 1839, p. 618; OD]. Subequally biconvex, suboval or subcircular shells, lenticular in profile, with or without dorsal fold and ventral sulcus; Delthyrium covered in apex by small concave plate, as in *Nucleospira*. Dental plates lacking; pedicle valve musculature essentially as in *Meifodia*; inner hinge plates conjunct; stout myophragm present in posterior; spiralia directed medially; jugum simple short, arising posteriorly. *L.Sil.(M.Llandovery)-M.Dev.*, Eu.——FIG. 525,4a,b. **G. obovata*, Wenlock, G.Brit.; *4a*, brachidium, ×2; *4b*, serial sec., ×6.5 (396, 744). ——FIG. 525,4c,d. *G. sulcata* SIEHL?, M.Dev., Boh.; *4c,d*, brach.v. int. mold, ped.v. int. mold, ×2 (53).

Lissatrypoidea BOUCOT & AMSDEN, 1958, p. 159 [**Nucleospira concentrica* HALL, 1859, p. 223

(*partim*, pl. 28B, fig. 16, =*Lissatrypa decaturensis* AMSDEN, 1949, p. 64) (*non* fig. 19)]. Subequally biconvex, suboval or subcircular shells, lenticular in profile, lacking well-developed fold and sulcus; beak incurved, foramen mesothyridid. Interior structures as in *Lissatrypa* except that hinge plates

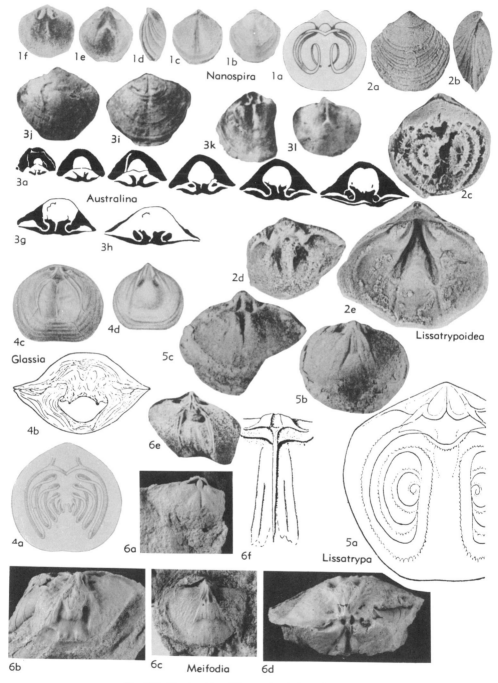

FIG. 525. Lissatrypidae (Lissatrypinae) (p. *H641-H644*).

bear stout, bulbous cardinal process. *U.Sil.(Wen-lock.-Ludlov.),* N.Am.——Fig. 525,2. **L. concentrica* (HALL), Brownsport F.; *2a,b,* ped.v., lat. views, ×3; *2c-e,* brach.v. int. (brachidium), brach.v. int. (cardinalia), ped.v. int., ×5 (28).

Meifodia WILLIAMS, 1951, p. 106 [**Hemithyris subundata* M'COY, 1851, p. 387; OD] [=*Tyro-*

thyris ÖPIK, 1953, p. 15 (type, *T. tyro*)]. Transversely suboval or elongate biconvex shells with brachial valve more convex; dorsal fold and ventral sulcus commonly well developed anteriorly; beaks small, strongly incurved. Dental plates very short, obsolescent or absent; pedicle-valve muscle area modified by adductor on small, raised, trans-

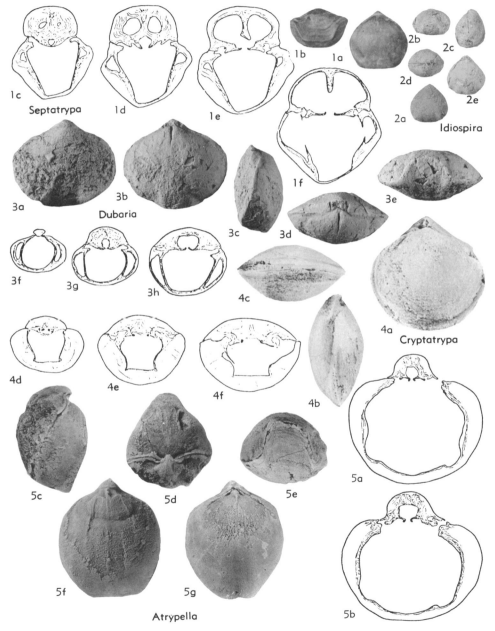

FIG. 526. Lissatrypidae (Septatrypinae) (p. *H644-H645*).

verse platform with steep posterior face and gently sloping anterior face; diductor tracks narrow, impressed; inner hinge plates disjunct, parallel medially, commonly supported by stout dorsal myophragm; dental sockets finely crenulate; spiralia directed dorsally. [Differs from *Glassia* principally by disposition of spiralia.]*L.Sil.(Llandovery)*, Eu. - Asia (Sib.) - S. Am.(Venez.)-Australia (Victoria).——FIG. 525,*6a-d*. *M.* sp., L. Sil., S.Am. (Venez.); *6a-d*, brach. int. mold, ped. int. mold, ped. int. mold, int. mold post. view, all ×2 (Boucot, Johnson, & Staton, n).——FIG. 525,*6e,f*. *M. ovalis supercedens* WILLIAMS, Wales; *6e*, ped.v. int. mold,×1.5; *6f*, brach.v. int., ×4 (870).

[ÖPIK inferred the presence of punctae in the shell of *Tyrothyris* from observation of external molds—not internal molds such as commonly provide evidence of punctation. Dr. J. A. TALENT (Geological Survey of Victoria) has examined numerous specimens of *T. tyro* without finding in them any indication of punctate shell structure and accordingly has concluded that *Tyrothyris* possesses an impunctate shell, with affinities to *Meifodia* and *Lissatrypa*. He has suggested (letter to J. G. JOHNSON) that the "punctation" illustrated by ÖPIK may correspond to the problematic *Cyclopuncta* described and figured by ELIAS (1958, Jour. Paleontology, v. 32, p. 50).]

Nanospira AMSDEN, 1949, p. 203 [**N. parvula*; OD]. Unequally biconvex or plano-convex, subcircular shells with shallow dorsal sulcus; ventral beak incurved. Dental plates lacking; teeth attached directly to wall of valve; pedicle valve with short myophragm that bifurcates and splays laterally; hinge plates disjunct, diverging slightly anteriorly; short dorsal myophragm may be present; jugum originating anteriorly, curves laterally to first volution of spiralia to join posteriorly; spiralia consisting of about 1.5 volutions directed submedially with slight dorsal inclination. [The only known species, *N. parvula,* is very small and may be an immature form of *Australina.* The spiralia are too poorly developed to determine the final disposition of their apices.] *U.Sil.(L.Ludlow),* N. Am. (Okla.). —— FIG. 525,*1*. **N. parvula,* Henryhouse F.; *1a*, brach.v. int. view (brachidium), ×10; *1b-f*, ped.v., brach.v., lat., ped.v. int., brach.v. int. views, ×5 (30).

Subfamily SEPTATRYPINAE Kozlowski, 1929

[Septatrypinae KOZLOWSKI, 1929, p. 176] [*emend.* BOUCOT, JOHNSON, & STATON, 1964, p. 812] [=Atrypellinae, Atrypopsinae POULSEN, 1943, p. 40]

Inner hinge plates absent. *M.Ord.-M.Dev.*

Septatrypa KOZLOWSKI, 1929, p. 176 [**S. secreta*; OD]. Unequally biconvex shells with brachial valve more convex, commonly transverse and slightly pentagonal in outline; dorsal fold and ventral sulcus commonly strongly developed anteriorly. Dental plates present; hinge plates disjunct, forming septalium supported by median septum; spiralia dorsomedially directed; jugum unknown. *L.Dev.,* Eu.-Asia.——FIG. 526,*1a,b*. **S. secreta,* Borszczów, Podolia; *1a,b*, brach.v., ant. views, ×1 (487).——FIG. 526,*1c-f*. *S. sapho* (BARRANDE), Lochkovian, Boh.; *1c-f,* serial secs., ×5.5 (744).

Atrypella KOZLOWSKI, 1929, p. 173 [**Atrypa prunum* DALMAN, 1828, p. 133; OD]. Strongly biconvex, elongate or rarely transverse shells with dorsal fold and ventral sulcus commonly distinctly developed anteriorly; ventral beak commonly strongly incurved. Dental plates lacking; pedicle umbonal cavity may be distinctly set off by transverse ridge or platform; ventral diductor tracks broadly divergent, separated by trapezoidal platform; hinge plates discrete, diverging anteriorly; dorsal myophragm may be present; spiralia dorsomedially directed. [May equal *Atrypoidea.*] *U.Sil.-L.Dev.* (Skala), Eu.-Asia (USSR)-N. Am. (Arctic-USA).——FIG. 526,*5a,b*. **A. prunum* (DALMAN), Sil., Sweden(Gotl.); *5a,b*, serial secs., ×3 (744). ——FIG. 526,*5c-g*. *A. carinata* JOHNSON, Sevy Dol., Nev.; *5c-g*, lat., post. int. mold, ant. ped.v. int. mold, brach.v. int. mold, ×1 (454).

Atrypoidea MITCHELL & DUN, 1920, p. 271 [**Meristina australis* DUN, 1904, p. 318; OD]. Externally like *Atrypella.* Pedicle valve interior unknown; cardinalia and jugum unknown; spiralia directed dorsomedially. [May equal *Atrypella.*] *?U.Sil.,* Australia.

Cryptatrypa SIEHL, 1962, p. 196 [**Terebratula philomela* BARRANDE, 1847, p. 387; OD]. Subequally biconvex, elongate or transversely oval, lenticular shells lacking well-developed fold and sulcus; ventral beak pointed, slightly incurved, foramen enclosed by deltidial plates. Short or obsolescent dental plates present or lacking; hinge plates discrete, commonly small, flat or concave; spiralia directed dorsally. *L.Sil.(U.Llandovery)-M.Dev.,* Eu.——FIG. 526,*4*. **C. philomela* (BARRANDE), M.Dev.(Greifensteiner Kalk), Ger.; *4a-c,* brach.v., ant., lat. views, ×3; *4d-f,* serial secs., ×6.5 (744).

Dubaria TERMIER, 1936, p. 1266 [**D. lantenoisi;* OD] [=*Atrypopsis* POULSEN, 1943, p. 44 (type, *A. varians*); *Rhynchatrypa* SIEHL, 1962, p. 199 (type, *Terebratula thetis* BARRANDE, 1847, p. 349)]. Unequally biconvex shells with brachial valve more convex, commonly transverse and slightly pentagonal in outline; dorsal fold and ventral sulcus commonly strongly developed anteriorly in most species; sulcus may bear several indistinct plications; ventral beak small, incurved. Dental plates present; hinge plates disjunct, diverging slightly anteriorly; dorsal myophragm may be present; apices of spiralia directed dorsally; jugum unknown. [Differs from the externally homeomorphous *Septatrypa* by absence of a dorsal median septum.] *U.Sil.-M.Dev.,* Eu.(Ger.-Czech.)-N.Afr.-Asia-N.Am.(N.Greenl.-Nev.). —— FIG. 526,*3a-e*. **D. lantenoisi,* ?U.Sil., N.Afr.; *3a-e,* ped.v., brach.v. int. mold, lat., post., ant. views, ×1.5 (113).——FIG. 526,*3f-h*. *D. thetis* (BARRANDE), M.Dev.(U.Eifel.), Ger.(Greifenstein); *3f-h*, ser. secs., ×4.5 (744).

Idiospira COOPER, 1956, p. 690 [**Camerella panderi* BILLINGS, 1859, p. 302; OD]. Subequally bicon-

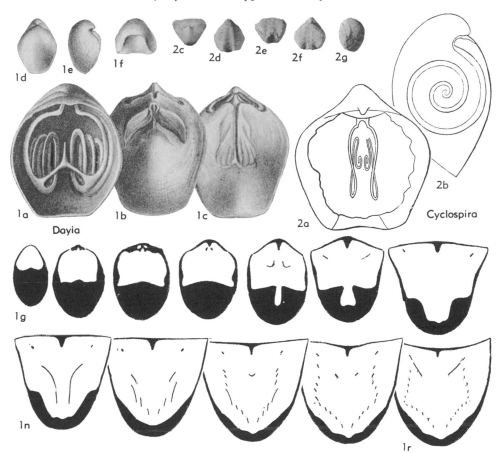

FIG. 527. Dayiidae (Cyclospirinae) *(2)*, (Dayiinae) *(1)* (p. *H645-H646*).

vex, commonly elongate shells with dorsal fold and ventral sulcus developed anteriorly; fold and sulcus may be faintly plicate at anterior margin; ventral beak small, incurved. Dental plates present or lacking; hinge plates disjunct; dorsal myophragm may be present; spiralia directed submedially with slight dorsal inclination; jugum short, simple, situated posteriorly. [Differs from *Dubaria* principally by the submedial disposition of apices of the spiralia.] *M.Ord.-L.Sil.*, N.Am.-Eu.(G.Brit.)-Asia(Sib.).——FIG. 526,2. **I. panderi* (BILLINGS), M.Ord.(Tyrone F.), Ky.; *2a-e*, ped.v., ant., lat., post., brach.v. views, ×1 (189).

Family and Subfamily UNCERTAIN

Loilemia REED, 1936, p. 116 [**L. proxima* REED, 1936, p. 116; OD]. Unequally biconvex or planoconvex shells, pedicle valve more convex, shallow dorsal sulcus. Hinge line very narrow, greatest width anterior to mid-length; pedicle valve with ponderous myophragm that widens anteriorly; brachial-valve interior unknown. *Sil.,* India.

Superfamily DAYIACEA Waagen, 1883

[*nom. transl.* RZHONSNITSKAYA, 1960, p. 264 (*ex* Dayiinae WAAGEN, 1883, p. 486)]

Spiralia directed ventrally, laterally, or planospiral parallel to median plane. *M. Ord.-M.Dev.*

Family DAYIIDAE Waagen, 1883

[*nom. transl.* RZHONSNITSKAYA, 1960, p. 264 (*ex* Dayiinae WAAGEN, 1883, p. 486)]

Smooth shells. *M.Ord.-L.Dev.*

Subfamily CYCLOSPIRINAE Schuchert, 1913

[*nom. transl.* BOUCOT, JOHNSON, & STATON, herein (*ex* Cyclospiridae SCHUCHERT, 1913, p. 410)]

Jugum lacking. *M.Ord., ?L.Sil.*

Cyclospira HALL & CLARKE, 1893, p. 146 [**Orthis bisulcata* EMMONS, 1842, p. 395; OD] [=*Triplecella* WILSON, 1932, p. 399 (type, *T. duplicata*)]. Unequally biconvex shells, pedicle valve more convex; dorsal sulcus and ventral fold modified in

la   lb

lc

ld   Anoplotheca   le

Fig. 528. Anoplothecidae (Anoplothecinae)
(p. *H648*).

some by low medial plication. Dental plates, if
present, buried in secondary shell material present in umbonal region; hinge plates conjunct;
median septum present; spiralia planospiral, with
few volutions, jugum lacking. [According to
SCHUCHERT & COOPER (729) the pedicle valve
musculature resembles that of *Dayia*.]. *M.Ord.,
?L.Sil.,* N.Am.-Eu.(Eng.).——FIG. 527,2. *C.
bisulcata* (EMMONS), M.Ord.(Trenton Ls.), N.Y.
(2a,b,); M.Ord.(Coburg F.), N.Y. *(2c-g);* *2a,b,*
brach. view (brachidium), lat. (brachidium), ×3;
2c-g, post., ped.v., ant., brach.v., lat., ×1.5 (189,
396).

Subfamily DAYIINAE Waagen, 1883

[*nom. correct.* SCHUCHERT, 1913, p. 409 (*pro* Dayinae
WAAGEN, 1883, p. 486, *nom. imperf.*)] [=Protozeugidae
TWENHOFEL, 1914, p. 29]

Jugum present. *U.Sil.(Wenlock.)-L.Dev.*

Dayia DAVIDSON, 1881, p. 291 [**Terebratula navicula* SOWERBY in MURCHISON, 1839, p. 611; OD]
[=*Daya* KOKEN, 1896, p. 240 (obj.) (*nom.
null.*)]. Unequally biconvex shells, pedicle valve
more convex; ventral fold and dorsal sulcus
developed most strongly anteriorly; ventral beak
strongly incurved; posterior of pedicle valve
thickened by secondary shell material. Dental
plates absent; teeth crenulate, fixed directly to
sides of valve; diductor scars strongly divergent,
situated near anterior edge of secondary shell
material; hinge plates disjunct, with bilobed cardinal process supported by stout myophragm (according to ALEXANDER, 9); spiralia directed laterally, jugum long, situated anteriorly, and projecting posteriorly with short stem. *U.Sil.-L.Dev.
(Skala),* Eu.-Asia-N.Afr.——FIG. 527,1a-f. **D.
navicula* (SOWERBY); U.Sil.(Ludlow), Br.Is.; *1a-c,*
brach.v. int. (brachidium), ped.v. int. mold, brach.
v. int. mold, ×3; *1d-f,* brach.v., lat., ant. views,
×1 (229, 396).——FIG. 527,1g-r. *D.* sp. cf.
**D. navicula;* *1g-r,* serial secs., ×3 (487).

Protozeuga TWENHOFEL, 1913, p. 51 [**Waldheimia? mawei* DAVIDSON, 1881, p. 145; SD TWEN-
HOFEL, 1914, p. 30]. Small, externally like *Dayia.*
[CLOUD (1942, p. 145) determined that the type-
species possesses laterally directed spiralia. Jugum
situated anteriorly, long and U-shaped, lacking
the stem present in *Dayia*.] *U.Sil.(Wenlock.),* Eu.,
(Eng.-Gotl.).

Subfamily AULIDOSPIRINAE Williams, 1962

[*nom. transl.* BOUCOT, JOHNSON, & STATON, 1964, p. 806 (*ex*
Aulidospiridae, WILLIAMS, 1962, p. 252)]

Primitive dayiids with rudimentary spi-
ralia coiled in plane parallel to median plane.
Shoe-lifter present, jugum lacking. *M.Ord.,
?U.Ord.*

Aulidospira WILLIAMS, 1962, p. 252 [**A. trippi;*
OD]. Unequally biconvex shells with pedicle valve
more convex; brachial valve with broad, shallow
sulcus. Dental plates present, attached to ventral
side of small, nearly flat, shoe-lifter process; hinge
plates discrete; myophragm present, median septum absent. *M.Ord., ?U.Ord.,* Eu.(G.Brit.-?Boh.)-
?N.Am.(Que.).

Family ANOPLOTHECIDAE Schuchert, 1894

[*nom. transl.* BOUCOT, JOHNSON, & STATON, 1964, p. 807 (*ex*
Anoplothecinae SCHUCHERT, 1894, p. 103)] [=Coelospiridae
HALL & CLARKE, 1895, p. 357]

Costate or plicate Dayiacea. *U.Sil.(Wenlock.)-M.Dev.*

Subfamily COELOSPIRINAE Hall & Clarke, 1894

[*nom. transl.* AMOS & BOUCOT, 1963, p. 441 (*ex* Coelospiridae
HALL & CLARKE, 1895, p. 357)]

Shells with bifurcating plications. *U.Sil.
(Wenlock.)-M.Dev.*

Coelospira HALL, 1863, p. 60 [**Leptocoelia concava*
HALL, 1857, p. 107; OD]. Small, unequally biconvex or plano-convex plicate shells, pedicle
valve more convex; anterior commissure rectimarginate or deflected slightly ventrally; plications may be flattened on their crests, bifurcating
anteriorly in some species. Dental plates lacking;
ventral myophragm present; hinge plates discrete,
arising directly from posterolateral shell margins;
small cardinal process may be slightly bilobate
posteriorly, situated between hinge plates; dorsal
myophragm may be present in front of cardinal
process; spiralia with few volutions and with
short axes directed sublaterally and slightly ventrally, jugum arising about midway between crural
bases and axes of spiralia, arching ventrally, and
joining to form simple stem. *U.Sil.(Wenlock.)-
M.Dev.,* ?Asia(Kazakh.)-N. Am.-S. Am.(Venez.).
——FIG. 529,1. *C. virginia* AMSDEN, L.Dev.(Haragan), USA(Okla.); *1a-e,* brach.v., ped.v., ant.,
post., lat. views, ×3 (33).——FIG. 529,5. *C.* sp.,
L.Dev.(Rabbit Hill Ls.), USA(Nev.); *5a,b,* ped.v.
int., brach.v. int., ×5 (Boucot, Johnson, & Staton,
n).

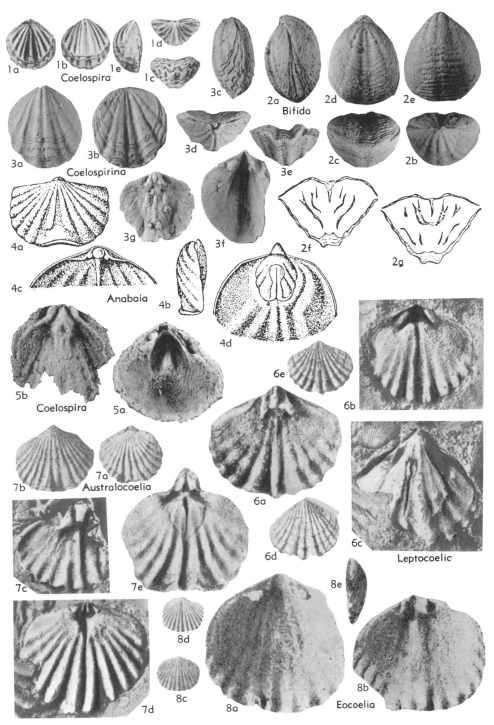

FIG. 529. Anoplothecidae (Coelospirinae) *(1,5)* (Anoplothecinae) *(2,3)*, Leptocoeliidae *(4, 6-8)* (p. *H646, H648-H649*).

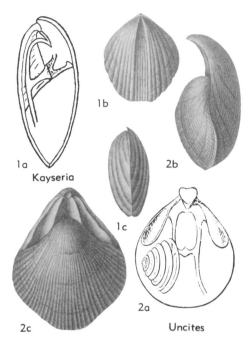

1b

1a

Kayseria

1c

2b

2a

2c   Uncites

FIG. 530. Kayseriidae *(1);* Uncitidae *(2)*
(p. *H649).*

Subfamily ANOPLOTHECINAE Schuchert, 1894

[Anoplothecinae SCHUCHERT, 1894, p. 103]

Shells with low, rounded plications crossed by lamellose growth lines. *L.Dev.-M.Dev.*

Anoplotheca SANDBERGER, 1855, p. 102 [*Terebratula venusta* SCHNUR, 1853, p. 180; OD (=*Productus lamellosus* SANDBERGER, 1850-56, p. 351)] [=*Hoplotheca* BIGSBY, 1878, p. 36 (type, *T. venusta*)]. Medium-sized lenticular, plano-convex shells with low, rounded plications and lamellose growth lines. Prominent, bilobed cardinal process present; muscle impressions and cardinalia otherwise essentially as in *Bifida.* Spiralia directed ventrally, inclined slightly laterally. *L.Dev.,* Eu.——FIG. 528,*1a-c.* *A. venusta* (SCHNUR), Ger.; *1a,b,* ped.v., brach.v., ×1 (396); *1c,* ped.v. int. mold, ×2 (113).——FIG. 528,*1d,e.* *A.* sp., U. Ems., Laubacher Schichten, Ger.; *1d,e,* brach.v. and ped.v. int. molds, ×3 (113).

Bifida DAVIDSON, 1882, p. 27 [*Terebratula lepida* D'ARCHIAC & DE VERNEUIL, 1842, p. 368; OD]. Small plano-convex or concavo-convex pauciplicate shells, pedicle valve strongly convex; low rounded plications crossed by lamellose growth lines. Dental plates absent; crural fossettes present on teeth; ventral myophragm present, cardinal process present, dorsal median septum present, attaining greatest height at mid-length; spiralia directed laterally,

jugal lamellae arising posterior to mid-length and joining over crest of median septum, giving rise to long stem that projects ventrally to touch pedicle valve; accessory lamellae arising from base of stem and projecting nearly to floor of brachial valve. *L.Dev.-M.Dev.,* Eu.-N.Afr.——FIG. 529,2. *B. lepida* (D'ARCHIAC & DE VERNEUIL), M. Dev.(Eifel., Rommersheimer Sh.), Ger.; *2a-e,* lat., post., ant., brach.v., ped.v., ×4; *2f,g,* ser. secs., ×6.5 (113, 744).

Coelospirina HAVLÍČEK, 1956, p. 586 [*C. modica;* OD]. Small, unequally biconvex shells like *Bifida,* but with stronger plications, numerous lamellose growth lines not developed; cardinal process present. *L.Dev.(U.Ems.),* Eu.(Czech.).——FIG. 529, 3. *C. modica,* Zlichov Ls.; *3a-g,* ped.v., brach.v., lat., post., ant., ped.v. int., brach.v. int, all ×3 (113).

Family LEPTOCOELIIDAE
Boucot & Gill, 1956

[Leptocoeliidae BOUCOT & GILL, 1956, p. 1174]

Shells with simple, strong, unbranched plications; lamellose growth lines present or absent; brachidium unknown. *L.Sil.-M. Dev.*

Leptocoelia HALL, 1857, p. 108 [*L. propria* (=*Atrypa flabellites* CONRAD, 1841, p. 55 =*Leptocoelia propria* HALL, 1857, p. 108); SD OEHLERT, 1887, p. 1324]. Subequally biconvex or plano-convex plicate shells, pedicle valve more convex; dorsal fold and ventral sulcus present; dorsal fold commonly bearing median groove; surface may or may not be covered with lamellose growth lines. Dental plates lacking, teeth crenulate; ventral beak incurved; pedicle-valve diductor scars flabellate, enclosing small, oval adductors; hinge plates flanking notothyrial platform bearing elevated, posteriorly trilobate cardinal process; dorsal myophragm present; brachidium unknown. *L.Dev.-M.Dev.,* N.Am.-S.Am.-Asia.—— FIG. 529,6. *L. flabellites* (CONRAD), Glenerie Ls., USA(N.Y.) *(6a,d,e);* Gaspé Ss., Can.(Quebec) *(6b,c);* *6a-c,* brach.v. int., brach.v. int. mold, ped.v. int. mold, ×2; *6d,e,* ped.v., brach.v., ×1 (111).

Anabaia CLARKE, 1893, p. 141 [*A. paraia;* OD]. Biconvex, plicate shells with dorsal fold and ventral sulcus, brachial valve commonly more convex; plications on fold and sulcus becoming obsolescent anteriorly on some shells. Musculature in pedicle valve essentially as in *Leptocoelia;* cardinal process present, consisting of simple rounded knob supported by dorsal myophragm and lying between disjunct hinge plates; brachidium unknown. *L.Sil.,* N.Am.-S.Am.(Brazil). —— FIG. 529,4. *A. paraia,* Brazil; *4a,b,* brach.v., lat., ×1; *4c,d,* brach.v. int. (cardinalia), ped.v. int. views, ×2 (396).

Australocoelia BOUCOT & GILL, 1956, p. 1174 [*A.

tourteloti; OD]. Unequally biconvex, plicate shells with dorsal fold and ventral sulcus, pedicle valve more convex; pedicle beak suberect. Dental plates lacking; hinge teeth stout, triangular in cross section; musculature in pedicle valve essentially as in *Leptocoelia*; cardinal process consisting of elevated median ridge, swollen terminally; cardinal process supported by dorsal myophragm; brachidium unknown. *L.Dev.,* S.Am.-S.Afr.-Australia. —— FIG. 529,7. **A. tourteloti,* Brazil-Arg.; *7a,b,* brach.v., ped.v., ×1; *7c-e,* brach.v. int., brach.v. int. mold, brach.v. int. mold, ×2 (111).

Eocoelia NIKIFOROVA, 1961, p. 252 [*Atrypa hemisphaerica* SOWERBY in MURCHISON, p. 637; OD]. Unequally biconvex or plano-convex, plicate shells, pedicle valve more convex; growth lines subdued; fold and sulcus lacking or subdued. Small, thin, dental plates present or lacking; hinge teeth bearing crural fossette on their median faces; discrete hinge plates bounding notothyrial cavity and notothyrial platform with or without knoblike cardinal process; brachidium unknown. *L.Sil.(L.Landovery)-U.Sil.(Wenlock.),* Eu.-Asia(Sib.)-N.Am.-S.Am.-Australia.——FIG. 529,8. **E. hemisphaerica* (SOWERBY), L.Sil., Sib.; *8a,b,* ped.v. int., brach.v. int., ×4; *8c-e,* ped.v., brach.v., lat. views, ×3 (602).

Family KAYSERIIDAE Boucot, Johnson, & Staton, 1964

[Kayseriidae BOUCOT, JOHNSON, & STATON, 1964, p. 807]

Accessory lamellae arising from the jugum and continuing intercoiled with the primary lamellae of the spiralia to their ends. *M.Dev.*

Kayseria DAVIDSON, 1882, p. 21 [*Orthis lens* PHILLIPS, 1841, p. 65; OD]. Biconvex, bisulcate, elongate-ovate costate shells, lenticular in profile, median costae finer than those on flanks, growth lines crossing ribs at irregular intervals. Ventral myophragm present; dental plates lacking; high triangular median septum extending along posterior part of brachial valve; jugum resting on median septum and produced ventrally as stem that contacts pedicle valve, jugal bifurcations arising from base of stem and accessory lamellae continuing intercoiled with primary volutions of spiralia to their ends. *M.Dev.,* Eu., Asia.——FIG. 530,1. **K. lens* (PHILLIPS), G.Brit.; *1a,* lat. (jugum), ×3; *1b,c,* brach.v., lat., views, ×2 (396).

Superfamily UNCERTAIN
Family UNCITIDAE Waagen, 1883

[*nom. transl.* SCHUCHERT & LEVENE, 1929 (*ex* Uncitinae WAAGEN, 1883, p. 494)]

Characters of *Uncites. M.Dev.*

Uncites DEFRANCE, 1825, p. 630 [*Terebratulites gryphus* SCHLOTHEIM, 1820, p. 259; OD] [=*Winterfeldia* SPRIESTERSBACH, 1942, p. 197 (type, *U. paulinae* WINTERFELD)]. Biconvex, elongate-oval,

costate shells, commonly lacking fold and sulcus; costae bifurcate, crossed by irregularly spaced growth lines; ventral beak attenuate, commonly twisted; foramen may be present behind concave deltidium. Dental plates and ventral myophragm present; cardinal plate supporting large, slightly bilobed cardinal process; hinge plates extending as long lobes along posterolateral margins; crura connecting directly with primary volutions of spiralia, jugum a simple band with low median projection. *M.Dev.,* Eu.-Asia.——FIG. 530,2. **U. gryphus* (SCHLOTHEIM), M.Dev., Eu.; *2a-c,* brach.v. int., lat., brach.v. ext., ×1 (229, 396).

Suborder RETZIIDINA Boucot, Johnson, & Staton, 1964

[*nom. correct.* BOUCOT, JOHNSON, & STATON, herein (*pro* suborder Retzioidea BOUCOT, JOHNSON, & STATON, 1964)] [Materials for this suborder prepared by A. J. BOUCOT, J. G. JOHNSON, and R. D. STATON]

Costate and plicate rhynchonelliform shells with crural loops directed medially connecting with laterally directed spiralia. *U.Sil.(Wenlock.)-Trias.*

Superfamily RETZIACEA Waagen, 1883

[*nom. transl.* BOUCOT, JOHNSON, & STATON, 1964, p. 813 (*ex* Retziinae WAAGEN, 1883, p. 486)]

Punctate shells with spiralia directed laterally. *U.Sil.(Wenlock.)-Perm.*

Family RETZIIDAE Waagen, 1883

[*nom. transl.* HALL & CLARKE, 1895, p. 358 (*ex* Retziinae WAAGEN, 1883, p. 486)]

Plications or costae developed evenly across shell. *L.Dev.-Perm.*

Retzia KING, 1850, p. 137 [*Terebratula adrieni* DE VERNEUIL, 1845, p. 471; OD] [=*Trigeria* BAYLE, 1878, pl. 13 (type, *T. adrieni*)]. Subequally biconvex, elongate-oval costate shells commonly without fold and sulcus; median costae same width as costae on flanks; deltidial plates conjunct; foramen permesothyridid. Thin, but prominent dental plates present; pedicle collar well developed; cardinal plate quadrilobate, flattened, posterior lobes extending into umbonal cavity of pedicle valve; cardinal plate supported by median septum; jugum joining and projecting posteroventrally as long stem that gives rise to short pronglike bifurcations. *L.Dev.,* Eu.-Asia(USSR). ——FIG. 531,2. **R. adrieni* (DE VERNEUIL), Fr.; *2a,* lat. (jugum), ×3; *2b,c,* brach.v., ped.v., ×1 (396).

Acambona WHITE, 1862, p. 27 [**A. prima*; OD]. Subequally biconvex, elongate oval, finely costate shells commonly lacking distinct fold and sulcus; costae may be interrupted by growth lines at irregular intervals toward anterior, costae on medial regions same width as those on flanks; deltidial plates conjunct. Foramen permesothyridid.

Dental plates lacking; pedicle collar present; cardinal plate bilobate posteriorly; jugum unknown. *L.Miss.*, N.Am.——Fɪɢ. 531,6. **A. prima*, Osagian

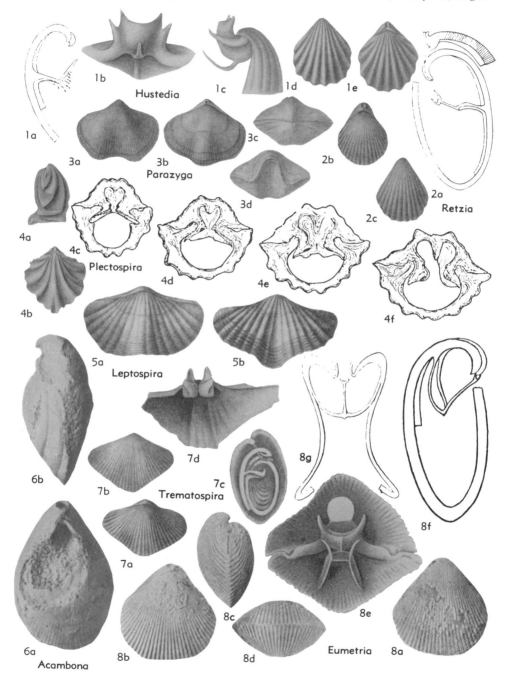

Fɪɢ. 531. Retziidae (p. *H649-H652*).

(Burlington Ls.), USA(Iowa); *6a,b*, brach.v., lat. views, ×1 (858).

Eumetria HALL, 1864, p. 54 [*Retzia vera* HALL, 1858, p. 704; OD]. Subequally biconvex, elongate-oval costate shells, commonly without fold and sulcus; median and flank costae same in width; deltidial plates conjunct; foramen permesothyridid. Dental plates lacking; cardinal plate crescent-shaped, with apices pointed posteroventrally; cardinal plate supported by transverse plate connecting crural plates; median septum lacking; limbs of jugum connecting medially and projecting backward as long stem that bifurcates into short stubs. *Miss.*, Eu.-N.Am.——FIG. 531, *8a-d.* **E. vera* (HALL), U.Miss.(Chester.), USA (Ill.); *8a-d*, ped.v., brach.v., lat., ant. views, ×1 (858).——FIG. 531,*8e-g.* *E. verneuiliana* (HALL), U.Miss.(St. Louis Ls.), USA(Ind.); *8e*, post. int.; ×10; *8f*, jugum (lat.), ×4; *8g*, jugum (ped. view), ×4 (396).

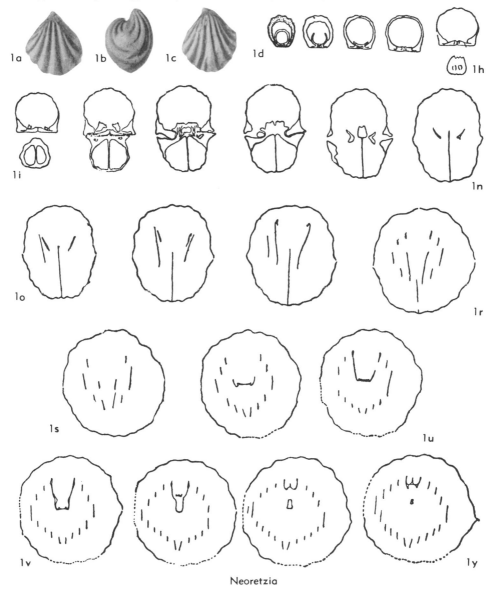

Neoretzia

FIG. 531A. Retziidae (p. *H652*).

Hustedia HALL & CLARKE, 1893, p. 120 [**Terebratula mormoni* MARCOU, 1858; OD]. Subequally biconvex, elongate-oval, costate shells, commonly lacking fold and sulcus; costae on medial regions same width as costae on flanks; deltidial plates conjunct; foramen permesothyridid. Dental plates lacking; pedicle collar present; cardinal process large, recurved posteriorly, bilobate on extremity, and bearing lateral pronglike crura projecting into umbonal cavity of pedicle valve; single pronglike stem, recurved backward, arising medially near base of cardinal plate; jugal lamellae rising anteroventrally to join; then projecting backward as simple stem; spines projecting backward from jugal stem. *Carb.-Perm.,* Eu.-N.Am.-S.Am.(Brazil-Peru)-Asia(India).——FIG. 531,*1.* **H. mormoni* (MARCOU), Penn., Mo.; *1a,* lat. (jugum), ×4; *1b,c,* ant. (cardinal plate), lat. (cardinal plate), ×5; *1d,e,* ped.v., brach.v., ×2 (396).

Leptospira BOUCOT, JOHNSON, & STATON, 1964, p. 814 [**Trematospira costata* HALL, 1859, p. 210; OD]. Externally shaped like *Trematospira,* but pauciplicate. Internally like *Rhynchospirina.* *L. Dev.,* ?*M.Dev.,* N.Am.——FIG. 531,*5.* **L. costata* (HALL), New Scotland F., USA(N.Y.); *5a,b,* brach.v., ped.v., ×1 (396).

Neoretzia DAGIS, 1963, p. 130 [**Retzia superbescens* BITTNER, 1890, p. 281; OD]. Biconvex shells with few subangular plications and radially grooved interspaces laterally; pedicle valve beak nearly straight, deltidial plates conjunct. Pedicle collar present, dental plates lacking; brachial valve with cardinal plate supporting bilobed cardinal process and supported by high, thin median septum; jugum wtih M-shaped median blade on ventral side and with saddle-shaped jugal stem; apices of spiralia directed laterally. *Trias.,* Eu.-Asia.——FIG. 531A,*1.* **R. superbescens* (BITTNER), Eu.(Crimea-Caucasus); *1a-c,* brach.v., lat., ped.v. views, ×1; *1d-y,* ser. transv. secs., ×3 (212a).

Parazyga HALL & CLARKE, 1893, p. 127 [**Atrypa hirsuta* HALL, 1857, p. 168; SD SCHUCHERT, 1897, p. 301]. Subequally biconvex, transversely oval or elongate costate shells, commonly with dorsal fold and ventral sulcus; costae simple, bearing fine spines; conjunct deltidial plates present or lacking. Dental plates present; incomplete pedicle collar present; cardinal plate quadrilobate as in *Trematospira* but less elevated and with anterior lobes only poorly defined by medial cleft; dorsal myophragm present; jugum as in *Trematospira.* *M.Dev.,* N.Am.——FIG. 531,*3.* **P. hirsuta* (HALL), M.Dev.(Hamilton.), USA(N.Y.); *3a-d,* ped.v., brach.v., post., ant. views, ×1 (396).

Plectospira COOPER, 1942, p. 288 [*pro Ptychospira* HALL & CLARKE, 1893, p. 112 (*non* SLAVIK, 1869)] [**Terebratula ferita* VON BUCH, 1834, p. 96; OD]. Subequally biconvex, pauciplicate shells of lenticular outline, dorsal fold commonly consisting of elevated median plication; deltidial plates conjunct. Dental plates absent; cardinal plate essentially as in *Homoeospira;* median septum present; jugal lamellae joined and projecting posteroventrally as simple stem. *L.Dev.-M.Dev.,* Eu.-Asia.——FIG. 531,*4.* **P. ferita* (VON BUCH), M. Dev., Ger.; *4a,b,* lat., brach.v. views, ×2; *4c-f,* serial secs., ×6.5 (396, 744).

Trematospira HALL, 1859, p. 207 [**Spirifer multistriatus* HALL, 1857, p. 59; SD HALL & CLARKE, 1893, p. 126]. Subequally biconvex, transverse-oval, costate shells, commonly with dorsal fold and ventral sulcus; costae subangular, approximately of same width in medial regions as on flanks, bifurcating anteriorly; deltidial plates conjunct; foramen mesothyridid. Short dental plates may be present; cardinal plate quadrilobate, posterior lobes projecting into pedicle cavity; small cardinal process may be developed between posterior lobes, as in *Rhynchospirina;* median septum absent; myophragm may be present; jugum essentially as in *Rhynchospirina.* *L.Dev.,* N.Am.——FIG. 531,*7.* **T. multistriata* (HALL), L.Dev.(New Scotland), USA(N.Y.); *7a-c,* brach.v., ped.v., lat. (jugum) views, ×1; *7d,* brach.v. int. (cardinalia), ×3 (396).

Family RHYNCHOSPIRINIDAE
Schuchert & LeVene, 1929 (1894)

[Rhynchospirinidae SCHUCHERT & LEVENE, 1929, p. 22 (1894); =Rhynchospirinae SCHUCHERT, 1894, p. 105] [The family-group name Rhynchospirinidae, based on the junior synonym of the replaced homonym *Rhynchospira,* is to be cited under its own author and date, but for purposes of priority takes the date of the replaced family name. Likewise, the type-genus, *Rhynchospirina,* takes the date of the name (*Rhynchospira* HALL) which it replaces (Code, Art. 39)]

Costae of second order originating medially in front of the beak or with medial costae finer than on flanks. *U.Sil.-L.Dev.*

Rhynchospirina SCHUCHERT & LEVENE, 1929, p. 121 (1859) [*pro Rhynchospira* HALL, 1859, p. 29 (*non* EHRENBERG, 1845)] [**Waldheimia formosa* HALL, 1857, p. 88; OD] [=*Rhyncospira* HALL, 1859, p. 213 (*nom. null.*); *Retziella* NIKIFOROVA, 1937, p. 57 (type, *R. weberi*)]. Externally like *Homoeospira;* deltidial plates conjunct; foramen mesothyridid to permesothyridid. Short dental plates may be present; incipient pedicle collar variably developed or absent; cardinal plate trapezoidal, projecting posteriorly into umbonal cavity of pedicle valve; crural bases stout and flattened, projecting ventrally; cardinal plate supported by median septum; jugal lamellae rising ventrally to join, then projects backward as short trough-shaped stem, or stem may be lacking. *U.Sil.-L.Dev.,* Eu.-N.Am.-Asia. —— FIG. 532,*1a.* **R. formosa* (HALL), L.Dev., USA(N.Y.); lat. view showing jugum, ×4 (396).——FIG. 532,*1b-e.* *R. maxwelli* AMSDEN, L.Dev., USA(Okla.); *1b-d,* brach.v., ped.v., ant. views, ×2; *1e,* brach.v. int., ×3 (33).

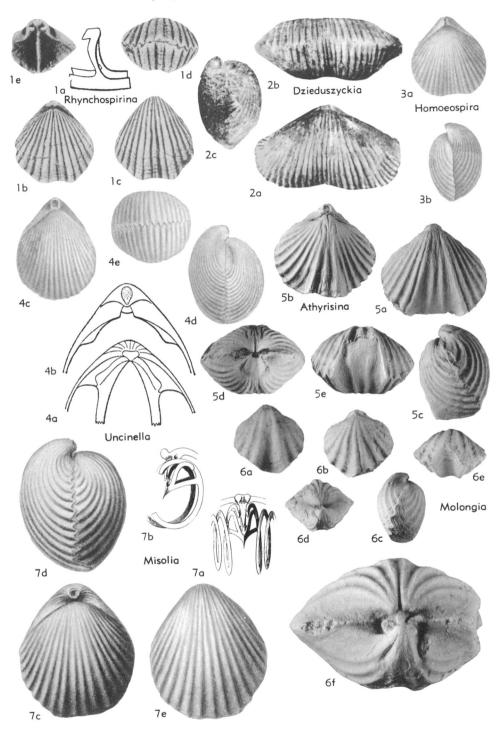

FIG. 532. Rhynchospirinidae *(1,3)*; Athyrisinidae *(2,4-7)* (p. *H652, H654*).

Homoeospira HALL & CLARKE, 1893, p. 112 [*Rhynchospira evax HALL, 1863, p. 213; SD SCHUCHERT, 1897, p. 231]. Subequally biconvex, elongate-oval costate shells, with or without poorly defined fold and sulcus, or shells may be faintly bisulcate; median costae narrower than costae on flanks; deltidial plates conjunct or disjunct, or delthyrium may be open; foramen submesothyridid to permesothyridid. Dental plates lacking; cardinal plate conjunct posteriorly, but disjunct, divergent anteriorly; small, linear cardinal process posteriorly; cardinal plate not extending into pedicle cavity, as in *Rhynchospirina,* supported by median septum that does not reach to anterior half of shell; jugum essentially as in *Rhynchospirina. U.Sil.(Wenlock.),* Eu.-N.Am.——FIG. 532,3. *H. evax* (HALL), U.Sil.(Waldron), USA(Ind.); *3a,b,* brach.v., lat. views, ×1 (396).

Superfamily ATHYRISINACEA Grabau, 1931

[*nom. transl.* BOUCOT, JOHNSON, & STATON, 1964, p. 814 (*ex* Athyrisininae GRABAU, 1931, p. 509)]

Impunctate. *M.Dev.-Trias.*

Family ATHYRISINIDAE Grabau, 1931

[*nom. transl.* BOUCOT, JOHNSON, & STATON, 1964, p. 815 (*ex* Athyrisininae GRABAU, 1931, p. 509)]

Characters of superfamily. *M.Dev.-Trias.*

Athyrisina HAYASAKA, 1920, p. 176 [*A. squamosa HAYASAKA, 1920, p. 176; OD]. Biconvex costate shells with dorsal fold and ventral sulcus. Short dental plates present; brachial median septum absent. Spiralia directed laterally; jugum unknown. *M.Dev.,* China.——FIG. 532,5. *A. minor* HAYASAKA; *5a-e,* ped.v., brach.v., lat., post., and ant. views of int. mold, ×1.5 (113).

Dzieduszyckia SIEMIRADZKI, 1909, p. 768 [*Terebratula (?) kielcensis ROEMER, 1866, p. 671; OD] [=Zigania NALIVKIN, 1937, opp. p. 112 (*nom. nud.*)]. Subequally biconvex, bisulcate, commonly transverse shells with crenulate, rectimarginate anterior commissure, or with commissure deflected slightly dorsally; costae coarse or fine, rounded, bifurcating anteriorly; ventral beak straight or slightly incurved. Short dental plates present; cardinal plate present, supported by median septum in posterior half of valve; crura forming loops subparallel to median plane; apices of spiralia directed toward posterolateral margins; jugum unknown. *U.Dev.(?Frasn.-Famenn.),* Eu.-N.Afr.——FIG. 532,2. *D. intermedia* TERMIER, Famenn., N.Afr.; *2a-c,* brach.v., ant., lat., ×1 (799).

Misolia VON SEIDLITZ, 1913, p. 172 [*M. misolica; OD]. Biconvex, elongate-oval, costate shells with or without dorsal fold and ventral sulcus; costae may bifurcate anteriorly; mesothyridid foramen present below conjunct deltidial plates. Cardinal plate not pierced apically, but bearing low bilobed

cardinal process on its posterior portion; jugum giving rise to dorsally directed bifurcations at point of joining, jugum pointed anteriorly, but lacking saddle. *Trias.,* Indonesia.——FIG. 532,7. *M. misolica*; *7a,b,* ped. and lat. views of jugum, ×2; *7c-e,* brach.v., lat., ped.v. views, ×1 (735).

Molongia MITCHELL, 1921, p. 546 [*M. elegans, p. 547; OD]. Small, subequally biconvex, pauciplicate shells; unplicate sulcus present in pedicle valve, medially grooved fold present in brachial valve; beak moderately incurved, foramen circular. Short thin dental plates present in pedicle valve, musculature not impressed; brachial valve with discrete hinge plates diverging anterolaterally; prominent myophragm present posteriorly; spiralia directed laterally. Shell substance impunctate. *U. Sil.* or *L.Dev.,* Australia (New S. Wales).——FIG. 532,6. *M. elegans*; *6a-e,* ped.v., brach.v., lat., ant. views, ×1.5; *6f,* post. view of int. mold, ×5 (113).

Uncinella WAAGEN, 1883, p. 494 [*U. indica; OD]. Biconvex, suboval, costate shells commonly lacking fold and sulcus; foramen as in *Misolia*; small deltidium present. Dental plates lacking; hinge plates discrete, diverging; spiralia as in *Retzia*; jugum unknown. *Perm.,* India.——FIG. 532,4. *U. indica*; *4a,b,* ped.v. beak, brach.v. beak, ×2; *4c-e,* brach.v. ext., lat., ant. views, ×1 (845).

Suborder ATHYRIDIDINA Boucot, Johnson, & Staton, 1964

[*nom. correct.* BOUCOT, JOHNSON, & STATON, herein (*pro* suborder Athyridoidea BOUCOT, JOHNSON, & STATON, 1964, p. 815)] [=suborder Rostrospiracea MOORE, 1952, p. 221] [Materials for this suborder prepared by A. J. BOUCOT, J. G. JOHNSON, and R. D. STATON]

Smooth or pauciplicate, impunctate, with narrow hinge line; pedicle-valve interarea obsolete or lacking; beak commonly truncated by circular foramen. Spiralia directed laterally or ventrally; crura project parallel to median plane; primary lamellae invariably united by more or less elaborate jugum; accessory lamellae arising from jugum commonly present. *U.Ord.-Jur.*

Superfamily ATHYRIDACEA M'Coy, 1844

[*nom. correct.* BOUCOT, JOHNSON, & STATON, 1964, p. 815 (*pro* Athyracea M'COY, 1844, *nom. transl.* WILLIAMS, 1956, p. 284 (*ex* Athyridae M'COY, 1844, p. 104, *emend.* DAVIDSON, 1881) *pro* Rostrospiracea SCHUCHERT & LEVENE, 1929 (invalid name not based on a family group)]

Spiralia directed laterally; crura united with primary lamellae by pair of S-shaped loops. *U.Ord.-Trias.*

Family MERISTELLIDAE Waagen, 1883

[*nom. transl.* HALL & CLARKE, 1895, p. 358 (*ex* Meristellinae WAAGEN, 1883, p. 449)]

Imperforate medially depressed cardinal

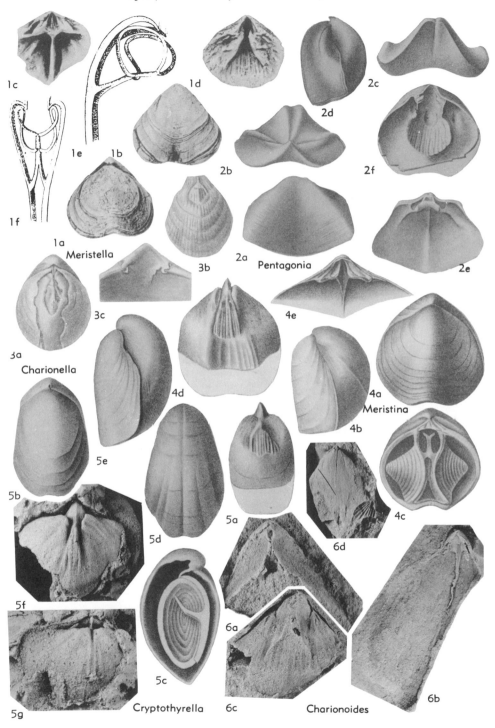

1c · 1d · 1e · 1b · 1f · 1a Meristella · 2c · 2d · 2f · 2b · 2a Pentagonia · 2e · 3b · 3c · 3a Charionella · 4e · 4d · 4a Meristina · 4b · 4c · 5e · 5b · 5d · 5a · 6d · 5f · 6a · 6b · 5g · 5c Cryptothyrella · 6c Charionoides

Fig. 533. Meristellidae (Meristellinae) (p. *H656*).

plate present, commonly forming septalium; jugal saddle not developed. Lamellar expansions at growth lines not developed. *U.Ord.-Miss.*

Subfamily MERISTELLINAE Waagen, 1883

[Meristellinae WAAGEN, 1883, p. 449]

Jugal stem present with bifurcations present or absent; pedicle-valve muscle field triangular, impressed, and longitudinally striate. *U.Ord.-U.Dev.*

Meristella HALL, 1859, p. 78 [*Atrypa laevis* VAN-UXEM, 1842, p. 120; SD S. A. MILLER, 1889, p. 354]. Unequally biconvex shells, commonly longer than wide; interarea obscure; ventral beak strongly incurved at maturity, commonly concealing foramen; deltidial plates may be exposed in early growth stages; dorsal fold and ventral sulcus may occur, or sulcation may affect only anterior commissure, or valves may be nonsulcate. Dental plates obsolescent; ventral muscle scar flaring strongly laterally, commonly deeply impressed into secondary shell material; cardinal plate strong, variable, triangular to subquadrate in outline; commonly concave on upper surface and depressed to form broad septalium; median septum originating beneath cardinal plate and extending part way to anterior margin; jugum produced backward as stem that bifurcates and recurves dorsally, then anteriorly, to reunite with stem. *L.Dev.*, N.Am.——FIG. 533,*1a-d*. *M. atoka* GIRTY, L.Dev.(Haragan), USA(Okla.); *1a,b*, brach.v., ped.v. views, ×1; *1c*, brach.v. int., ×3; *1d*, ped. v. int., ×2 (33).——FIG. 533,*1e,f*. *M. walcotti* HALL & CLARKE, L.Dev., Can.(Ont.); *1e,f*, lat. and ped.v. views showing jugum, ×3 (396).

Charionella BILLINGS, 1861, p. 148 [*Atrypa scitula* HALL, 1843, p. 171; OD]. External features, pedicle-valve interior, and jugum as in *Meristella*; cardinal plate simple, imperforate, depressed, and sessile; dorsal myophragm may be present; differs from *Meristella* by absence of dorsal median septum. *L.Dev.(L.Ems.)-U.Dev.(Tully)*, N.Am.——FIG. 533,*3*. *C. scitula* (HALL), M.Dev., Can. (Ont.); *3a,b*, brach.v. (shell partly removed), ped. int. mold, ×1; *3c*, brach.v. int. (cardinal plate), ×3 (396).

Charionoides BOUCOT, JOHNSON, & STATON, 1964, p. 817 [*Meristella doris* HALL, 1860, p. 84; OD]. Biconvex, elongate shells commonly lacking well-defined fold and sulcus; ventral beak with round foramen posterior to triangular delthyrium; cardinal shoulders angular, defining broad palintrope; ventral beak slightly incurved. Short dental plates present; ventral muscle field not strongly impressed; brachial valve bearing posteriorly sessile cruralium elevated on short median septum anteriorly; spiralia and jugum unknown. Differs from *Charionella* in shape of ventral beak and in presence of dorsal median septum. *L.Dev.(Ems.)-*

M.Dev., E.N.Am.——FIG. 533,*6*. *C.* sp. cf. *C. doris* (HALL), L.Dev., Tomhegan F., USA(Maine); *6a*, brach.v. view (ped. beak), ×3; *6b,c*, brach.v. int., ped.v. int. mold, ×2; *6d*, brach. int. mold, ×1 (113).

Cryptothyrella COOPER, 1942, p. 233 [*Whitfieldella quadrangularis* FOERSTE, 1906, p. 327; OD]. Large, elongate, smooth shells, subequally biconvex or with pedicle valve more convex, ventral beak incurved, dorsal fold and ventral sulcus absent or only poorly developed. Ventral interior with long subparallel dental plates bounding deeply impressed muscle field; umbonal cavities may bear chevron-like corrugations; brachial valve with sockets set widely apart, large flat cardinal plate present with sessile septalium medially, prominent myophragm present, no median septum; jugum united and projecting backward as simple stem. *U. Ord. (Ashgill.)-up. L. Sil.(C₂, up. Lland-over.)*, N.Am.-S.Am.(Venez.), Eu.-Asia(Sib. platform).——FIG. 533,*5a-e*. *C. cylindrica* (HALL), L.Sil. (*Platymerella* Z.); *5a-e*, ped.v. int. mold, brach.v., lat. int., ped.v., lat. views, all ×1 (396). ——FIG. 533,*5f,g*. *C.* sp., L.Sil., S.Am.(Venez.); *5f,g*, ped.v. int. mold, ×2, brach.v. int. mold, ×3 (Boucot, Johnson, & Staton, n).

Meristina HALL, 1867, p. 299 [*Meristella maria* HALL, 1863, p. 212; OD] [=*Whitfieldia* DAVID-SON, 1882, p. 107 (type, *Atrypa tumida* DALMAN, 1828, p. 134)]. External configuration and internal shell structures similar to *Meristella* but with narrow ventral muscle field and dental plates extending forward as distinct ridges that bound muscle area; jugum united and projecting backward as stem, bifurcating into 2 short stubs that may or may not recurve to rejoin jugum. *Sil.-U. Dev.(Tully)*, cosmop.——FIG. 533,*4*. *M. maria* (HALL), U.Sil.(Waldron), USA(Ind.); *4a-d*, brach.v., lat., brach.v. int. (brachidium), ped.v. int. mold, ×1; *4e*, brachial int. (cardinal plate), ×2 (396).

Pentagonia COZZENS, 1846, p. 158 [*P. peersii* (=*Atrypa unisulcata* CONRAD, 1841, p. 56); OD] [=*Goniocoelia* HALL, 1861, p. 101 (type, *G. uniangulata*, =*Atrypa unisulcata*)]. Biconvex shells of pentagonal outline; pedicle valve with very broad sulcus and abrupt lateral slopes; brachial valve with broad rounded fold commonly with narrow medial groove. Pedicle valve muscle impressions essentially as in *Meristella*; short dental plates present; dorsal median septum continuing posteriorly as faint median ridge on cardinal plate which arises vertically from bottom of valve so as to present erect, concave anterior face, top of cardinal plate extended posteriorly as scoop-shaped concavity; spiralia conforming with contracted interior cavity of shell; nature of jugal bifurcations unknown. *L.Dev.(Ems.)-M.Dev.*, N. Am.-S. Am.(Colom.-Venez.).——FIG. 533,*2*. *P. unisulcata* (CONRAD), M.Dev.(Onondaga), USA

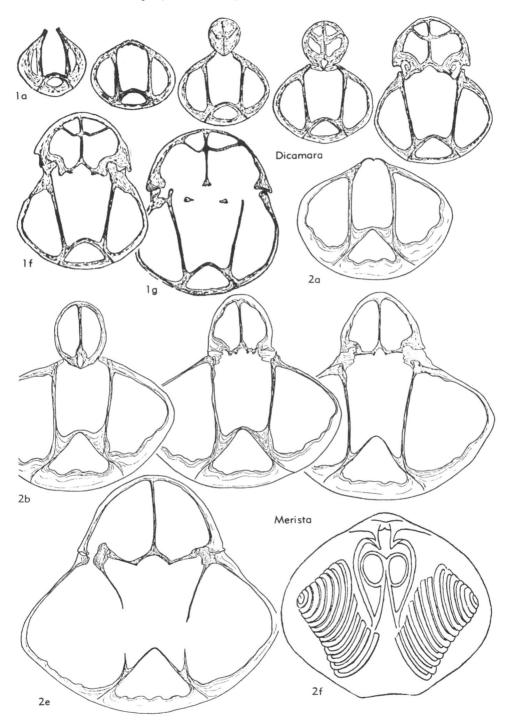

FIG. 534. Meristellidae (Meristinae) (p. *H*658).

(N.Y.); *2a-f,* ped.v., post., ant., lat., brach.v. int., ped.v. int., ×1 (396).

Subfamily MERISTINAE Hall & Clarke, 1895

[*nom. transl.* SCHUCHERT & LEVENE, 1929, p. 22 (*ex* Meristidae HALL & CLARKE, 1895, p. 345)]

Jugal bifurcations present; shoe-lifter process present in pedicle valve. *L.Sil.-M. Dev.*

Merista SUESS in DAVIDSON, 1851, p. 150 [**Terebratula herculea* BARRANDE, 1847, p. 382; OD] [*=Camarium* HALL, 1859, p. 486 (type, *C. typum*)]. Biconvex, elongate or transverse shells with dorsal fold and ventral sulcus commonly developed anteriorly. Dental plates short or may be produced anteriorly as thickened ridges that unite with outer part of medially placed shoelifter process along its lateral edges, shoe-lifter process with form of posteriorly plunging roof-shaped plate; cardinal plate depressed to form septalium supported by median septum; jugum united to form stem, then bifurcating posteriorly and recurving to rejoin in front of jugal stem, loops thus differing from *Meristella* by not rejoining at stem. *U.Sil.(Wenlock.)-M.Dev.,* Eu.-N. Am.-S. Am.(Venez.).——FIG. 534,*2f. M. typa* HALL, U.Sil.(L.Kayser), USA(Md.); brach.v. int. (jugum), ×2 (396).——FIG. 534,*2a-e.* **M. herculea* (BARRANDE), L.Dev., Boh.; *2a-e,* serial secs., ×4.5 (744).

Dicamara HALL & CLARKE, 1893, p. 73 [**Atrypa plebeia* SOWERBY, 1840, pl. 56, fig. 12,13 (*=Terebratula scalprum* ROEMER, 1844, p. 68); OD]. Externally like *Merista.* Pedicle valve with shoelifter process as in *Merista;* brachial valve with shoe-lifter divided and pierced by median septum. [Differs mainly from *Merista* in presence of brachial-valve shoe-lifter.] *M.Dev.,* Eu.——FIG. 534,*1.* **M. plebeia* (SOWERBY), Eifel(Gerolstein), Ger.; *1a-g,* serial secs., ×4.5 (744).

Subfamily CAMAROPHORELLINAE Schuchert & LeVene, 1929

[Camarophorellinae SCHUCHERT & LEVENE, 1929, p. 22]

Jugal bifurcations present, spondylium in pedicle valve. *M.Dev.-Miss.*

Camarophorella HALL & CLARKE, 1893, p. 215 [**Pentamerus lenticularis* WHITE & WHITFIELD, 1862, p. 295; OD]. Transversely subovate or elongate, biconvex shells, with or without dorsal fold and ventral sulcus; growth lines may be crossed by irregular, fine radial lines. Dental plates converging to form spondylium which rests on short median septum and is supported laterally by mystrochial plates; cardinal plate deeply concave, supported by thickened median septum; adductor muscle platform raised above floor of brachial valve in form of shoe-lifter penetrated by median septum, as in *Dicamara;* jugum consisting of inverted troughlike structure resting on median septum, projecting backward as stem that

bifurcates and recurves dorsally, then anteriorly to rejoin near its base. *U.Dev.(Louisiana Ls.)-Miss.,* N.Am.——FIG. 535,*3. C. mutabilis* HYDE, L.Miss., USA(Ohio); *3a,b,* ped.v., lat. views showing jugum, ×4; *3c,d,* umbonal region int., ped.v. int., ×2 (440).

Camarospira HALL & CLARKE, 1893, p. 82 [**Camarophoria eucharis* HALL, 1867, p. 368; OD] [*=Rowleyella* WELLER, 1911, p. 448 (type, *Terebratula fabulites* ROWLEY, 1900, p. 265)]. Externally like *Merista.* Dental plates converging to form spondylium which rests on short median septum; cardinal plate supported by median septum; jugum unknown. *M.Dev.-Miss.,* N.Am.——FIG. 535,*2.* **C. eucharis* (HALL), M.Dev., USA (Ind.); *2a-d,* brach.v., ped.v., lat. views, transv. sec., ×1 (396).

Subfamily HINDELLINAE Schuchert, 1894

[Hindellinae SCHUCHERT, 1894, p. 106 (*emend.* BOUCOT, JOHNSON, & STATON, 1964, p. 818]

Inner hinge plates in plane of commissure, separated by narrow fissure in most species; hinge plates not depressed to form septalium, as in other meristellid subfamilies; jugal bifurcations absent; pedicle-valve diductor muscle field poorly impressed; longitudinal striations lacking. *U.Ord.-L.Dev.*

Hindella DAVIDSON, 1882, p. 130 [**Athyris umbonata* BILLINGS, 1862, p. 144; OD]. Subcircular, transverse, elongate-ovate, subequally biconvex shells, with or without dorsal fold and ventral sulcus. Pedicle valve with long subparallel dental plates; cardinal plate depressed medially; long low median septum present; jugum originating in front of axis of spiralia and projecting backward at low angle, joining to project as short stem. *L.Sil.(Ellis Bay),* N.Am.——FIG. 535,*4.* **H. umbonata* (BILLINGS), L.Sil., Can.(Anticosti Is.);*4a,* brach.v. int. and brachidium, ×3 (229); *4b-d,* ped.v., brach.v., lat. views, ×1 (818).

Hyattidina SCHUCHERT, 1913, p. 415 [*pro Hyattella* HALL & CLARKE, 1893, p. 61 (*non* LENDENFELD, 1889)] [**Atrypa congesta* CONRAD, 1842, p. 265; OD]. Biconvex shells with or without ventral sulcus and dorsal fold that may be strongly accentuated by bounding furrows. Short dental plates occupying apex of pedicle valve; diductor tracks linear, impressed, slightly divergent; hinge plates triangular, medially divided by narrow fissure or may be anteriorly conjunct; median septum lacking; jugum united and projecting backward, without jugal stem. *?U.Ord.(English Head F.), L.Sil.(Clinton.)-U.Sil.(Greenfield Dol.),* N.Am.-Eu.(Eng.).——FIG. 535,*1.* **H. congesta* (CONRAD), L.Sil.(Clinton.), USA(N.Y.); *1a-c,* ped.v. int. mold, post. int. mold, brach.v. int. showing brachidium, ×2; *1d,* brach.v. int. showing hinge plates, ×5 (396).

Whitfieldella HALL & CLARKE, 1893, p. 58 [**Atrypa nitida* HALL, 1843; OD]. Biconvex, elongate trigonal, with or without faint sulcus on both valves. Short dental plates; pedicle-valve musculature only slightly impressed; inner hinge plates partly covering septalium supported by median septum that projects only short distance anteriorly; jugum united and projecting backward as simple stem. *Sil., ?L.Dev.,* Eu.-N.Am.——Fig. 535,5a-d. *W. upsilon* (BARRANDE), Sil., Boh.; *5a-d,* serial secs., ×4.5 (744).——Fig. 535,5e-h. *W. nitida* (HALL), U.Sil.(Waldron Sh.), USA(Ind.); *5e-h,* ped.v., brach.v., lat., ant. views, ×1 (Tillman, n).

Family ATHYRIDIDAE M'Coy, 1844

[*nom. correct.* BOUCOT, JOHNSON, & STATON, 1964, p. 817 (*pro* Athyridae M'COY, 1844, p. 104, *emend.* DAVIDSON, 1881, p. 4)] [Athyridae proposed by PHILLIPS, 1841, p. 54 for *Producta* and *Calceola* is not an available group name under Article 11e of the International Code]

Smooth or pauciplicate shells commonly with cardinal plate pierced apically, not depressed to form septalium; jugal saddle may occur and lamellar expansions may be developed at growth lines. *U.Sil.(Wenlock.)-Trias.*

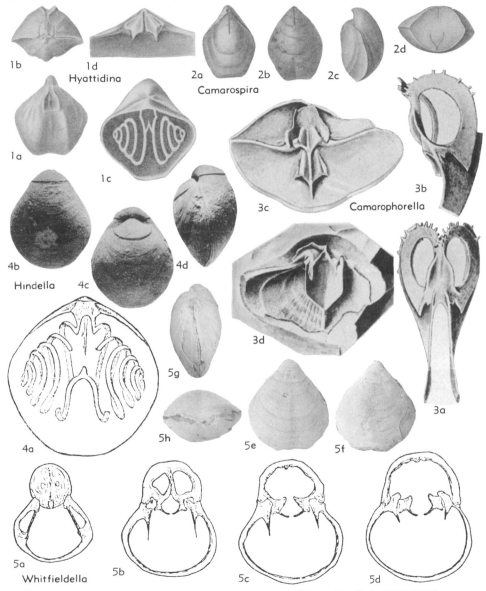

FIG. 535. Meristellidae (Camarophorellinae) *(2-3),* (Hindellinae) *(1,4-5)* (p. H658-H659).

Subfamily PROTATHYRIDINAE Boucot, Johnson, & Staton, 1964

[Protathyridinae BOUCOT, JOHNSON, & STATON, 1964, p. 819]

Primitive athyridids lacking jugal saddle; cardinal plate pierced apically. *U.Sil.(Wenlock.)-M.Dev.*

Protathyris KOZLOWSKI, 1929, p. 223 [**P. praecursor*; OD]. Subequally biconvex, elongate shells, with or without dorsal fold and ventral sulcus. Dental plates present, confined to apex; ventral muscle scars only faintly impressed; hinge plates divided apically by small fissure, conjunct anteriorly; dorsal myophragm may be present; jugum joined, to form stem that projects backward, bifurcating and extending into proximity with primary lamellae. *?U. Sil. (?Wenlock.), L. Dev., Eu.-N.Am.*——FIG. 536,*1a. P. "didyma"* (DALMAN), U.Sil., Eng.; ped.v. int. mold, ×2 (229). ——FIG. 536,*1b-l. *P. praecursor*, L.Dev.(Czortków), Podolia; *1b*, brach.v. view, ×1; *1c-k*, serial secs., ×3; *1l*, brachidium, ×6 (487).

Buchanathyris TALENT, 1956, p. 36 [**B. westoni*; OD]. Subequally biconvex, transversely oval or

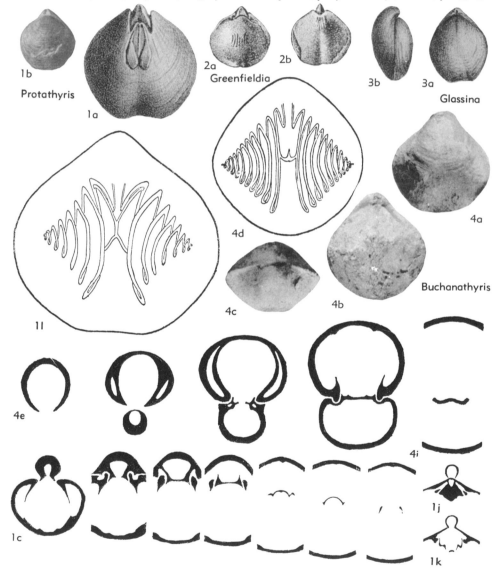

FIG. 536. Athyrididae (Protathyridinae) (p. *H660-H662*).

elongate shells, with or without faintly developed dorsal fold and ventral sulcus. Dental plates present; cardinal plate essentially flat, perforated, as in *Athyris*; jugum united and pointed backward, lacking jugal stem, saddle, or bifurcations. *L.Dev. (Ems.) - M.Dev.,* Australia.——Fig. 536,4. **B. westoni;* 4a-c, ped.v., brach.v., post. views, ×1.5; 4d, brachidium, ×3; 4e-i, serial secs., ×5 (796).

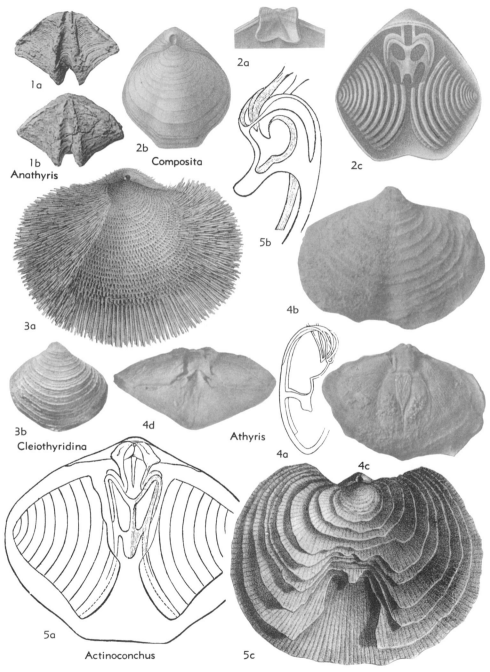

Fig. 537. Athyrididae (Athyridinae) (p. *H662-H663*).

Glassina HALL & CLARKE, 1893, p. 98 [*Terebratula laeviuscula SOWERBY in MURCHISON, 1839, p. 631; OD]. Small, elongate, subequally biconvex shells resembling Protathyris. Jugum taking form of backward inclined X, lacking jugal stem present in Protathyris. U.Sil.(Wenlock.-L.Ludlow), Eu.——FIG. 536,3. *G. laeviuscula (SOWERBY); Eng.; 3a,b, brach.v., lat. views, ×2 (229).

Greenfieldia GRABAU, 1910, p. 148 [*G. whitfieldi; OD]. Subequally biconvex, elongate shells, with or without dorsal fold and ventral sulcus. Dental plates present; confined to apex; muscle scars only faintly impressed; hinge plate medially divided by narrow fissure, or may be anteriorly conjunct; jugum unknown. [May equal Protathyris.] U.Sil., N.Am.——FIG. 536,2. *G. whitfieldi, Greenfield Dol., USA(Mich.); 2a,b, brach.v. int. mold, ped. v. int. mold, ×1 (364).

Subfamily ATHYRIDINAE M'Coy, 1844

[nom. correct. BOUCOT, JOHNSON, & STATON, 1964, p. 819 (pro Athyrinae M'COY, 1844, nom. transl. WAAGEN, 1883, p. 450 (ex Athyridae M'COY, 1844, p. 104, emend. DAVIDSON, 1881, p. 4)]

Jugal saddle present, jugal bifurcations terminating between 1st and 2nd volutions of spiralia. L.Dev.-Trias.

Athyris M'COY, 1844, p. 146 [*Terebratula concentrica VON BUCH, 1834, p. 123; SD KING, 1850, p. 136] [=Cliothyris AGASSIZ, 1846, p. 90; Spirithyris QUENSTEDT, 1868, p. 30; Euthyris QUENSTEDT, 1869, p. 442 (type, T. concentrica); Cleidothyris PAETEL, 1875, p. 45]. Biconvex, transverse or elongate shells, with or without dorsal fold or ventral sulcus; dorsal fold may bear median furrow and flanks may bear pair of broad plications; broad lamellar expansions which may be developed at growth lines, may bear fine radially arranged spines. Dental plates present; ventral myophragm may occur; cardinal plate of variable shape, pierced apically, free, flat, concave, or medially crested; dorsal myophragm may be present; jugum united to form saddle-shaped plate that projects forward, saddle narrowing posteriorly into stem with bifurcations that arise beneath stem or at its posterior terminus and recurve dorsally, terminating between 1st and 2nd volutions of spiralia. L.Dev.-Trias., cosmop.——FIG. 537,4a. A. vittata HALL, M.Dev., USA(N.Y.); 4a, lat. (jugum), ×3 (396).——FIG. 537,4b-d. A. lamellosa (LÉVEILLÉ), L.Miss., USA(Mo., Ill.); 4b-d, ped.v. ext., ped.v. int. mold, posterior int. mold, ×1 (858).

Actinoconchus M'COY, 1844, p. 149 [*Spirifera planosulcata PHILLIPS, 1836, p. 220 (=A. paradoxus M'COY, 1844, p. 149); OD]. Subequally biconvex, transversely oval, commonly without well-developed dorsal fold and ventral sulcus; broad lamellar expansions developed at growth lines, lamellae traversed by close-set fine radial grooves, but spines not developed. Dental plates

present; ventral myophragm may be present; brachial valve internal structures essentially as in Cleiothyridina. Carb., Eu.——FIG. 537,5. *A. planosulcata (PHILLIPS), Eng.; 5a,b, brach.v. int. (brachidium), ×2, lat. (jugum), ×3 (229); 5c, brach.v. ext., ×1 (229).

Anathyrella KHALFIN, 1961, p. 476 [*Anathyris ussovi KHALFIN, 1933; OD]. Like Anathyris externally but with widely flaring flanks, sulcus in pedicle valve and fold in brachial valve. Apically perforate cardinal plate reportedly present in brachial valve. U.Dev., USSR(Kuznetzk Basin).

Anathyris VON PEETZ, 1901, p. 134 [*Spirifera phalaena PHILLIPS, 1841, p. 71; SD SCHUCHERT & LEVENE, 1929, p. 29]. Biconvex plicate shells with lamellose growth lines; ventral sulcus present, dorsal fold deeply plicate; lateral slopes bearing 1 or 2 broad, rounded troughs which may be separated by narrow cusplike plications. Dental plates present; cardinal plate concave; dorsal myophragm may be present; jugum unknown. L.Dev.-U.Dev., Eu.-Asia(USSR). —— FIG. 537,1. A. ezquerrai (DE VERNEUIL & D'ARCHIAC), L.Dev., Eu.; 1a,b, ped.v., brach.v. views, ×1 (171). [=Plicathyris KHALFIN, 1946 (type, Athyris sibirica V. KHALFIN, 1940).]

Cleiothyridina BUCKMAN, 1906, p. 324 [pro Cleiothyris KING, 1850, p. 137 (non PHILLIPS, 1841)] [*Spirifer de roissyi LÉVEILLÉ, 1835, p. 39 (=Athyris royssii DAVIDSON, 1860, p. 84); OD] [=Cliothyris HALL & CLARKE, 1893, p. 90 (non AGASSIZ, 1846)]. Subequally biconvex, transversely suboval shells, commonly with dorsal fold and ventral sulcus developed anteriorly; surface covered with broad lamellar expansions at growth lines, lamellae projecting anteriorly as long flat spines, which are broader and flatter than in Athyris. Dental plates present; cardinal plate small, triangular, pierced apically; jugum essentially as in Athyris. U.Dev.-Perm., cosmop.—— FIG. 537,3a. *C. deroissii (LÉVEILLÉ), Carb., Eu. (Eng.); brach.v. ext., ×1.5 (229).——FIG. 537, 3b. C. pectinifera (D'ORBIGNY), U.Perm., Greenl.; ped.v. view, ×1 (269).

Composita BROWN, 1849, p. 131 [*Spirifer ambiguus SOWERBY, 1823, p. 105; OD] [=Seminula HALL & CLARKE, 1893, p. 93 (non M'COY, 1844, p. 158)]. Biconvex shells with dorsal fold and ventral sulcus, or fold may bear medial furrow; lateral slopes may bear indistinct low plications; growth lines not lamellose. Short dental plates present; cardinal plate perforate apically, and may develop posteriorly extended flanges; dorsal myophragm may be present; jugum essentially as in Athyris. U.Dev.(Famenn.)-Perm., Eu.-N.Am.-Australia.——FIG. 537,2. C. subtilita (HALL), U. Miss.(Chester.), USA(Ky.) (2a); U.Penn.(Mo.), USA(Mo.-Iowa) (2b,c); 2a, brach.v. int. showing card. plate, ×5; 2b, brach.v. view, ×1; 2c, ped.v. int. showing brachidium, ×3 (396). [=?Inia-

thyris BESNOSSOVA, 1963, p. 312 (type, *I. topkensis*); ?*Pseudopentagonia* BESNOSSOVA, 1963, p. 315 (type, *P. injensis*).]

Leptathyris SIEHL, 1962, p. 212 [**L. gryphis*; OD]. Subequally biconvex, small, subcircular, elongate or transverse shells, commonly faintly bisulcate; lamellar expansions at growth lines not known. Internally, like *Athyris*, except that cardinal plate is deeply depressed and medially crested; stout myophragm may be developed. Differs from *Meristospira* in presence of depressed cardinal plate. *L. Dev.(Ems.)-M. Dev.,* Eu.-N. Am.(Nev.). ——FIG. 538,4. **L. gryphis,* M.Dev., Ger.; *4a-c,* brach.v., lat., ant. views, ×3; *4d-f,* serial secs., ×6.5 (744).

Meristospira GRABAU, 1910, p. 158 [**M. michiganense*; OD]. Small, subcircular or transversely oval biconvex shells with or without faintly developed dorsal fold and ventral sulcus; growth lines not lamellose; few low plications may be present. Short dental plates present; cardinal plate free, pierced apically; dorsal myophragm present; spiralia and jugum unknown. Differs from *Athyris* principally in presence of free cardinal plate. [May equal *Composita.*] *M.Dev.,* N.Am. ——FIG. 538,2. **M. michiganense,* Amherstberg Dol., USA(Mich.); *2a,* post. int. mold both valves, ×3; *2b,c,* lat., ped.v. int. mold, ×2 (364).

Spirigerella WAAGEN, 1883, p. 450 [**S. derbyi*; SD OEHLERT, 1887, p. 1300] [=*Athyrella* RENZ, 1913, p. 620 *(nom. nud.)*]. Biconvex, elongate, or transversely suboval shells with dorsal fold and ventral sulcus, depth of broad sulcus tending to reduce convexity of pedicle valve; surface lacking lamellar expansions or spines. Dental plates, if present, buried in secondary shell material that is strongly developed in umbonal cavities; cardinal plate large, free, recurved posteriorly, and apically perforated; median septum lacking; jugum as in *Athyris,* but with median septum on saddle extending backward as far as jugal stem. *Perm.,* India-S.Am.——FIG. 538,3. **S. derbyi,* India; *3a,b,* lat. (jugum), brach.v. int., ×3; *3c-f,* brach.v., ped.v., lat., ant. views, ×1 (845).

Triathyris COMTE, 1938, p. 45 [**Terebratula mucronata* DE VERNEUIL, 1850, p. 171; SD BOUCOT, JOHNSON, & STATON, herein]. External configuration like *Anathyris* but with lateral slopes nonplicate, median plication on each valve raised to form anterior projection. Interior structures unknown. *L. Dev.(U. Ems.),* Eu.(Spain). —— FIG. 538,1. **T. mucronata* (DE VERNEUIL); *1a,b,* ped. v., brach.v. views, ×1 (171).

Subfamily DIPLOSPIRELLINAE Schuchert, 1894

[*nom. correct.* SCHUCHERT, 1913, p. 418 (*pro* Diplospirinae SCHUCHERT, 1894, p. 106, *nom. imperf.*)]

Jugal bifurcations continuing intercoiled with primary volutions of spiralia to their ends. *Trias.*

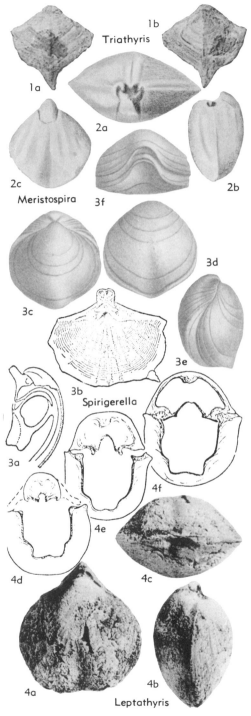

FIG. 538. Athyrididae (Athyridinae) (p. *H663*).

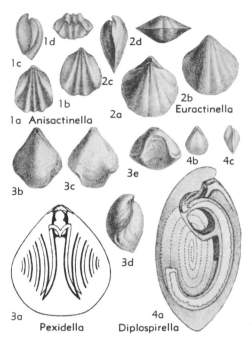

FIG. 539. Athyrididae (Diplospirellinae) (p. *H664*).

Diplospirella BITTNER, 1890, p. 297 [*Terebratula wissmanni* MÜNSTER, 1841, p. 64; OD]. Small, biconvex, transversely oval or pentagonal shells, with or without faintly sulcate brachial valve. Jugal lamellae united and projecting backward as stem that bifurcates, accessory lamellae produced by bifurcation continuing between primary volutions of spiralia to their ends. *Trias.*, Eu.(Aus.).——FIG. 539,4. *D. wissmanni* (MÜNSTER); *4a*, lat. (jugum), ×4; *4b,c*, brach.v., lat. views, ×1 (76).

Anisactinella BITTNER, 1890, p. 302 [*Terebratula quadriplecta* MÜNSTER, 1841, p. 58; OD]. Small, biconvex shells with broad, plicate dorsal fold and ventral sulcus; flanks smooth. Cardinal plate well developed; jugum as in *Dioristella*, but accessory lamellae continuing intercoiled with primary spiral volutions to their ends. *Trias.*, Eu. (Aus.).——FIG. 539,1. *A. quadriplecta* (MÜNSTER); *1a-d*, brach.v., ped.v., lat., ant. views, ×1 (76).

Euractinella BITTNER, 1890, p. 302 [*Terebratula contraplecta* MÜNSTER, 1841, p. 59; OD]. Biconvex, suboval shells lacking fold and sulcus; low rounded plications may be developed in corresponding position on each valve. Accessory lamellae of jugum continuing intercoiled wtih primary spiral volutions to their ends. *Trias.*, Eu. (Aus.).——FIG. 539,2. *T. contraplecta* (MÜN-

STER); *2a-d*, brach.v., ped.v., lat., ant. views, ×2 (76).

Pexidella BITTNER, 1890, p. 300 [*Spirifer strohmayeri* SUESS, p. 27; OD]. Small, biconvex, elongate-oval shells, commonly with dorsal fold and ventral sulcus; may be thickened by secondary shell material in umbonal region. Jugum situated posteriorly, giving rise directly to accessory lamellae which continue between primary volutions of spiralia to their ends. *Trias.*, Eu.(Aus.).——FIG. 539,3. *P. strohmayeri* (SUESS); *3a*, ped.v. view showing brachidium, ×2; *3b-e*, brach.v., ped.v., lat., ant. views, ×1 (76).

Subfamily UNCERTAIN

Amphitomella BITTNER, 1890, p. 298 [*Terebratula hemisphaeroidica* KLIPSTEIN, 1843, p. 222; OD]. Smooth, biconvex shells, with ventral median septum; cardinal plate strong, divided; dorsal median septum strongly developed to meet ventral median septum; jugum as in *Athyris*, but lacking saddle, jugal bifurcations terminating anteriorly, between 1st and 2nd volutions of spiralia. *Trias.*, Eu.(Aus.).——FIG. 540,8. *A. hemisphaeroidica* (KLIPSTEIN); lat. (jugum), ×4 (76).

Anomactinella BITTNER, 1890, p. 300 [*Terebratula flexuosa* MÜNSTER, 1841, p. 59; OD]. Finely plicate, biconvex shells, with plications in corresponding position on each valve, plications strong anteriorly, umbones smooth. Jugal bifurcations short, not extending beyond origin of 1st volution of spiralia. *Trias.*, Eu.(Aus.).——FIG. 540,3. *A. flexuosa* (MÜNSTER); *3a,b*, brach.v., ped.v. views, ×1 (76).

Dioristella BITTNER, 1890, p. 299 [*Terebratula indistincta* VON BEYRICH, 1862, p. 34; SD HALL & CLARKE, 1895, p. 775]. Smooth, biconvex, elongate-oval shells, commonly nonsulcate. Jugum extending ventrally to join, then long jugal stem projecting abruptly backward, jugal bifurcations curving dorsally, then bending abruptly to rejoin jugal stem at its origin *Trias.*, Eu.(Aus.).——FIG. 540,2. *D. indistincta* (VON BEYRICH); *2a*, lat. (jugum), ×3; *2b-e*, ped.v., lat., brach.v., ant. views, ×1 (76).

Janiceps FRECH, 1901, p. 551 [*Athyris peracuta* STACHE, 1878, pl. 6, fig. 4; SD SCHUCHERT & LEVENE, 1929, p. 70]. Small trigonal shells, pointed at posterior and anterolateral extremities, each valve bearing medially plicate fold. Interior unknown. *U.Perm.*, Eu.(Aus.).——FIG. 540,4. *J. peracuta* (STACHE); *4a,b*, ped.v., brach.v. views, ×1 (312).

Pentactinella BITTNER, 1890, p. 300 [*Terebratula quinquecostata* MÜNSTER, 1841, p. 59; SD HALL & CLARKE, 1895, p. 783]. Plicate, biconvex shells, with plication in corresponding position on each valve, medial part of each valve with plication rather than furrow. Interior unknown. *Trias.*, Eu.(Aus.).——FIG. 540,1. *P. quinquecostata* (MÜNSTER); *1a,b*, brach.v., ant. views, ×1 (76).

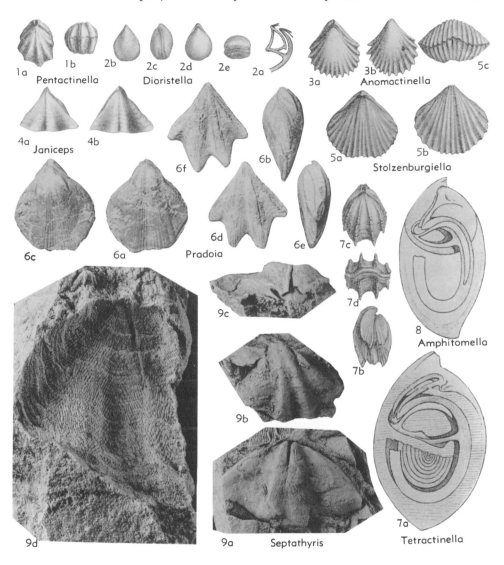

FIG. 540. Athyrididae (Subfamily Uncertain) (p. *H664-H666*).

Pradoia COMTE, 1938, p. 43 [**Terebratula torenoi* DE VERNEUIL & D'ARCHIAC, 1845, p. 460; SD BOUCOT, JOHNSON, & STATON, herein]. Externally like *Athyris* or *Anathyris* but lacking strong regular concentric lamellae, surface covered instead by fine radiating lines which may bifurcate anteriorly. Dental plates and perforate cardinal plate present; spiralia directed laterally. *L.Dev.(U.Ems.),* Eu.(Spain).——FIG. 540,*6a-c.* **P. torenoi* (DE VERNEUIL & D'ARCHIAC); *6a-c,* ped.v., lat., brach. v. views, ✕1 (171).——FIG. 540,*6d-f. P. collettei* (DE VERNEUIL); *6d-f,* ped.v., lat., brach.v. views, ✕1 (171).

Septathyris BOUCOT, JOHNSON, & STATON, 1964, p. 819 [**Athyris aliena* DREVERMANN, 1904, p. 258; OD]. Resembles *Anathyris* in external configuration but lacks lamellose growth lines, ornament consisting of fine irregular lines that diverge from plications and join in interspaces to form zigzag pattern, or intersection of lines may not meet in the furrows. Short dental plates present; cardinal plate broad, slightly concave, not known to be pierced apically; cardinal plate supported by median septum that does not continue to anterior; spiralia and jugum unknown. *L.Dev.(Siegen.-L. Ems.),* Eu.-N.Afr.——FIG. 540,*9.* **S. aliena*

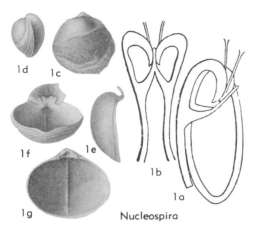

FIG. 541. Nucleospiridae (p. *H666*).

(DREVERMANN), Siegen., Ger.; *9a,* brach. int. mold, ×1; *9b,* ped. int. mold, ×1.5; *9c,* brach. int. mold post. view, ×2; *9d,* brach. ext. mold, ×3 (113).

Stolzenburgiella BITTNER, 1903, p. 508 [**S. bukowskii;* OD]. Biconvex, elongate or transverse, plicate shells resembling *Tetractinella,* but with fine secondary plications developed between primary plications in corresponding position on each valve; secondary plications may increase anteriorly by implantation; dorsal fold and ventral sulcus may be present. Spiralia with few volutions. *Trias.,* Eu.(Aus.).——FIG. 540,5. **S. bukowskii;* 5a-c, brach.v., ped.v., ant. views, ×2 (80).

Tetractinella BITTNER, 1890, p. 300 [**Terebratulites trigonellus* VON SCHLOTHEIM, 1820, p. 271; SD HALL & CLARKE, 1895, p. 783]. Small, biconvex shells externally resembling *Anathyris,* but lacking regularly lamellose growth lines, commonly with 4 narrow cusplike plications in corresponding positions on each valve. Dental plates present; jugum joined, forming rudimentary saddle, jugal stem projecting ventrally, bifurcating to form short accessory lamellae. *Trias.,* Eu.(Aus.).——FIG. 540,7. **T. trigonella* (VON SCHLOTHEIM); 7a, lat. (jugum), ×4; 7b-d, lat., brach.v., ant. views, ×1 (76).

Family NUCLEOSPIRIDAE Davidson, 1881

[Nucleospiridae DAVIDSON, 1881, p. 4]

Smooth athyridaceans with free, imperforate, cardinal plate; jugum lacking bifurcations. *U.Sil.-L.Carb.(Miss.).*

Nucleospira HALL, 1859, p. 24 [**Spirifer ventricosus* HALL, 1857, p. 57; OD]. Small, subequally biconvex, subcircular shells; surface may be covered by numerous fine, irregularly spaced spinules; narrow, poorly defined dorsal fold and ventral sulcus may be present, or sulcation may be ex-

pressed only by deflection of anterior commissure; interarea narrow and low, commonly obscured by small incurved pedicle-valve beak; small concave plate in apex of delthyrium. Dental plates lacking; low, thin, long median septum commonly present in each valve; ventral diductor scars flabellate, feebly impressed, enclosing elongate adductor scars; cardinal plate imperforate, originating normal to plane of commissure and bending backward into delthyrial cavity of pedicle valve; jugum united and projecting backward as simple stem. *U.Sil.-L.Carb.(Miss.),* cosmop.—— FIG. 541,1. **N. ventricosa* (HALL), L.Dev.(New Scotland), USA(N.Y.); 1a,b, lat., ped.v. views showing jugum, ×4; 1c-g, brach.v., lat., lat. (brach.v. only), brach.v. int. (ant. view), ped.v. int., ×2 (396).

Superfamily KONINCKINACEA Davidson, 1853

[*nom. transl.* BOUCOT, JOHNSON, & STATON, 1964, p. 820 (*ex* Koninckidae DAVIDSON, 1853, p. 92)]

Spiralia directed ventrally; jugum giving rise to bifurcations that are intercoiled with primary volutions of spiralia to their ends. *Trias.-Jur.*

Family KONINCKINIDAE Davidson, 1853

[Koninckinidae DAVIDSON, 1853, p. 92]

Concavo-convex shells. *Trias.-Jur.*

Koninckina SUESS in DAVIDSON, 1853, p. 92 [**Productus leonhardi* WISSMANN, 1841, p. 18; OD]. Concavo-convex, smooth, alate shells. Spiralia lacking marginal spinules. *Trias.,* Eu.(Aus.).—— FIG. 542,4a-i. *K. leopoldi austriae* BITTNER; 4a, ped.v. view (brachidium), ×2; 4b-i, ser. secs., ×2 (76).——FIG. 542,4j,k. **K. leonhardi* (WISSMANN); 4j,k, ped.v., post. views, ×1 (76) [p. *H904*].

Amphiclina LAUBE, 1865, p. 28 [**Producta dubia* MÜNSTER, 1841, p. 68; OD]. Concavo-convex, smooth shells with narrow cardinal margin; anterior and lateral margins thickened. *Trias.,* Eu. (Aus.).——FIG. 542,2. *A. amoena* BITTNER; 2a-c, ped.v., lat., ant. views, ×1 (76).

Amphiclinodonta BITTNER, 1888, p. 288 [**A. liasina;* OD]. Concavo-convex, smooth shells with narrow cardinal margin, resembling *Amphiclina,* but bearing submarginal rows of thickened articulating tubercles as in *Koninckodonta.* Spiralia bearing marginal spines. *Trias.-Jur.,* Eu.(Aus.). ——FIG. 542,1. *A. carnica* BITTNER, Trias., Aus.; 1a-c, ped.v., lat., post. views, ×1 (76).

Koninckella MUNIER-CHALMAS, 1880, p. 280 [**Leptaena liasina* BOUCHARD, 1847, fig. 2; OD]. Concavo-convex, smooth, alate shells resembling *Koninckina,* but bearing marginal spinules on spiralia; cardinal process present. *Trias.-Jur.,* Eu.

(Aus.).——FIG. 542,*3. K. fastigata* BITTNER; Trias., Aus.; *3a-d,* brach.v., lat., ped.v., ant. views, ×1 (76).

Koninckodonta BITTNER, 1893, p. 137 [**K. fuggeri*; OD]. Concavo-convex, smooth, alate shells resembling *Koninckina,* but bearing submarginal rows of thickened articulating tubercules, as in *Amphiclinodonta. Trias.,* Eu.(Aus.).——FIG. 542, 5. **K. fuggeri*; brach.v. int. (brachidium), ×2 (78).

SPIRIFERIDINA

[Materials for this suborder prepared by CHARLES W. PITRAT]

In the Lower Silurian two groups of impunctate spiriferoids make their appearance nearly simultaneously. The Eospiriferinae, which seem to have appeared slightly earlier and which are judged to be more primitive structurally, are characterized by long crural plates, a nonstriate cardinal process, and micro-ornament of rather prominent radial striae. The somewhat later Acrospiriferinae may have been derived from the Eospiriferinae or they may have arisen separately from a primitive group of spire-bearers. In any case they possess a distinctly different sort of brachial valve, with shorter crural plates and a longitudinally striate cardinal process, as well as fimbriate micro-ornament.

The superfamily Cyrtiacea, as here interpreted, includes the Eospiriferinae and their impunctate derivatives, the Cyrtiinae and the Ambocoeliidae. The Eospiriferinae and Cyrtiinae are thought to be very closely related, since the two groups are substantially the same except for overall shell shape and modifications of the delthyrium. The position of the Ambocoeliidae is not quite so certain. However, the presence of a nonstriate cardinal process and well-developed crural plates points to the Eospiriferinae as their progenitors. The highly variable micro-ornament of the Ambocoeliidae neither confirms nor denies the postulated relationship.

Nearly half the recognized genera of the Spiriferidina are placed in the superfamily Spiriferacea, which was initiated in the Early Silurian with appearance of the Acrospiriferinae. As evolution of the Spiriferacea unfolded, most of the characters so distinctly expressed in the ancestral Acrospiriferinae became modified and specializations appeared. The primitive fimbriate micro-

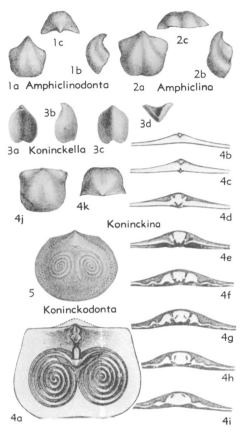

FIG. 542. Koninckinidae (p. *H666-H667*).

ornament gave rise to a bewildering variety of radial striae, concentric growth lamellae, granules, pustules, spines, and various combinations of them. The crural plates, short or absent in the Acrospiriferinae, became so variable that apparently they are useless in later groups for classification above the generic level. Even the typically impunctate shell developed punctae in one group (Syringothyridinae). Aside from radial macro-ornament, which occurs also in many of the Cyrtiacea, and the spiral brachidium universally common to spiriferoids, the only feature of importance found in all Spiriferacea is the longitudinally striate cardinal process. Indeed, the assumption that this feature is of critical importance is the main basis for the suprageneric classification here proposed.

The spiriferoids with smooth shells are especially difficult to deal with confidently at the higher levels of classification. One group of them, the Ambocoeliidae, is placed in the Cyrtiacea on the reasonably firm grounds of their retention of well-developed crural plates and possession of a nonstriate cardinal process. The others, which are segregated, perhaps artificially, in the superfamily Reticulariacea, have a longitudinally striate cardinal process which bespeaks a spiriferacean ancestry. The Reticulariidae display a micro-ornament of concentric growth lamellae upon which uniramous spines or granules are generally superposed. This micro-ornament, together with observation that the Recticulariidae appear in the Upper Silurian or Lower Devonian, suggests an ancestor among the Delthyrididae. The Elythidae, which appeared in the Middle Devonian, comprise a rather closely knit group of smooth spiriferoids, characterized by a micro-ornament of fine double-barreled spines. They are probably derived from the Reticulariidae. The remaining group of essentially smooth spiriferoids, the Martiniidae, possess a micro-ornament of surficial pits. It seems likely that they were derived from the Reticulariidae in Early Carboniferous time, although the possibility of other (even polyphyletic) origin cannot be discounted.

Punctate shell structure is found in three distinct groups of spiriferoids. One occurrence, in late Paleozoic Syringothyridinae, has already been mentioned, this group being classed with the Spiriferacea on the basis of overall similarity of the shell with those of the impunctate Spinocyrtiidae and Licharewiinae. Another group, the Suessiacea, which appeared in the Silurian, evidently either arose from the Cyrtiidae or was independently derived from some primitive spire-bearer, as evidenced by the nonstriate cardinal process. The origin of the third group, the Spiriferinacea, which appeared in the Early Carboniferous, is also somewhat problematical. The main evidence for the general assumption that the Spiriferinacea were derived from the Suessiacea appears to be the presence in both groups of punctation—a character of dubious reliability. Such a relationship has the disadvantage of requiring separate deriva-

tion of the striate cardinal process in the Spiriferinacea and the Spiriferacea. It seems more likely that the Spiriferinacea were developed from some Late Devonian or Early Carboniferous spiriferacean. Of all the known spiriferaceans, the mucrospiriferid *Tylothyris,* on the basis of general shell form and possession of a well-developed ventral median septum, seems to be the most plausible candidate for recognition as ancestor.

Suborder SPIRIFERIDINA Waagen, 1883

[*nom. correct.* P̲ɪ̲ᴛʀᴀᴛ, herein (*pro* suborder Spiriferacea Wᴀᴀɢᴇɴ, 1883, p. 447)]

Smooth, plicate, or costate, generally transverse Spiriferida with rather long hinge line; pedicle valve interarea generally well developed, bearing delthyrium which is usually either fully open or constricted by disjunct deltidial plates. Spiralia directed laterally or posterolaterally, with primary lamellae directed parallel and close to sagittal plane, either separate or connected by simple jugum; area of diductor attachment on brachial valve smooth in primitive forms, deeply striate longitudinally in advanced and more typical forms. Shell substance either punctate or impunctate. *L.Sil.-L.Jur.*

Superfamily CYRTIACEA Frederiks, 1919 (1924)

[*nom. transl.* P̲ɪ̲ᴛʀᴀᴛ, herein (*ex* Cyrtiinae Fʀᴇᴅᴇʀɪᴋs, 1919 (1924), p. 312)]

Form, macro- and micro-ornament highly variable. Pedicle valve interior with or without dental plates, generally lacking median septum; brachial valve interior with crural plates and nonstriate cardinal process. Shell substance impunctate. *L.Sil.-Perm.*

Family CYRTIIDAE Frederiks, 1919 (1924)

[*nom. transl.* Iᴠᴀɴᴏᴠᴀ, 1959, p. 55 (*ex* Cyrtiinae Fʀᴇᴅᴇʀɪᴋs, 1919 (1924), p. 312) [=Eospiriferinae Sᴄʜᴜᴄʜᴇʀᴛ & LᴇVᴇɴᴇ, 1929, p. 20]

Form variable, fold and sulcus present, generally smooth, some genera with minor ribbing; lateral slopes smooth, plicate, or costate, micro-ornament consisting of prominent capillae crossed by growth lines, providing intersections which rarely are nodose but never spinose; delthyrium provided with deltidial plates in varying stages of development. Interior of pedicle valve with well-developed, generally long dental plates; brachial valve interior with long crural plates. *L. Sil. (Llandover.)-M. Dev. (Couvin.).*

Subfamily CYRTIINAE Frederiks, 1919 (1924)
[*emend.* Boucot, 1963, p. 701]

Pedicle valve pyramidal; interarea very high; delthyrium narrow, in small specimens occupied by disjunct deltidial plates, in larger specimens by conjunct deltidial plates with or without third plate at base, delthyrium of very large specimens completely closed by deposits of secondary material. *L.Sil.(U.Llandover.)-Dev.(Ems.-?Couvin.).*

Cyrtia DALMAN, 1828, p. 92 [**Anomites exporrectus* WAHLENBERG, 1821, p. 64; SD DAVIDSON, 1853, p. 83]. Entire shell smooth. *Sil.(U.Llandover.-Ludlov.),* N. Am.; *Sil. (U. Llandover.) - Dev. (Ems.-?Couvin.),* Eu.——FIG. 543,*1*. **C. exporrecta* (WAHLENBERG), U. Llandover., Sweden (Gotl.); *1a-e,* brach.v., ped.v., post., ant., lat. views, ×2 (104).

Plicocyrtia BOUCOT, 1963, p. 704 [**Cyrtia petasus* BARRANDE, 1848, p. 183; OD]. Similar to *Cyrtia,* but with 1 to 3 low rounded lateral plications separated by shallow U-shaped depressions. *Sil.(U. Wenlock.-?Ludlov.),* Eu.-Asia.——FIG. 543,*3*. **P. petasus* (BARRANDE), Czech.; *3a-c,* brach.v., lat., ped.v. views, ×1; *3d-e,* brach.v. and ped.v. int. molds, ×2 (104).

?Tannuspirifer IVANOVA, 1960, p. 267 [**Spirifer pedaschenkoi* CHERNYSHEV, 1937, p. 51; OD]. Small; shape as in *Cyrtia*; lateral slopes plicate; fold and sulcus profound, bald; micro-ornament consisting of radial striae only; data on interior wanting. *Sil.,* USSR.——FIG. 543,*2*. **T. pedaschenkoi* (CHERNYSHEV); *2a,b,* ped.v. ant., ped.v. post. views, ×1.5 (448).

Subfamily EOSPIRIFERINAE Schuchert & LeVene, 1929
[Eospiriferinae SCHUCHERT & LEVENE, 1929, p. 20 (*emend.* BOUCOT, 1963, p. 685)]

Pedicle valve convex but not pyramidal; interarea not abnormally high; delthyrium triangular, not unusually narrow, occupied by deltidial plates which are generally disjunct. *L.Sil.(Llandover.)-M.Dev.(Couvin.).*

Eospirifer SCHUCHERT, 1913, p. 411 [**Spirifer radiatus* J. DE C. SOWERBY, 1840, p. 245 (=*Spirifer lineatus* J. DE C. SOWERBY, 1825, p. 151) (*non Conchyliolites anomites lineatus* MARTIN, 1809, =*Terebratula? lineata* J. SOWERBY, 1822); OD]. Medium-sized, moderately transverse, wholly nonplicate; hinge line ranging from one-half to almost equal maximum width; fold and sulcus well developed. *Sil.(Llandover.-Ludlov.),* cosmop.; *L.*

FIG. 543. Cyrtiidae (Cyrtiinae) (p. H669).

Dev.(Gedinn.-Ems.), Eu.-Asia.——Fig. 544,*1.* **E. radiatus* (Sowerby), Sil., Sweden(Gotl.); *1a-e,* ant., lat., post., brach.v., ped.v. views, ×1; Sil.,

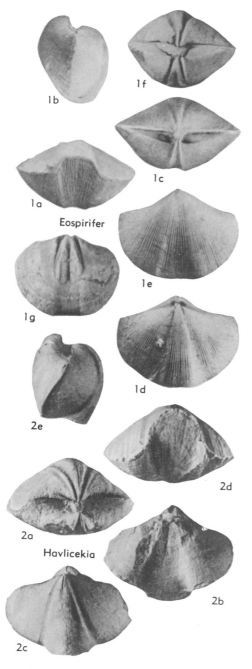

Fig. 544. Cyrtiidae (Eospiriferinae) (p. *H669-H670*).

USA(Ill.); *1f-g,* post. int. mold, ped.v. int. mold, ×1 (104).

[The name *Conchyliolites anomites lineatus* was published by Martin (1809) for a form now included in *Phricodothyris.* J. Sowerby (1822) referred to this species as *Terebratula? lineata* (Martin). Somewhat later J. de C. Sowerby (1825) published the name *Spirifer lineatus* for the fossil which is now the type of *Eospirifer.* By 1840 he recognized that Martin's species was a spire-bearer, and therefore, according to practice of the time, belonged in *Spirifer.* Thus, *Spirifer lineatus* J. de C. Sowerby, 1825, became a junior homonym of *Spirifer lineatus* (Martin), 1809, a situation which was corrected in 1840 when Sowerby substituted *Spirifer radiatus* for the invalid *Spirifer lineatus* J. de C. Sowerby, 1825. In 1848 the I.C.Z.N. invalidated the work of Martin. Subsequently the Commission (1956) ruled that the fossil which Martin had named *Conchyliolites anomites lineatus* should be credited to Sowerby in the original combination *Terebratula? lineata* J. Sowerby, 1822, thereby retaining the priority of this fossil over *Spirifer lineatus* J. de C. Sowerby, 1825. Although *Terebratula? lineata* J. Sowerby, 1822, and *Spirifer lineatus* J. de C. Sowerby, 1825, are not now regarded as belonging to the same genus, the fact that they were once so treated requires permanent suppression of *Spirifer lineatus* J. de C. Sowerby, 1825, and the substitution of *Spirifer radiatus* J. de C. Sowerby, 1840 (Code, Art. 59c)]

Havlicekia Boucot, 1963, p. 693 [*Spirifer secans Barrande, 1848, p. 168; OD]. Lateral slopes weakly plicate in early growth stages, smooth thereafter; fold and sulcus smooth, wide, and in late growth stages very deep, resulting in markedly uniplicate anterior commissure. *Sil.(Wenlock.)-Dev.(Ems.),* cosmop.——Fig. 544,*2. H.* sp., Lochkov Ls., Czech.; *2a-e,* post., ped.v., brach.v., ant., lat. views, ×1 (104).

Janius Havlíček, 1957, p. 247 [*Spirifer nobilis Barrande, 1848, p. 184; OD]. Lateral slopes with about 3 anteriorly bifurcating plications, separated by U-shaped furrows; fold and sulcus either smooth or with single median costa in sulcus and single median groove in fold; micro-ornament of radial striae and fine concentric growth lines with nodose intersections. *Sil.(U.Wenlock.-Ludlov.),* N.Am.-Eu.; *Sil.(Ludlov.) - Dev.(Couvin.),* Asia. ——Fig. 545,*4.* **J. nobilis* (Barrande), U.Sil. (Ludlov.), Boh.; *4a-b,* brach.v., ped.v. views, ×1 (411).

Macropleura Boucot, 1963, p. 690 [**Delthyris macropleurus* Conrad, 1840, p. 207; OD]. Large, rather transverse, with hinge line almost equal to maximum width; lateral slopes with 3 to 6 simple plications separated by broad, rounded depressions; plications very strong posteriorly, in some shells tending to weaken anteriorly; fold and sulcus broad, smooth, rather low, but producing strongly uniplicate anterior commissure. *Sil. (Wenlock.)-Dev.(Gedinn.),* N.Am.; *Sil.(Llandover.) - Dev.(Couvin.),* Eu.-Asia.——Fig. 545,*3.* **M. macropleura* (Conrad), L.Dev.(Helderberg.), USA(N.Y.) *(3a-e)*; L.Dev. (Birdsong F.), USA (Tenn.) *(3f)*; L.Dev.(Helderberg.), USA(Md.) *(3g)*; *3a-e,* brach.v., ant., post., ped.v., lat. views, ×0.7; *3f,* ped.v. int., ×1; *3g,* brach.v. int., ×2 (104).

Nikiforovaena Boucot, 1963, p. 697 [**Spirifer (Eospirifer) ferganensis* Nikiforova, 1937, p. 48; OD]. Lateral costae rounded, separated by U-shaped depressions; fold with one or more promi-

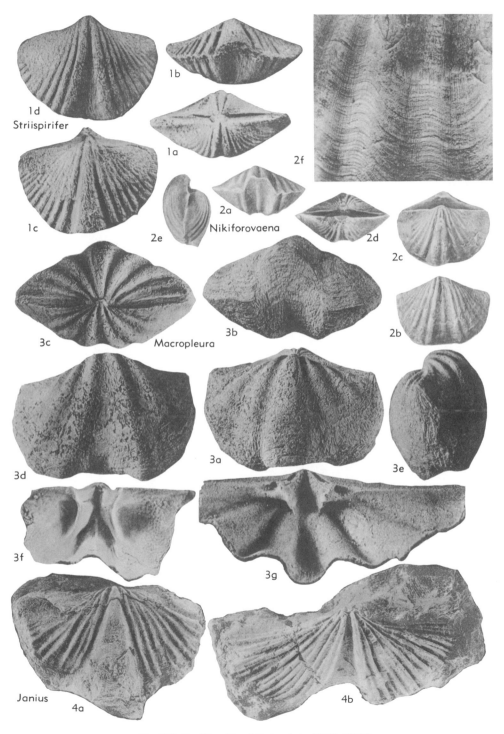

FIG. 545. Cyrtiidae (Eospiriferinae) (p. *H670, H672*).

nent grooves; sulcus with one or more prominent plications; otherwise similar to *Strüspirifer*. *Sil.* (*?Wenlock.-Ludlov.*), Asia-Australia.——Fig. 545, 2. **N. ferganensis* (Nikiforova), USSR; *2a-e*, ant., ped.v., brach.v., post., lat. views, ✕1; *2f*, surface, ✕5 (104).

Striispirifer Cooper & Muir-Wood, 1951, p. 195 [*pro Schuchertia* Frederiks, 1926, p. 406 (*non* Gregory, 1899)] [**Delthyris niagarensis* Conrad, 1842, p. 261; OD]. Like *Eospirifer* except lateral slopes with numerous well-defined costae separated by narrow V-shaped grooves; fold and sulcus smooth. *Sil.(U.Llandover.-Ludlov.)*, N.Am.; *Sil. (Llandover.) - M. Dev.(Couvin.)*, Eu.——Fig. 545,1. **S. niagarensis* (Conrad), M.Sil.(Clinton.), USA(N.Y.); *1a-d*, post., ant., brach.v., ped.v. views, ✕2 (104).

Family AMBOCOELIIDAE George, 1931

[*nom. transl.* Ivanova, 1959, p. 56 (*ex* Ambocoeliinae George, 1931, p. 42)]

Generally small; unequally biconvex to plano-convex; generally approximately equidimensional to moderately transverse; hinge line commonly slightly less than maximum width; fold and sulcus generally present but variably expressed, bald; lateral slopes generally bald, less commonly plicate; pedicle valve interior commonly without dental plates; brachial valve interior with crural plates, and generally with nonstriate cardinal process. *?U.Sil., L.Dev.-Perm.*

Ambocoelia Hall, 1860, p. 71 [**Orthis umbonata* Conrad, 1842, p. 264; OD]. Small; pedicle valve strongly convex, with prominent incurved beak, large interarea with open delthyrium; brachial valve semicircular, weakly convex; hinge line slightly less than maximum width; pedicle valve with narrow sulcus; brachial valve without fold; anterior commissure rectimarginate; macro-ornament lacking; micro-ornament consisting of concentric growth lamellae and very weak capillae. Pedicle valve interior without dental plates; brachial valve interior with adductor scars placed anteriorly; crural plates distinct; cardinal process rather large, bifid, nonstriate. *L.Dev.-Miss.*, cosmop.——Fig. 546,1. **A. umbonata* (Conrad), M.Dev., USA(N.Y.); *1a-c*, brach.v., lat., ped.v. views, ✕1.5; *1d,e*, ped.v. int., brach.v. int., ✕3 (396).

?Alaskospira Kirk & Amsden, 1952, p. 61 [**A. dunbari*; OD]. Small; unequally biconvex; moderately transverse, with hinge line almost equal to maximum width; macro-ornament wanting; micro-ornament consisting of concentric growth lamellae and radial striae; pedicle valve interior with prominent median ridge or platform extending nearly to anterior margin, lacking dental plates; brachial valve interior with low median

ridge appearing just anterior to beak, extending forward several millimeters, then replaced by shallow depression. *U.Sil.*, Alaska.——Fig. 546,2. **A. dunbari*; *2a-m*, transv. secs., ✕2; *2n-q*, lat., ped.v., brach.v., ant. views, ✕2 (476).

Ambothyris George, 1931, p. 42 [**Spirifera infima* Whidbourne, 1893, p. 108; OD]. Pedicle valve with feebly curved beak; micro-ornament lacking or consisting of radial striae only; otherwise seemingly similar to *Ambocoelia*, but interior unknown. *M.Dev.*, Br.I.——Fig. 546,5. **A. infima* (Whidbourne); *5a-d*, ped.v., brach.v., ant., lat. views, ✕4.5 (334).

Attenuatella Stehli, 1954, p. 343 [**A. texana*; OD]. Pedicle valve narrow, long, markedly convex; beak greatly attenuated, strongly incurved. Pedicle valve interior with strong, rather wide median ridge; otherwise seemingly similar to *Crurithyris* but brachial valve interior unknown. *L.Perm.*, W.Tex.——Fig. 546,4. **A. texana*; *4a,b*, ped.v. int., ped.v. ext., ✕4.5 (773).

Bisinocoelia Havlíček, 1953, p. 7 [**B. bisinuata*; OD]. Brachial valve interior with erect, rodlike crural bases set far apart and embedded in thickened valve floor which bears prominent Y-shaped ridge; otherwise similar to *Crurithyris*. *L.Dev.*, Czech.——Fig. 546,3. **B. bisinuata*, *3a*, brach.v. int., ✕8; *3b*, brach.v. ext., ✕3 (411).

Crurithyris George, 1931, p. 42 [**Spirifer urei* Fleming, 1828, p. 376; OD]. Both valves with very weak sulci; anterior commissure rectimarginate; micro-ornament consisting of concentric growth lamellae only; cardinal process triangular, tuberculate; brachial valve interior with adductor scars in normal position; otherwise similar to *Ambocoelia*. *Dev.-Perm.*, cosmop.——Fig. 546,6. **C. urei* (Fleming), L.Carb., Br.I.; *6a-d*, ped.v., brach.v., lat. views and brach.v. int. mold, ✕9 (334).

Echinocoelia Cooper & Williams, 1935, p. 844 [**E. ambocoelioides*; OD]. Cardinal process stout, bilobed; micro-ornament consisting of concentric growth lamellae, each bearing row of short, fine spines; otherwise similar to *Crurithyris*. *M.Dev.*, E.N.Am.——Fig. 547,5. **E. ambocoelioides*, Tully, USA(N.Y.); *5a-c*, brach.v. ext., brach.v. int., ped.v. views, ✕2 (198).

Emanuella Grabau, 1923, p. 192 [**Nucleospira takwanensis* Kayser, 1883, p. 86; OD]. Both valves markedly convex, pedicle valve somewhat more so; beaks feebly incurved; anterior commissure weakly uniplicate. *M.Dev.-U.Dev.*, cosmop.—— Fig. 547,8. **E. takwanensis* (Kayser), M.Dev., China; *8a-c*, brach.v., lat., ant. views, ✕3 (358).

?Ilmenia Nalivkin, 1941, p. 186 [**I. altovae*; OD]. Micro-ornament consisting of both capillae and concentric growth lamellae; pedicle valve interior with distinct dental plates; otherwise similar to *Crurithyris*. *M.Dev.-U.Dev.*, Eu.-Asia.—— Fig. 547,1. **I. altovae*, U.Dev., USSR; *1a-d*, ped. v., brach.v., ant., lat. views, ✕1 (594).

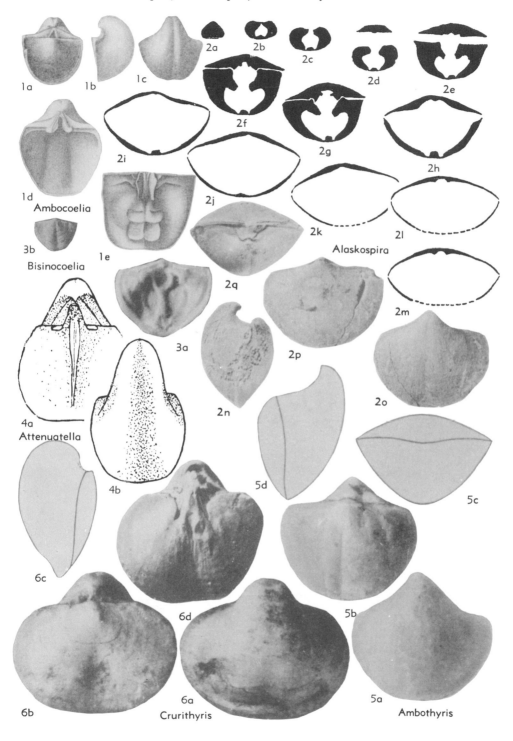

FIG. 546. Ambocoeliidae (p. *H672*).

?Ilmenispina HAVLÍČEK, 1959, p. 180 [**I. hanaica*; OD]. Transverse, with rounded cardinal extremities; crural plates large, dorsally convergent, in some forming sessile septalium; micro-ornament consisting of fine, irregularly placed, radially elongated spines; otherwise similar to *Ilmenia*. *M.Dev.* *(Givet.),* Czech.——FIG. 547,2. **I. hanaica*; *2a-e,* brach.v. int., ped.v., brach.v., ant., lat. views, ×2 (411).

Ladjia VEEVERS, 1959, p. 125 [**L. saltica*; OD]. Anterior commissure rectimarginate to weakly uniplicate; micro-ornament consisting of conspicu-

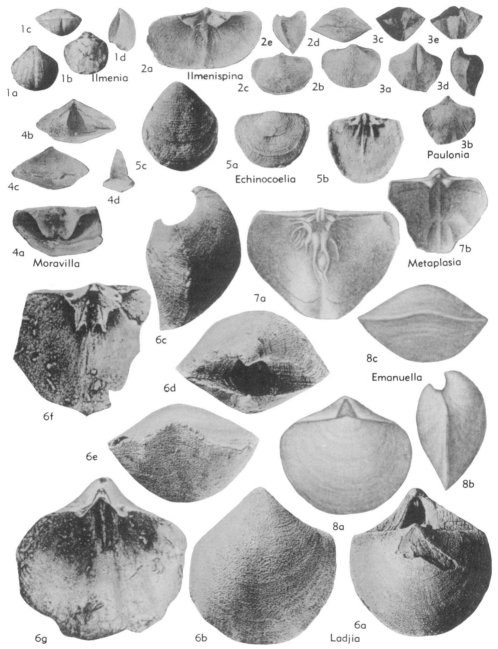

FIG. 547. Ambocoeliidae (p. *H672, H674-H675*).

ous radial striae and concentric growth lamellae; brachial valve interior with low, triangular, possibly striate cardinal process; otherwise similar to *Crurithyris*. *U.Dev.(Frasn.)*, W.Australia.——Fig. 547,6. *L. saltica; 6a-e,* brach.v., ped.v., lat., post., ant. views, ×3; *6f,g,* brach.v. int., ped.v. int., ×4.5 (838).

Metaplasia HALL & CLARKE, 1894, p. 56 [*Spirifer pyxidatus* HALL, 1859, p. 428; OD]. Rather small; moderately transverse, with hinge line approximately equal to maximum width; fold and sulcus weak; lateral slopes generally with 1 or 2 broad, low plications; micro-ornament consisting of concentric growth lamellae only; pedicle valve interior lacking dental plates; brachial valve interior with crural plates which reach floor of valve near mid-line; cardinal process bilobed. *Dev. (Siegen.-Ems.)*, N.Am.——Fig. 547,7. *M. pyxidata* (HALL), Oriskany, USA(N.Y.); *7a-b,* brach. v. int., ped.v. int., ×3 (396).

?Moravilla HAVLÍČEK, 1953, p. 4 [*M. ficneri;* OD]. Extremely transverse, with hinge line equal to maximum width; pedicle valve hemipyramidal; brachial valve weakly convex; crural plates very long, extending almost to anterior margin, strongly divergent dorsally and meeting valve floor close to mid-line; otherwise similar to *Ilmenia*. *M.Dev. (Givet.)*, Czech.——Fig. 547,4. *M. ficneri; 4a,* brach.v. int., ×3; *4b-d,* post., ant., lat. views, ×1.5 (411).

?Paulonia NALIVKIN, 1925, p. 267 [*Spirifer ranovensis* PEETZ, 1893, p. 53; OD]. Rather small; unequally convex; slightly transverse, with rounded cardinal extremities; fold and sulcus distinct; entire surface with numerous, fine costae; micro-ornament consisting of papillae; pedicle valve interior lacking dental plates and median septum; brachial valve interior unknown. *L.Carb.*, USSR.——Fig. 547,3. *P. ranovensis* (PEETZ); *3a-e,* ped.v., brach.v., post., lat., ant. views, ×1 (752).

Plicoplasia BOUCOT, 1959, p. 19 [*P. cooperi;* OD]. Pedicle valve with strongly biplicate fold; brachial valve with sulcus bearing median plication; lateral plications very strong; otherwise similar to *Metaplasia*. *L.Dev.*, N. Am.-S. Am.-Afr.——Fig. 548,3. *P. cooperi,* Oriskany, USA(N.Y.); *3a-f,* brach. v. int., lat., ant., ped.v., brach.v., ped.v. int., ×3 (99).

?Prosserella GRABAU, 1910, p. 139 [*P. modestoides;* SD SCHUCHERT & LEVENE, 1929, p. 101] [=?*Rhynchospirifer* PAULUS, 1957, p. 51 (type, *R. halleri*)]. Biconvex; approximately equidimensional; pedicle valve beak prominent, somewhat incurved; interarea rather high with open delthyrium; anterior commissure rectimarginate to weakly uniplicate; macro-ornament lacking; micro-ornament a reticulate network of concentric and radial elements. Pedicle valve interior with long, parallel, closely spaced dental plates; brachial valve interior with crural plates which, in some species, converge, forming septalium supported

by median septum. *L.Dev.(Ems.)-M.Dev.(Givet.)*, Eu.-N.Am.——Fig. 548,4a-c. *P. modestoides,* L.Dev., USA(Mich.); *4a-c,* ped.v. post. int. mold; ped.v. int. mold, ped.v. lat. int. mold, ×1 (364).——Fig. 548,4d. P. subtransversa (GRABAU), L. Dev., USA(Mich.); brach.v. int., ×2 (364).—— Fig. 548,4e,f. P. halleri (PAULUS), M.Dev., Ger.; *4e,* brach.v. int., ×3; *4f,* ped.v. int. mold, ×1.5; *4g-i,* brach.v., post., lat., ×2.2 (637).

Pustulatia COOPER, 1956, p. 769 [*pro Vitulina* HALL, 1860, p. 72 (*non* SWAINSON, 1840); *Pustulina* COOPER, 1942, p. 228 (*non* QUENSTEDT, 1857)] [*Vitulina pustulosa* HALL, 1860; OD]. Micro-ornament consisting of radial striae which may be interrupted to form radially aligned pustules; otherwise similar to *Plicoplasia* except for slightly more numerous lateral plications. *Dev.*, N.Am.-S.Am.-Afr.——Fig. 548,2. *P. pustulosa* (HALL), M.Dev., USA(N.Y.); *2a,* ped.v., ×2; *2b,* brach.v., ×3; *2c,* brach.v. int., ×2 (178).

?Quasimartinia HAVLÍČEK, 1959, p. 179 [*Q. rectimarginata;* OD]. Exterior similar to *Ambocoelia*, but seemingly lacking micro-ornament; pedicle valve interior with neither dental plates nor median septum; brachial valve interior lacking crural plates. *L.Dev.*, Boh.——Fig. 548,1. *Q. rectimarginata; 1a-d,* ped.v., brach.v., ant., lat., ×2 (411).

Spinoplasia BOUCOT, 1959, p. 18 [*S. gaspensis;* OD]. Micro-ornament consisting of growth lamellae bearing fine spines; cardinal process simple; otherwise similar to *Metaplasia*. *L.Dev.*, E.Can. (Gaspé).——Fig. 548,5. *S. gaspensis; 5a,b,* brach.v. int. mold, brach.v. mold, ×10 (99).

Superfamily SUESSIACEA
Waagen, 1883

[*nom. transl.* PITRAT, herein (*ex* Suessiinae WAAGEN, 1883, p. 498)]

Pedicle valve hemipyramidal, with high interarea; brachial valve weakly convex; radial macro-ornament absent, confined to lateral slopes or present over entire shell; pedicle valve interior with median septum which either supports spondylium or is expanded at its posterodorsal margin; brachial valve interior with nonstriate cardinal process; shell substance commonly punctate. *Sil.-L.Jur.*

Family CYRTINIDAE Frederiks, 1912

[*nom. transl.* STEHLI, 1954, p. 350 (*ex* Cyrtininae FREDERIKS, 1912)]

Generally rather small and equidimensional, with weakly convex brachial valve and hemipyramidal pedicle valve; interarea large, triangular; delthyrium generally open or with convex pseudodeltidium; costae or plications either absent, present only on

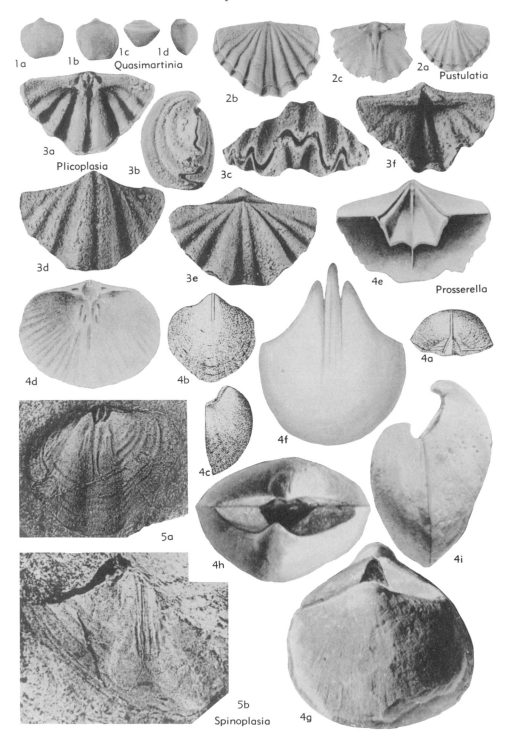

Fig. 548. Ambocoeliidae (p. *H675*).

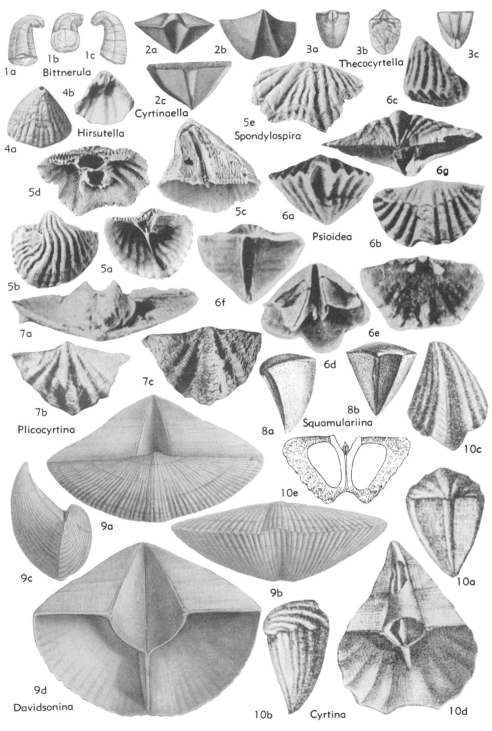

FIG. 549. Cyrtinidae (p. *H678-H679*).

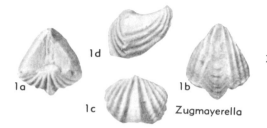

1a 1d 1b 1c Zugmayerella

Fig. 549A. Cyrtinidae (p. *H679*).

lateral slopes, or present over entire shell; micro-ornament highly variable; pedicle valve interior with dental plates and median septum uniting to form spondylium, with or without tichorhinum; brachial valve interior with nonstriate cardinal process, complete jugum; shell substance generally punctate. *Sil.-U.Trias.*

Cyrtina DAVIDSON, 1858, p. 66 [*Calceola heteroclita* DEFRANCE, 1828, p. 306; SD HALL & CLARKE, 1894, p. 44] [=*Spinocyrtina* FREDERIKS, 1916, p. 18 (type, *Cyrtia hamiltonensis* HALL, 1857); *Cyrtinaellina* FREDERIKS, 1926, p. 414 (type, *Cyrtina acutirostris* SHUMARD, 1855)]. Small to medium-sized; almost equidimensional; pedicle valve hemipyramidal, in some deformed; brachial valve weakly convex; interarea very high; delthyrium covered by convex pseudodeltidium bearing large foramen near apex; lateral slopes with several distinct plications; fold and sulcus bald; micro-ornament consisting of concentric growth lamellae in various stages of expression; pedicle valve interior with prominent spondylium with tichorhinum; shell substance punctate. *Sil.-Perm.*——Fig. 549,*10*. *C. heteroclita* (DEFRANCE), M. Dev., W.Eu.; *10a-c*, ant., lat., ped.v., ×1.5; *10d*, ped.v. int., ×3 (229); *10e*, transv. sec., ?×2 (613). [Fixation of type-species is usually credited to DALL (1877, p. 24), though he merely stated *Cyrtina heteroclita* (DEFRANCE) to be the "first species." The first unequivocal designation of the type-species seems to be that of HALL & CLARKE (1894, p. 44).]

Bittnerula HALL & CLARKE, 1894, p. 764 [*Cyrtina zitteli* BITTNER, 1890, p. 117; OD]. Pedicle valve highly deformed; lateral slopes bald; otherwise seemingly similar to *Cyrtina*, but detailed structure of spondylium not known. *U.Trias.*, Eu.——Fig. 549,*1*. *B. zitteli* (BITTNER); *1a-c*, post., brach.v., lat., ×1 (76).

Cyrtinaella FREDERIKS, 1916, p. 18 [*Cyrtina biplicata* HALL, 1857, p. 165; OD] [=?*Pyramidalia* NALIVKIN, 1947, p. 124 (type, *Spirifera simplex* PHILLIPS, 1841, p. 71)]. Lateral slopes bald; micro-ornament consisting of capillae and

concentric growth lamellae; otherwise similar to *Cyrtina*. Dev., N.Am., ?Eu.——Fig. 549,*2*. *C. biplicata* (HALL), L.Dev., USA(N.Y.); *2a-c*, ant., ped.v., post., ×1 (392).

Davidsonina SCHUCHERT & LEVENE, 1929, p. 120 [*pro Cyrtinopsis* FREDERIKS, 1916, p. 17 (*non* SCUPIN, 1896); *Davidsonella* FREDERIKS, 1926, p. 413 (*non* MUNIER-CHALMAS, 1880; *nec* WAAGEN, 1885)] [*Spirifera septosa* PHILLIPS, 1836, p. 216; OD]. Rather large, transverse; interarea not unusually high; delthyrium open; both lateral slopes, fold, and sulcus with numerous bifurcating costae; pedicle valve interior with spondylium, lacking tichorhinum; otherwise similar to *Cyrtina*. *L.Carb.*, Eu.-Asia.——Fig. 549,*9*. *D. septosa* (PHILLIPS), Visean, Br.I.; *9a-d*, brach.v., ant., lat., ped.v. int., ×0.7 (229).

?Hirsutella COOPER & MUIR-WOOD, 1951, p. 195 [*pro Hirsutina* KIRCHNER, 1933, p. 106 (*non* TUTT, 1909)] [*Spirifer? hirsutus* ALBERTI, 1864, p. 156; OD]. General shape cyrtiniform; delthyrium partly closed by disjunct deltidial plates; fold and sulcus weak; entire shell costate, costae on fold and sulcus being somewhat weaker than those on lateral slopes; pedicle valve interior with spondylium; shell substance impunctate. *M.Trias.*, Ger.——Fig. 549,*4*. *H. hirsuta* (ALBERTI); *4a,b*, ped.v. ext., ped.v. int., ×1 (475).

Plicocyrtina HAVLÍČEK, 1956, p. 608 [*Cyrtina (Plicocyrtina) sinuplicata* HAVLÍČEK, 1956; OD]. Lateral plications few, very strong; sulcus narrow, deep, bald posteriorly, but developing single, prominent, median plication anteriorly; fold considerably broader than sulcus, bald posteriorly, bearing shallow depression anteriorly; micro-ornament consisting of very strong imbricating growth lamellae; otherwise similar to *Cyrtina*. *Dev.(?Couvin.)*, Eu.-?N.Afr.——Fig. 549,*7*. *P. sinuplicata*, Bohem.; *7a-c*, brach.v. post., ped.v., brach.v., ×1.5 (408).

Psioidea HECTOR, 1879, p. 538 [*Spiriferina nelsonensis* TRECHMANN, 1918, p. 223; SD MARWICK, 1953, p. 39] [=?*Lepismatina* WANG, 1955, p. 353 (type, *L. hsui*)]. Moderately to markedly transverse; fold and sulcus sharp, narrow, bald; delthyrium open; otherwise similar to *Cyrtina*. *Trias.*, N.Am.-Australia-?Asia-?Eu.——Fig. 549,*6g*. *P. nelsonensis* (TRECHMANN), Carn., N.Z.; post., ×1 (536).——Fig. 549,*6a-f*. *P. hsui* (WANG), M.Trias., China; *6a-f*, ant., brach.v., lat., ped.v. int., brach.v. int., post., ×1 (852).

Spondylospira COOPER, 1942, p. 232 [*S. reesidei*; OD]. Fold and sulcus plicate; delthyrium open; hinge line denticulate; descending lamellae of spiralia supported by calcareous net; otherwise similar to *Cyrtina*. *Trias.*, W.N.Am.——Fig. 549, *5*. *S. reesidei*, U.Trias., USA(Idaho); *5a-e*, ped.v. int., ped.v., ant., brach.v. int., brach.v., ×1 (178).

Squamulariina FREDERIKS, 1916, p. 19 [*Cyrtina parva* GÜRICH, 1896, p. 266; OD]. Micro-ornament of fine papillae; otherwise seemingly similar to *Cyrtinaella,* but internal features poorly known. *M.Dev.,* Pol.——Fig. 549,*8*. **S. parva* (Gürich); *8a,b,* lat., post., ×2.5 (373).

Thecocyrtella Bittner, 1892, p. 15 [*pro Cyrtotheca* Bittner, 1890, p. 116 (*non* Hicks, 1872)] [**Cyrtotheca ampezzana* Bittner, 1890, p. 116; OD]. Very small; pedicle valve very high, curved; micro-ornament consisting of concentric growth lamellae only; delthyrium completely closed by pseudodeltidium; shell substance probably impunctate; otherwise similar to *Cyrtinaella. U.Trias.,* W. Eu.——Fig. 549,*3*. **T. ampezzana* (Bittner); *3a-c,* ped.v., ped.v. int., post., ×3 (76).

Zugmayerella Dagis, 1963, p. 99 [**Spiriferina koessenensis* Zugmayer, 1882; OD]. Fold and sulcus smooth; otherwise similar to *Spondylospira. U.Trias.,* Eu.——Fig. 549A,*1*. **Z. koessenensis* (Zugmayer); *1a-d,* post., ped.v., brach.v., lat., ×1 (212a).

Family SUESSIIDAE Waagen, 1883

[*nom. transl.* Pitrat, herein (*ex* Suessiinae Waagen, 1883, p. 498)]

Shells externally cyrtiniform, with hemipyramidal pedicle valve and weakly convex brachial valve; pedicle valve interior with prominent median septum which is horizontally expanded at its posterodorsal margin; brachial valve interior with very large hinge plate, apparently bearing adductor muscle scars; shell substance seemingly impunctate. *L.Jur.*

Suessia Deslongchamps, 1854, p. 6 [**S. costata;* SD Davidson, 1854, p. 28]. Entire shell plicate; delthyrium open; pedicle valve interior with dental plates reduced to teeth ridges. *L.Jur.,* Eu.—— Fig. 550,*1*. **S. costata; 1a-c,* post., ped.v. post., brach.v. int., ×? (397).

Superfamily SPIRIFERACEA King, 1846

[*nom. transl.* Schuchert, 1896, p. 333 (*ex* Spiriferidae King, 1846, p. 28)]

Shell form variable but generally rather transverse with either angular or narrowly rounded cardinal extremities and hinge line equal to or slightly less than maximum width; lateral slopes invariably costate or plicate; fold and sulcus generally present, tending to be bald in earlier forms, costate or plicate in later ones; pedicle valve interior with or without dental plates, commonly without median septum; brachial valve interior with striate cardinal process, with or without crural plates; shell substance generally impunctate. *L.Sil.-U.Perm.*

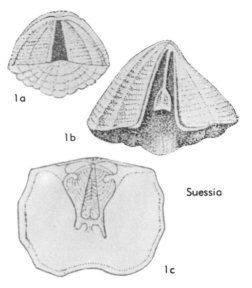

Fig. 550. Suessiidae (p. *H679*).

Family DELTHYRIDIDAE Waagen, 1883

[*nom. correct.* Pitrat, herein (*pro* Delthyridae Waagen, 1883, *nom. transl.* Ivanova, 1959, p. 56, *ex* Delthyrinae Waagen, 1883, p. 507)] [Formerly the family group taxon Delthyrididae has been attributed to Phillips, 1841 (p. 54). However, under the International Code (1961), Article 11e, in order to be available a family group name when first published must be based on the name then valid for a contained genus. Phillips, 1841 (p. 68), synonymized *Spirifera* Phillips, 1836, and *Delthyris* Dalman, 1828, and retained the former name in his description of species. However, *Spirifera* Phillips, 1836, is an "unjustified emendation" of *Spirifer* Sowerby, 1816, and is hence a junior objective synonym. Thus effectively Phillips synonymized *Spirifer* and *Delthyris* and although both names were available, only *Spirifer* was valid. Consequently Delthyridae Phillips, 1841, is not an available family group name, for the conditions of Article 11e are not satisfied.]

Biconvex, weakly to strongly transverse; lateral slopes plicate or costate; fold and sulcus smooth or with median rib in sulcus and median groove on fold; micro-ornament typically consisting of distinct growth lamellae upon which are superposed capillae that become fimbriate at anterior margin of each growth lamella, more rarely consisting of capillae only or of teardrop-shaped granules; interior of pedicle valve generally with well-developed dental plates, some with distinct median septum but none with delthyrial plate; brachial-valve interior with area of diductor attachment deeply striate longitudinally, resulting in comblike cardinal process; crural plates absent or short; shell substance impunctate. *L.Sil.(Llandover.)-M.Dev.(Couvin.).*

Subfamily DELTHYRIDINAE Phillips, 1841

[*nom. correct.* PITRAT, herein (*ex* Delthyrinae WAAGEN, 1883, p. 507)]

Lateral slopes with few, strong plications;

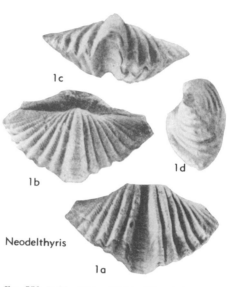

FIG. 552. Delthyrididae (Delthyridinae) (p. *H681*).

micro-ornament consisting of subdued growth lamellae and capillae which become fimbriate at anterior margin of each growth lamella; interior of pedicle valve with well-developed median septum, and commonly strong dental plates; brachial-valve interior with or without crural plates. *U.Sil.(Wenlock.)-M.Dev.(Couvin.).*

Delthyris DALMAN, 1828, p. 120 [**D. elevata*; OD] [=*Quadrifarius* FUCHS, 1929, p. 195 (type, *Delthyris loculata* FUCHS, 1923, p. 854)]. Pedicle-valve interior with well-developed dental plates and high, narrow median septum; interior of brachial valve with short crural plates. *U.Sil. (Wenlock.)-L.Dev.(Gedinn.),* Eu.——FIG. 551,2. **D. elevata,* Sil., Br.I.; *2a-c,* brach.v., post., lat., ×1.5 (229).

?Howittia TALENT, 1956, p. 34 [**Spirifer howitti* CHAPMAN, 1905, p. 18; OD]. Rather small; moderately transverse with slightly rounded cardinal extremities; lateral slopes with several rather strong simple plications; fold with prominent median groove; sulcus with from 1 to 3 plications; micro-ornament consisting of growth lamellae and fine capillae with papillae; pedicle valve interior with strong dental plates and short, high median septum; brachial valve interior lacking crural plates. *?M.Dev.(Couvin.),* Australia. ——FIG. 551,4. **H. howitti* (CHAPMAN); *4a,* transv. sec. ped.v., ×3; *4b,* transv. sec. brach.v., ×10; *4c-e,* brach.v., ant., post., ×1.5 (796).

Ivanothyris HAVLÍČEK, 1957, p. 438 [**Spirifer gibbosus* BARRANDE, 1879, p. 99; OD]. Brachial-valve interior without crural plates; pedicle-valve interior with median septum thickened posteriorly,

FIG. 551. Delthyrididae (Delthyridinae)
(p. *H680-H681*).

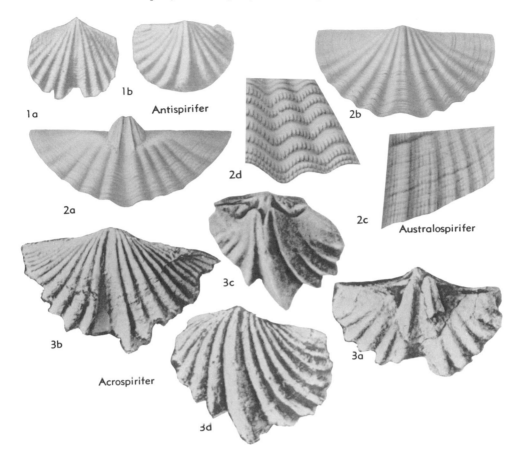

Fig. 553. Delthyrididae (Acrospiriferinae) (p. *H681-H683*).

and muscle field deeply impressed owing to deposition of secondary material in umbonal cavities; otherwise like *Delthyris*. *?L.Dev.(Gedinn.)*, Boh.——Fig. 551,*3*. **I. gibbosa* (BARRANDE); *3a-d*, ped.v., brach.v., lat., ant., ×1 (411).

?**Neodelthyris** HOU, 1963, p. 413 [**N. sinensis*; OD]. Large; extremely transverse; plications sharp, rather numerous, confined to lateral slopes; pedicle valve interior with thick dental plates, very short median septum; brachial valve lacking crural plates and cardinal prcoess. *M.Dev.(Couvin.)*, China.——Fig. 552,*1*. **N. sinensis*; *1a-d*, ped.v., brach.v., ant., lat., ×1 (433c).

Uralospirifer HAVLÍČEK, 1959, p. 142 [**Spirifer (Delthyris) mansy* KHODALEVICH, 1951, p. 96; OD]. Like *Delthyris* except pedicle valve with dental plates reduced to teeth ridges and brachial valve without crural plates. *Dev.(Coblenz.-Couvin.)*, USSR.——Fig. 551,*1*. **U. mansy* (KHODALEVICH); *1a*, transv. sec., ?mag. (411); *1b-e*, ped.v., brach.v., ant., lat., ×1 (467).

Subfamily ACROSPIRIFERINAE Termier & Termier, 1949

[Acrospiriferinae TERMIER & TERMIER, 1949, p. 96]

Lateral plications generally few, strong, angular in early forms, tending to become weaker and more numerous in later forms; micro-ornament primitively as in Delthyridinae, in advanced forms consisting of capillae or teardrop-shaped granules; interior of pedicle valve with well-developed dental plates, without median septum; brachial-valve interior generally without crural plates. *L.Sil.(Llandover.)-U.Dev.(Frasn.)*.

Acrospirifer HELMBRECHT & WEDEKIND, 1923, p. 952 [**Spirifer primaeva* STEININGER, 1853, p. 72; SD WEDEKIND in SALOMON, 1926, p. 202]. Lateral plications fairly numerous, very strong, angular; micro-ornament as in *Howellella*; umbonal cavities of pedicle valve filled with secondary material, muscle field deeply impressed, dental plates short; brachial-valve interior without crural plates. *L.Dev.(Siegen.-Ems.),* cosmop.——FIG. 553,3.

A. primaevus (STEININGER), Siegen., Ger.; *3a-d*, ped.v. int. mold, ped.v., brach.v. int. mold, brach. v., ×1 (528).

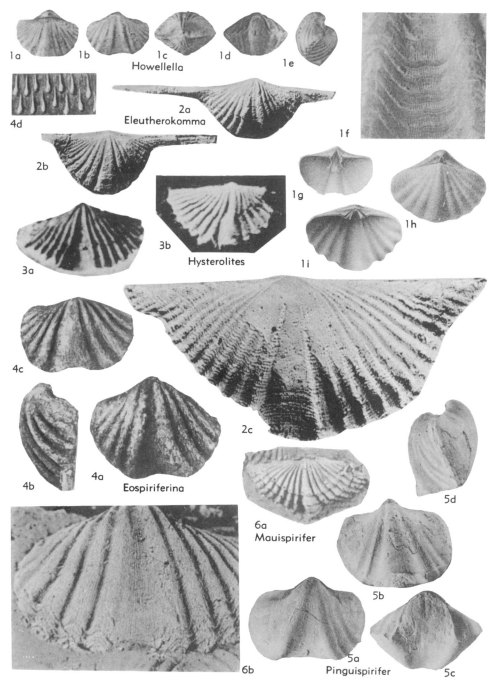

FIG. 554. Delthyrididae (Acrospiriferinae) (p. *H*683).

Antispirifer WILLIAMS, 1916, p. 114 [*A. harroldi; OD]. Pedicle valve essentially flat; brachial valve convex; otherwise like *Acrospirifer*. *L.Dev.*, USA (Maine).——FIG. 553,1. *A. harroldi; 2a,b,* ped. v., brach.v., ×1.5 (883).

Australospirifer CASTER, 1939, p. 159 [*Spirifer kayserianus* CLARKE, 1913, p. 252; OD]. Lateral plications few, distinct, rounded; micro-ornament in 2 stages, initially as in *Howellella*, later with radial striae only; interiors of both valves as in *Acrospirifer*. *L.Dev.*, S.Am.——FIG. 553,2. *A. kayserianus* (CLARKE), Brazil; *2a,b,* ped.v. int. mold, brach.v., ×0.7; *2c,d,* mature ornament, immature ornament, ×5 (165).

?Eleutherokomma CRICKMAY, 1950, p. 219 [*E. hamiltoni; OD]. Small to medium-sized; biconvex; extremely transverse with acuminate cardinal extremities; lateral slopes with rather numerous well-defined plications; fold and sulcus distinct, bald or with single median costa in sulcus; micro-ornament consisting of prominent closely spaced concentric growth lamellae and capillae; pedicle valve interior with strong dental plates, lacking median septum. *M.Dev.(Givet.)-U.Dev. (Frasn.)*, N.Am.——FIG. 554,2a,b. *E. hamiltoni,* Waterways F., Alta.; *2a,b,* ped.v., brach.v., ×2 (202).——FIG. 554,2c. E. beardi, ?Frasn., Alta.; brach.v. ext. mold, ×6 (202).

?Eospiriferina GRABAU, 1931, p. 494 [*Spiriferina (Eospiriferina) lachrymosa; OD]. Lateral plications coarse, few; micro-ornament consisting of posteriorly tapering teardrop-shaped pustules; pedicle valve interior with well-developed, divergent dental plates, lacking median septum; presence of punctae not established. *M.Dev.*, China.——FIG. 554,4. *E. lachrymosa* (GRABAU); *4a-c,* ped.v., lat., brach.v., ×2; *4d,* micro-ornament, ×10 (358).

Howellella KOZLOWSKI, 1946, p. 295 [*pro Crispella* KOZLOWSKI, 1929, p. 190 (*non* GRAY, 1870)] [*Terebratula crispus* HISINGER, 1826, pl. 7, fig. 4 (*non Anomia crispa* LINNÉ, 1758), =*Delthyris elegans* MUIR-WOOD, 1925, p. 89; OD]. Lateral plications few; micro-ornament of capillae becoming fimbriate at anterior edges of growth lamellae; pedicle-valve interior with well-developed dental plates; brachial-valve interior with short crural plates. *L.Sil.(U.Llandover.)-Dev.(L. Gedinn.)*, cosmop.——FIG. 554,1g-i. H. elegans (MUIR-WOOD), M.Sil. (Niagaran), E.USA; *1g-i,* ped.v. int., brach.v., brach.v. int., ×2 (178).—— FIG. 554,1a-f. H. angustiplicatus KOZLOWSKI, U. Sil.(Ludlov.), Pol.; *1a-e,* brach.v., ped.v., post., ant., lat., ×1; *1f,* surface, ×7 (487).

[Formerly both *Anomia crispa* LINNÉ, 1758, and *Terebratula crispus* HISINGER, 1826, were regarded as belonging to *Delthyris* MUIR-WOOD (1925, p. 89-91) proposed *Delthyris elegans* as a substitute for *Delthyris crispa* (HISINGER). According to the Zoological Code (Art. 59,c) the substitution stands, even though the two species are not now considered congeneric.]

Hysterolites VON SCHLOTHEIM, 1820, p. 247 [*H. hystericus; SD DALL, 1877, p. 38]. Pedicle-valve interior with rather long, thin dental plates, umbonal cavities not filled with secondary material, muscle field not deeply impressed; otherwise seemingly similar to *Acrospirifer*, but interior of brachial valve poorly known. *L.Dev.(Siegen.),* Ger.—— FIG. 554,3. *H. hystericus; 3a,b,* ped.v. int. mold, brach.v., ×1.5 (528).

Mauispirifer ALLAN, 1947, p. 445 [*M. hectori; OD]. Micro-ornament consisting of capillae, interrupted only locally by growth lamellae; otherwise similar to *Acrospirifer*. *L.Dev.*, N.Z.——FIG. 554,6. *M. rectori; 6a,* brach.v., ×1; *6b,* ped.v., ×3 (27).

?Pinguispirifer HAVLÍČEK, 1957, p. 246 [*Spirifer infirmus* BARRANDE, 1879, p. 47; OD]. Medium-sized to large; rather transverse, with rounded cardinal extremities; fold and sulcus well developed, bald; lateral slopes with several low plications; micro-ornament consisting of fine capillae only; pedicle valve interior with dental plates short or obsolete; brachial valve interior lacking crural plates. *L.Dev.-M.Dev.*, Boh.——FIG. 554,5. *P. infirmus* (BARRANDE); *5a-d,* ped.v., brach.v., ant., lat., ×1 (411).

Spinella TALENT, 1956, p. 21 [*S. buchanensis; OD]. Lateral slopes with numerous, simple, rather low, rounded plications; micro-ornament consisting of very numerous teardrop-shaped granules; interiors of both valves similar to those of *Acrospirifer*. *M. Dev. (?Couvin.)*, Australia.——FIG. 555,1. *S. buchanensis; 1a-d,* ped.v., brach.v., post., ant., ×1.5; *1e,* surface, ×10.5; *1f,* transv. sec., ×1; *1g,* cardinalia, enl. (796).

Subfamily KOZLOWSKIELLININAE Boucot, 1957

[*nom. correct.* BOUCOT, 1958, p. 1031 (*pro* Kozlowskiellinae BOUCOT, 1957, p. 317)]

Lateral slopes with few, very strong plications; micro-ornament consisting of strong growth lamellae which tend to bend outward and become frilly at their anterior margins, and capillae which become fimbriate at edges of frills; interior of pedicle valve with well-developed dental plates, and generally median septum; brachial-valve interior with short crural plates. *U.Sil. (Wenlock.)-L.Dev.(Ems.).*

Kozlowskiellina BOUCOT, 1957 (1958), p. 1031 [*pro Kozlowskiella* BOUCOT, 1957, p. 317 (*non* PRIBYL, 1953)] [*Kozlowskiella strawi* BOUCOT, 1957, p. 318; OD] [=*Megakozlowskiella* BOUCOT, 1957, p. 322 (type, *Spirifer perlamellosus* HALL, 1857, p. 57); *Megakozlowskiellina* AMSDEN & VENTRESS, 1963, p. 114 (*nom. van.*)]. Pedicle valve with well-developed median septum; brachial valve with deeply striate, bilobed cardinal process. [Although *Kozlowskiellina* was proposed in 1958 as a replacement for the preoccupied *Kozlowskiella* BOUCOT, 1957, it takes the

FIG. 555. Delthyrididae (Acrospiriferinae) (p. *H683*).

date 1957, because it is the basis for a subfamily (Code, Art. 39,a).] *U.Sil.-L.Dev.,* N.Am.; *Sil.* (*Wenlock.-Ludlov.*), Eu.; *L.Dev.,* Australia.——— FIG. 556,*1a-e.* **K. strawi* (BOUCOT), Wenlock., Br.I.; *1a-e,* post., ant., lat., brach.v., ped.v., ×4 (97).———FIG. 556,*1f,g. K.* sp. (BOUCOT), L.Dev. (Haragan), USA(Okla.); brach.v., brach.v. int., ×3 (97).———FIG. 556,*1h. K. raricosta* (CONRAD), L.Dev.(Onondaga), USA(N.Y.); oblique post. int., ×2 (97).

Hedeina BOUCOT, 1957, p. 323 [*Anomia crispa LINNÉ, 1758, p. 702; OD]. Pedicle valve without median septum; brachial valve with simple, striate cardinal process; otherwise like *Kozlowskiellina. U. Sil.* (*Wenlock. - Ludlov.*), N.Am.-Eu. ——— FIG. 556,*2.* **H. crispa* (LINNÉ), Ludlov., Gotl.; *2a-e,* post., ant., lat., brach.v., ped.v., ×3 (97).

Subfamily PARASPIRIFERINAE Pitrat, n.subfam.

Lateral plications very numerous, generally low, but distinct, simple or bifurcating; micro-ornament as in Delthyridinae;

interior of pedicle valve with dental plates, but without median septum; brachial-valve interior without crural plates. *L.Dev.* (*Siegen.*)-*M.Dev.*(*Couvin.*).

Paraspirifer WEDEKIND in SALOMON, 1926, p. 198 [**Spirifer cultrijugatus* ROEMER, 1844, p. 70; OD]. Large, slightly transverse to equidimensional, with maximum width at mid-length; brachial valve highly convex, pedicle valve less so; lateral costae numerous, low, straplike, mostly simple, but those near fold or sulcus bifurcating anteriorly; fold smooth, broad, very strong, carinate; sulcus smooth, wide, V-shaped. *L.Dev.(Ems.)-M.Dev.* (*Couvin.*), cosmop.———FIG. 557,*1a,b.* **P. cultrijugatus* (ROEMER), Ems., Ger.; *1a,b,* ped.v., post., ×0.7 (374).———FIG. 557,*1c-e. P. acuminatus* (CONRAD), L.Dev. (Onondaga), E.USA; *1c-e,* brach.v., ant., lat., ×0.7 (178).

Brachyspirifer WEDEKIND in SALOMON, 1926, p. 198 [**Spirifer carinatus* SCHNUR, 1853, p. 202; OD]. Medium-sized, subequally biconvex, transverse; hinge line slightly less than maximum width; in-

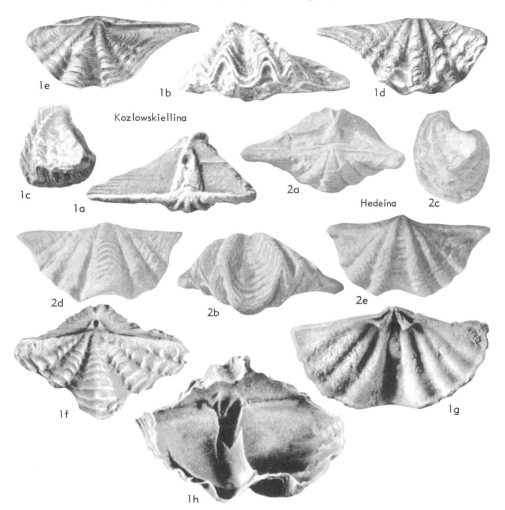

FIG. 556. Delthyrididae (Kozlowskiellininae) (p. *H683-H684*).

terarea high with wide delthyrium; lateral plications numerous, distinct, simple; fold and sulcus well developed, smooth; fold carinate; sulcus V-shaped. *L.Dev.(Siegen.-Ems.),* Eu.——FIG. 557,2. **B. carinatus* (SCHNUR), Siegen., Ger.; *2a-c,* brach. v. int. mold, lat. int. mold, ped.v. int. mold, ×1 (721).

Euryspirifer WEDEKIND in SALOMON, 1926, p. 202 [**Terebratulites paradoxus* VON SCHLOTHEIM, 1813, p. 28; OD] [=*?Rostrospirifer* GRABAU, 1931, p. 407 (type, *Spirifer tonkinensis* MANSUY, 1908, p. 41)]. Extremely transverse with highly acuminate cardinal extremities; interarea low; otherwise similar to *Brachyspirifer. L.Dev.(Ems.)-M.Dev.,* Eu.-Asia.——FIG. 557,3. **E. paradoxus* (VON SCHLOTHEIM), Ems., Ger.; *3a-c,* brach.v. int. mold, ped.v. int. mold, post. int. mold, ×1 (721).

Subfamily CYRTINOPSINAE Boucot, 1957

[*nom. transl.* PITRAT, herein (*ex* Cyrtinopsidae BOUCOT, 1957, p. 38)]

Lateral slopes plicate; micro-ornament consisting of very prominent growth lamellae crossed by short radial crenulations; pedicle valve interior with dental plates converging to form spondylium supported by 3-layered septum. *M.Dev.(Couvin.).*

Cyrtinopsis SCUPIN, 1896, p. 247 [*non* FREDERIKS, 1916, p. 17] [**Spirifer undosus* SCHNUR, 1853, p. 204; OD]. Medium-sized; transverse; markedly and unequally biconvex with highly arched beak and rather high interarea on pedicle valve; lateral slopes with numerous, rounded plications; fold and sulcus distinct, broad, lacking macro-ornament; brachial valve interior with short

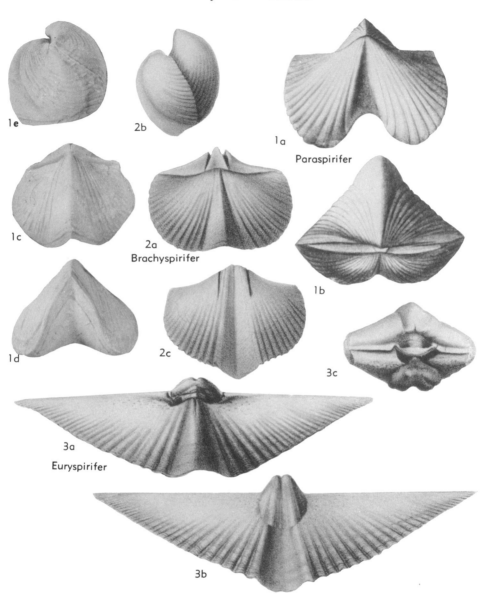

FIG. 557. Delthyrididae (Paraspiriferinae) (p. *H684-H685*).

crural plates. *M.Dev.(Couvin.),* W.Eu.——FIG. 558,*1.* **C. undosa* (SCHNUR); *1a-d,* lat., ant., brach.v., post., ×1; *1e,* ornament, ×3; *1f,* transv. sec. showing spondylium, ×5 (96).

Family MUCROSPIRIFERIDAE Pitrat, n.fam.

Biconvex, generally strongly transverse; lateral slopes plicate; fold and sulcus smooth or with median ridge in sulcus and median groove on fold; micro-ornament consisting of strong, imbricating growth lamellae; interior of pedicle valve with short dental plates or teeth ridges, rarely with distinct median septum, commonly with apical callus; interior of brachial valve generally with comblike cardinal process; crural plates lacking or short; shell substance impunctate. *L.Dev.(Ems.)-L.Carb.(Visean).*

Mucrospirifer GRABAU, 1931, p. 408 [**Delthyris mucronata* CONRAD, 1841, p. 54; OD] [*=Lamelli-*

spirifer NALIVKIN, 1937, p. 87, obj.]. Highly transverse; cardinal extremities commonly mucronate; lateral plications numerous; fold and sulcus bald or with single median ridge in sulcus; pedicle valve interior with short dental plates; median septum wanting. *M.Dev.(Couvin.-Givet.).* cosmop.——FIG. 559,6. *M. mucronatus* (CONRAD), Hamilton, USA(N.Y.); *6a,b,* ped.v., brach. v., ×1 (178).

Amoenospirifer HAVLÍČEK, 1957, p. 436 [*Spirifer*

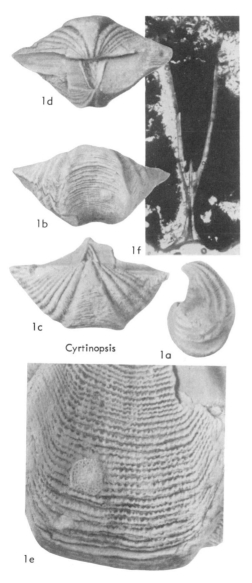

1d

1b

1f

1c

Cyrtinopsis

1a

1e

FIG. 558. Delthyrididae (Cyrtinopsinae) (p. *H685-H686*).

thetidis BARRANDE, 1848, p. 24; OD]. Moderately transverse; pedicle valve interior with dental plates reduced to teeth ridges; otherwise similar to *Brevispirifer. M.Dev.(Couvin.),* Boh.——FIG. 559,3. *A. thetidis* (BARRANDE); *3a-d,* ped.v., brach.v., lat., ant., ×1.5 (411).

Brevispirifer COOPER, 1942, p. 231 [*Spirifer gregaria* CLAPP, 1857, p. 127; OD]. Length and width subequal; lateral plications few; fold and sulcus bald or with median ridge in sulcus and median groove on fold; otherwise similar to *Mucrospirifer. L. Dev.(Ems.)-M. Dev.(Couvin.),* N.Am.——FIG. 559,4. *B. gregarius* (CLAPP), Onondaga, USA (N.Y.); *4a-c,* lat., brach.v., brach.v. int., ×1 (178).

Strophopleura STAINBROOK, 1947, p. 324 [*Spirifer notabilis* KINDLE, 1909, p. 26; OD]. Small, rather transverse; lateral plications numerous, unusually strong and almost perpendicular to hinge line near cardinal extremities; fold and sulcus strong, narrow, bald except for median ridge in sulcus and median groove on fold of some specimens; pedicle valve interior with teeth ridges, without median septum. *U.Dev.(Ouray Ls.),* USA(Colo.-N.Mex.); *?Tournais.,* Australia.——FIG. 559,1. *S. notabilis* (KINDLE), U.Dev., Colo.; *1a,b,* post., brach.v., ×2 (469).

Tylothyris NORTH, 1920, p. 195 [*Cyrtia laminosa* M'COY, 1844, p. 137; OD] [=*Welleria* MAILLIEUX, 1931, p. 35 (*nom. nud.*) (*non* ULRICH & BASSLER, 1923; *nec* ROTAI, 1941); *?Bouchardopsis* MAILLIEUX, 1933, p. 80 (type, *Spirifer bouchardi* MURCHISON, 1840, p. 253) (*nom. nud.*)]. Rather transverse; lateral plications numerous, distinct; fold and sulcus bald or with median ridge in sulcus and median groove on fold; pedicle valve interior with distinct dental plates and well-developed median septum. [*Welleria* MAILLIEUX, 1931, is an unavailable name, for MAILLIEUX failed to designate a type-species. *Bouchardopsis* MAILLIEUX, 1933, was not accompanied by a statement of differentiating characters or by a bibliographic reference to such a statement and is therefore a *nom. nud.*]*?M.Dev.(Givet.), U.Dev.(Frasn.)-L. Carb.(Visean),* cosmop.——FIG. 559,2. *T. laminosa* (M'COY), Tournais., Br.I.; *2a,b,* ped.v., brach. v., ×1; *2c,* ped.v. int., ×1.5; *2d,* transv. sec., *?*×1 (all 607).

Family FIMBRISPIRIFERIDAE Pitrat, n.fam.

Biconvex; weakly to moderately transverse; fold and sulcus distinct; entire surface covered with rather numerous anteriorly bifurcating costae; micro-ornament consisting of numerous concentric growth lamellae, each bearing fringe of minute spines; shell substance impunctate. *L.Dev.(Ems.)-M.Dev.(Givet.).*

Fimbrispirifer COOPER, 1942, p. 231 [*Spirifer venustus* HALL, 1860, p. 82; OD]. Pedicle valve interior with strong dental plates; brachial valve interior with short crural plates. *L.Dev.(Ems.)-M.Dev.(Givet.),* N.Am.——FIG. 559,5. **F. venustus* (HALL), M.Dev., USA(N.Y.); *5a,* microornament, ×2; *5b,* brach.v., ×1 (178).

Family SPINOCYRTIIDAE Ivanova, 1959

[*nom. transl.* PITRAT, herein (*ex* Spinocyrtiinae STRUVE, 1963, *nom. correct. pro* Spinocyrtinae IVANOVA, 1959, p. 59)] [=Guerichellinae PAECKELMANN, 1931, p. 24 *(partim)*]

Biconvex, moderately to strongly transverse; lateral slopes generally with numer-

FIG. 559. Mucrospiriferidae *(1-4,6)*; Fimbrispiriferidae *(5)* (p. *H686-H688*).

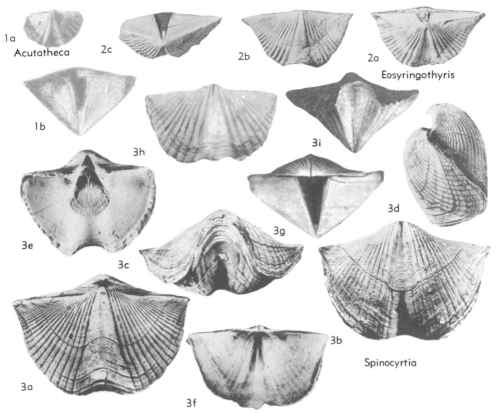

FIG. 560. Spinocyrtiidae (p. *H689, H691*).

ous distinct nonbifurcating plications; fold and sulcus commonly bald, but rarely with incipient plication; micro-ornament variable; interior of pedicle valve wtih distinct dental plates and delthyrial plate, without median septum; brachial valve interior lacking crural plates; shell substance impunctate. *L.Dev.(Ems.)-U.Dev.(Frasn.).*

Guerichellinae is invalid by virtue of failure to satisfy provisions of the Zoological Code (Art. 11e) which states that a family-group name "must, when first published, be based on the name then valid for a contained genus. . . ." The genus *Guerichella* PAECKELMANN, 1931, is, and has always been, a junior objective synonym of *Adolfia* GÜRICH, 1909.

Spinocyrtia FREDERIKS, 1916, p. 18 [*Delthyris granulosa* CONRAD, 1839, p. 65; SD FREDERIKS, 1926, p. 411] [=?*Platyrachella* FENTON & FENTON, 1924, p. 158 (type, *Spirifera macbridei* CALVIN, 1883, p. 433)]. Generally large, transverse; lateral plications numerous; fold and sulcus bald;

micro-ornament consisting of rather weak concentric growth lamellae and stronger capillae from summits of which rise minute teardrop-shaped granules; interarea of pedicle valve consisting of central area bearing horizontal and vertical striae, and marginal areas with horizontal striae only. *M. Dev. (Couvin.)-U. Dev. (Frasn.),* cosmop.——FIG. 560,*3a-f*; 561,*4*. *S. granulosa* (CONRAD), M.Dev.(Hamilton), USA(N.Y.); 560, *3a-f*, brach.v., ped.v., ant., lat., ped.v. int., brach. v. int., ×0.7; 561,*4*, ornament, ×5 (272).—— FIG. 560,*3g-i*. *S. macbridei* (CALVIN), U.Dev., USA(Iowa); *3g-i*, post., brach.v., ant., ×1 (296).

?**Acutatheca** STAINBROOK, 1945, p. 55 [*A. propria*; OD]. Small; moderately transverse; pedicle valve hemipyramidal with high interarea ornamented as in *Spinocyrtia*, and narrow delthyrium closed apically by delthyrial cover; lateral slopes with several low, rounded plications; fold and sulcus distinct, bald; micro-ornament consisting of concentric growth lamellae and capillae; pedicle valve interior with short, divergent dental plates, seemingly lacking delthyrial plate. *U.Dev.,* N. Am.——FIG. 560,*1*. *A. propria*, USA(Iowa); *1a*, ped.v., ×1.5; *1b*, post., ×3 (768).

Adolfia Gürich, 1909, p. 136 [*Spirifer deflexus* Roemer, 1843, p. 13; SD Schuchert & LeVene, 1929, p. 27] [=*Guerichella* Paeckelmann, 1913, p. 299 (obj.); *Gürichia* Wedekind in Salomon, 1926, p. 198 *(nom. null.)*; ?*Plectospirifer* Grabau, 1931, p. 379 (type, *Spirifer (Plectospirifer) heimi)*]. Generally medium-sized, moderately transverse; lateral plications fairly numerous; fold and sulcus bald or with incipient plication; micro-ornament as in *Spinocyrtia* except for fanlike divergence of radial striae in some. *L.Dev.(Ems.)- U.Dev.(Frasn.),* N.Am.-Eu.-Asia.——Fig. 561,*1.* *A. deflexa* (Roemer), Frasn., Ger.; *1a-d,* ped.v., brach.v., post., lat., ×1 (831).

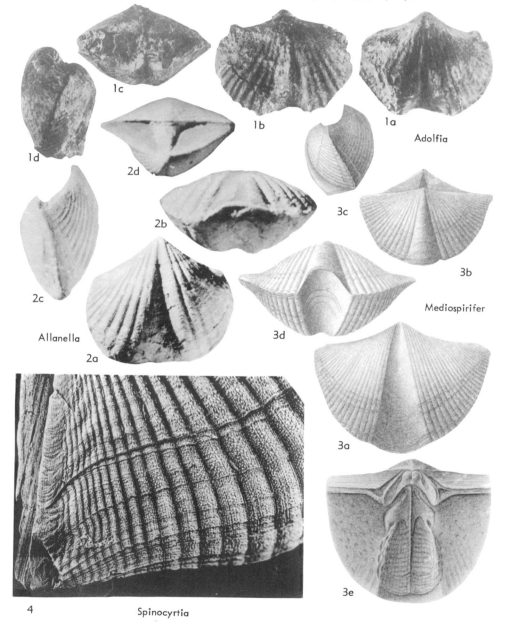

Fig. 561. Spinocyrtiidae (p. *H689-H691*).

[In erecting the genus *Adolfia,* GÜRICH included the species *A. deflexa* and *A. bifida,* but did not choose a type. PAECKELMANN (1913) proposed *Guerichella* expressly as a replacement for *Adolfia* which he erroneously believed to be a junior homonym; again no type was designated. FREDERIKS (1926) accepted the name *Guerichella* and stated the type to be *G. angustistellata,* an action which is invalid because *G. angustistellata* was not mentioned in GÜRICH's original description of *Adolfia.* SCHUCHERT & LEVENE (1929) listed both genera, indicated their synonymy, but then, unaccountably, stated that the type of *Adolfia* is *A. deflexa* and that of *Guerichella* is *G. angustistellata.* Despite the ambiguity of their entries, SCHUCHERT & LEVENE were the first to designate a legally eligible type-species for *Adolfia, A. deflexa,* and that species must also be regarded as the type of *Guerichella.*]

Alatiformia (*see* p. H903).

Allanella CRICKMAY, 1953, p. 5 [**Spirifer allani* WARREN, 1944, p. 123; OD] [=*Allanaria* CRICKMAY, 1953 (obj.)]. Rather small, weakly transverse; lateral plications fairly numerous; fold and sulcus bald; micro-ornament of concentric growth lamellae and capillae. *M.Dev.(Givet.),* W.Can. ——FIG. 561,2. **A. allani* (WARREN); *2a-d,* ped. v., ant., lat., post., ×2 (206).

[In the text of CRICKMAY's article (1953, p. 5) the genus *Allanella* was erected. To the flyleaf of the publication is affixed a section entitled "Addendum" in which CRICKMAY pointed out the existence of the genus *Allanella* BOUČEK, 1936, which he erroneously believed invalidated his use of *Allanella.* In this same section CRICKMAY proposed substitution of *Allanaria* for *Allanella.* According to the Rules (Art. 24) choice of one of two or more names simultaneously published for the same taxon is to be made by the first reviser. To the present time there has not been a "first reviser" in the meaning of the Rules (Art. 24,a,i). Therefore, as such first reviser, I now choose the name *Allanella,* relegating *Allanaria* to the status of a junior objective synonym.]

Chimaerothyris (*see* p. H903).

Eosyringothyris STAINBROOK, 1943, p. 431 [**Spirifera aspera* HALL, 1858, p. 508; OD]. Pedicle valve hemipyramidal with very high interarea; delthyrial plate terminating anterodorsally in short spine; otherwise similar to *Spinocyrtia. M.Dev.,* N.Am.——FIG. 560,2. **E. aspera* (HALL), Cedar Valley Ls., USA(Iowa); *2a-c,* ped.v., brach.v., post., ×1 (767).

Mediospirifer BUBLICHENKO, 1956, p. 102 [**Delthyris medialis* HALL, 1843, p. 208 (=**Delthyris audacula* CONRAD, 1842, p. 262); OD]. Micro-ornament of widely spaced growth lamellae only; otherwise similar to *Spinocyrtia. M.Dev.(Hamilton),* USA(N.Y.).——FIG. 561,3. **M. audacula* (CONRAD); *3a-d,* ped.v., brach.v., lat., ant., ×1; *3e,* brach.v. int., ×2 (396).

Family SYRINGOTHYRIDIDAE
Frederiks, 1926

[*nom. correct* PITRAT, herein (*pro* Syringothyridae FREDERIKS, 1926 (*nom. imperf.*) *nom. transl.* IVANOVA, 1959, p. 55, *ex* Syringothyrinae FREDERIKS, 1926, p. 411)]

Typically biconvex, large, transverse, with high interarea on pedicle valve; lateral slopes with numerous nonbifurcating plications; fold and sulcus commonly bald, rarely with weak plication; micro-ornament variable; interior of pedicle valve generally with well-developed dental plates and delthyrial plate, in some with syrinx; shell substance

FIG. 562. Syringothyrididae (Syringothyridinae) (p. H692).

punctate in earlier forms, becoming impunctate in later ones. *U.Dev.-U.Perm.*

Subfamily SYRINGOTHYRIDINAE Frederiks, 1926

[*nom. correct.* PITRAT, herein (*pro* Syringothyrinae FREDERIKS, 1926, p. 411)]

Generally rather large, transverse; interarea of pedicle valve commonly high, consisting of central area with horizontal and vertical markings, and marginal areas with horizontal markings only; lateral slopes with numerous nonbifurcating plications; fold and sulcus distinct, generally bald; micro-ornament somewhat variable, typically consisting of a textile-like pattern of intersecting capillae and concentric growth lamellae, in some complicated by pustules; pedicle valve interior generally with well-developed dental plates; delthyrial plate commonly present, in some bearing syrinx on its dorsal surface; shell substance normally punctate. *U.Dev.-Perm.*

Syringothyris WINCHELL, 1863, p. 6 [**S. typa* (=**Spirifer carteri* HALL, 1857, p. 170); SD ICZN Opinion 100, 1928, p. 377] [=*Syringopleura* SCHUCHERT, 1910 (type, *Spirifer randalli* SIMPSON, 1890, p. 441); *Prosyringothyris* FREDERIKS, 1916, p. 51 (type, *P. northi*); *Protosyringothyris* FREDERIKS, 1918, p. 88 *(nom. null.)*]. Fold and sulcus bald; micro-ornament of minute pustules; pedicle valve interior with long dental plates, delthyrial plate and syrinx, lacking median septum. *U.Dev.(Famenn.)-Miss.,* cosmop.——FIG. 563,*1a,b.* **S. carteri* (HALL), Burlington Ls., USA (Iowa); *1a,b,* post., ant., ×0.7 (858).——FIG. 563,*1c-f. S. texta* (HALL), Keokuk F., USA(Ind.); *1c-f,* ped. v. int., brach.v., ped.v. int. mold, transv. sec., ×0.7 (396).

Asyrinx HUDSON & SUDBURY, 1959, p. 46 [**A. haushensis;* OD]. Pedicle valve interior with dental plates reduced to teeth ridges, lacking delthyrial plate and syrinx; delthyrial cavity with thick callus deposits simulating dental plates. *L. Perm.,* Arabia.——FIG. 562,*1.* **A. haushensis; 1a,* transv. sec., ×1; *1b-d,* ped.v., lat., post., ×0.7 (438).

Asyrinxia CAMPBELL, 1957, p. 80 [**Spirifera lata* M'COY, 1847, p. 223; OD]. Sulcus with several very weak plications; interior of pedicle valve with neither delthyrial plate nor syrinx; otherwise similar to *Syringothyris. L.Carb.(U.Tournais.),* Australia(New S. Wales)-?Japan. —— FIG. 563,*3.* **A. lata* (M'COY), Australia; *3a-c,* ped.v. int. mold, ped.v. interarea, brach.v. post., ×1 (140).

?Plicatosyrinx MINATO, 1952, p. 168 [**P. singulare;* OD]. Fold and sulcus plicate; pedicle valve with syrinx, lacking dental plates and median septum; shell substance impunctate. [The genus is based on a single specimen which is so badly deformed that the cardinal extremities have been shoved against the beak. Morphology as well as systematic position are in doubt.] *L.Carb.,* Japan.

Pseudosyringothyris FREDERIKS, 1916, p. 51 [**P. karpinskii;* OD]. Pedicle valve interior with syrinx incompletely developed, consisting of pair of longitudinally directed parallel thickenings on underside of delthyrial plate; otherwise similar to *Syringothyris. L.Perm.,* USSR.——FIG. 563,*2.* **P. karpinskii; 2a,* ped.v. post., ×0.7; *2b,* transv. sec., ×2 (314).

Pseudosyrinx WELLER, 1914, p. 404 [**P. missouriensis;* OD]. Pedicle valve interior lacking syrinx; otherwise similar to *Syringothyris. Miss.,* cosmop.; *?Perm.,* Arabia.——FIG. 563,*4.* **P. missouriensis,* L.Miss.(Burlington Ls.), USA(Mo.); *4a,b,* ant., post., ×1 (858). [*?=Verkhotomia* SOKOLSKAYA, 1963, p. 280 (type, *V. plenoides*).]

Septosyringothyris VANDERCAMMEN, 1955, p. 2 [**S. demaneti;* OD]. Pedicle valve with conspicuous median septum; otherwise similar to *Syringothyris. L.Carb.,* Eu.-S.Am.——FIG. 562,*2.* **S. demaneti,* Belg.; *2a,* ped.v., ×1; *2b,* ped.v. int., ×3; *2c,d,* transv. secs., ×5 (828).

Subfamily LICHAREWIINAE Slusareva, 1958

[Licharewiinae SLUSAREVA, 1958, p. 582]

Biconvex, generally large, transverse, with high interarea; lateral slopes with numerous, distinct, nonbifurcating plications; fold and sulcus generally bald, rarely with weak plication; micro-ornament of concentric growth lamellae with or without capillae; interior of pedicle valve generally with dental plates and delthyrial plate, lacking syrinx; shell substance impunctate. *U. Carb.-U.Perm.*

Some members of this group are morphologically indistinguishable from some of the Spinocyrtiidae. They are excluded from the Spinocyrtiidae because of the long hiatus between disappearance of this family in the late Devonian and appearance of similar Licharewiinae in the Late Carboniferous. It is considered likely that the Licharewiinae were derived from the Syringothyridinae, being merely heterochronous homeomorphs of the Spinocyrtiidae, but it must be admitted that future work may reveal a direct connection between Spinocyrtiidae and Licharewiinae, in which case the two groups would have to be merged.

Licharewia EINOR, 1939, p. 69 [**Spirifer stuckenbergi* NETSCHAJEW, 1900, p. 18; OD] [=*Permospirifer* KULIKOV, 1950, p. 5 (type, *Spirifer keyserlingi* NETSCHAJEW, 1911, p. 84); *Rugulatia* SOKOLSKAYA, 1952, p. 187 (type, *Spirifer rugula-*

tus KUTORGA, 1842, p. 22)]. Fold and sulcus generally bald, but rarely with several weak costae; micro-ornament largely concentric, but rarely with weak capillae; interior of pedicle valve with strong, thick, anteriorly diverging den- tal plates and delthyrial plate, lacking median septum. *U.Perm.*, USSR.——FIG. 565,4. *L. stuckenbergi* (NETSCHAJEW); 4a,b, ped.v., brach.v., ×1 (448).——FIG. 564,2a,b. *L. keyserlingi* (NETSCHAJEW), 2a,b, ped.v., brach.v., ×1 (598).——

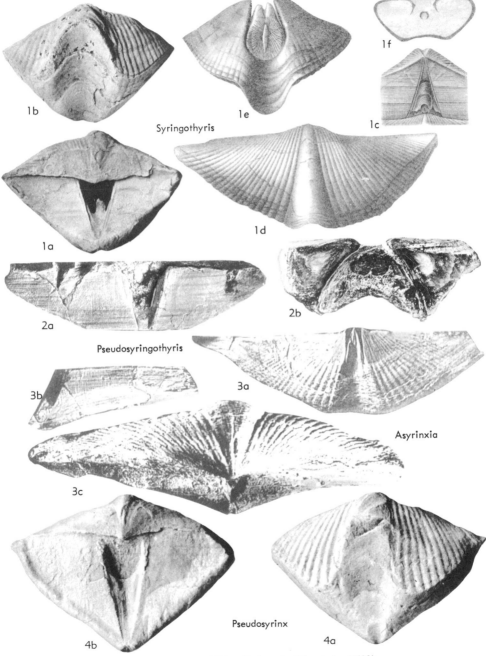

FIG. 563. Syringothyrididae (Syringothyridinae) (p. *H692*).

FIG. 564. Syringothyrididae (Licharewiinae) (p. *H692-H693, H696*).

Fig. 564,*2c-e*. *L. rugulatus* (Kutorga); *2c-e*, ped. v., brach.v., lat., ×1 (598).

?**Alispirifer** Campbell, 1961, p. 434 [**A. laminosus*; OD]. Highly transverse; hinge line denticulate; lateral slopes with distinct rounded plications; fold and sulcus well developed, bald except

for median plication in sulcus of some; microornament consisting of concentric growth lamellae and unusually distinct capillae; pedicle valve interior with dental plates almost obscured with callus; median septum lacking; presence of delthyrial plate not established. *U.Carb.*, Australia-

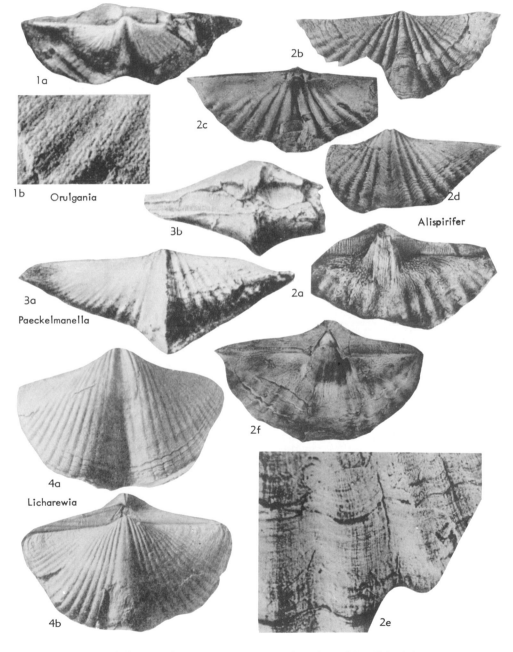

Fig. 565. Syringothyrididae (Licharewiinae) (p. H692-H693, H695-H696).

Arg.——Fig. 565,2. **A. laminosus*, New S. Wales; *2a-d,f*, ped.v. int. mold, brach.v., brach.v. int., ped.v., ped.v. int., ×1; *2e*, micro-ornament, ×7 (143).

Cyrtella FREDERIKS, 1919 (1924), p. 312 [**Cyrtia kulikiana* FREDERIKS, 1916, p. 43; OD]. Fold with narrow median groove; sulcus bald; otherwise similar to *Licharewia*. *L.Perm.*, USSR.—— Fig. 564,4. **C. kulikiana* (FREDERIKS); *4a,b*, ped.v. ant., brach.v. post., ×0.7 (314).

Darvasia LIKHAREV, 1934, p. 212 [**D. edelsteini*; OD]. Pedicle valve with very high interarea and narrow, open delthyrium; interior of pedicle valve with well-developed dental plates connected near valve floor by delthyrial plate; median septum present, almost reaching delthyrial plate. *L.Perm.*, E.USSR.——Fig. 564,3. **D. edelsteini*; *3a-c*, ped. v., post., lat., ×1.5 (448).

Orulgania SOLOMINA & CHERNYAK, 1961, p. 61 [**O. naumovi*; OD]. Fold and sulcus bald; dental plates thin, very long, tending to converge anteriorly; otherwise similar to *Licharewia*. *U.Carb.*, N.USSR.——Fig. 565,1. **O. naumovi*; *1a*, ped.v., ×1; *1b*, ornament, ×10 (759).

Paeckelmanella LIKHAREV, 1934, p. 212 [**Spirifer dieneri* CHERNYSHEV, 1902, p. 535; OD]. Highly transverse, with moderately high interarea; fold bald, carinate; sulcus with weak median costa; pedicle valve interior with well-developed dental plates and delthyrial plate and long, high median septum; hinge line denticulate. *L.Perm.*, USSR. ——Fig. 565,3. **P. dieneri*; *3a,b*, brach.v., post., ×1 (448).

Pterospirifer DUNBAR, 1955, p. 128 [**Spirifer alatus* VON SCHLOTHEIM, 1813, p. 58; OD]. Fold bald; sulcus with weak median costa; lateral plications prominent; micro-ornament of very conspicuous growth lamellae and obscure capillae; delthyrial plate weak or absent; otherwise similar to *Licharewia*. *U.Perm.*, Eu.-Greenl.——Fig. 564,5. **P. alatus* (VON SCHLOTHEIM), Greenl.; *5a,b*, ped. v., brach.v., ×1.5; *5c*, ornament, ×7.5 (269).

?**Subansiria** SAHNI & SRIVASTAVA, 1956, p. 212 [**S. ranganensis*; OD]. Pedicle valve interior lacking delthyrial plate; otherwise similar to *Licharewia*. *?U.Carb.*, India.——Fig. 564,1. **S. ranganensis*; *1a-c*, brach.v., ped.v., post., ×0.7 (701).

Family COSTISPIRIFERIDAE
Termier & Termier, 1949

[*nom. transl.* PITRAT, herein (*ex* Costispiriferinae TERMIER & TERMIER, 1949, p. 98)] [=Theodossiinae IVANOVA, 1959, p. 61 (*partim*)]

Biconvex, weakly to moderately transverse; lateral slopes with moderately to very numerous simple costae; fold and sulcus generally indistinct, poorly delineated from lateral slopes, provided with costae which are indistinguishable from lateral costae except for tendency toward bifurcation; micro-ornament of distinct capillae and weak

growth lamellae in some; pedicle valve interior with dental plates, lacking delthyrial plate and median septum; brachial valve interior with longitudinally striated area of diductor attachment, with or without crural plates; shell substance impunctate. *L.Dev. (Siegen.)-U.Dev.(Frasn.), ?L.Carb.*

Costispirifer COOPER, 1942, p. 232 [**Spirifer arenosus planicostatus* SWARTZ, 1929, p. 56; OD]. Costae moderately numerous, flat-topped; delthyrium covered by short, flat pseudodeltidium; pedicle valve interior with short, thick dental plates, and dense callus deposits in delthyrial and umbonal cavities; brachial valve lacking crural plates. *L.Dev.(Oriskany)*, N.Am. —— Fig. 566,3. *C. arenosus* (CONRAD), USA(Pa.); *3a,b*, ped.v., brach.v., ×1 (178).

?**Eudoxina** FREDERIKS, 1929, p. 382 [**Spirifer medius* LEBEDEV, 1912, p. 18; OD]. Fold and sulcus weak but distinct; entire shell with very numerous, fine costae; pedicle valve interior with dental plates reduced to teeth ridges, lacking median septum; brachial valve interior unknown. *L.Carb.*, USSR.——Fig. 566,4. **E. media* (LEBEDEV); brach.v., ×1.5 (448).

Lazutkinia RZHONSNITSKAYA, 1951, p. 151 [**Spirifer (Yavorskiella) mamontoviensis* LAZUTKIN *in* YAVORSKY, 1940, p. 44; OD] [=*Yavorskiella* LAZUTKIN *in* YAVORSKY, 1940, p. 44 (*nom. nud.*)]. Strongly biconvex with prominent, curved beak on pedicle valve; hinge line very short; fold and sulcus poorly defined, narrow; entire shell provided with rather numerous costae; micro-ornament consisting of both capillae and concentric growth lamellae; pedicle valve interior with short dental plates, lacking median septum; brachial valve interior with distinct septalium. *M. Dev.*, USSR.——Fig. 566,1. **L. mamontoviensis* (LAZUTKIN); *1a-d*, ant., brach.v., ped.v., lat., ×1 (465).

Theodossia NALIVKIN, 1925, p. 267 [**Spirifer anossofi* DE VERNEUIL, 1845, p. 153; OD] [=*Vandergrachtella* CRICKMAY, 1953, p. 7 (type, *V. arcuum*)]. Costae very numerous, rounded; delthyrium open; pedicle valve interior generally with thin, anteriorly diverging dental plates. *M.Dev. (Givet.)*, USSR; *U.Dev.*, cosmop.——Fig. 566, *2a,b*. **T. anossofi* (DE VERNEUIL), U.Dev., W. USSR; *2a,b*, ped.v., brach.v., ×1.5 (448).—— Fig. 566,*2c-f*. *T. hungerfordi* (HALL), U.Dev., USA(Iowa); *2c-f*, post., lat., brach.v., ped.v., ×1 (296).——Fig. 566,*2g-i*. *T. arcuum* (CRICKMAY), U.Dev., W.Can.; *2g-i*, brach.v., ped.v., lat., ×1 (206).

?**Urella** RZHONSNITSKAYA, 1960, p. 402 [**U. asiatica*; OD]. Medium-sized; equidimensional or elongate oval; hinge line very short; delthyrium with disjunct deltidial plates; fold and sulcus weak to absent; macro-ornament consisting of rounded, simple or branching costae; micro-ornament comprising capillae only; pedicle valve interior with long, thin dental plates. *M.Dev.,* USSR.

Family CYRTOSPIRIFERIDAE
Termier & Termier, 1949

[*nom. transl.* BESNOSSOVA, 1958 (*ex* Cyrtospiriferinae TERMIER & TERMIER, 1949, p. 99)]

Biconvex, generally transverse; lateral slopes commonly with numerous simple costae, more rarely with fewer or bifurcating costae; fold and sulcus generally distinct, with bifurcating costae; micro-ornament commonly consisting of distinct capillae and concentric growth lamellae, some with pustules; pedicle valve interior with dental plates and delthyrial plate, lacking median septum; shell substance impunctate. *?M. Dev., U.Dev.(Frasn.)-L.Carb.(Visean).*

Cyrtospirifer NALIVKIN in FREDERIKS, 1919 (1924), p. 312 [*Spirifer verneuili* MURCHISON, 1840, p. 252; OD] [=*Sinospirifer* GRABAU, 1931, p. 241

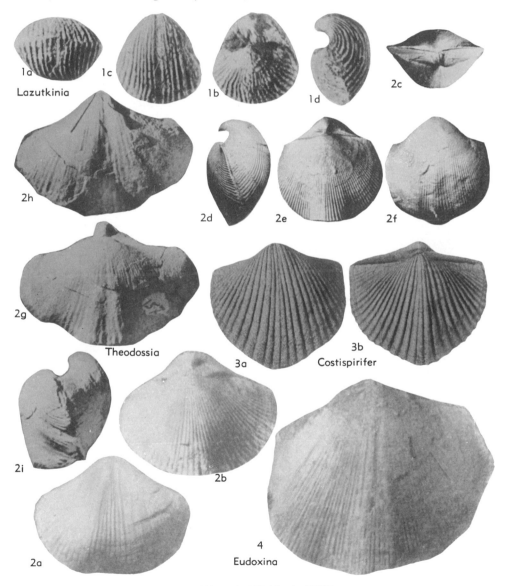

FIG. 566. Costispiriferidae (p. *H696*).

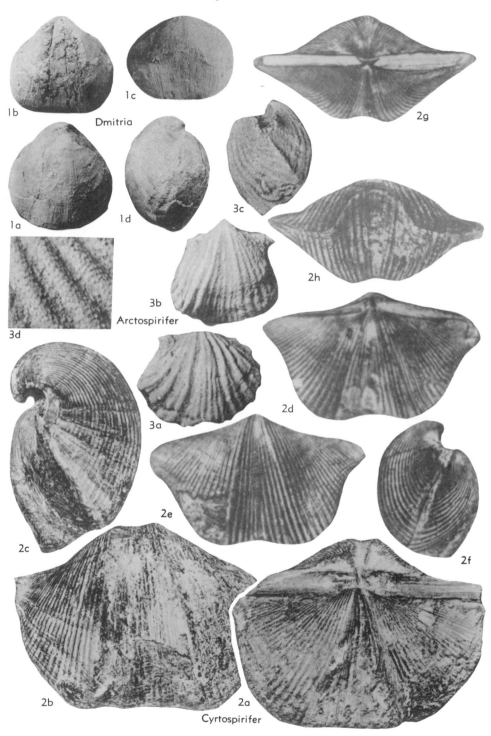

1b

1c

Dmitria

1a

1d

3c

3d

3b

Arctospirifer

2g

2h

2d

3a

2e

2c

2f

2b

2a

Cyrtospirifer

FIG. 567. Cyrtospiriferidae (p. *H*697, *H*699-*H*700).

(type, *Spirifer (Sinospirifer) sinensis* GRABAU, 1931, =*Spirifer chinensis* GRABAU, 1923); *Centrospirifer* TIEN, 1938, p. 111 (type, *Spirifer (Sinospirifer) chaoi* GRABAU, 1931); *Hunanospirifer* TIEN, 1938 (type, *Spirifer (Hunanospirifer) wangi* TIEN, 1938); *Deothossia* GATINAUD, 1949, p. 488 (type, *Spirifer (Sinospirifer) anossofioides* GRABAU, 1931); *Eurytatospirifer* GATINAUD, 1949, p. 487 (type, *Spirifer disjunctus* SOWERBY, 1840); *Grabauispirifer* GATINAUD, 1949, p. 413 (type, *Spirifer (Sinospirifer) archiaciformis* GRABAU, 1931); *Iubagraspirifer* GATINAUD, 1949, p. 487 (*nom. vet.*); *Iugrabaspirifer* GATINAUD, 1949, p. 487 (*nom. vet.*); *Lamarckispirifer* GATINAUD, 1949, p. 489 (type, *Spirifer (Sinospirifer) hayasakai* GRABAU, 1931); *Martellispirifer* GATINAUD, 1949, p. 487 (*nom. vet.*); *Mirtellaspirifer* GATINAUD, 1949, p. 488 (*nom. vet.*); *Yrctospirifer* GATINAUD, 1949, p. 488 (*nom. vet.*); *Liraspirifer* STAINBROOK, 1950, p. 381 (type, *L. tricostatus* STAINBROOK, 1950); *Regelia* CRICKMAY, 1952, p. 3 (type, *Cyrtospirifer glaucus* CRICKMAY, 1952)].
Generally transverse, with hinge line equal to maximum width; fold and sulcus strong; costae very numerous on lateral slopes and on fold and sulcus, simple on lateral slopes, bifurcating on fold and sulcus; micro-ornament consisting of rather weak growth lamellae and radial striae; delthyrium generally open; pedicle valve interior with well-developed, anteriorly diverging dental plates, delthyrial plate, lacking median septum; brachial valve interior without crural plates. *U. Dev.-L.Miss.*, cosmop.——FIG. 567,*2a-c*. **C. verneuili* (MURCHISON), Frasn., Belg.; *2a-c*, brach.v., ped.v., lat., ×1 (834).——FIG. 567,*2d-h*. *C. chinensis* (GRABAU), U.Dev., China; *2d-h*, brach. v., ped.v., lat., post., ant., ×1.5 (358).

[In a series of papers which reported results of a statistical study of the bifurcation patterns of costae in the sulci of several spiriferoids, GATINAUD (1949) erected numerous new taxa, ostensibly at the genus level. Eleven genera, one of them new, were divided into subgenera, sections, and subsections according to the scheme below.

Genus *Sinospirifer* GRABAU, 1931
Genus *Cyrtospirifer* NALIVKIN, 1919 (1924)
  Subgenus *Grabauispirifer* GATINAUD, 1949
    Section *Grabauispirifer, s.s.*
      Subsection *Grabauispirifer, s.s.*
      Subsection *Iubagraspirifer* GATINAUD, 1949
    Section *Iugrabaspirifer* GATINAUD, 1949
  Subgenus *Eurytatospirifer* GATINAUD, 1949
  Subgenus *Cyrtospirifer, s.s.*
    Section *Martellispirifer* GATINAUD, 1949
      Subsection *Martellispirifer, s.s.*
      Subsection *Mirtellaspirifer* GATINAUD, 1949
    Section *Cyrtospirifer, s.s.*
      Subsection *Cyrtospirifer, s.s.*
      Subsection *Yrctospirifer* GATINAUD, 1949
Genus *Deothossia* GATINAUD, 1949
Genus *Theodossia* NALIVKIN, 1925
  Subgenus *Platyspirifer* GRABAU, 1931
  Subgenus *Theodossia, s.s.*
Genus *Tenticospirifer* TIEN, 1938
  Subgenus *Tenticospirifer, s.s.*
  Subgenus *Lamarckispirifer* GATINAUD, 1949
Genus *Hunanospirifer* TIEN, 1938
Genus *Cyrtiopsis* GRABAU, 1923
  Subgenus *Cyrtiopsis, s.s.*
    Section *Cyrtiopsis, s.s.*
    Section *Alphacyrtiopsis* GATINAUD, 1949
    Section *Betacyrtiopsis* GATINAUD, 1949
    Section *Paracyrtiopsis* GATINAUD, 1949

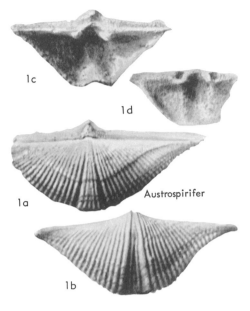

1c

1d

Austrospirifer

1a

1b

FIG. 568. Cyrtospiriferidae (p. *H*700).

  Subgenus *Grabauicyrtiopsis* GATINAUD, 1949
  Subgenus *Sinocyrtiopsis* GATINAUD, 1949
Genus *Spirifer* SOWERBY, 1816
Genus *Neospirifer* FREDERIKS, 1919 (1924)
  Section *Neospirifer, s.s.*
  Section *Alphaneospirifer* GATINAUD, 1949
  Section *Betaneospirifer* GATINAUD, 1949
Genus *Choristites* FISCHER, 1825
  Section *Choristites, s.s.*
  Section *Alphachoristites* GATINAUD, 1949
  Section *Betachoristites* GATINAUD, 1949
Genus *Spiriferella* CHERNYSHEV, 1902

In my opinion, GATINAUD's single-minded attention to the details of sulcal costae has resulted in a fragmentation of spiriferoid genera which is taxonomically absurd. Hence, all taxa introduced by GATINAUD which are available nomenclaturally are here placed in subjective synonymy with previously established genera. This is done with full realization that some of the taxa are likely to be revived in future and may be shown to be taxonomically valid on the basis of numerous characters not considered by GATINAUD.——¶Several of the GATINAUD taxa appear to contravene provisions of the International Code of Zoological Nomenclature (1961). Article 42 (a) states that the "genus group, which is next below the family-group and next above the species-group in the hierarchy of classification, includes the categories genus and subgenus." The Code does not explicitly forbid the use of additional genus-group categories, but no provision is made for them. The case against recognition of GATINAUD's sections and subsections becomes stronger when one compares the wording quoted above for the genus-group with parallel wording for the family-group. According to Art. 35 (a), "The family group includes the categories tribe, subfamily, family, and superfamily and any supplementary categories required." Presumably, the phrase "any supplementary categories required" would have been included in the paragraph on the genus-group, had there been any intention of allowing categories other than genus and subgenus. Art. 42(d) also bears upon the problem. It states that "a uninominal name proposed for a primary subdivision of a genus, even if the subdivision is designated by a term such as 'section' or 'division,' has the status in nomenclature of a subgeneric name, provided the name satisfies relevant provisions of Chapter IV." This provision would appear to rescue several of GATINAUD's taxa (e.g., *Alphaneospirifer*) which are primary subdivisions of genera. However, in several cases, GATINAUD's sections and subsections are secondary and tertiary subdivisions of genera, not primary ones. The clear implication is that these secondary and tertiary subdivisions are not to be regarded as subgenera. In summary, the Code seems to outlaw any genus-group name other than genus and subgenus, and to imply that several of GATINAUD's taxa cannot be regarded either as genera or subgenera. Under the circumstances the following taxa are considered to be unavailable names: *Iubagraspirifer, Iugrabaspirifer, Martellispirifer, Mirtellaspirifer, Yrctospirifer, Alphacyrtiopsis, Betacyrtiopsis, Paracyrtiopsis.*]

Arctospirifer STAINBROOK, 1950, p. 382 [**A. constrictus*; OD]. Hinge line short in most specimens, but in some with earlike extensions; lateral plications few, very strong; plications of fold and sulcus smaller, more numerous; micro-ornament consisting of radially aligned pustules; pedicle valve interior with dental plates and delthyrial plate; brachial valve interior unknown. *?U.Dev.(Aplington F.),* Iowa.——FIG. 567,*3.* **A. constrictus*; *3a-c,* brach.v., ped.v., lat., ×1.5; *3d,* ornament, ×6 (770).

Austrospirifer GLENISTER, 1955, p. 58 [**A. variabilis*; OD]. Rather small, very transverse; delthyrium almost completely closed by convex delthyrial cover; pedicle valve interior with weak dental plates or teeth ridges; brachial valve interior with crural plates; otherwise similar to *Cyrtospirifer.* *U.Dev.(Frasn.),* W.Australia.——FIG. 568,*1.* **A. variabilis*; *1a-d,* brach.v., ped.v., ped.v. int., brach.v. int., ×2 (352).

Cyrtiopsis GRABAU, 1923, p. 194 [**C. davidsoni*; SD GRABAU, 1931, p. 424] [=*Alphacyrtiopsis* GATINAUD, 1949, p. 490 *(nom. vet.)*; *Betacyrtiopsis* GATINAUD, 1949, p. 490 *(nom. vet.)*; *Grabauicyrtiopsis* GATINAUD, 1949, p. 490 (type, *Cyrtiopsis graciosa* GRABAU, 1923); *Paracyrtiopsis* GATINAUD, 1949, p. 490 *(nom. vet.)*; *Sinocyrtiopsis* GATINAUD, 1949, p. 491 (type, *Cyrtiopsis transversa* GRABAU, 1931); *Uchtospirifer* LYASHENKO, 1957, p. 885 (type, *U. nalivkini*)]. Very strongly biconvex; hinge line somewhat less than maximum width; interarea of pedicle valve rather high with delthyrium closed by prominent, convex pseudodeltidium; pedicle valve interior with long dental plates; otherwise similar to *Cyrtospirifer. U.Dev.,* cosmop.——FIG. 569,*3.* *C. intermedia* GRABAU, China; *3a-e,* brach.v., ped.v., lat., post., ant., ×3 (358). [For comments on genera erected by GATINAUD (1949), see note following *Cyrtospirifer,* p. H697.]

Dmitria SIDYACHENKO, 1961, p. 80 [**Spirifer (Cyrtospirifer) romanowskii* NALIVKIN, 1930, p. 127; OD]. Rather large, inflated; length and width approximately equal; hinge line much less than maximum width; fold and sulcus weak, generally reflected only in slight uniplication of anterior margin; lateral costae numerous, fine, simple; costae of fold and sulcus very numerous, fine, tending to bifurcate; interior similar to *Cyrtospirifer. U.Dev.(Famenn.),* USSR.——FIG. 567, *1.* **D. romanowskii*; *1a-d,* brach.v., ped.v., ant., lat., ×0.7 (591).

?Indospirifer GRABAU, 1931, p. 359 [**Spirifer padaukpinensis* REED, 1908, p. 101; OD]. Me-

dium-sized; moderately transverse, with slightly rounded cardinal extremities; fold and sulcus distinct; entire surface covered with strong, generally simple plications; micro-ornament consisting of distinct capillae which originate in grooves and diverge fan-wise onto plications; pedicle valve interior with strong dental plates, lacking median septum; presence of delthyrial plate not established. *M.Dev.-U.Dev.,* cosmop.——FIG. 569,*1.* *I. padaukpinensis maoerhchuanensis* GRABAU, M.Dev., China; *1a-e,* brach.v., ped.v., lat., post., ant., ×1.5; *1f,* micro-ornament, ×9 (358).——FIG. 570,*6.* *I. varians* (FENTON), U. Dev., Iowa; micro-ornament, ×10 (295).

Platyspirifer GRABAU, 1931, p. 355 [**Schizophoria paronai* MARTELLI, 1902, p. 365; OD]. Length and width subequal; cardinal extremities rounded, yielding rather short hinge line; fold and sulcus weak; lateral slopes, fold, and sulcus with very numerous, fine costae; otherwise seemingly similar to *Cyrtospirifer,* but interior unknown. *?U. Dev.,* China.——FIG. 569,*2.* **P. paronai* (MARTELLI); *2a-e,* brach.v., ped.v., lat., post., ant., ×1.5 (358).

Prospira MAXWELL, 1954, p. 35 [**P. typa*; OD]. Similar to *Cyrtospirifer* except for narrow fold and sulcus with costae tending toward obsolescence. *L.Carb.(U.Tourn.-Visean),* Australia-Japan-?Eu.-?N.Am.——FIG. 570,*2.* **P. typa,* Queensl.; ped.v., ×1 (541).

?Schizospirifer GRABAU, 1931, p. 353 [**Spirifer aperturatus* var. *latistriatus* FRECH, 1911, p. 53; OD]. Somewhat resembling *Cyrtospirifer,* but with coarser costae, some of which branch on lateral slopes; interior unknown. *?M.Dev.,* China. ——FIG. 570,*5.* **S. latistriatus* (FRECH); *5a-e,* brach.v., ped.v., lat., post., ant., ×1.5 (358).

Sphenospira COOPER, 1954, p. 330 [**Spirifera alta* HALL, 1866, p. 246; OD]. Very similar to *Syringospira* but with longer dental plates, less elaborate frill; blisters lacking. *U.Dev.,* E.N.Am.——FIG. 570,*7.* **S. alta* (HALL); post., ×1 (183).

Sulcatospirifer MAXWELL, 1954, p. 11 [**S. primus*; OD]. Fold divided by prominent, rather wide median groove; costae of fold and sulcus very obscure; micro-ornament consisting of concentric growth lamellae, capillae, and pustules. *U. Dev.(Famenn.),* Australia.——FIG. 570,*1.* **S. primus*; ped.v. mold, ×2 (539).

Syringospira KINDLE, 1909, p. 28 [**S. prima*; OD]. Pedicle valve hemipyramidal; lateral slopes with simple costae; fold and sulcus present, costate; micro-ornament consisting of radially elongate pustules; shell substance overgrown so as to produce prominent frill or flange which serves to greatly increase size of interarea; pedicle valve interior with short dental plates, prominent delthyrial plate and stegidium in old specimens; umbonal cavities filled with blister-like plates. *U.Dev. (Percha Sh.),* W.N.Am.——FIG. 570,*4.* **S. prima*; *4a-c,* ant., lat., post., ×2; *4d,* post., ×2 (183).

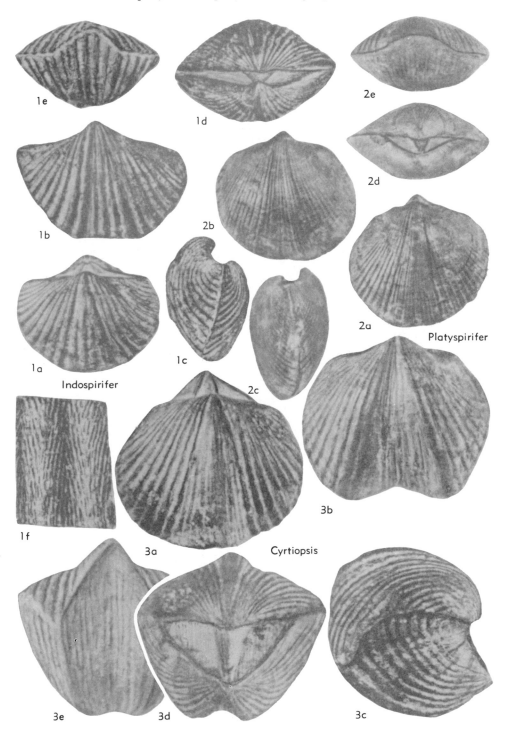

FIG. 569. Cyrtospiriferidae (p. *H*700).

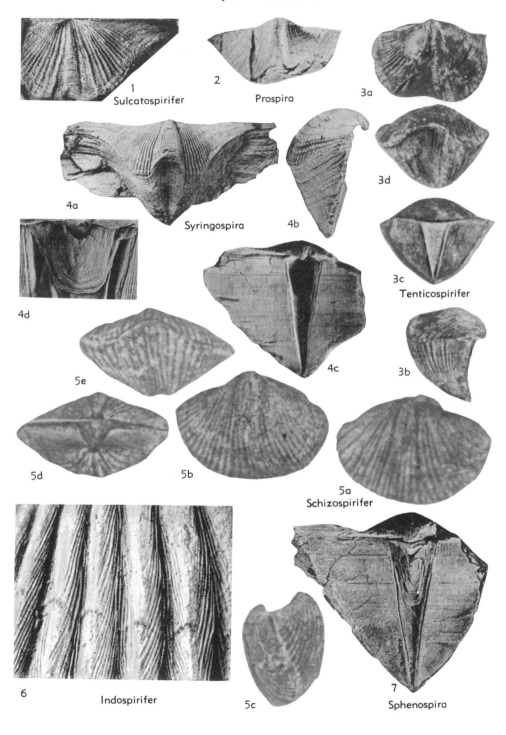

1 Sulcatospirifer

2 Prospira

3a

4a

Syringospira

4b

3d

3c

Tenticospirifer

4d

4c

3b

5e

5d

5b

5a

Schizospirifer

6 Indospirifer

5c

7 Sphenospira

Fig. 570. Cyrtospiriferidae (p. *H*700, *H*704).

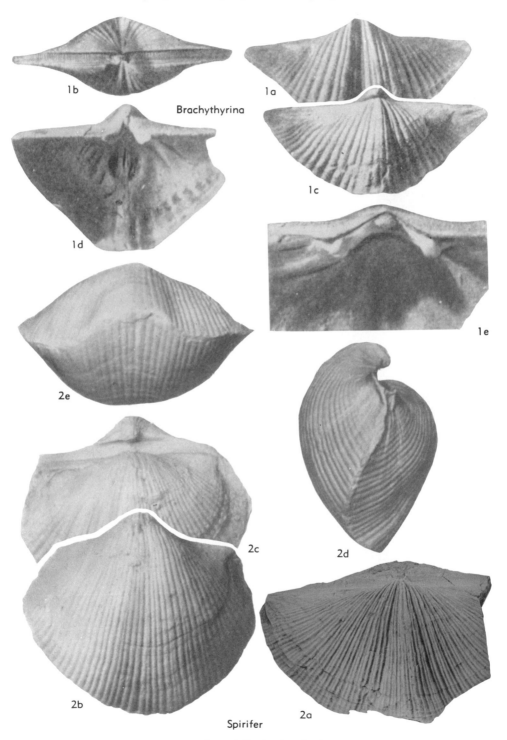

FIG. 571. Spiriferidae (p. *H*704).

FIG. 572. Spiriferidae (p. H704).

Tenticospirifer TIEN, 1938, p. 113 [*Spirifer tenticulum* DE VERNEUIL, 1845, p. 159; OD]. Pedicle valve hemipyramidal, with large, high, essentially equilateral interarea; otherwise similar to *Cyrtospirifer*. ?M.Dev., U.Dev., cosmop.——FIG. 570,3. *T. tenticulum* (DE VERNEUIL), U.Dev., USSR; *3a-d*, ped.v., lat., post., ant., ×1 (811).

Family SPIRIFERIDAE King, 1846
[Spiriferidae KING, 1846, p. 28]

Biconvex; slightly to markedly transverse; cardinal extremities rounded in some, making hinge line somewhat less than maximum width, more typically angular, with hinge line equal to maximum width, hinge commonly denticulate; lateral plications few to numerous, rarely simple, more commonly bifurcating, in some becoming fasciculate; fold and sulcus commonly distinct, plicate; pedicle valve interior generally with dental plates, lacking median septum; delthyrial plate weakly developed or lacking; brachial valve interior rarely with crural plates; shell substance impunctate. *L.Carb.-Perm.*

Spirifer SOWERBY, 1816, p. 41 [*Conchyliolithus (Anomia) striatus* MARTIN, 1793, pl. 23; SD ICZN Opinion 100, 1928, p. 377] [=*Spiriferus* DE BLAINVILLE, 1827, p. 291 (nom. van.); *Spirifera* PHILLIPS, 1836, p. 216 (nom. van.); ?*Lytha* FREDERIKS, 1919 (1924), p. 298 (type, *Spiriferella? tschernyschewiana* FREDERIKS, 1916)]. Biconvex; almost equidimensional to moderately transverse; cardinal extremities generally rounded, providing hinge line somewhat less than maximum width; lateral plications numerous, generally bifurcating adjacent to fold and sulcus, elsewhere generally simple, rarely somewhat fasciculate; fold and sulcus with numerous bifurcating plications; micro-ornament typically consisting of obscure concentric growth lamellae and capillae; pedicle valve interior with short, stout dental plates, lacking median septum and delthyrial plate; brachial valve interior without crural plates. *Carb.*, cosmop.——FIG. 571,2a. *S. striatus* (MARTIN), L.Carb., Br.I.; brach.v., ×0.7 (Sadlick, n). ——FIG. 571,2b-e. *S. gregeri* WELLER, Miss. (Chouteau Ls.), USA(Mo.); *2b-e*, ped.v., brach.v., lat., ant., ×1.5 (858).

Anthracospirifer LANE, 1963, p. 387 [*A. birdspringensis*; OD]. Lateral costae strong, few; otherwise similar to *Unispirifer*. *Penn.*, N.Am. ——FIG. 572,1. *A. birdspringensis*, Morrow., Nev.; *1a-d*, brach.v. int., brach.v., ped.v. int., ped.v., ×2 (502a).

Brachythyrina FREDERIKS, 1929, p. 385 [*pro Anelasma* IVANOV, 1925, p. 33 (*non* DARWIN, 1851; *nec* SOERENSEN, 1873; *nec* COSSMANN, 1889)] [*Spirifer strangwaysi* DE VERNEUIL, 1845, p. 164; OD] [=*Anelasmina* SEMIKHATOVA, 1939, p. 324 (obj.); *Elinoria* COOPER & MUIR-WOOD, 1951, p. 195 (*pro Elina* FREDERIKS, 1924, p. 321, *non* BLANCHARD, 1852, *nec* FERRARI, 1878) (type, *Spirifer rectangulus* KUTORGA, 1844, p. 90)]. Micro-ornament consisting of rather weak concentric growth lamellae; pedicle valve interior with dental plates reduced to teeth ridges; otherwise similar to *Unispirifer*. *L.Carb.-Perm.*, Eu.-Asia.——FIG. 571,1a-c. *B. strangwaysi* (DE VERNEUIL), U.Carb. (Moscov.), USSR; *1a-c*, ped. v., post., brach.v., ×1.5 (448).——FIG. 571,1d,e. *B. strangwaysi lata* CHAO, U.Carb. (Moscov.), USSR; *1d,e*, ped.v. int., ×1.5; brach.v. int., ×6 (448).

Fusella M'Coy, 1844, p. 128 [*Spirifera fusiformis* PHILLIPS, 1836, p. 217; OD]. Extremely transverse; lateral slopes with about 8 rounded plications; fold with about 3 weaker plications; sulcus bald; otherwise seemingly similar to *Unispirifer*. *L.Carb.*, Br.I.——Fig. 573,2. *F. fusiformis* (PHILLIPS); *2a,b*, ant., post. (oblique), ×1.5 (640).

[The type-species of *Fusella* is so poorly known that the generic characters cannot be regarded as well established. Until this situation is remedied, it seems best to confine use of the name *Fusella* to the type-species. Other species commonly placed in *Fusella* seem assignable to *Unispirifer*.]

Grantonia BROWN, 1953, p. 60 [*G. hobartensis*; OD]. Generally similar to *Neospirifer* but fasciculate plications stronger, less numerous, and shell greatly thickened with callus. *Perm.*, Tasmania.——Fig. 573,3. *G. hobartensis*; *3a-c*, post. (oblique), ped.v. post., ped.v. int., ×1 (123).

Imbrexia NALIVKIN, 1937, p. 105 [*Spirifer imbrex* HALL, 1858, p. 601; OD]. Micro-ornament consisting of imbricate growth lamellae without capillae; otherwise seemingly similar to *Unispirifer*, but type-species poorly known. *L.Carb.*, cosmop.——Fig. 573,5. *I. imbrex* (HALL), Miss. (Burlington Ls.), USA(Iowa); brach.v., ×1.5 (858).

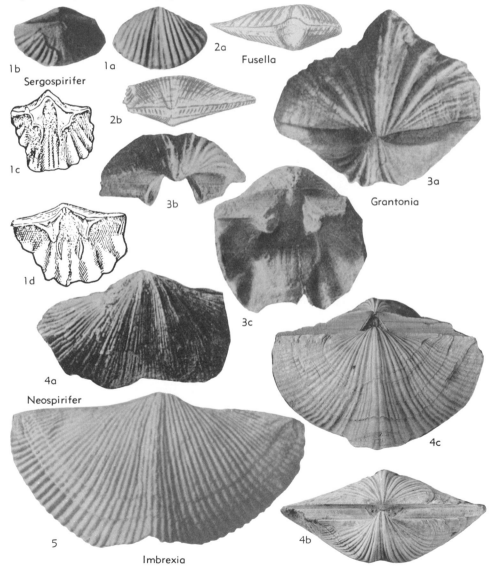

1b 1a 2a
Sergospirifer Fusella
1c
2b
1d
3b
3a
Grantonia
3c
4a
Neospirifer
5 4c
Imbrexia 4b

FIG. 573. Spiriferidae (p. H705-H706).

Neospirifer FREDERIKS, 1919 (1924), p. 311 [*non* NIKITIN, 1900, p. 385, *nom. nud.*] [**Spirifer fasciger* KEYSERLING, 1846, p. 231; OD] [=*Alphaneospirifer* GATINAUD, 1949, p. 491 (type, *Spirifer mahaensis* HUANG, 1933); *Betaneospirifer* GATINAUD, 1949, p. 491 (type, *Spirifer moosakhailensis* DAVIDSON, 1862)]. Generally rather large, transverse; cardinal extremities commonly angular; hinge line equal or almost equal to maximum

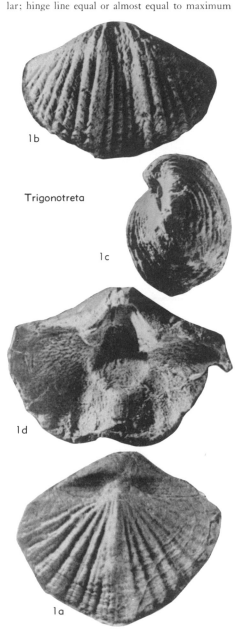

1b

Trigonotreta

1c

1d

1a

FIG. 574. Spiriferidae (p. *H706*).

shell width, typically denticulate; lateral slopes with numerous fasciculate plications; fold and sulcus distinct, plicate; micro-ornament consisting of distinct concentric growth lamellae and rather obscure capillae; pedicle valve interior with short dental plates; brachial valve interior without crural plates. [For comment on taxa of GATINAUD, *see* entry following *Cyrtospirifer.*] *U.Carb.-Perm.,* cosmop.——FIG. 573,*4a.* **N. fasciger* (KEYSERLING), L.Perm., USSR; ped.v., ×1 (123).—— FIG. 573,*4b,c.* N. *cameratus* (MORTON), M.Penn., USA(Ohio); *4b,c,* post., brach.v., ×1 (178).

Sergospirifer IVANOVA in SARYCHEVA & SOKOLSKAYA, 1952, p. 190 [**Spirifer okensis* NIKITIN, 1890, p. 28; OD]. Rather small; moderately transverse; cardinal extremities rounded; hinge line slightly less than maximum width; lateral slopes with 6 to 8 prominent, simple costae; fold and sulcus rather obscure, bearing 3 or 4 bifurcating costae; pedicle valve interior with long, thin, widely separated, essentially parallel dental plates; brachial valve interior with very well-developed crural plates. *U.Carb.,* USSR.——FIG. 573,*1.* **S. okensis* (NIKITIN); *1a,b,* ped.v., brach.v., ×1; *1c,d,* ped.v. int., brach.v. int., ×1.5 (711).

Trigonotreta KOENIG, 1825, p. 3 [**T. stokesi*; SD HALL & CLARKE, 1894, p. 8]. Lateral plications strong, few; each lateral plication tending to bifurcate once, doubled plications remaining in genetic pairs; otherwise similar to *Neospirifer.* *Perm.,* Tasmania.——FIG. 574,*1.* **T. stokesi, 1a-d,* brach.v., ped.v., lat., ped.v. int., ×1 (123).

Unispirifer CAMPBELL, 1957, p. 67 [**Spirifer striatoconvolutus* BENSON & DUN, 1920, p. 350; OD] [=?*Grandispirifer* YANG, 1959, p. 116 (type, *G. mylkensis*)]. Biconvex; rather strongly transverse; hinge line equal to maximum shell width, denticulate; lateral costae numerous, mostly simple, but some bifurcating, never fasciculate; fold and sulcus narrow, with several distinct, bifurcating costae; micro-ornament consisting of distinct capillae; pedicle valve interior with stout dental plates; brachial valve interior lacking crural plates. *L.Carb.(Tournais.-Visean),* cosmop. ——FIG. 575,*1a-c.* **U. striatoconvolutus* (BENSON & DUN), Tournais., Australia(New S. Wales); *1a-c,* post., ped.v., brach.v., ×1 (140).——FIG. 575,*1d-f.* U. *mylkensis* (YANG), Visean, China; *1d-f,* ped.v., brach.v., post., ×4.7 (898).

Family BRACHYTHYRIDIDAE
Frederiks, 1919 (1924)

[*nom. transl. et correct.* PITRAT, herein (*ex* Brachythyrinae FREDERIKS, 1919 (1924), p. 316)]

Shell markedly biconvex; hinge line generally short, commonly denticulate; interarea generally distinct, triangular or trapezoidal; fold and sulcus almost always present, with bifurcating costae or plications which in some shells are obscure; lateral

FIG. 575. Spiriferidae (p. *H706*).

slopes invariably with costae or plications, generally bifurcating, but more rarely simple; micro-ornament variable; pedicle valve interior with dental plates or teeth ridges; lacking delthyrial plate and median septum; brachial valve interior with or without crural plates; shell substance impunctate. *?U.Dev., L.Carb.-Perm.*

Brachythyris M'COY, 1844, p. 128 [**Spirifera ovalis* PHILLIPS, 1836, p. 219; OD] [*=Ovalia* NALIVKIN, 1937, p. 107 (obj.)]. Biconvex; hinge much less than maximum shell width, nondenticulate; interarea triangular; lateral plications few, broad, low, simple; fold and sulcus distinct, with well-developed plications in some, more typically with plications weak or obsolete; pedicle valve interior with teeth ridges, lacking dental plates. *?U.Dev., L.Carb.,* cosmop.——FIG. 576,*3a-c. *B. ovalis* (PHILLIPS), Visean, Br.I.; *3a-c,* brach.v., ped.v., lat., ×1 (229).——FIG. 576,*3d-f. B. subcardiiformis* (HALL), Miss.(Salem Ls.), USA(Ill.); *3d-f,* ped.v., lat., brach.v., ×1 (858).

?Cancellospirifer CAMPBELL, 1953, p. 10 [**C. maxwelli*; OD]. Biconvex; equidimensional to slightly transverse, with rounded cardinal extremities; hinge line less than maximum shell width; lateral slopes with about 7 distinct, rounded, simple costae; fold and sulcus well defined, bearing 2 or 3 very weak costae; micro-ornament consisting of imbricate growth lamellae and capillae, together producing cancellated effect; pedicle valve interior with dental plates; brachial valve interior lacking crural plates. *Perm.(Ingelara Beds),* Australia(Queensl.).——FIG. 576,*4;* 577,*3. *C. maxwelli;* 576,*4a,b,* ped.v., brach.v., ×1.5; 577,*3,* ornament, ×10 (139).

Choristitella IVANOV & IVANOVA, 1937, p. 163 [**Choristites podolskensis* IVANOV, 1926, p. 17; OD]. Interarea triangular; dental plates short, commonly greatly thickened with callus; otherwise similar to *Choristites. U.Carb.,* USSR.——FIG. 576,*5. *C. podolskensis* (IVANOV); *5a,b,* ped.v., ped.v. post., ×1 (447).

Choristites FISCHER DE WALDHEIM, 1825, p. 7 [**C. mosquensis*; SD BUCKMAN, 1908, p. 30] [*=Neomunella* OZAKI, 1931, p. 24 (type, *Spirifer (Neomunella) chaoi* OZAKI, 1931); *Yatsengina* SEMIKHATOVA, 1936, p. 216 (type, *Y. plana*); *Alphachoristites* GATINAUD, 1949, p. 492 (type, *Choristites bisulcatiformis* SEMIKHATOVA, 1934); *Betachoristites* GATINAUD, 1949, p. 492 (type, *Choristites kschemyschensis* SEMIKHATOVA, 1941); *Jatsengina* IVANOVA, 1960, p. 270 *(nom. null.)*]. Strongly biconvex; hinge line generally slightly less than maximum width; beak of pedicle valve strong, curved; interarea trapezoidal; hinge line denticulate; lateral costae generally very numerous, straplike, with narrow interspaces, those nearest fold and sulcus bifurcating, marginal ones simple; fold and sulcus rather shallow, marginal, with poorly defined margins, bearing numerous bifurcating costae; micro-ornament of capillae and concentric growth lines; pedicle valve interior with rather long, close-set, almost parallel dental plates which intersect muscle field; brachial valve interior without crural plates. [For comments on genera of GATINAUD (1949), see note following *Cyrtospirifer* (p. *H697*).] *L.Carb.-L.Perm.,* cosmop.——FIG. 577,*1. *C. mosquensis,* U.Carb. (Moscov.), USSR; *1a-c,* ped.v., brach.v., lat., ×1 (448).——FIG. 576,*6. *C.* sp.; ped.v. int., ×1 (447).

Ectochoristites CAMPBELL, 1957, p. 71 [**E. wattsi*;

OD]. Pedicle valve interior with short, thick dental plates; brachial valve interior with short crural plates; otherwise similar to *Choristites*. *L.Carb.* (*Tournais.*), *?U.Carb.(Namur.)*, Australia - ?N.

Am.-?Eu.——Fig. 577,2. **E. wattsi*, Tournais., Australia; *2a-d*, ped.v., brach.v., lat., ant., ×1 (140).

Eliva FREDERIKS, 1919 (1924), p. 319 [**Spirifer*

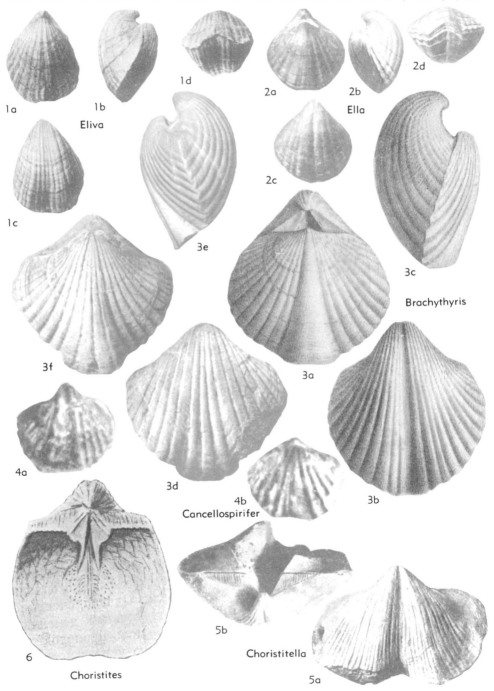

Fig. 576. Brachthyrididae (p. *H707-H709*)

1a

1b

Choristites

1c

2d

2b

2a

Ectochoristites

2c

Cancellospirifer

3

FIG. 577. Brachythyrididae (p. *H707-H708*).

lyra KUTORGA, 1844, p. 92; OD]. Similar to *Spiriferella* but plications more numerous and dental plates reduced to teeth ridges. *L.Perm.*, USSR.——FIG. 576,1. *E. lyra* (KUTORGA); *1a-d*, brach.v., lat., ped.v., ant., ×1 (158).

Ella FREDERIKS, 1918, p. 87 [*Martinia simensis* CHERNYSHEV, 1902, p. 569; OD]. Small; lateral costae few, simple; dental plates reduced to teeth ridges; otherwise similar to *Purdonella*. *L.Perm.*, USSR.——FIG. 576,2. *E. simensis* (CHERNYSHEV); *2a-d*, brach.v., lat., ped.v., ant., ×1 (158).

Eochoristites CHU, 1933, p. 28 [*E. neipentaiensis*; OD]. Moderately biconvex; fold and sulcus weak, rather narrow; lateral costae generally simple; interarea triangular; hinge nondenticulate; brachial valve interior with short, thin crural plates; otherwise similar to *Choristites*. *L.Carb.*, Asia.——FIG. 578,3. *E. neipentaiensis*, Kinling Ls., SE.China; *3a-e*, ped.v., brach.v., lat., post., ant., ×1 (161).

Palaeochoristites SOKOLSKAYA, 1941, p. 26 [*Spirifer cinctus* KEYSERLING, 1847, p. 229; OD]. Fold

and sulcus lacking; brachial valve interior with rather long crural plates; otherwise similar to *Eochoristites*. *L.Carb.*, USSR.——FIG. 578,4. *P. cinctus* (KEYSERLING); *4a-e*, ped.v., brach.v., post., ant., lat., ×1 (752).

Purdonella REED, 1944, p. 218 [*pro Munella* FREDERIKS, 1919 (1924), p. 314 (*non* BONNIER, 1896)] [*Spirifer nikitini* CHERNYSHEV, 1902, p. 542; OD]. Interarea triangular; hinge nondenticulate; otherwise similar to *Choristites*. *Perm.*, Asia-Arctic.——FIG. 578,5. *P. nikitini* (CHERNYSHEV), L. Perm., USSR; *5a-d*, brach.v., lat., ped.v., ant., ×0.7(158).

Spiriferella CHERNYSHEV, 1902, p. 121 [*Spirifer saranae* DE VERNEUIL, 1845, p. 169; OD] [=*Elivina* FREDERIKS, 1919 (1924), p. 315 (=*Dienerina* OZAKI, 1931, p. 25; type, *Spirifer tibetanus* DIENER, 1897); *Blasispirifer* KULIKOV, 1950, p. 6 (type, *Spirifer blasii* DE VERNEUIL, 1845)]. Pedicle valve highly convex; brachial valve less so; hinge generally very short, nondenticulate; lateral plica-

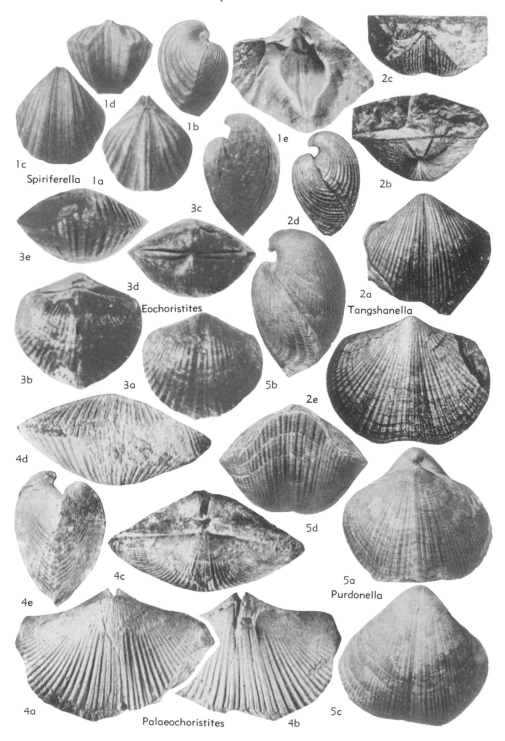

Fig. 578. Brachythyrididae (p. *H*709, *H*711).

tions few, very strong, bifurcating; fold and sulcus ranging from well defined and almost bald to poorly defined with strong plications; pedicle valve interior with well-developed dental plates, commonly thickened with callus, deeply impressed muscle field; brachial valve interior unknown. *U.Carb.-Perm.,* cosmop.——Fig. 579,2. **S. saranae* (DE VERNEUIL), L.Perm., USSR; *2a,* ped.v., ×1; *2b,* ornament, ×15 (448).——Fig. 578,*1a-d. S. tibetana* (DIENER), L.Perm., Tibet; *1a-d,* brach.v., lat., ped.v., ant., ×1 (158).—— Fig. 578,*1e. S. keilhavii* (VON BUCH), U.Perm., Greenl.; ped.v. int., ×1 (269).

Tangshanella CHAO, 1929, p. 57 [**T. kaipingensis;* OD]. Pedicle valve interior with dental plates reduced to teeth ridges; otherwise similar to *Choristites. U.Carb.,* China.——Fig. 578,*2. *T. kaipingensis,* Penchi Series; *2a-e,* ped.v., post., ant., lat., brach.v., ×1 (154).

Family UNCERTAIN

Thomasaria STAINBROOK, 1945, p. 57 [**T. altumbona;* OD]. Rather small; weakly transverse, with slightly rounded cardinal extremities; brachial valve moderately convex; pedicle valve hemipyramidal, with high interarea; macro-ornament lacking; micro-ornament consisting of fine growth lamellae which give rise to fine spines; delthyrium high and narrow, constricted by pair of plates similar to conjunct or disjunct deltidial plates but occupying position of delthyrial plate; pedicle valve interior with long dental plates which diverge anteriorly and ventrally; brachial valve interior with short crural plates, striate cardinal process. *U.Dev.(Frasn.),* N.Am.(Iowa-N. Mex.).——Fig. 579,*1. *T. altumbona,* Iowa; *1a-f,* ped.v., brach.v., post., ant., lat., transv. sec. ped.v., ×1 (768).

Superfamily SPIRIFERINACEA Davidson, 1884

[*nom. transl.* IVANOVA, 1959, p. 57 (*ex* subfam. Spiriferinidae DAVIDSON, 1884, p. 354)]

Shell rather variable, but typically small, rather transverse, biconvex, with distinct fold and sulcus and plicate lateral slopes; micro-ornament variable, consisting of growth lamellae, capillae, spines, and granules in various combinations; interior of pedicle valve with well-developed median septum, and commonly with dental plates; shell substance generally, but not invariably, punctate, *L.Carb.-L.Jur.*

Family SPIRIFERINIDAE Davidson, 1884

[*nom. transl.* IVANOVA, 1959, p. 57 (*ex* subfam. Spiriferinidae DAVIDSON, 1884, p. 354)]

Characters of superfamily. *L.Carb.-L.Jur.*

FIG. 579. Brachythyrididae *(2)*; Family Uncertain *(1)* (p. *H*711).

Spiriferina D'ORBIGNY, 1847, p. 268 [**Spirifer walcotti* SOWERBY, 1823, p. 106; SD DALL, 1877, p. 64]. Shell small to medium-sized, equidimensional to moderately transverse; cardinal extremities rounded; lateral slopes ranging from smooth to coarsely plicate; fold and sulcus generally distinct in plicate forms, commonly obscure or wanting in smooth forms; micro-ornament of growth lamellae and numerous fine, tubular spines; punctate. *Trias.-L.Jur.,* cosmop.——Fig. 580,*5. *S. walcotti* (SOWERBY), Lias., Br.I.; *5a,b,* brach.v., ant., ×1 (229).

[No nominal species were assigned to *Spiriferina* when the genus was erected by D'ORBIGNY (1847). Later D'ORBIGNY (1849) assigned 11 species to the genus, including *Spirifer walcotti* SOWERBY, 1823, but not including *Terebratulites rostratus* VON SCHLOTHEIM, 1822; no type-species was designated. DALL (1877) stated the type to be *S. rostratus* (SCHLOTHEIM) =*S. walcotti* (SOWERBY), attributing this information to DAVIDSON (1856, p. 161). I have been unable to locate the DAVIDSON reference, and I believe it to be erroneous, inasmuch as both before and after 1856, DAVIDSON maintained that *S. walcotti* and *S. rostratus* were separate species. In any case, DALL's designation of *S. rostratus* as type and his synonymizing of it with *S. walcotti,* one of the eligible species, makes the latter the type, despite the fact that the synonymy appears to be in error (Code, Art. 69a,iii,iv)]

?Acanthospirina SCHUCHERT & LEVENE, 1929, p. 119 [*pro Acanthospira* WELLER, 1914, p. 418 (*non* REINSCH, 1877)] [**Spirifer aciculifera* ROWLEY, 1893, p. 307; OD]. Very small; moderately

FIG. 580. Spiriferinidae (p. H711, H713-H714).

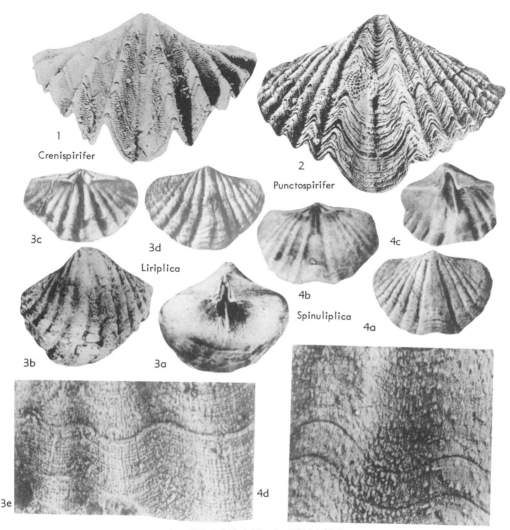

FIG. 581. Spiriferinidae (p. *H713-H714*).

transverse, with rounded cardinal extremities; micro-ornament consisting of fine, radially arranged spines; pedicle valve interior with dental plates, lacking median septum; neither presence nor absence of punctation firmly established. *Miss.*, N.Am.——FIG. 580,6. **A. aciculifera* (ROWLEY), L.Miss.; *6a,b*, brach.v., ped.v., ×12 (683).

Altiplecus STEHLI, 1954, p. 349 [**A. cooperi*; OD]. Small, highly transverse, triangular in outline; fold and sulcus deep, rather narrow, smooth; lateral plications few, low, rounded; micro-ornament of strong spines, generally in 1 or 2 concentric rows on irregular growth lamellae; punctate. *L. Perm.*, N.Am.——FIG. 580,2. **A. cooperi*, USA (Tex.); *2a*, ant., ×1.5; *2b*, ped.v., ×3 (773).

Callispirina COOPER & MUIR-WOOD, 1951, p. 195 [*pro Maia* FREDERIKS, 1919 (1924), p. 298 (*non*

LAMARCK, 1801; *nec* REICHENBACH, 1850); *pro Maya* RAKUSZ, 1932, p. 77 (*non* BLATTNY, 1925); *pro Mansuyella* REED, 1944, p. 249 (*non* ENDO, 1937)] [**Spiriferina ornata* WAAGEN, 1883, p. 505; OD] [=?*Paraspiriferina* REED, 1944, p. 252 (type, *Spiriferina (Paraspiriferina) ghundiensis*)]. Small, essentially equidimensional; cardinal extremities rounded; fold and sulcus distinct, smooth; lateral slopes with 3 or 4 very strong, angular plications; micro-ornament of numerous growth lines only; punctate. *M.Perm.*, India.——FIG. 580,*1*. **C. ornata* (WAAGEN); *1a-d*, brach.v., ped.v., lat., ant., ×1.5; *1e*, surface, ×? (845).

Crenispirifer STEHLI, 1954, p. 347 [**Spiriferina angulata* KING, 1930, p. 122; OD]. Small, transverse, subequally biconvex; cardinal extremities slightly rounded; fold and sulcus narrow, smooth;

lateral plications few, very high, angular; micro-ornament of exceedingly numerous fine spines; punctate. *U.Penn.-U.Perm.*, N.Am.——Fig. 581,*1*; 580,*3*. *C. angulata* (King), L.Perm., USA(Tex.); 581,*1*, ped.v., ×2; 580,*3a-e*, brach.v., brach.v. int., ant., ped.v. int., lat., ×1.5 (773).

?**Dimegelasma** Cooper, 1942, p. 232 [*Spirifer neglectus* Hall, 1858, p. 643; OD]. Strongly biconvex; approximately equidimensional; cardinal extremities rounded, producing hinge somewhat less than maximum width; interarea rather high, with delthyrium closed by flat pseudodeltidium; fold and sulcus pronounced, bald; lateral slopes with several, well-developed, rounded plications; pedicle valve interior with long, rather thin dental plates, lacking median septum; brachial valve interior with hinge plate supported by median septum; shell substance punctate. *Miss(L.Carb.)*, N.Am.-Australia.——Fig. 580,*4*. *D. neglectum* (Hall), Miss. (Keokuk Ls.); *4a-d*, ped.v., lat., brach.v., ant., ×1 (858); *4e,f*, ped.v. int., brach.v. int., ×0.7 (178).

Liriplica Campbell, 1961, p. 440 [*L. alta*; OD]. Small to medium-sized, equidimensional to moderately transverse, strongly biconvex; fold and sulcus smooth except for small median fold in sulcus and small median groove in fold; lateral slopes with about 6 distinct, rounded plications; micro-ornament of prominent growth lamellae and discontinuous capillae; pedicle valve interior with thick umbonal callus; punctate. *U. Carb.(?Westphal.)*, Australia (New S.Wales).—— Fig. 581,*3*, *L. alta*; *3a-d*, ped.v. int. mold, ped.v. mold, brach.v. int. mold, brach.v. mold, ×1; *3e*, surface, ×10 (143).

?**Mentzeliopsis** Trechmann, 1918, p. 229 [*M. spinosa*; OD]. Equidimensional to weakly transverse; cardinal extremities slightly rounded; fold and sulcus distinct, low, wide, bald; lateral slopes with several rather weak plications; micro-ornament consisting of prominent imbricating growth lamellae and large tubular spines; pedicle valve interior with dental plates and prominent median septum; shell substance seemingly impunctate. *Trias.*, N.Z.——Fig. 582,*2*. *M. spinosa*; *2a-c*, brach.v., ped.v., int. mold, ×1 (816).

Odontospirifer Dunbar, 1955, p. 154 [*O. mirabilis*; OD]. Small, transverse, strongly biconvex; fold narrow, smooth; sulcus with single median plication; lateral plications strong; micro-ornament of strong growth lamellae and faint capillae; hinge line denticulate; dental plates weak; impunctate. *U.Perm.*, Greenl.——Fig. 582,*4*. *O. mirabilis*; *4a-d*, brach.v., ped.v., lat., ant., ×3; *4e-f*, ped.v., ped.v. int., ×7.5 (269).

Punctospirifer North, 1920, p. 212 [*P. scabricosta*; OD]. Small to medium-sized, transverse; fold and sulcus distinct, wide, smooth; lateral plications strong and rather numerous; micro-ornament of strong, imbricate growth lamellae and distinct capillae; punctate. *L. Carb. - Perm.*, cosmop.——Fig. 581,*2*; 582,*3*. *P. scabricosta*, L. Carb.(Visean), Br.I.; 581,*2*, ped.v., ×3.3; 582, *3a-d*, brach.v., post., ant., lat., ×2 (142).

?**Punctothyris** Hyde, 1953, p. 288 [*P. argus*; OD]. Small; slightly transverse; cardinal angles rounded, resulting in hinge line less than maximum width; lateral slopes with about 10 gently rounded costae; fold and sulcus rather obscure, bearing weaker costae; micro-ornament consisting of closely spaced growth lamellae; pedicle valve interior with short dental plates, lacking median septum; shell substance presumably punctate. [The features interpreted by Hyde as punctae may be merely spine bases, in which case *Punctothyris* should be reassigned to the *Spiriferidae*.] *Miss.*, N.Am.——Fig. 582,*1*. *P. argus*, USA(Ohio); *1a-e*, lat., brach.v., ped.v., ant., post., ×2; *1f*, micro-ornament, ×15 (441).

Rastelligera Hector, 1879, p. 538 [*R. elongata* Hector in Thomson, 1913, p. 50; SD Thomson, 1913, p. 50]. Medium-sized, highly transverse; lateral slopes with 4 or 5 distinct but gently rounded plications; hinge line denticulate; shell structure not known, probably punctate. *U.Trias.(Rhaet.)*, N.Z.——Fig. 583,*2*. *R. elongata*; *2a-c*, brach.v., ped.v. int., ant., ×1 (806).

Reticulariina Frederiks, 1916, p. 16 [*Spirifer spinosus* Norwood & Pratten, 1855, p. 71; OD]. Rather small, transverse; fold and sulcus rather narrow, smooth; lateral slopes with 5 or 6 distinct plications; micro-ornament of few large spines which leave elliptical scars; punctate. *Miss.*, N.Am. ——Fig. 583,*6*. *R. spinosa* (Norwood & Pratten), Chester., USA(Ill.); *6a*, brach.v., ×5; *6b-e*, ped.v., ant., post., lat., ×2 (142).

Sinucosta Dagis, 1963, p. 104 [*Spirifer emmrichi* Suess, 1854; OD] [?=*Guseriplia* Dagis, 1963, p. 107 (type, *G. multicostata*)]. Medium-sized, unequally biconvex, slightly transverse, with narrowly to broadly rounded cardinal extremities; lateral slopes with few to many distinct costae; fold and sulcus generally rather obscure, bearing costae similar to those on lateral slopes. *U.Trias.*, Eu.-Asia.——Fig. 583A,*1*. *S. emmrichi* (Suess); Rhaet., Alps; *1a-d*, brach.v., ped.v., post., lat., ×1 (212a).

Spinuliplica Campbell, 1961, p. 442 [*S. spinulosa*; OD]. Medium-sized, moderately transverse, with rounded cardinal extremities; fold distinct, smooth; sulcus with single median costa; lateral slopes with 6 or 7 rounded plications; micro-ornament of closely spaced, distinct growth lamellae and very numerous, fine, anteriorly directed spines; punctate. *U.Carb.(?Westphal.)*, Australia (New S. Wales).——Fig. 581,*4*. *S. spinulosa*; *4a-c*, brach.v., brach.v. int., ped.v. int., ×1; *4d*, surface, ×10 (143).

Spiriferellina Frederiks, 1919 (1924), p. 299 [*Terebratulites cristatus* von Schlotheim, 1816, p. 28; OD] [=*Tylotoma* Grabau, 1934, p. 100 (obj.)]. Small, moderately transverse, with

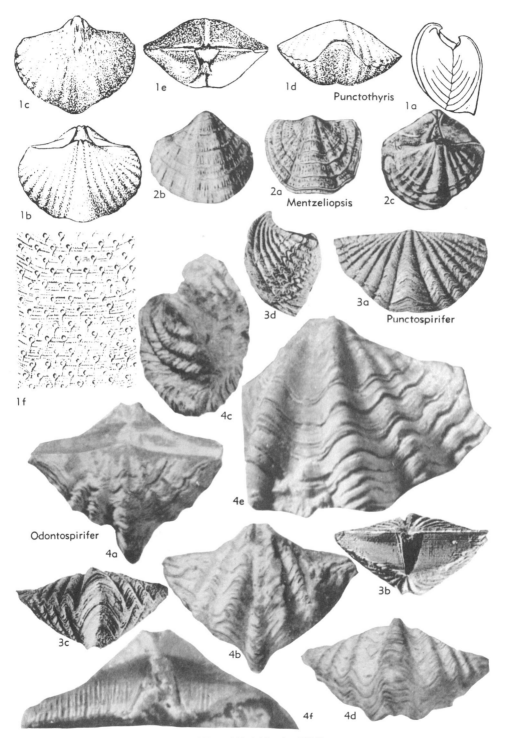

FIG. 582. Spiriferinidae (p. *H*714).

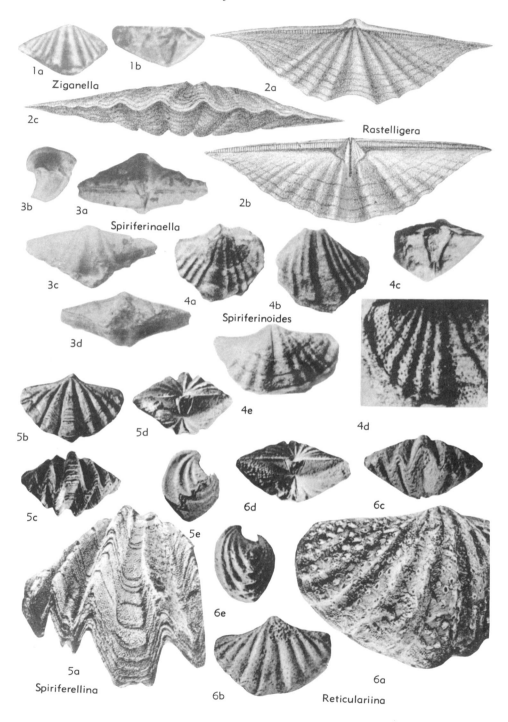

Fɪɢ. 583. Spiriferinidae (p. *H714, H717*).

rounded cardinal extremities; fold and sulcus deep, narrow, smooth; lateral slopes with 3 to 6 angular plications; micro-ornament of imbricate growth lamellae and very numerous fine granules; punctate. *Perm.*, Eu.-Asia-N.Am.——FIG. 583,5. **S. cristata* (VON SCHLOTHEIM), U.Perm., Ger.; *5a*, ped.v., ✕5; *5b-e*, brach.v., ant., post., lat., ✕2 (142).

Spiriferinaella FREDERIKS, 1926, p. 407 [**Spirifer artiensis* STUCKENBERG, 1898, p. 266; OD]. Like *Odontospirifer* except lateral plications somewhat weaker and more numerous, and dental plates stronger. *L.Perm.*, USSR.——FIG. 583,3. **S. artiensis* (STUCKENBERG); *3a-d*, post., lat., ped.v., ant., ✕1 (158).

Spiriferinoides TOKUYAMA, 1957, p. 101 [**S. sakawanus* KOBAYASHI & TOKUYAMA in TOKUYAMA, 1957; OD]. Small, rather transverse, with rounded cardinal extremities; pedicle valve semiconical; brachial valve nearly flat; fold and sulcus well developed, smooth; lateral slopes with about 6 distinct plications; micro-ornament of imbricate growth lamellae and numerous, fine, tubular spines; dental plates weak, commonly reduced to teeth ridges; impunctate. *M.Trias.-U.Trias.*, Eu.-Asia.——FIG. 583,4a-d. **S. sakawanus*, U.Trias. (Carn.), Japan; *4a-c*, brach.v. int. mold, ped.v. int. mold, post. int. mold, ✕2; *4d*, surface mold, ✕5 (812).——FIG. 583,4e. *S. yeharai* KOBAYASHI & TOKUYAMA, U.Trias.(Carn.), Japan; ped.v. int. mold, ✕2 (812).

?Ziganella NALIVKIN in IVANOVA, 1960, p. 280 [**Z. ziganensis*; OD]. Like *Punctospirifer* but lacking median septum in pedicle valve. *L.Carb.* (*Tournais.*), USSR.——FIG. 583,1. **Z. ziganensis*; *1a,b*, ped.v., post., ✕1.5 (448).

Superfamily RETICULARIACEA Waagen, 1883

[*nom. transl.* PITRAT, herein (*ex* Reticulariinae WAAGEN, 1883, p. 538)]

Generally biconvex; equidimensional to slightly transverse; cardinal extremities rounded, resulting in rather short hinge line; fold and sulcus generally present, tending to be rather weak; macro-ornament generally lacking or consisting of low, weak plications; micro-ornament highly variable; pedicle valve interior with or without dental plates and median septum; brachial valve interior with striate cardinal process, with or without crural plates; shell substance impunctate. *?U.Sil., L.Dev.-Perm., ?Trias.*

Family RETICULARIIDAE Waagen, 1883

[*nom. transl.* IVANOVA, 1959, p. 56 (*ex* Reticulariinae WAAGEN, 1883, p. 538)]

Biconvex; equidimensional to slightly transverse, with rounded cardinal extremi-

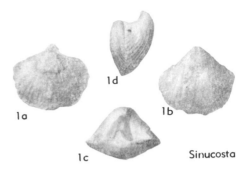

FIG. 583A. Spiriferinidae (*p. H714*).

ties and hinge line much less than maximum width; lateral slopes generally bald, more rarely weakly plicate; never costate; fold and sulcus generally present, commonly weak, bald; micro-ornament consisting of concentric growth lamellae, generally in combination with uniramous spines, papillae or granules; pedicle valve interior with or without dental plates and median septum; brachial valve interior generally lacking crural plates; cardinal process longitudinally striate; shell substance impunctate. *?U.Sil., L.Dev.-L.Carb., ?U.Carb.-?Trias.*

Reticularia M'COY, 1844, p. 142 [**Terebratula? imbricata* SOWERBY, 1822, p. 10; SD DAVIDSON, 1882, p. 80] [=*?Sinothyris* MINATO, 1953, p. 68 (type, *Spirifer maureri* HOLZAPFEL, 1896)]. Unequally biconvex; slightly transverse, with rounded cardinal extremities; fold and sulcus very weak to absent; lateral slopes bald; micro-ornament consisting of conspicuous concentric growth lamellae and uniramous spines; pedicle valve interior with well-developed dental plates and median septum; brachial valve interior lacking crural plates. *?Dev., L.Carb.*, Eu.-Asia.——FIG. 584,8a-d. **R. imbricata* (SOWERBY), L.Carb., Br. I.; *8a-d*, brach.v., ped.v., ant., lat., ✕1 (335).——FIG. 584,8e-j. *R. maureri* (HOLZAPFEL), Dev., China; *8e-i*, brach.v., ped.v., lat., post., ant., ✕1; *8j*, ornament, ✕5 (358).

[MINATO diagnosed *Sinothyris* as being "without dental plates, but with a median septum in the ventral valve." The diagnosis was based on Chinese forms referred to *Spirifer maureri* HOLZAPFEL, 1896, by GRABAU (1931, p. 394-96). Evidently MINATO based the diagnosis on GRABAU's statement that "no indications of dental plates have been seen in these specimens, but a median septum is indicated in one of them. . . ." The specimens to which GRABAU referred were not the whole suite of specimens which he had before him, but rather three exceptionally well-preserved specimens which evidently were not sectioned. In a later paragraph GRABAU (p. 396) mentioned a different specimen, the beak of which was broken, and wrote that "close-set dental lamellae are seen, much thickened by stereoplasm. . . ." In the light of this last statement it would appear that the specimens do possess dental plates, in which case no grounds are seen for separation of the genus based upon them from *Reticularia*.]

?**Ambikella** Sahni & Srivastava, 1956, p. 207 [**A. fructiformis*; OD]. Pedicle valve with prominent fold, brachial valve with distinct sulcus; anterior commissure sulcate; micro-ornament of concentric growth lamellae; pedicle valve interior with thin dental plates and median septum; brachial valve

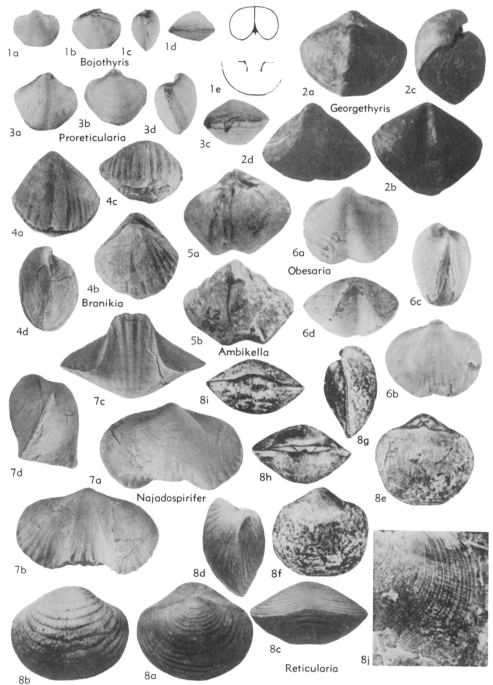

Fig. 584. Reticulariidae (p. *H717-H719*).

interior with long crural plates and median septum. *?U.Carb.*, India.——FIG. 584,5. **A. fructiformis; 5a,b,* brach.v., ped.v., ×1 (701).

?**Bojothyris** HAVLÍČEK, 1959, p. 147 [**B. nikiforovae;* OD]. Micro-ornament consisting of concentric growth lamellae only; pedicle valve interior with spondylium simplex; otherwise similar to *Reticularia. L.Dev.(Siegen.),* Boh.——FIG. 584, *1.* **B. nikiforovae; 1a-d,* ped.v., brach.v., lat., ant., ×1; *1e,* transv. sec., ×3 (411).

?**Branikia** HAVLÍČEK, 1957, p. 437 [**Spirifer ascanius* BARRANDE, 1879, p. 99; OD]. Approximately equidimensional, equally biconvex; fold and sulcus absent; macro-ornament consisting of rather low plications on entire valve surface, microornament of growth lines only; pedicle valve interior with short, thick dental plates, lacking median septum; brachial valve interior lacking crural plates. *L.Dev.(Ems.),* Boh.——FIG. 584,4. **B. ascania* (BARRANDE); *4a-d,* ped.v., brach.v., ant., lat., ×1.5 (411).

?**Elythina** RZHONSNITSKAYA, 1952, p. 61 [**E. salairica;* OD]. Weakly transverse with slightly rounded cardinal extremities; macro-ornament consisting of rather well-developed plications on lateral slopes; micro-ornament consisting of concentric growth lamellae and fine, long spines; pedicle valve interior with thick dental plates, lacking median septum. *M.Dev.(Couvin.-Givet.),* USSR.——FIG. 586,2. **E. salairica; ?a-d,* lat., brach.v., ped.v., ant., ×1 (465).

Georgethyris MINATO, 1953, p. 68 [**Reticularia alexandri* GEORGE, 1932, p. 553; OD]. Pedicle valve interior lacking median septum; fold and sulcus very strong, producing markedly uniplicate anterior commissure; otherwise similar to *Reticularia. L.Carb.(Visean),* Br.I.——FIG. 584,2. **G. alexandri* (GEORGE); *2a-d,* brach.v., ped.v., lat., ant., ×1 (335).

Najadospirifer HAVLÍČEK, 1957, p. 246 [**Spirifer najadum* BARRANDE, 1848, p. 171; OD] [=*Naiadospirifer* HAVLÍČEK, 1957, p. 246 *(nom. null.)*]. External appearance typically reticulariid in early growth stages but later developing strong angular fold and sulcus and plications over entire shell; plications very strong on fold and sulcus; pedicle valve interior with dental plates, lacking median septum; brachial valve interior with short crural plates, lacking median septum. *L.Dev.(Ems.),* Eu.——FIG. 584,7. **N. najadum* (BARRANDE); *7a-d,* ped.v., brach.v., ant., lat., ×1 (411).

Obesaria HAVLÍČEK, 1957, p. 438 [**Spirifer indifferens* var. *obesa* BARRANDE, 1848, p. 159; OD]. Anterior commissure strongly uniplicate; micro-ornament consisting of concentric rows of granules; pedicle valve interior lacking dental plates and median septum; otherwise similar to *Reticularia. M.Dev.(Couvin.),* Boh.——FIG. 586,1; 584, 6. **O. obesa* (BARRANDE); *586,1,* micro-ornament, ×12; *584,6a-d,* ped.v., brach.v., lat., ant., ×1.5 (411).

?**Proreticularia** HAVLÍČEK, 1957, p. 247 [**Spirifer carens* BARRANDE, 1879, p. 218; OD]. Anterior commissure rectimarginate to weakly uniplicate; macro-ornament lacking; micro-ornament consisting of densely crowded growth lamellae bearing papillae on their anterior terminations; pedicle valve interior lacking dental plates and median septum; brachial valve interior without crural plates. *U.Sil.(Ludlov.),* Bohemia.——FIG. 584,3. **P. carens* (BARRANDE); *3a-d,* ped.v., brach.v., ant., lat., ×1 (411).

?**Quadrithyrina** HAVLÍČEK, 1959, p. 136 [**Q. ivanovae;* OD]. Rather transverse, with distinct fold and sulcus, uniplicate anterior commissure; micro-ornament consisting of concentric growth lamellae with or without capillae; pedicle valve interior with strong, high median septum, lacking dental plates; otherwise similar to *Reticularia. M. Dev.(Couvin.),* Boh.——FIG. 585,1. **Q. ivanovae; 1a-e,* ped.v., brach.v., ant., lat., transv. sec., ×1 (411).

?**Quadrithyris** HAVLÍČEK, 1957, p. 437 [**Spirifer robustus* BARRANDE, 1848; OD]. Markedly transverse, with strong fold and sulcus and uniplicate anterior commissure; lateral slopes generally bald, rarely with weak plications; otherwise similar to *Reticularia. L.Dev.(Ems.)-M.Dev.(Couvin.),* Eu.-N.Afr.——FIG. 585,4. **Q. robustus* (BARRANDE), Ems., Boh.; *4a-d,* ped.v., brach.v., lat., ant., ×1.5 (411).

Reticulariopsis FREDERIKS, 1916, p. 17 [**Spirifer (Reticularia) dereimsi* OEHLERT, 1901, p. 236; SD FREDERIKS, 1918, p. 87] [=*Eoreticularia* NALIVKIN in FREDERIKS, 1919 (1924), p. 314 (type, *Spirifer indifferens* BARRANDE, 1847); *Tingella* GRABAU, 1931, p. 407 (type, *T. reticularioides*)]. Lateral slopes bald or rarely with very weak plications; fold and sulcus distinct; pedicle valve interior with well-developed dental plates, lacking median septum; brachial valve interior with crural plates, lacking median septum; otherwise similar to *Reticularia. L.Dev.-M.Dev., ?U. Dev.,* Eu.-Asia-Afr.-?Australia.——FIG. 585,2a-f. **R. dereimsi* (OEHLERT), L.Dev.(Ems.), Spain; *2a-f,* transv. sec., transv. sec., brach.v., ant., ped.v., lat., ×1 (613).——FIG. 585,2g-l. *R. reticularioides* (GRABAU), M.Dev., China; *2g-j,* brach.v., ped.v., ant., lat., ×3; *2k,* ornament, ×6; *2l,* int. mold post., ×4.5 (358).

[FREDERIKS (1916, p. 17) diagnosed *Reticulariopsis* as "*Reticularia*-like, but without a septum." He based the genus on three species, *Spirifer elliptica* PHILLIPS, 1836, *Delthyris fimbriata* CONRAD, 1842, and *Spirifer (Reticularia) dereimsi* OEHLERT, 1901, all of which were said to be "type specimens." In 1918 FREDERIKS (p. 87) designated *R. dereimsi* (OEHLERT) as type, but apparently changed his mind and in 1926 (p. 404) designated *R. elliptica* (PHILLIPS) as type. Subsequent authors have overlooked the earlier designation and have taken *R. elliptica* as type-species. GEORGE (1932, p. 525) reinvestigated *R. elliptica* and found it possesses a median septum; on that basis he synonymized *Reticulariopsis* with *Reticularia*. Later workers have tended to follow GEORGE's lead. However, *R. dereimsi*, which must be regarded as the type-species of *Reticulariopsis*, does not possess a median septum, and therefore *Reticulariopsis* cannot properly be placed in the synonymy of *Reticularia*.]

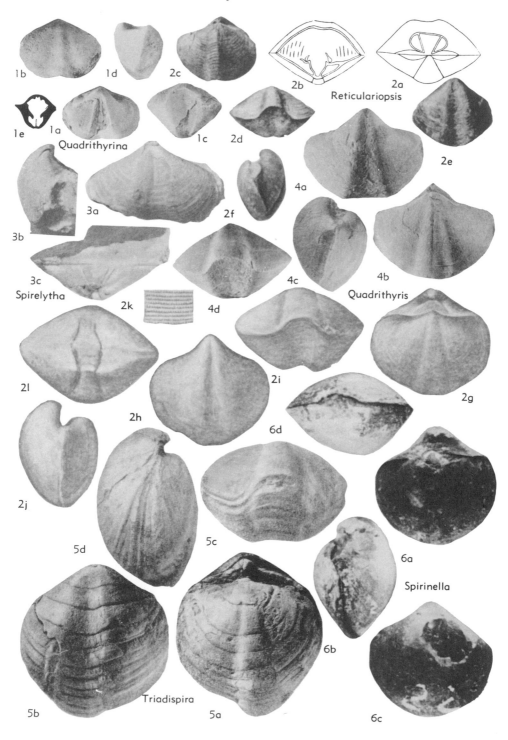

1b   1d   2c

2b

2a
Reticulariopsis

1e   1a
Quadrithyrina

1c   2d

2e

3a

2f

4a

3b

4b

3c
Spirelytha

2k

4d

4c

Quadrithyris

2l

2i

2h

6d

2g

2j

5d

5c

6a
Spirinella

6b

Triadispira

5b   5a   6c

FIG. 585. Reticulariidae (p. *H719, H721*).

?**Spirelytha** FREDERIKS, 1919 (1924), p. 304 [*Spirifer schei* CHERNYSHEV, 1916, p. 69; OD]. Pedicle valve reticulariid in shape with well-developed sulcus; macro-ornament lacking; micro-ornament consisting of concentric growth lamellae and small, irregular tubercles; pedicle valve interior with dental plates, lacking median septum; brachial valve not known. *Perm.*, Arctic.——FIG. 585, 3. *S. schei* (CHERNYSHEV); *3a-c*, ped.v., lat., post., ×1 (160).

?**Spirinella** JOHNSTON, 1941, p. 161 [*S. caecistriata*; OD]. Rather small; biconvex; essentially equidimensional; anterior commissure weakly uniplicate; macro-ornament lacking; micro-ornament consisting of concentric growth lamellae and obscure capillae; pedicle valve interior with dental plates, lacking median septum; brachial valve interior with neither median septum nor crural plates. ?*U.Sil.(Ludlov.),* Australia (New S. Wales).——FIG. 585,6. *S. caecistriata*; *6a-d*, brach.v., lat., ped.v., ant., ×2 (454a).

?**Triadispira** DAGIS, 1961, p. 457 [*T. caucasica*; OD]. Large; markedly biconvex; longer than wide; fold and sulcus distinct, forming markedly uniplicate anterior commissure; micro-ornament consisting of concentric growth lamellae and radially elongate papillae; otherwise similar to *Reticularia. Trias.*, USSR.——FIG. 585,5. *T. caucasica*, U.Trias.; *5a-d*, brach.v., ped.v., ant., lat., ×0.7 (212).

?**Undispirifer** HAVLÍČEK, 1957, p. 439 [*Spirifer undiferus* ROEMER, 1844, p. 73; OD]. Fold and sulcus distinct, bald; lateral slopes with low, rather obscure plications; pedicle valve interior with dental plates, lacking median septum; otherwise similar to *Reticularia. M.Dev.(Givet.)-U.Dev. (Frasn.),* Eu.——FIG. 587,3a. *U. undiferus* (ROEMER), Ger.; brach.v., ×2.5 (411).——FIG. 587,3b-e. *U. transiens* (BARRANDE), Givet., Boh.; *3b-e*, ped.v., brach.v., ant., lat., ×1 (411).

Warrenella CRICKMAY, 1953, p. 596 [*W. eclectea*; OD] [=*Minatothyris* VANDERCAMMEN, 1957, p. 1 (type, *Spirifer euryglossus* SCHNUR, 1851)]. Fold and sulcus distinct, producing markedly uniplicate anterior commissure; pedicle valve interior with dental plates, lacking median septum; otherwise similar to *Reticularia. M.Dev.-U.Dev.,* cosmop. ——FIG. 587,1a-f. *W. eclectea*, U.Dev., Alta.; *1a-e*, brach.v., ped.v., post., ant., lat., ×1.5; *1f*, transv. sec., ×2 (205).——FIG. 587,1g-j. *W. euryglossus* (SCHNUR), U.Dev., Ger.; *1g-j*, brach. v., ped.v., lat., ant., ×1.5 (830).

Xenomartinia HAVLÍČEK, 1953, p. 6 [*X. monosepta*; OD]. Micro-ornament consisting of concentric growth lamellae and capillae, commonly with papillae at intersections; pedicle valve interior lacking dental plates; otherwise similar to *Reticularia. L.Dev.(Ems.)-M.Dev.(Couvin.),* Eu.——FIG. 587,2. *X. monosepta*, Boh.; *2a-d*, ped.v., brach.v., ant., lat., ×2; *2e-f*, ped.v. int., brach.v. int., ×3 (411).

FIG. 586. Reticulariidae (p. H719).

Family ELYTHIDAE Frederiks, 1919 (1924)

[*nom. transl.* PITRAT, herein (*ex* Elythinae FREDERIKS, 1919 (1924), p. 304] [=Phricodothyrinae CASTER, 1939]

Markedly and almost evenly biconvex; equidimensional to moderately transverse, with broadly rounded cardinal extremities; fold and sulcus generally weak or absent, yielding rectimarginate to weakly uniplicate anterior commissure; macro-ornament lack-

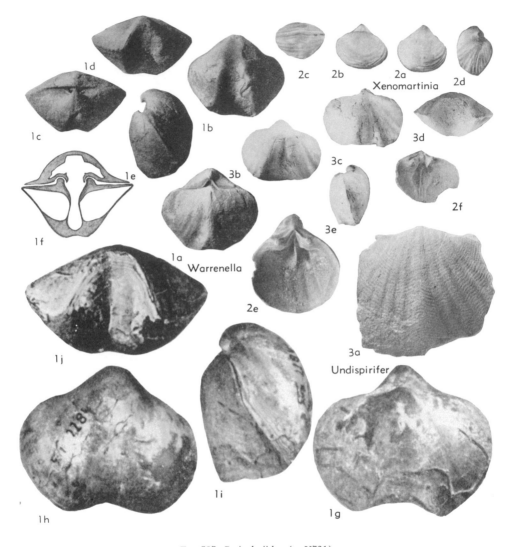

FIG. 587. Reticulariidae (p. *H721*).

ing, or less commonly consisting of rounded lateral plications; micro-ornament consisting of concentric growth lamellae, each terminating anteriorly in row of fine double-barreled spines; internal structures highly variable. *M.Dev.-Perm.*

Elita FREDERIKS, 1918, p. 87 [**Delthyris fimbriata* CONRAD, 1842, p. 263; OD] [=*Elytha* FREDERIKS, 1919 (1924), p. 304 *(nom. van.)*; *Elyta* IVANOVA, 1960, p. 277 *(nom. null.)*]. Lateral slopes with several low, rounded plications; pedicle valve interior with stout dental plates and median septum; brachial valve interior with short crural plates. *M.Dev.-U.Dev.*, cosmop.——FIG. 588,*4*. **E.*

fimbriata (CONRAD), Hamilton, USA(N.Y.); *4a,b,* brach.v., lat., ✕1 (178).

Kitakamithyris MINATO, 1951, p. 374 [**Torynifer (Kitakamithyris) tyoanjiensis* MINATO, 1951; OD]. Brachial valve lacking median septum; otherwise similar to *Torynifer*. *L.Carb.*, Japan-Australia.—— FIG. 588,*6*. **K. tyoanjiensis*, Tournais., Japan; *6a,* post., ✕1; *6b,* ornament, ✕2 (561).

Martinothyris MINATO, 1953, p. 70 [**Terebratula? lineata* J. SOWERBY, 1822, p. 39 (=*Conchyliolites (Anomites) lineatus* MARTIN, 1809) (ICZN Opinion 420, p. 132, 1956); OD]. Pedicle valve interior with dental plates; brachial valve interior with crural plates; otherwise similar to *Phricodothyris*. *L.Carb.*, Br.I.——FIG. 588,*1*. **M. lineatus*

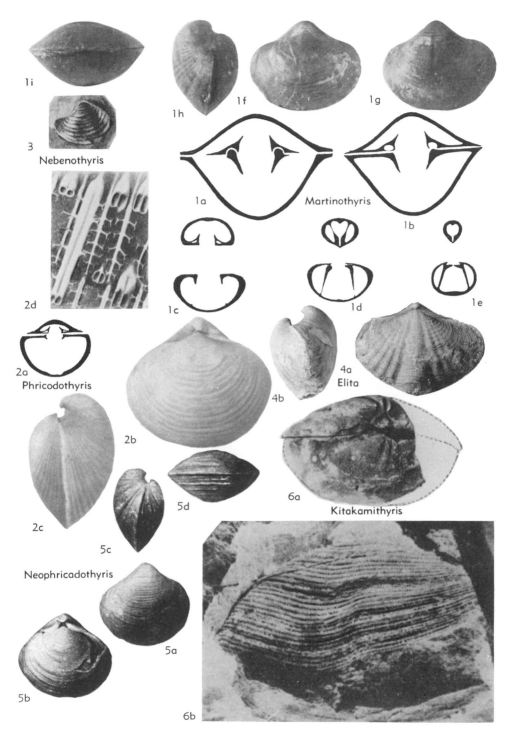

Fig. 588. Elythidae (p. *H722, H724*).

ld

lc

le

lb

Torynifer

la

FIG. 589. Elythidae (p. H724).

(SOWERBY); *1a-e,* transv. sec., ×2; *1f-i,* brach.v.,
ped.v., lat., ant., ×1 (335).
Nebenothyris MINATO, 1953, p. 72 [**N. lineatus;*
OD]. Pedicle valve interior with median septum;
brachial valve interior with median septum; otherwise seemingly similar to *Phricodothyris;* poorly
known. [MINATO deliberately used NEBE's (1911,
p. 447) misidentification of *Terebratula? lineata*
J. SOWERBY as type-species for *Nebenothyris.* According to the *Code* (Art. 70,b,i) the type-species
is to be based on the NEBE material, but is to be
known as *Nebenothyris lineatus* MINATO, 1953.]

L.Carb., Ger.——FIG. 588,*3.* **M. lineatus;* ped.v.,
×1 (596).
Neophricadothyris LIKHAREV, 1934, p. 214
[**Squamularia asiatica* CHAO, 1929, p. 91; OD]
[=*Neophricodothyris* IVANOVA, 1960, p. 277
(nom. null.)]. Spiralia directed posterolaterally;
otherwise similar to *Phricodothyris. Perm.,* cosmop.
——FIG. 588,*5.* **N. asiatica* (CHAO), L.Perm.,
China; *5a-d,* ped.v., brach.v., lat., ant., ×1 (154).
Phricodothyris GEORGE, 1932, p. 524 [**P. lucerna;*
OD] [?=*Squamularia* GEMMELLARO, 1899, p.
189 *(non* ROTHPLETZ, 1896) (type, *S. rotundata);*
Condrathyris MINATO, 1953, p. 69 (type, *Spirifer*
perplexa McCHESNEY, 1860)]. Macro-ornament
wanting; pedicle valve interior typically lacking
dental plates and median septum; brachial valve
interior with laterally directed spiralia, generally
lacking crural plates and median septum. [The
name *Squamularia* was first used by ROTHPLETZ
(1896) for a fossil which could be either animal
or plant. Its use by GEMMELLARO (1899) is here
regarded as invalid.] *L.Carb.-Perm.,* cosmop.——
FIG. 588,*2a.* **P. lucerna,* L.Carb., Br.I.; transv.
sec., ×2 (335).——FIG. 588,*2b-d.* *P. perplexa*
(McCHESNEY), Penn., USA; *2b,c,* brach.v., lat.,
×1.5; *2d,* spines, ×8 (485).
Torynifer HALL & CLARKE, 1894, pl. 84 [**T. criti-*
cus (=*Spirifer pseudolineatus* HALL, 1858, p.
645); OD]. Pedicle valve interior with dental
plates and median septum; brachial valve interior
with low median septum; otherwise similar to
Phricodothyris. U.Dev.-Miss., cosmop. —— FIG.
589,*1.* **T. pseudolineatus* (HALL), Meramec.,
USA; *1a-c,* ped.v., brach.v., lat., ×1.5 (858);
1d,e, ped.v. int., brach.v. int., ×2 (178).

Family MARTINIIDAE Waagen, 1883

[*nom. transl.* IVANOVA, 1959, p. 56 *(ex* Martiniinae
WAAGEN, 1883, p. 524)]

Biconvex; approximately equidimensional,
with broadly rounded cardinal extremities
and rather short hinge line; fold and sulcus
generally present, but commonly weak;
macro-ornament generally lacking, rarely
with lateral plications or costae; micro-ornament consisting of concentric growth lamellae and surficial pits, some shells with
capillae; pedicle valve interior with or
without dental plates and median septum;
brachial valve interior with low, longitudinally striated cardinal process, with or without crural plates. *L.Carb.-Perm., ?Trias.*

Martinia M'COY, 1844, p. 128 [**Spirifer glaber* J.
SOWERBY, 1820, p. 123; SD ICZN Opinion 421,
1956, p. 171] [=*Pseudomartinia* LEIDHOLD, 1928,
p. 82 (obj.); *?Paramartinia* REED, 1949, p. 471
(type, *Martinia (Paramartinia) lingulata* REED,
1949)]. Fold and sulcus distinct, forming uni-

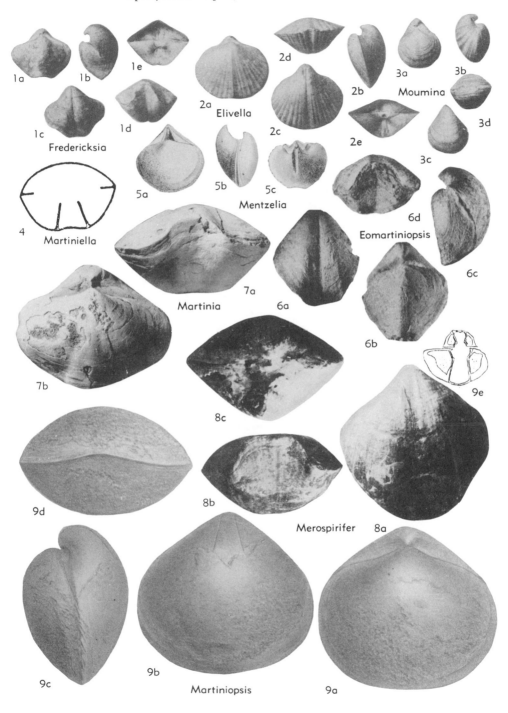

FIG. 590. Martiniidae (p. *H724, H726-H727*).

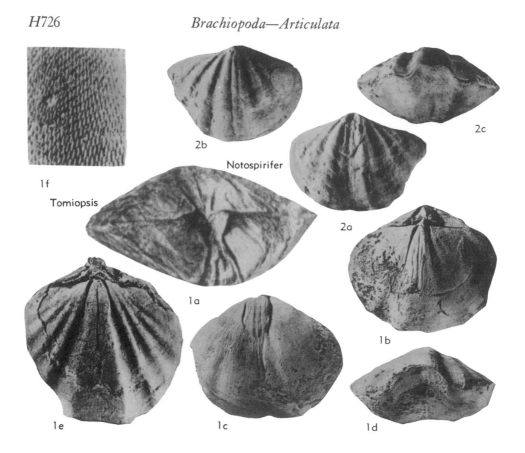

1f

Tomiopsis

2b

Notospirifer

2c

2a

1a

1b

1e

1c

1d

FIG. 591. Martiniidae (p. *H727*).

plicate anterior commissure; macro-ornament lacking; micro-ornament consisting of concentric growth lamellae and surficial pits; pedicle valve interior lacking dental plates and median septum; brachial valve interior lacking crural plates. *L.Carb., ?U.Carb.,* cosmop.——FIG. 590,7. *M. glabra* (SOWERBY), L.Carb., Br.I.; *7a,b,* ant., brach.v., ×0.5 (581).

Elivella FREDERIKS, 1919 (1924), p. 316 [*Martiniopsis baschkirica* CHERNYSHEV, 1902, p. 558; SD FREDERIKS, 1926, p. 403]. Lateral slopes with rather numerous, low, rounded plications; micro-ornament consisting of capillae and surficial pits; pedicle valve interior with dental plates, lacking median septum; otherwise similar to *Martinia. L.Perm.,* USSR.——FIG. 590,2. *E. baschkirica* (CHERNYSHEV); *2a-e,* brach.v., lat., ped.v., ant., post., ×1 (158).

Eomartiniopsis SOKOLSKAYA, 1941, p. 78 [*E. elongata;* OD]. Fold and sulcus distinct, producing strongly uniplicate anterior commissure; crural plates very short; otherwise similar to *Martiniopsis. L.Carb.,* USSR.——FIG. 590,6. *E. elongata; 6a-d,* ped.v., brach.v., lat., ant., ×1 (752).

Fredericksia PAECKELMANN, 1931, p. 48 [*pro Munia* FREDERIKS, 1918, p. 88 (*non* HODGE, 1836)] [*Spiriferina (Mentzelia) simensis* CHERNYSHEV, 1902, p. 514; OD]. Fold and sulcus strong; micro-ornament consisting of concentric growth lamellae, capillae, and surficial pits; pedicle valve interior with dental plates and median septum; otherwise similar to *Martinia. L.Perm.,* USSR.——FIG. 590,1. *F. simensis* (CHERNYSHEV); *1a-e,* brach.v., lat., ped.v., ant., post., ×1 (158).

?Martiniella GRABAU & TIEN in GRABAU, 1931, p. 420 [*M. nasuta;* OD]. Micro-ornament consisting of capillae; presence of surficial pits not established; pedicle valve interior with dental plates, lacking median septum; otherwise similar to *Martinia. Carb.,* China.——FIG. 590,4. *M.* sp. GRABAU; transv. sec., ×1 (359).

Martiniopsis WAAGEN, 1883, p. 524 [*M. inflata;* SD ETHERIDGE, 1892, p. 238]. Fold and sulcus weak to absent; pedicle valve interior with well-developed dental plates; brachial valve interior with long, thin crural plates; otherwise similar to *Martinia. U.Carb.-Perm.,* Eu.-Asia.——FIG. 590,9. *M. inflata,* L.Perm., India; *9a-d,* brach.v.,

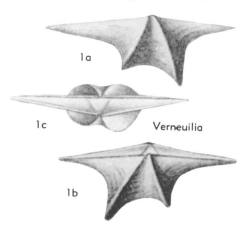

FIG. 592. Family Uncertain (p. *H727*).

ped.v., lat., ant., ×1 (845); *9e*, transv. sec., ×1 (122).

?**Mentzelia** QUENSTEDT, 1871, p. 522 [**Spirifer medianus* QUENSTEDT, 1852, p. 482 (=**Spirifer mentzeli* DUNKER, 1851, p. 287); OD]. Shape typically martiniid, but fold and sulcus wanting; anterior commissure rectimarginate; macro-ornament lacking; micro-ornament poorly known; pedicle valve interior with dental plates, strong median septum, and delthyrial plate; brachial valve interior lacking crural plates. *?Perm., Trias.,* Eu.-Asia.——FIG. 590,5. **M. mentzeli* (DUNKER), M.Trias., Ger.; *5a-c*, brach.v., lat., ped.v. int., ×? (651).

?**Merospirifer** REED, 1948 (1949), p. 467 [**Martinia (Merospirifer) insolita*; OD]. Macro-ornament consisting of very numerous, very weak costae; micro-ornament including concentric growth lamellae and granules, evidently lacking surficial pits; pedicle valve interior with dental plates and median septum. *L.Carb.,* Br.I.——FIG. 590,8. **M. insolita*; *8a-c*, ped.v., ant., post., ×1 (665).

Moumina FREDERIKS, 1919 (1924), p. 321 [**Martinia incerta* CHERNYSHEV, 1902, p. 569; OD]. Small; micro-ornament consisting of concentric growth lamellae, capillae and surficial pits; otherwise similar to *Martinia. L.Perm.,* USSR.—— FIG. 590,3. **M. incerta* (CHERNYSHEV); *3a-d*, brach.v., lat., ped.v., ant., ×1 (158).

Notospirifer HARRINGTON, 1955, p. 115 [**Spirifer darwini* MORRIS, 1845, p. 279; OD]. Fold and sulcus distinct; lateral slopes with 3 to 6 moderately strong plications; brachial valve interior with crural plates short or lacking; otherwise similar to *Martiniopsis. Perm.,* Australia.——FIG. 591,2. **N. darwini* (MORRIS); *2a-c*, ped.v., brach. v., ant., ×1 (141).

Tomiopsis BENEDIKTOVA, 1956, p. 169 [**Brachythyris kumpani* YANISHEVSKIY, 1935, p. 68; OD]

[=*Ingelarella* CAMPBELL, 1959, p. 340 (type, *I. angulata*)]. Fold and sulcus distinct; lateral slopes with several very weak plications; otherwise similar to *Martiniopsis. ?L.Carb., U.Carb.-Perm.,* Asia-Australia.——FIG. 591,1a. **T. kumpani* (YANISHEVSKIY), *Carb.(?Visean-?Namur.),* USSR; post. int. mold, ×1.5 (448).——FIG. 591,1b-e. *T. angulata* (CAMPBELL), Perm., Australia; *1b-e*, brach.v. int. mold, ped.v. int. mold, ant. int. mold, ×0.7, brach.v. int. mold, ×1 (141).——FIG. 591,1f. *T.* sp. CAMPBELL, Perm., Australia; micro-ornament, ×5 (141).

Superfamily and Family UNCERTAIN

Verneuilia HALL & CLARKE, 1894, p. 58 [**Spirifer cheiroptyx* D'ARCHIAC & DE VERNEUIL, 1842, p. 370; SD HALL & CLARKE, 1895, p. 762]. Medium-sized; very transverse with sharply pointed cardinal extremities; both brachial and pedicle valves with strong median sulci bounded by very pronounced, flaring ridges; lateral slopes markedly concave; macro-ornament lacking; micro-ornament consisting of concentric growth lines only; interior unknown. *Dev., ?L.Carb.,* Eu.-?Asia.—— FIG. 592,1. **V. cheiroptyx* (D'ARCHIAC & DE VERNEUIL), M.Dev., Ger.; *1a-c*, ped.v., brach.v., post., ×1.5 (396).

Suborder, Superfamily, and Family UNCERTAIN

Clavigera HECTOR, 1879, p. 538 [**C. bisulcata* HECTOR in THOMSON, 1913, p. 50; SD MARWICK, 1953, p. 45] [=*Hectoria* TRECHMANN, 1918, p. 233 (obj.) (*non* TEPPER, 1889, *nec* CASTELNAU, 1873); *Hectorina* FINLAY, 1927, p. 533 (obj.); *Clavigerina*

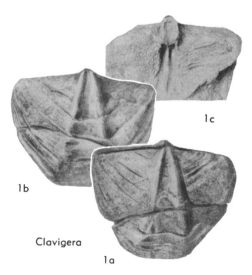

FIG. 593. Family Uncertain (p. *H727-H728*).

Marwick, 1953, p. 45 *(nom. null.)*]. Large, gently biconvex; length and width subequal; each valve with shallow median sulcus bordered by a distinct ridge on either side; anterior commissure rectimarginate; macro-ornament lacking; micro-ornament consisting of growth lamellae only; pedicle valve interior with distinct dental plates and low median ridge; brachial valve interior with laterally directed spiralia and very large, spoon-shaped, non-striate cardinal process projecting far into pedicle valve; shell substance im-punctate. *U.Trias.-L.Jur.,* N.Z.——Fig. 593,*1a,b*. **C. bisulcata,* U.Trias. (Rhaet.); *1a,b,* brach.v., ped.v., ×0.67 (816).——Fig. 593,*1c*. *C. tumida* Hector in Thomson, 1913, U.Trias. (Rhaet.); brach.v. int., ×0.67 (816).

Cryptospirifer Grabau, 1931, p. 405 [**C. lochengensis*; OD] [=*Lochengia* Grabau, 1931, p. 478 (type, *L. holoensis*) *(nom. nud.)*]. Original generic description is so all-encompassing as to be meaningless. Type-species has never been illustrated. *Carb.,* China.

TEREBRATULIDA

By H. M. Muir-Wood, F. G. Stehli, G. F. Elliott, and Kotora Hatai

[British Museum (Natural History); Western Reserve University; Iraq Petroleum Company, Ltd.; Tohoku University]

Brachiopods classed as terebratulids are a long-ranging, distinctive group of the phylum, differentiated by persistent characters of external form, by the punctate nature of their shell, and by internal features, chief of which is the relatively simple calcareous loop extending from the beak region of the brachial valve for support of the lophophore. In outline, the shells most commonly are teardrop-shaped, pointed at the posterior extremity, near which the valves are hinged and where a generally rounded foramen provides for egress of the pedicle, and rounded anteriorly where the valves open widest. The shell surface commonly is smooth, but it may be finely to somewhat coarsely plicate and marked by more or less prominent radially disposed rounded depressions (sulci) and elevations (folds) extending from the beak to the anterior shell margin.

This assemblage of brachiopods ranging in age from Early Devonian to Recent is assigned to the Order Terebratulida which contains three suborders—the Centronellidina, ranging from Lower Devonian to Permian, most primitive and earliest to appear; the Terebratulidina, well represented in the Paleozoic but primarily post-Paleozoic; and the Terebratellidina, sparsely represented in the Paleozoic but important in post-Paleozoic time.

The description of the terebratuloid brachiopods has been divided among authors in such a way that neither taxonomically bounded nor stratigraphically limited segments are separable readily according to authorship without undesirable overlap and offlap. Hence, an editorial organization which assigns precedence to the subordinal groups is adopted, with treatment of stratigraphically differentiated groups held secondary. The contributions of some authors are thus broken into parts, and cross references between chapters are needed in order to avoid repetition. These features should cause little inconvenience, however, to students of the Terebratulida who consult the *Treatise.*

TEREBRATULIDA— MAIN GROUPS

By H. M. Muir-Wood & F. G. Stehli

A brief history of the classification of Brachiopoda given by Muir-Wood (1955) indicates subordinal and family-group classification (p. 85-93) extant at the time of her writing. Whereas genera distinguished as terebratulids were divided among four families (Centronellidae, Stringocephalidae, Terebratulidae, Terebratellidae) by Schuchert in 1913, all placed in the single superfamily Terebratulacea, Muir-Wood recorded 15 families. Of these, eight (Centronellidae, Rhipidothyridae, Stringocephalidae, Meganteridae, Dielasmatidae, Terebratulidae, Cancellothyridae, Orthotomidae) were included in the superfamily Terebratulacea (also suborder Terebratuloidea) and seven (Zeilleriidae, Megathyridae, Platidiidae, Kraussinidae, Dallinidae, Laqueinidae, Terebratellidae) were assigned to the superfamily Terebratellacea (also suborder Terebratelloidea). All Paleozoic genera were classified as belonging to the Terebratulacea, which therefore was considered to contain the root stocks of the whole assemblage; the exclusively Mesozoic and Cenozoic Terebratellacea evidently comprised forms derivative from terebratulaceans. Descendants of Paleozoic families are an abundant component also of the Terebratulacea in Mesozoic and Cenozoic parts of the geologic column.

A systematic description of the whole assemblage as now distinguished follows.

Order TEREBRATULIDA
Waagen, 1883

[*nom. correct.* MOORE in MOORE, LALICKER, & FISCHER, 1952, p. 220 (*pro* order Terebratulacea KUHN, 1949, p. 105, *nom. transl. ex* suborder Terebratulacea WAAGEN, 1883, p. 447)]

Punctate articulate brachiopods with functional pedicle, delthyrium more or less closed by deltidial plates or some similar structure; adult loop highly variable but basically centronelliform, terebratuliform, terebratelliform, or some derivative of one of these plans, and arising from cardinalia, or in part from median septum; dental plates present or absent; lophophore trocholophous, schizolophous, ptycholophous, spirolophous, or plectolophous; mantle canals with 2 or 4 main trunks in each mantle; small internal calcareous spicules in some families; Recent forms with pair of nephridia. *L.Dev.-Rec.*

The Terebratulida now are divided into three major groups ranked as suborders. As previously noted, these are named Centronellidina, Terebratellidina, and Terebratulidina. Four groups (Centronellidae, Stringocephalidae, Rhipidothyridae—*recte* Rhipidothyrididae—, Meganteridae—*recte* Meganterididae) formerly classed as belonging within the Terebratuloidea (=Terebratulidina) are the main representatives of the suborder Centronellidina. The Meganterididae, however, are treated as a subfamily of the Centronellidae. An additional family (Mutationellidae) completes the assemblage as currently defined. No genera of the Centronellidina are known to occur in deposits younger than Permian.

Suborder CENTRONELLIDINA
Stehli, n. suborder

Archaic terebratuloids which primitively are characterized by centronelliform loop and possibly by crural plates but which, in advanced forms, bear more complex types of loop and may lack crural plates. *L.Dev.-Perm.*

The Centronellidina contain primarily the forms involved in the initial adaptive radiation of the Terebratulida during the Early and Middle Devonian. It includes also a few persistent but relatively unsuccessful stocks that failed to survive the Paleozoic. As appears commonly to be the case, successful descendent lineages which supplanted the Centronellidina became distinct during the Early Devonian, soon after beginning of the radiation. Both the Terebratulidina and Terebratellidina appear to have arisen from the family Mutationellidae, some members of which were characterized by extraordinary diversity of the loop.

The removal of the Centronellidina from the old, more inclusive Terebratuloidea makes the remnant Terebratulidina improved in cohesive attributes and focuses attention on comparing them with the Terebratellidina. Relationships of the two groups, which comprise the entirety of Mesozoic and Cenozoic terebratuloid faunas and their respective lines of descent are too poorly known to warrant any firm conclusions. If classificatory assignments given by STEHLI are accepted, the origins of families in each of the three suborders date from the Early Devonian. The seeming disappearance before Triassic time of all genera of the Centronellidina and of the single Paleozoic family (Cryptonellidae) of the Terebratellidina, as contrasted with survival of at least one Paleozoic family (Dielasmatidae) of the Terebratulidina to the Late Triassic (numerous genera) or even Early Jurassic (*Propygope, Pseudokingena*), lacks compelling phylogenetic implications. In the *Treatise,* the Terebratulidina are placed next after the Centronellidina partly because this accords with traditional taxonomic arrangement, preceding the Terebratellidina, and partly because Paleozoic representatives of the former are much more numerous than those of the latter.

Suborder TEREBRATULIDINA Waagen, 1883

[*nom. correct.* MUIR-WOOD & STEHLI, herein (*pro* suborder Terebratulacea WAAGEN, 1883, p. 447)]

Brachial loop primitively and persistently short, of terebratulid, terebratulinid or chlidonophorid type in most forms but neotenously centronelliform or with complex derivative thereof in some Paleozoic forms, loop developed directly from cardinalia; median septum normally lacking; internal spicules usually developed in mantle, body wall, lophophore, and filaments of Recent species. *L.Dev.-Rec.*

The Terebratulidina contain four families (19 genera) which are confined to the Paleozoic (Cranaenidae, Heterelasminidae, Labaiidae, Notothyrinidae). The Dielasmatidae is the only family assemblage known to transgress the Paleozoic-Mesozoic boundary, containing 9 described Paleozoic genera and 15 Mesozoic genera. Eight families are exclusively Mesozoic or Mesozoic-Cenozoic (Terebratulidae, Pygopidae, Cancellothyrididae, Orthotomidae, Cheniothyrididae, Dictyothyrididae, Tegulothyrididae, Dyscoliidae). Genera of this latter group are overwhelmingly Mesozoic in distribution (1 Triassic, 48 Jurassic, 24 Cretaceous—total 73—as compared with 17 recorded from Cenozoic deposits); in addition, the Mesozoic dielasmatids include 14 Triassic forms and one or two Lower Jurassic genera.

Suborder TEREBRATELLIDINA Muir-Wood, 1955

Loop developed in connection with both cardinalia and median septum. *L.Dev.-Rec.*

As presently classified, the Terebratellidina contain the Paleozoic family Cryptonellidae, the Mesozoic superfamily Zeilleriacea, and the Mesozoic-Cenozoic superfamily Terebratellacea. STEHLI considers the Cryptonellidae as probably ancestral zeilleriaceans, but because of doubts expressed by others, this family is not here included in the superfamily. The Zeilleriacea contain the Zeilleriidae (21 genera) and Eudesiidae (1 genus). Assigned to the Terebratellacea are the six families Megathyrididae, Platidiidae, Kraussinidae, Dallinidae, Laqueidae, and Terebratellidae. Genera of this

superfamily are chiefly Cenozoic (59 genera); described Mesozoic genera number 26.

For the Terebratulida as a whole, we still lack essential information concerning internal characters, especially of the Triassic genera, and consequently it is not now possible to shape a sound classification or to give a satisfactory outline of evolution. It is noteworthy that no hard and fast break in the sequence of these brachiopods is found at the end of the Paleozoic. Many of the Triassic forms seem to be related more closely to Paleozoic genera than to those of the Jurassic and Cretaceous, but much larger collections of Triassic species and genera need to be made and serial sections of them prepared.

PALEOZOIC TEREBRATULIDA
By F. G. STEHLI

[Contribution number 12 of the Department of Geology, Western Reserve University]

INTRODUCTION

The relatively diverse and abundant living representatives of the Terebratulida are related by a long and excellent fossil record to their first clearly recognizable progenitors in the Early Devonian. The general excellence of the fossil record devolves from a number of fortunate circumstances. First, the Terebratulida, like other articulate brachiopods, construct their shells of relatively stable low-magnesium calcite. Secondly, the two valves do not easily become separated; and in some cases, the hinge teeth are actually swollen within the sockets so that disarticulation without breakage of the hinging mechanism is impossible. Thirdly, these animals appear to have rather continuously occupied widespread and well-preserved environments—though there are numerous exceptions. The Paleozoic portion of the record of the Terebratulida encompasses about 40 percent of their total history and appears to include the origin of the three subordinal groups. Thus, during the Paleozoic, the character of the group and the general course of its future evolution seemingly was established.

Despite the length and excellence of the record of this order of brachiopods and despite the possibilities for comparison offered by still-living forms, it has been the subject of relatively few phyletic studies.

This situation is particularly evident among the Paleozoic representatives of the group. The descriptive studies which must provide the basis for phyletic interpretation have, however, proceeded apace and now provide much of the information necessary for useful phyletic work. The masterly study of CLOUD (1942) stands out as an example of what may be accomplished by a vertical, or phyletic, rather than a horizontal, or faunal, approach to the study of fossils. This work has made known the general course of evolution in the Terebratulida through the end of the Devonian and, thus, treats the primary adaptive radiation of the group. Studies by BOUCOT (1959, 1960), BOUCOT & PANKIWSKYI (1962), and BOUCOT, CASTER, IVES, & TALENT (1963) have materially added to our knowledge of Devonian forms and their geographic and stratigraphic occurrence. Studies by CAMPBELL (1957, 1964) have advanced our knowledge of upper Paleozoic forms from Australia. A series of papers by STEHLI (1956, 1961, 1962) has added information on other Paleozoic forms, especially the relatively stable groups evolved from the primary Devonian radiation and the secondary radiations which occurred within them.

All these studies and a host of others permit us to see with some clarity the general course of evolution within the Terebratulida during the Paleozoic. Major gaps in our knowledge still remain, however, and these relate primarily to three intervals: (1) the Late Silurian, during which time the order probably differentiated from a spiriferoid parental stock (879); (2) the Pennsylvanian, during which our information is markedly less complete than in either the Mississippian or Permian; and (3) the Upper Permian and Lower and Middle Triassic, a time interval relating to which study has been so incomplete that the course of evolution from well-established Paleozoic stocks into those dominating the Mesozoic is frequently unclear.

Classification of any group of fossils is an intensely subjective matter, since one attempts to express phyletic relationships which the available data are usually, if not always, inadequate to support fully. If, within a group, only two genera are known,

one is likely to regard them as two points on an evolutionary straight line. As additional genera are recognized, it soon becomes evident that few of the new control points fall on the straight line, which must now be modified into a curve best fitting a scatter of points. With still more information, it is apparent that the best-fitting curve is as woefully inadequate to show the true course of evolution as was the straight line and that a family of curves must be substituted. Our knowledge of evolution within the Paleozoic Terebratulida is not sufficiently complete, as yet, and may never be sufficiently detailed to indicate the true course of evolution from one form to another at the specific and perhaps even the generic level. We can, however, frequently see the general drift of genetic change. Depending upon the group in question, a summation of our data may be as simple and naïve as a straight line between two points, a more realistic fitting of a curve to a scatter of points, or as complex—yet doubtless, still naïve—as a family of curves.

The classification here suggested deals primarily with genera and their arrangement into subfamilial, familial, and superfamilial groupings. Since the genus is a subjective and abstract unit considerably removed from the species or breeding population which forms the reservoir of genetic change, the choice of such a unit is open to attack on genetic grounds. It is justified, however, on practical grounds, since paleontological species are not always well defined or objectively recognized, owing to the fortunes of preservation; the subtlety of many specific characteristics; the arbitrary definition of species as segments of a continuously evolving lineage; the rarity of individuals of some species and the resulting lack of understanding of the range of variation; and perhaps, we must admit, to the description of poor or unrecognizable specimens as the types of new species. The more inclusive generic units characterized by more pronounced characteristics are, at our present state of knowledge, more objectively recognized as units for phyletic study and are thus worked with more easily.

Classification, to be useful, must be flexible and susceptible to change as new data or more logical interpretations indicate the necessity for it. It is clearly recognized that the system here developed for the inclusion of the Paleozoic Terebratulida should and will be ephemeral. Some parts of the classification are regarded as more likely to change than others, because the data upon which they are based are fewer or less readily interpreted. Most likely to change is the concept, here expressed, of relationships between Paleozoic and Mesozoic groups. Not only is information poor for the critical time, but diverse opinions regarding probable relationships have been expressed in the literature (283, 775, 875).

BASES OF CLASSIFICATION

In a group as large and diverse as the Terebratulida, it is inevitable that the skeletal features, most useful in classification, will evolve at different rates in different lineages; and that, even within a single lineage, a given feature may, at different times, be conservative or change rapidly. It is, thus, not possible to present a simple classification based throughout on morphologic features, the variation of which can be assigned significance at any fixed taxonomic level. The features of terebratuloid shells, which have been used in this study, are discussed in some detail below; and some probable evolutionary changes which reflect variation in rates of change in these features are outlined.

EXTERNAL FEATURES

A note of warning should be injected with regard to the study and description of terebratuloid shells. External variation is limited, and both parallelism and convergence are common. Thus, it is absolutely mandatory that the internal structures of all shells be studied, if inordinate taxonomic confusion is to be avoided.

Notwithstanding the risks involved in making taxonomic assignments on the basis of external features alone, there is considerable variation in the shape, size, and ornamentation of terebratuloid shells which is useful in their study. Differences in size can be important, but are notoriously subject to error of interpretation, since environmental factors both during life and after death may have influenced size distributions. Before size may be used as a taxo-

nomic criterion, ontogenetic sequences and the range of variation within and between populations must be known with some certainty.

Variations in shell shape are quite likely to express a true genotypic variation and are, thus, more likely to be taxonomically significant than size alone. Because of the nature of growth of brachiopod shells (686, 875), most changes in shape are simply changes in growth rate or in proportion. These changes generally affect the relations between length, width, and thickness of the shell but also commonly affect the length or attitude of the beak, changes in shell curvature, and the development of major folding (as distinct from a more superficial pattern of radial ornamentation, even though the latter may affect the whole thickness of the shell). Since folding (666b, 686) is apparently related to nature and disposition of the lophophore within the shell and the efficient control of incurrent and excurrent water streams, it may reflect important changes in soft parts and be of considerable taxonomic significance.

External ornamentation of the shell is frequently significant at a surprisingly high taxonomic level, though it can rarely by itself be used with safety. Some of the finer kinds of ornamentation in brachiopod shells reflect the number and pattern of insertion of sensory setae in the mantle margin (Fig. 594,3a,b). Change in pattern may, thus, indicate change in an important element of the sensory system of the animal and could be expected to have considerable taxonomic significance. Some of the coarser kinds of ornamentation do not seem to be related to sensory bristles nor can they yet be related with assurance to other soft parts and, thus, may not be significant beyond the lower taxonomic levels (Fig. 594,5).

Though the pigments involved in brachiopod coloration are contained in the protein matrix so generously distributed through the shell (Fig. 594,2), color patterns are not uncommonly preserved in Paleozoic terebratulids. Color patterns seem significant on the specific level (e.g., among species of *Cranaena* illustrated by CLOUD, 1942), but it appears doubtful that they are important in recognizing higher taxa.

The pedicle foramen is of interest because

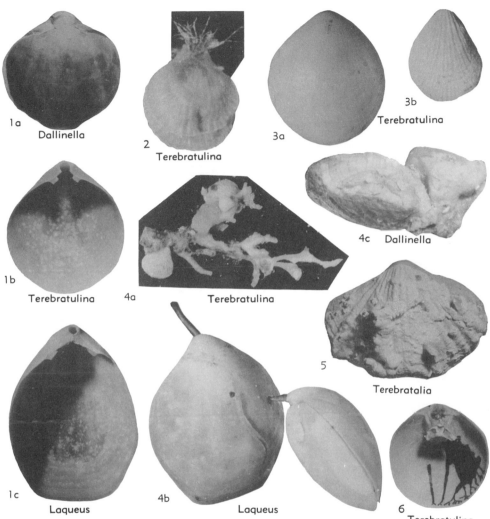

Fig. 594. Morphological features shown by Recent terebratulids from California (Stehli, n). *1.* Pedicle valves showing features of pedicle foramina; *1a, Dallinella occidentalis,* with abraded pedicle foramen seemingly characteristic of terebratulids living closely appressed to surfaces by short attached pedicles, ×1; *1b, Terebratulina crossei,* belonging to a different suborder, also showing effect of characteristic very short pedicle attachment, ×1; *1c, Laqueus californianus,* showing type of pedicle foramen characterizing forms with long pedicle which holds animal well above object of attachment, ×1.——*2.* Completely decalcified specimen of *Terebratulina unguicula* showing threadlike attachment strands of pedicle and extensive organic matrix of shell (pigments responsible for color patterns in brachiopods contained within this organic matrix), ×2.——*3.* Exterior of pedicle valves showing patterns of ornamentation in *Terebratulina*; *3a, T. crossei,* with fine ornamentation of calcareous shell reflecting very fine tactile bristles in mantle, ×1; *3b, T. unguicula,* with relatively coarse pattern of surface which also reflects position of insertion and nature of addition of tactile bristles at edge of mantle, ×1.——*4.* Modes of shell attachment; *4a,* several specimens of *Terebratulina unguicula* attached to coral and showing closely appressed attachment resulting in distinctively abraded pedicle foramen (as in *T. crossei*), ×1; *4b,* specimens of *Laqueus californianus* with long flexible pedicles which appear to correlate with unabraded pedicle foramina such as shown in Fig. *1c,* ×1; *4c,* typical closely appressed attachment of *Dallinella occidentalis,* here attached to a stone, which results in abraded beaks as seen in Fig. *1a,* ×1.——*5.* Exterior of pedicle valve of *Terebratalia transversa caurina* showing abundant growth of calcareous algae on specimen dredged alive (such association very common in this species), ×1.——*6.* Interior of brachial valve of *Terebratulina crossei* showing character of portion of pallial sinus system, ×1.

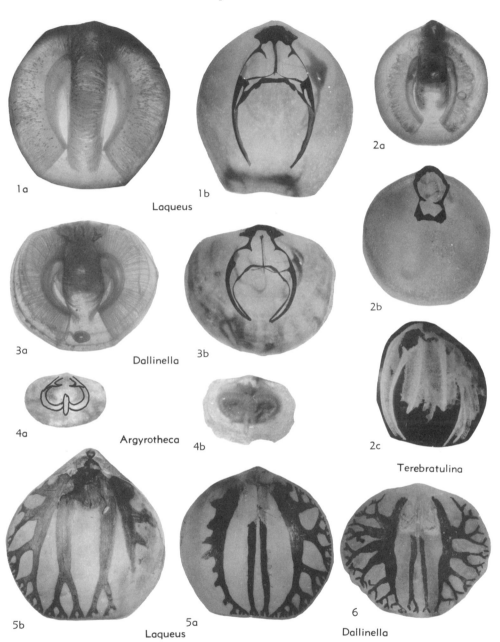

1a
1b

Laqueus

2a

2b

3a
3b

Dallinella

4a
4b

Argyrotheca

2c

Terebratulina

5b
5a

Laqueus

6

Dallinella

Fig. 595. Morphological features shown by Recent terebratulid brachiopods from California and Bermuda (Stehli, n).——*1. Laqueus californianus* (Koch) dredged alive off the coast of southern California; *1a,* brachial valve of preserved specimen showing nature and disposition of plectolophous lophophore within shell, its 2 lateral arms each consisting of upper and lower bands with filaments so disposed as to form partially closed tube at either side of shell through which incurrent water streams enter; median, spiral arm also consisting of 2 bands with filaments in form of closed spiral tube of decreasing diameter; excurrent water stream leaves valve at anterior margin in front of spiral arm and between lateral arms (photographed under water and in transmitted light, ×1);*1b,* brachial valve with soft parts removed in order to reveal calcareous brachidium or "long" terebratelliform loop which supports lophophore; mouth of animal

of its relative stability in the Terebratulidina and Terebratellidina of the Paleozoic, its variability in the Centronellidina, and its reflection of the mode of attachment in modern forms. Among Paleozoic and most conservative later Terebratulidina, the foramen is of the type seen in *Lowenstamia* (Fig. 616,5); that is, mesothyridid to permesothyridid and labiate to marginate. This type of foramen (almost invariably associated with an internal pedicle collar) remains unchanged in upper Paleozoic terebratuloid lineages, whereas the usually conservative loop undergoes drastic changes and thus the foramen provides means for recognizing affinities despite alteration of the loop. Among the Paleozoic Terebratellidina the foramen is rarely, if ever, likely to be confused with that of the conservative Terebratulidina and accordingly this furnishes a rather accurate means of telling from the exterior whether a shell will bear a long or a short loop. Among the Centronellidina, which were undergoing a ma-

jor adaptive radiation in the Devonian, the form of the pedicle foramen was not stabilized and varies greatly (167).

An examination of living brachiopods from the Pacific Coast of the United States suggests that forms having long pedicles (similar to those of *Laqueus*, Fig. 594,4b) maintained the shell in a position some distance above the attachment object. Presumably, this type of attachment gives the animal considerable flexibility. On the other hand, shells in which the beak is abraded are found to be closely appressed to the attachment object during life (Fig. 594,4c). Paleozoic Terebratulidina and Terebratellidina having pedicle openings suggestive of long pedicles and a *Laqueus*-like attachment are common. No forms with pedicles suggestive of closely appressed attachment are known to me from Paleozoic rocks. There remains the large assemblage of Centronellidina with their varied pedicle openings, about which little can, as yet, be said.

located between crural points along which main bands of lophophore lie, with both main and recurved bands receiving support from brachidium throughout their extent in the lateral arms (brachidium, cardinal plate and median septum artificially darkened, ×1). [Note that no similar supporting structure exists for the spiral arms which receive their only support from the transverse band, characteristically sculptured at its points of contact with the spiral.]——2. *Terebratulina crossei* (DAVIDSON) dredged alive off the coast of southern California; *2a*, brachial valve showing nature and disposition of plectolophous lophophore in shell (spiral arm only slightly developed in this species), terminal ends of filaments damaged, being considerably longer in life; current system in this brachiopod same as that described above for *Laqueus* (photographed under water and in transmitted light, ×1); *2b*, brachial valve after removal of soft parts showing "short" or terebratuliform type of loop, mouth located between crural points on which main bands rest, no solid calcareous support for lateral or spiral arms but modification of transverse band clearly indicating its relationship to spiral arm (loop and cardinal structures artificially darkened, ×1); *2c*, spicular supporting skeleton of lophophore as revealed following solution of that organ with hypochloric acid (resembling long loop in development of supporting structures for bands of lateral arms but surpassing long loop in development as well as support for spiral arms), entire structure organically united to solid calcareous loop and lost upon decay of soft parts, ×1.——3. *Dallinella occidentalis* (DALL) dredged off coast of southern California; *3a*, brachial valve of preserved specimen showing nature and disposition of plectolophous lophophore within shell (photographed under water and in transmitted light, ×1); *3b*, brachial valve following removal of soft parts, with "long" loop and structural relationships of its parts to soft anatomy as described for *Laqueus* (loop reflecting broad short nature of lophophore and close approach of anterior ends of 2 sides of loop showing that spiral arm was only moderately large, not extending between front ends of lateral arms as in *Laqueus*) (loop, septum, and cardinalia artificially darkened, ×1).——4. *Argyrotheca bermudana* (DALL) collected off Bermuda; *4a*, brachial valve after removal of soft parts, loop (here outlined in ink) of type called centronelliform and while occurring in *Argyrotheca*, as result of paedomorphosis, in all probability very similar in gross morphology to centronelliform loops of ancient terebratulid brachiopods; reduced median septum of *Argyrotheca* occupying position of vertical plate in ancient centronelliform types, ×10; *4b*, brachial valve of preserved specimen showing nature and distribution of schizolophous lophophore which is associated with centronelliform loop in this species (distribution of incurrent and excurrent water streams in forms with schizolophous lophophores not yet understood), schizolophous loop similar to that of *Argyrotheca* probably having been associated with centronelliform loops of ancient terebratuloids (photographed under water in reflected light, ×10).——5. Pallial sinus system in *Laqueus californianus*; *5a*, brachial valve showing complete system of right side, system injected with ink and specimen then dried and course of trunks artificially darkened, ×1; *5b*, pedicle valve showing complete system of left side (prepared as above), ×1.——6. Pallial sinus system of *Dallinella occidentalis*; brachial valve showing correctly disposition of main trunks and their major branches but inaccurately representing fine terminations which are actually similar to those of *Laqueus*, ×1.

Fig. 596. Evolution of the cardinal plate in Paleozoic Terebratulidina (Stehli, n). In each lettered pair the left figure represents a plan view and the right figure a cross section of the cardinal plate. The sequence *A-I-J-K-L* represents the evolutionary sequence in the Heterelasminidae. Presumably, though not demonstrably, this group arose from the Cranaeninae *(A)* by means of a hypothetical step *(I)*; as first seen in *Afilasma (J)* the basic pattern is established and is altered by simplification in *Beecheria (K)* and *Jisuina (L)*. A somewhat parallel sequence unites the Cranaeninae *(A-E)* with the Dielasmatinae *(F-G, F-H)*.

INTERNAL FEATURES

The internal features of shells of Paleozoic terebratuloids are of extraordinary significance in determining phylogeny and, thus, taxonomic assignments (875). This importance devolves from the intimate relationship which exists between these internal calcareous structures and the associated soft parts. Quite commonly the internal structures are of more significance in phylogeny than any, or even all external features of the shell.

Foremost in terms of interest among the internal structures of the terebratuloid shell is the loop, or calcareous support for the feeding organ, the lophophore. An understanding of the relationship of the loop and lophophore in living terebratuloids sometimes makes it possible to infer from the loops of fossil forms the probable nature of the lophophore in Paleozoic terebratuloids, though a unique conclusion is difficult to reach and diverse opinions exist (686, 775, 879). The loops of Paleozoic terebratuloids fall naturally into three large classes: (1) the short or terebratuliform loop, which characterizes all conservative members of the suborder Terebratulidina (Fig. 595,*2b*); (2) the long terebratelliform (e.g., cryptonelliform) loop, which characterizes normal members of the suborder Terebratellidina (Fig. 595,*1,3*); and (3) the centrelliform loop, which characterizes many members of the Centronellidina, as well as appearing in the ontogeny of Paleozoic members of the Terebratulidina (57a, 191, 775) (Fig. 595,*4a*).

Both the long and the short loop found in living forms can support a very similar (plectolophous) lophophore (Fig. 595,*1-3*), but the degree of support afforded to the lophophore by each kind of loop varies (283, 775, 879). The short loop supports the lophophore only at its base, while additional (probably nonpreservable) support may be furnished by calcitic spicules within the tissues of the lophophore itself (Fig. 595,*2c*). The long loop provides a greater degree of support, since both the ascending and descending bands of the lophophore rest on a calcareous structure. This support may substitute in part for the spicular framework which is its structural equivalent in some short-loop forms but it does not achieve the same level of support since apparently there is no calcareous support of any kind for the spiral arms.

While it is not possible to examine the relationship of the loop and lophophore in the extinct Centronellidina, some paedomorphic living genera of the Terebratellidina have an essentially centronelliform loop. *Argyrotheca* (Fig. 595,*4*) is such a genus, and its centronelliform loop is associated with a bilobed trocholophe (283, 775, 875). It has been assumed on what seem to be reasonable grounds (283, 775, 879), that centronelliform-looped Paleozoic forms also possessed trocholophous or schizolophous lophophores.

During the primary adaptive radiation of the Terebratulida during the Early and Middle Devonian, there was considerable "experimentation" with various arrangements of the loop and lophophore, as may be seen by reference to *Cimicinella* (Fig. 613,*4*) or *Meganteris* (Fig. 607,*2b*). Most genera for which the loop is known, however, seem to have retained a basically centronelliform type. Each of the "successful" stocks which arose from the Centronellidina to found new suborders had evolved a loop which seems designed to support a plectolophous lophophore, though WILLIAMS & WRIGHT (879) agree only insofar as the early Terebratellidina are concerned.

The nature of the transition between the centronelliform loop with its schizolophous or even simpler lophophore and the long or short loop with its possibly plectolophous lophophore can be seen in the ontogeny and phylogeny of various terebratuloids (283, 775, 875, 879). Once the long loop and

---

Following full sessility of the central portion of the cardinal plate in *Dielasma (F)* one lineage proceeds in the direction of its reconstruction as in *Dielasmina (H)* while another proceeds toward its further reduction as in *Lowenstamia (G)*. Also derived from the Cranaeninae *(A)* are several lesser lines of modification *(A-B, A-C, A-D)*. Development of a median septum and suppression of the apical foramen resulted in the initiation of the Girtyellinae *(B)*. Virtual stability of the cardinal plate characterizes the Notothyridinae *(C)* despite development of extreme neoteny in the loop. In the Labaiidae *(D)* the central portion of the cardinal plate becomes obsolete, leading to the type of cardinalia found in many Mesozoic Terebratulidina.

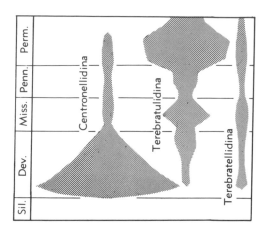

Fig. 597. Generic time-diversity graphs for suborders of Terebratulida during Paleozoic time, based on the known fossil record (Stehli, n).

plectolophous lophophore had become established in the Terebratellidina in the Early Devonian, there was apparently no further change during Paleozoic time. Quite a different situation prevails within the Terebratulidina, however. In this group, it has been shown that the ontogeny of the loop in typical Paleozoic forms includes a centronelliform step, though the adult loop is of the normal short or terebratuliform type (57a). Late in the Paleozoic, the centronelliform loop seemingly was paedomorphically carried on into the adult stage in the members of the family Notothyridae. From the centronelliform adult loop of early members of this family, there arose genera (*Gefonia*, Fig. 619,5*b*; *Timorina*, Fig. 619,4*b*) which seemingly began once more to evolve in the direction of a more complex lophophore, though it appears unlikely that they progressed beyond a zygolophous or incipiently plectolophous stage (775). It is of interest that during this second evolution of the complex lophophore, the supporting loop was developing toward the long, rather than the short, type characteristic of normal members of this suborder. In the case of this interesting family, the characteristics of the pedicle foramen furnish the surest criterion of subordinal affinity, while the usually conservative loop is characteristic only at the family level or below.

Secondly, in terms of interest and significance only to the loop are the structures con-

stituting other parts of the "cardinalia." The more interesting of these structures in the most primitive Terebratulida are as follows: (1) the socket ridges, bounding the sockets on the interior edge; (2) the cardinal plate, which extends between the socket ridges (e.g., *Cranaena*, Fig. 614,1*b*) and functions for the insertion of the dorsal pedicle muscles and commonly bears an apical perforation of unknown function; (3) and the crural plates, which are paired, and which, when present, appear to support the cardinal plate (e.g., *Etymothyris*, Fig. 604,2*b*). The true purpose of the crural plates is not known; and in many cases structures described under this name in the literature are either ridges peripheral to the adductor muscles scars, or structures homologous to the cardinal plate, rather than crural plates. The cardinal plate undergoes interesting modifications in three terebratuloid families—the Dielasmatidae, Heterelasminidae, and Labaiidae. These changes can be interpreted in terms of modification in the place of insertion of the dorsal pedicle muscles and in the case of the Dielasmatidae and Heterelasminidae a striking parallelism is evidenced. Figure 596 shows the course of evolution of the cardinal plate in each of the three families in a diagrammatic fashion. In the Labaiidae, the insertion point of these two muscles moves laterally to a final insertion between

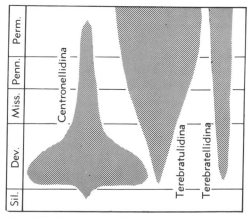

Fig. 598. Generic time-diversity graphs for suborders of Terebratulida during Paleozoic time, smoothed to indicate the probable true nature of the development of each group (Stehli, n).

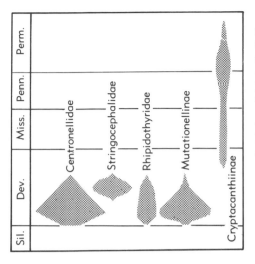

Fig. 599. Generic time-diversity graph showing the development of families of the Centronellidina (Stehli, n). The Mutationellidae is represented by its two important subfamilies, the Devonian Mutationellinae and the post-Devonian Cryptacanthiinae.

Among members of the Dielasmatidae, the picture is more complex, and the evolutionary sequence must first be picked up in the ancestral family, Cranaenidae. In *Cranaena* itself (Fig. 614,*1b*), the cardinal plate is shown in its primitive condition, extending unsupported between the socket ridges. *Hamburgia* shows the next step, in which the cardinal plate has become apically sessile. In *Dielasma* the cardinal plate is completely sessile in its medial portions and is separated into two sloping plates extending from the socket ridges to the floor of the brachial valve (Fig. 617,*1a,c*). On exceptionally well-preserved silicified material from the Guadalupe Mountains, one can detect the insertion scars of the dorsal pedicle muscles on the floor of the brachial valve between the now separate halves of the cardinal plate. Once the dorsal muscles had become inserted on the valve floor, the two parts of the cardinal plate served only to support the crura (Fig. 617,*1c*) and might be expected to have become obsolete either between the crura and the valve floor, as they seem to in *Lowenstamia,* or between the crura and the socket ridges. Interestingly enough, a reversal of evolutionary direction occurs in *Dielasmina* and some other

the crural bases and the socket ridges, whereupon the apparently functionless central portion of the cardinal plate between the crural bases is lost, as in *Pseudodielasma* itself (Fig. 615,*2b*).

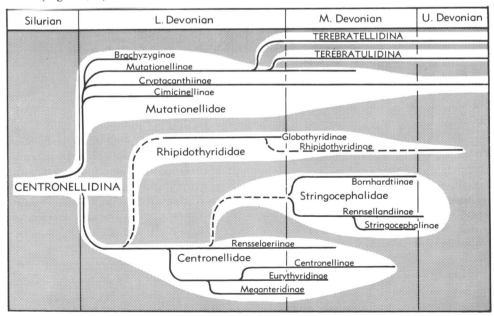

Fig. 600. Phylogeny of the Centronellidina as presently interpreted (Stehli, n).

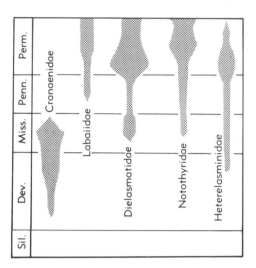

Fig. 601. Generic time-diversity graph showing development of the families of the Terebratulidina during the Paleozoic (Stehli, n).

genera (Fig. 617,7) in which the sessile and separated halves of the cardinal plate become reunited to form a kind of median septum, raising the medially depressed and secondarily entire cardinal plate above the valve floor and carrying with it the insertion of the dorsal pedicle muscles.

Among the Heterelasminidae, a strikingly parallel development occurs, though it leads to a somewhat different end. The genus *Afilasma,* the earliest known, though surely not the earliest member of the sequence, has what was probably once a continuous cardinal plate supported by crural plates, though in this genus it has become obsolete between the crural bases and the socket ridges. In *Beecheria* (Fig. 620,2), the portion of the cardinal plate between the crural bases has become medially sessile, much as it did in *Dielasma,* and though no material well enough preserved to show muscle scars has been examined, it is supposed that the dorsal pedicle muscles had, likewise, become inserted on the valve floor so that the remaining portions of the cardinal plate no longer served a useful function. *Jisuina* represents the last structural grade in this sequence (Fig. 620,5c), and here it is seen that the remnants of the cardinal plate have, indeed, disappeared, leav-

ing the crural bases now supported only by crural plates.

Internal structures, impressed into the shell, such as the mantle-canal system, which is quite distinctive in modern (Fig. 595, 5a-c) and many fossil terebratuloid brachiopods (WILLIAMS, 875) are poorly preserved in most Paleozoic Terebratulida and appear to be of little use at the present time in classification.

In the pedicle valve, a pedicle collar is consistently present in the Terebratulidina and absent in other forms (except *Globithyris* among the Centronellidina). Dental plates are commonly suppressed, and it is doubtful if, in most groups, their presence or absence is of more than generic importance.

The above-noted morphological features of the calcareous shells of terebratuloids have been used in recognizing relationships within the Paleozoic members of the order. As in most other things, however, success has been the criterion which has resulted in the award of higher taxonomic status. Thus, while the Centronellidina are an extraordinarily diverse assemblage, the suborder is short-lived and seems quite clearly to represent an early adaptive radiation. This radiation consists of numerous short-lived, adaptive "experiments," many of which lead to forms morphologically more divergent from the earliest Centronellidina than forms placed in other suborders. These "adaptive experiments" which did not encounter lasting success are left to be contained within the Centronellidina. Only those forms which adopted the successful features of the two still-living suborders have received this elevated taxonomic status (Fig. 597-601).

Suborder CENTRONELLIDINA
Stehli, n. suborder

A diagnosis of this assemblage is given in the section on Terebratulida-Main Groups (see p. *H729*).

Superfamily
STRINGOCEPHALACEA
King, 1850

[*nom. transl.* STEHLI, herein (*ex* Stringocephalidae *nom. correct.* DAVIDSON, 1853, p. 51, *pro* Stringocephalidae KING, 1850, p. 141)]

Characters of suborder. *L.Dev.-U.Perm.*

Family CENTRONELLIDAE Waagen, 1882

[*nom. transl.* HALL & CLARKE, 1895, p. 356 (*ex* Centronellinae WAAGEN, 1882, p. 331)]

Externally variable; cardinal plate supported by crural plates and perforate or sessile, loop typically centronelliform but quite variable. *L.Dev.-M.Dev.*

Subfamily CENTRONELLINAE Waagen, 1882

[Centronellinae WAAGEN, 1882, p. 331]

Moderate-sized, smooth, concavo-convex to biconvex; deltidial plates discrete or conjunct; anterior commissure sulcate; without dental plates; hinge teeth large; cardinal plate much thickened, sessile, medially depressed so as to appear almost as 2 plates; ridgelike cardinal process common; loop centronelliform. *L.Dev.-MDev.*

Centronella BILLINGS, 1859, p. 131 [**Rhynchonella glans-fagea* HALL. 1857, p. 125; SD HALL, 1863, p. 45]. Small to medium-sized; concavo-convex to nearly plano-convex, adult shells not known to be biconvex; cardinal plate sessile and so deeply depressed as to appear medially divided; cardinal process small, apically located (380). *L.Dev.-M. Dev.,* N. Am.——FIG. 602,*1.* **C. glansfagea* (HALL), composite figures; *1a,* brach.v. view, ×1.85; *1b,c,* lat. and brach.v. int., ×1.5 (167).

Oriskania HALL & CLARKE, 1893, p. 269 [**O. navicella*; OD]. Moderate-sized, concavo-convex to biconvex, pedicle valve deeper than brachial valve; cardinal plate moderately concave; cardinal process large, elongate (396). *L.Dev.,* N.Am.——FIG. 602,*2.* **O. navicella,* Oriskany, USA(N.Y.); *2a-c,* brach.v., lat., brach.v. int. views, ×2.1, ×1.9, ×1.9 (167).

Subfamily RENSSELAERIINAE Raymond, 1923

[Rensselaeriinae RAYMOND, 1923, p. 467] [includes Amphigeniinae CLOUD, 1942, p. 77]

Small to large, more or less radially ornamented; generally rather strongly biconvex; dental plates obsolete, distinct and separate or distinct and united in spondylium; cardinal plate perforate, supported by crural plates or partly sessile; loop variable but typically centronelliform though modified in some shells by loss of median plate and development of transverse band. *L.Dev.-M.Dev.*

Rensselaeria HALL, 1859, p. 39 [**Terebratula ovoides* EATON, 1832, p. 45 (*non* SOWERBY, 1812, p. 227) (=**Atrypa elongata* CONRAD, 1839, p. 65); SD HALL & CLARKE, 1893, p. 257]. Large, entire shell costellate; elongately subovate to subcircular; subequally biconvex; pedicle valve deeper than brachial valve; anterior commissure recti-

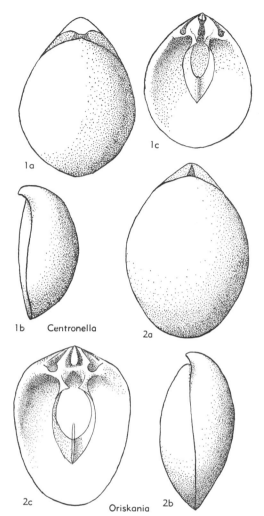

FIG. 602. Centronellidae (Centronellinae) (p. *H741*).

marginate; umbonal region of both valves thickened; dental plates obsolescent to obsolete; cardinal plate thickened, sessile except at anterior edge; crural plates much thickened; loop centronelliform, main bands broadly united anteriorly (172). *L.Dev.,* N.Am.——FIG. 603,*1.* **R. elongata* (CONRAD); *1a,b,* brach.v. view, brach.v. int., ×0.9; *1c,* post. int., ×1 (167).

[CLOUD (1942, p. 54) is correct in pointing out that the commonly cited type-species of *Rensselaeria* given as *R. ovoides* (EATON) lacks validity. This is because *Terebratula ovoides* EATON, 1832, must be rejected as a junior primary homonym of *T. ovoides* SOWERBY, 1812, a very different Cretaceous brachiopod. "A species-group name that is a junior primary homonym must be permanently rejected" (Zool. Code, 1961, Art. 59,a). *Atrypa elongata* CONRAD, 1839, is a nominal species included among those originally assigned to *Rensselaeria* by HALL (1859, p. 38) and referred to as "the more common form from the Oriskany

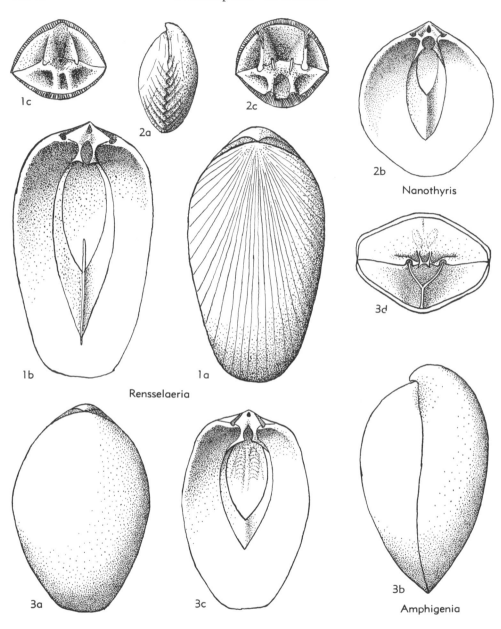

1c

2a

2c

2b

Nanothyris

3d

1b

1a

Rensselaeria

3a

3c

3b

Amphigenia

Fig. 603. Centronellidae (Rensselaeriinae) (p. *H741-H743*).

sandstone" *(loc. cit.)*. Yet he did not definitely indicate an opinion that *A. elongata* is a synonym of Eaton's *T. ovoides*, described as "one of the most common species in the Oriskany sandstone" *(loc. cit.)*, figured as *Rensselaeria ovoides* (Hall, 1859, p. 41). Schuchert (1897, p. 341), Cloud (1942, p. 56) and others have recognized the two nominal species as subjective synonyms and since *A. elongata* is the oldest available synonym, it must be adopted under its own authorship and date as the type-species of *Rensselaeria* (Art. 60,a).]

Amphigenia Hall, 1867, p. 163 [*Pentamerous elongata* Vanuxem, 1842, p. 132; OD]. Large, entire shell costellate; elongate outline; biconvex; anterior commissure rectimarginate to broadly sulcate; dental plates united adventrally to form spondylium; cardinal plate somewhat thickened, posteriorly sessile; crural plates somewhat thickened; loop centronelliform, union of main bands short (835). *M.Dev.*, N.Am.-S.Am.——Fig. 603, 3. **A. elongata* (Vanuxem), USA(N.Y.); *3a-c,* brach.v. and lat. views, brach.v. int., ×0.7; *3d,* post. int., ×1.3 (167).

Etymothyris Cloud, 1942, p. 59 [**Rensselaeria ovoides gaspensis* Clarke, 1909, p. 238; OD]. Large, entire shell costellate; outline linguloid, commonly deeper than wide; anterior commissure rectimarginate; umbones not thickened; dental plates distinct and separate, subparallel to slightly convergent adventrally; cardinal plate free of valve floor, supported by unthickened crural plates; loop unknown (162). *L.Dev.-M.Dev.,* N.Am.—— Fig. 604,2. **E. gaspensis* (Clarke), composite figure; *2a,* brach.v. view, ×0.8; *2b,* post. int., ×1.7; *2c,* brach.v. int., ×1.1 (167).

Nanothyris Cloud, 1942, p. 45 [**Meganteris mutabilis* Hall, 1857, p. 97; OD]. Small to moderate-sized; umbones smooth, shell margin radially ornamented; anterior commissure rectimarginate, dental plates distinct and separate; cardinal plate thickened slightly or unthickened; loop centronelliform, extending ¾ length of valve; main bands broadly united (380). *L.Dev.,* N.Am.-Eu.-Afr.—— Fig. 603,2. **N. mutabilis* (Hall), composite figure; *2a,* lat. view, ×4.7; *2b,* brach.v. int., ×6.3; *2c,* post. int., ×8 (167).

Rensselaerina Dunbar, 1917, p. 466 [**R. medioplicata;* OD]. Medium-sized to moderately large; radial ornamentation mainly on anteromedial region; umbones smooth; anterior commissure rectimarginate; both valves with internally thickened umbones; dental plates weakly developed or obsolete; cardinal plate thickened, partially to almost completely sessile; crural plates secondarily thickened and ill-defined; loop variable, ranging from typical centronelliform pattern to modification in which small transverse band develops (268). *L. Dev.,* N.Am.——Fig. 604,1. **R. medioplicata,* composite figure; *1a,b,* brach.v. and lat. views, ×1.9; *1c,* brach.v. int., ×1.8; *1d,* loop, ×2 (167).

Subfamily EURYTHYRIDINAE Cloud, 1942

[*nom. correct.* Stehli, herein (*pro* Eurythyrinae Cloud, 1942, p. 60)]

Small to moderate-sized, more or less radially ornamented; typically wider than thick; anterior commissure rectimarginate; lateral margins introverted; deltidial plates conjunct; dental plates obsolescent or obsolete; cardinal plate perforate or not and more or less thickened, almost or completely sessile; with or without pronounced cardinal process; crural plates present or lost by sessility of cardinal plate and secondary thickening; loop centronelliform, long, heart-shaped to subtriangular in outline; main bands broadly united. *L.Dev.*

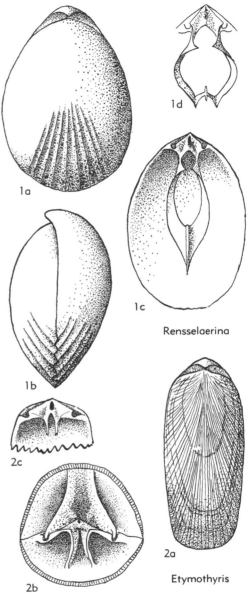

1d

1a

1c

Rensselaerina

1b

2c

2a

2b

Etymothyris

Fig. 604. Centronellidae (Rensselaeriinae) (p. *H743*).

Eurythyris Cloud, 1942, p. 63 [**Oriskania lucerna* Schuchert in Schuchert & Maynard, 1913, p. 390; OD]. Small, much wider than thick; plano-convex, brachial valve almost flat; anteriorly costellate or smooth; dental plates obsolete; cardinal plate imperforate, swollen, sessile; low, linear cardinal process present; crural plates lost due to sessility or faintly visible (732). *L.Dev.,* N.Am.

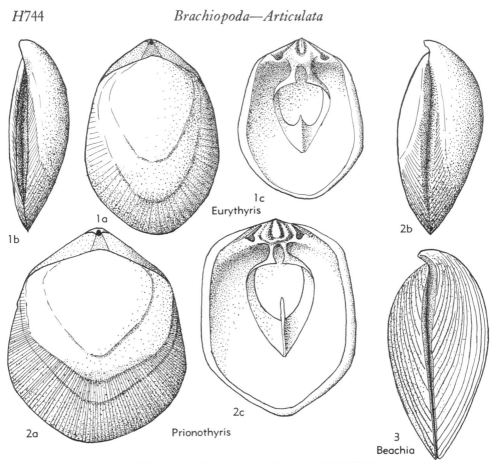

FIG. 605. Centronellidae (Eurythyridinae) (p. *H743-H744*).

——FIG. 605,*1*. **E. lucerna* (SCHUCHERT), Oriskany, USA(N.Y.); *1a,b,* brach.v. and lat. views, ×2; *1c,* brach.v. int., ×1.8 (167).

Beachia HALL & CLARKE, 1893, p. 260 [**Meganteris suessana* HALL, 1857, p. 100; OD (M)]. Moderate-sized; subequally biconvex; all or partly costellate; dental plates obsolescent to obsolete; cardinal plate perforate, more or less swollen and sessile, distinct cardinal process lacking; crural plates evident (380). *L.Dev.,* N.Am.——FIG. 605, *3*; 606,*1*. **B. suessana* (HALL), composite figure; 605,*3*, lat. view, ×2.2; 606,*1a,b,* ped.v. and brach.v. ints., ×4; 606,*1c,* ped.v. ext., ×2.7 (Stehli, n).

Prionothyris CLOUD, 1942, p. 66 [**P. perovalis;* OD]. Moderate-sized; anteriorly costellate or smooth; biconvex, wider than thick; dental plates obsolete; cardinal plate imperforate, thick, completely sessile, bearing ponderous, erect cardinal process; crural plates faintly developed or not visible (167). *L.Dev.,* N.Am.-S.Am.-?N.Z.——FIG. 605,*2*. **P. perovalis,* composite figure; *2a,b,* brach.v. and lat. views, ×2.2, ×2; *2c,* brach.v. int., ×2 (167).

Subfamily MEGANTERIDINAE
Schuchert & LeVene, 1929

[*nom. correct.* STEHLI, herein (*pro* Meganterinae, *nom. transl.* BOUCOT, 1959, p. 766, *ex* Meganteridae SCHUCHERT & LEVENE, 1929, p. 23)]

Moderate-sized to large, subequally biconvex, subcircular to subelliptical in outline; anterior commissure rectimarginate; pedicle foramen mesothyridid; lateral margins not introverted; dental plates present or absent; hinge teeth large, triangular in cross section; cardinal plate perforate or imperforate, posteriorly to entirely sessile; crural plates present. *L.Dev.*

Meganteris SEUSS, 1855, p. 51 [**Terebratula archiaci* DE VERNEUIL, 1850, p. 40; OD] [=*Megalanteris* OEHLERT, 1887, p. 1319; *Vltavothyris* HAVLÍČEK, 1956]. Large, smooth; dental plates becoming obsolete in adults; cardinal plate sessile but crural plates evident in immature specimens; cardinal process large; loop long, more or less cryptonelliform but with crural points extended probably to support spiral arms of lophophore

(843). *L.Dev.*, Eu.——FIG. 607,2. **M. archiaci* (DE VERNEUIL), composite figure; *2a*, lat., ×0.7; *2b*, brach.v. int., ×1.4 (167).

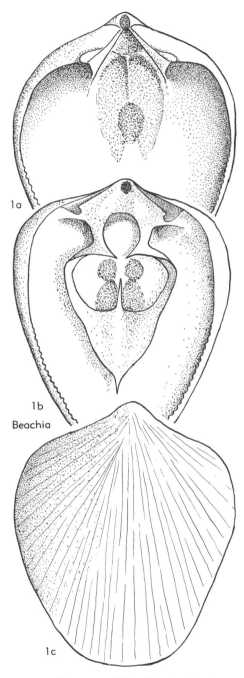

FIG. 606. Centronellidae (Eurythyridinae) (p. *H744*).

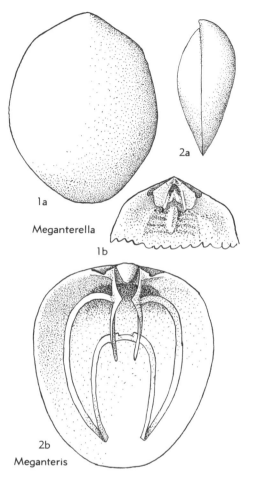

FIG. 607. Centronellidae (Meganteridinae) (p. *H744-H745*).

Meganterella BOUCOT, 1959, p. 767 [**M. finksi*; OD]. Moderate-sized, smooth; biconvex, pedicle valve more convex than brachial valve; dental plates present but short; cardinal plate posteriorly sessile; crural plates present; small cardinal process present in large specimens; loop unknown (101). *L.Dev.*, N.Am.——FIG. 607,*1*. **M. finksi*, Esopus F., USA(N.Y.); *1a*, ped.v. ext., ×2; *1b*, brach.v. int., ×1.7 (101).

Family STRINGOCEPHALIDAE King, 1850

[*nom. correct.* DAVIDSON, 1853, p. 51 (*pro* Strigocephalidae KING, 1850, p. 141)]

Large, thick-shelled, some forms asymmetrical, generally unornamented; deltidial plates discrete or conjunct; pedicle foramen hypothyridid to permesothyridid; dental

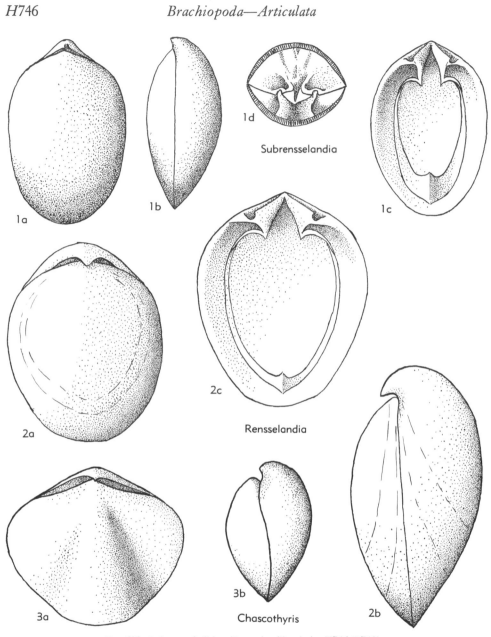

FIG. 608. Stringocephalidae (Rensselandiinae) (p. *H746-H748*).

plates obsolescent to obsolete in adults; hinge plates discrete; loop long, centronelliform. *M.Dev.*

Subfamily RENSSELANDIINAE Cloud, 1942

[Rensselandiinae CLOUD, 1942, p. 92]

Moderate-sized to large, symmetrical; deltidial plates discrete; ventral palintrope and beak relatively inconspicuous; cardinal process and median septa absent; hinge plates discrete and (except in *Subrensselandia*) unsupported by crural plates. *M. Dev.*

Rensselandia HALL, 1867, p. 385 [*Rensselaeria? johanni* HALL, 1867, p. 385; SD SCHUCHERT, 1897, p. 271] [=*Newberria* HALL, 1891, p. 236; *Denckmannia* HOLZAPFEL, 1912, p. 115 (*non* BUCKMAN, 1898); *Denckmannella* SCHUCHERT &

LeVene, 1929, p. 120]. Moderate-sized, biconvex, elongate-subovate to subcircular in outline; anterior commissure rectimarginate; hinge plates discrete, crural plates absent; loop long, anteriorly broad; dental plates obsolescent or obsolete. *M. Dev.*, N.Am.-Eu.——Fig. 608,2. **R. johanni*

(Hall), Cedar Valley Ls., USA(Iowa); *2a,b,* brach.v. and lat. views, ×0.8, ×1.3; *2c,* brach.v. int., ×1.3 (167). [*See* p. H904.]

Chascothyris Holzapfel, 1895, p. 234 [**C. barroisi;* SD Schuchert & LeVene, 1929, p. 40]. Large, generally transverse but exceptionally sub-

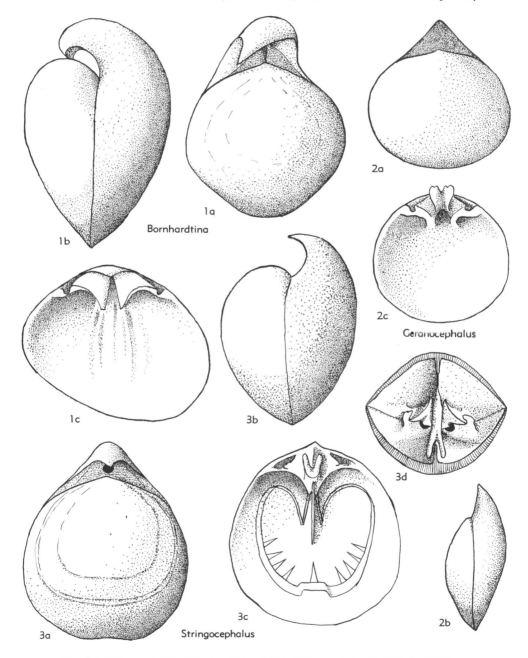

Fig. 609. Stringocephalidae (Bornhardtininae) *(1),* (Stringocephalinae) *(2-3)* (p. H748).

circular; ventral beak short; ventral sulcus and dorsal fold present; hinge plates discrete; crural plates lacking; loop unknown (432). *M.Dev.*, Eu.———Fig. 608,3. **C. barroisi,* Stringocephalus Ls., Ger.; *3a,b,* brach.v. and lat. views, ×0.7, ×0.6 (167).

Subrensselandia Cloud, 1942, p. 92 [**Newberryia claypolii* Hall, 1891, p. 97; OD]. Moderate-sized; anterior commissure rectimarginate; hinge plates discrete, supported by crural plates; dental plates thin, short, obsolescent (395). *M.Dev.*, N.Am.-Eu.———Fig. 608,*1.* **S. claypolii* (Hall), Montebello Ss., USA(Pa.); *1a,b,* brach.v. and lat. views, ×0.9, ×0.8; *1c,* brach.v. int., ×0.85; *1d,* apical int., ×0.7 (167).

Subfamily BORNHARDTININAE Cloud, 1942

[Bornhardtininae Cloud, 1942, p. 100]

Large, strongly asymmetrical; ventral beak large and conspicuous; hinge plates discrete, not supported by crural plates; cardinal process and median septa lacking. *M.Dev.*

Bornhardtina Schulz, 1914, p. 363 [**B. uncitoides;* SD Cloud, 1942, p. 101] [=*Rauffia* Schulz, 1914, p. 371]. Size and shape variable; ventral beak large, conspicuous; asymmetrical; deltidial plates conjunct; pedicle foramen hypothyridid; anterior commissure rectimarginate; dental plates obsolete; loop incompletely known, but apparently like that of *Stringocephalus* (733). *M.Dev.*, Eu.———Fig. 609,*1.* **B. uncitoides,* Ger.(Gerolstein); *1a,b,* brach.v. and lat. views, ×0.8, ×0.95; *1c,* brach.v. int., ×0.7 (167).

Subfamily STRINGOCEPHALINAE King, 1850

[*nom. transl.* Cloud, 1942, p. 104 (*ex* Stringocephalidae King, 1850, p. 51)]

Large, slightly asymmetrical; ventral beak large, conspicuous; ventral palintrope well developed; deltidial plates conjunct; pedicle foramen hypothyridid; cardinal process and median septa present; crural plates are present and well developed or suppressed. *M. Dev.*

Stringocephalus Defrance in de Blainville, 1825, p. 511 [*nom. subst.* Sandberger, 1842, p. 386 (*pro Strygocephale* Defrance in de Blainville, 1825, p. 511) (ICZN pend.)] [**Terebratula Burtini* Defrance in de Blainville, 1825, p. 511; OD (M)] [=*Strygocephalus* Defrance in de Blainville, 1827, pl. 53, fig. 1 (obj.)] *Strigocephalus* Sowerby, 1839, pl. 56, fig. 10 (obj.)]. Subglobular to transversely sublenticular; ventral beak large, sharp, conspicuous, slightly asymmetrical; pedicle foramen hypothyridid; ventral palintrope large; dental plates obsolete; both valves with median septum; cardinal process large, rodlike, terminally bifid, united with primitively discrete crural plates; crural points pronounced;

loop long, centronelliform, with posteriorly directed spines (246). *M.Dev.*, Eu.-N.Am.-Asia.——— Fig. 609,*3.* **S. burtini* (Defrance), Ger.; *3a,b,* brach.v. and lat. views, ×0.75; *3c,* brach.v. int., ×0.65; *3d,* apical int., ×1 (*3a-c,* 167; *3d,* Stehli, n).

[The supposition expressed by Davidson (1865, p. 12) that Defrance intended to derive the generic name *Strygocephale* from the Greek words for screech owl (*strix*) and head (*cephala*) but transliterated the components of the name incorrectly cannot be substantiated, though one may agree that either a Greek-derived *Stringocephala* or its Latinized equivalent *Stringocephalus* are properly formed. Unquestionably, *Strygocephale* has priority over the synonyms *Strygocephalus* Defrance, 1827, *Strigocephalus* Sowerby, 1839, and *Stringocephalus* Sandberger, 1842, all of which constitute "unjustified emendations" classed as invalid subsequent spellings according to the Zoological Code (Art. 33,a,ii). It happens that all but universal usage has established *Stringocephalus* as the preferred name in nomenclature calls for retaining it. This cannot be done within stipulations of the Code by invoking Art. 33,a,i, which allows correction of an incorrect original spelling, because Art. 32,a,ii excludes incorrect transliteration as emendable. Thus, Cloud's (1942, p. 106) justification for *Stringocephalus* as a corrected error of transcription must be rejected.———¶*Stringocephalus* may be validated under the Code in two ways, both of which call for appeal to ICZN. (1) Because *Strygocephale, Strygocephalus,* and *Strigocephalus* are all "forgotten names" (*nomina oblita*), unused in more than 50 years, Art. 23,b provides that they may be placed on the Official List of Rejected Generic Names in Zoology, reference to the Commission being made for this purpose. Then, *Stringocephalus* Sandberger, 1842, automatically gains place as oldest name for this genus. (2) A more direct and desirable course is to seek action under the plenary powers of ICZN (Art. 78), because this could establish *Stringocephalus* with Defrance, 1825, as author and similarly could reject *Strygocephale, Strygocephalus,* and *Strigocephalus* from possibility of being resurrected by anyone. Accordingly, an application to stabilize nomenclature in this way has been filed, for usage alone lacks force in legalizing zoological names.—R. C. Moore]

Acrothyris Hou, 1963, p. 419, 427 [**A. kwangsiensis;* OD]. Medium-sized to large, oval or elongate-oval in outline; beak protruding, foramen in apex. Pedicle valve with short divergent dental plates; brachial valve with massive bilobed cardinal process. [Data from Hou furnished by M. Rowell.] *M.Dev.*, S.China.

Geranocephalus Crickmay, 1954, p. 157 [**G. inopinus;* OD]. Large, smooth, biconvex; ventral beak large, erect, with broad palintrope; deltidial plates conjunct; pedicle foramen hypothyridid; dental plates present; hinge plates discrete or posteriorly united by large bifid cardinal process; crural plates present; median septa lacking; loop unknown (207). *M.Dev.*, N.Am.———Fig. 609,2. **G. inopinus;* composite figure; *2a,b,* brach.v. and lat. views, ×0.8; *2c,* brach.v. int., ×1 (207).

Family RHIPIDOTHYRIDIDAE Cloud, 1942

[*nom. correct.* Stehli, herein (*pro* Rhipidothyridae Cloud, 1942, p. 80)]

Small to large, costate to smooth; anterior commissure rectimarginate to gently plicate; pedicle foramen submesothyridid to hypothyridid; dental plates well developed to obsolescent; primitively with discrete hinge plates and crural plates but common-

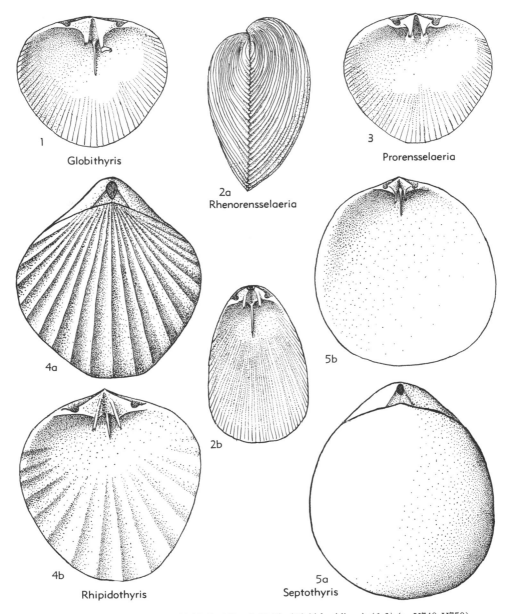

1 Globithyris

2a Rhenorensselaeria

3 Prorensselaeria

4a

2b

5b

4b Rhipidothyris

5a Septothyris

FIG. 610. Rhipidothyrididae (Rhipidothyridinae) *(4-5)*, (Globithyridinae) *(1-3)* (p. *H749-H750*).

ly with these structures united to form septalium; nature of loop unknown. *L.Dev.-M.Dev.*

Subfamily RHIPIDOTHYRIDINAE Cloud, 1942

[*nom. correct.* STEHLI, herein (*pro Rhipidothyrinae* CLOUD, 1942, p. 87)]

Small, lenticular, smooth or costate; pedicle foramen submesothyridid; cardinal plate concave, supported by a median septum; dental plates present. *M.Dev.*

Rhipidothyris COOPER & WILLIAMS, 1935, p. 847 [**R. plicata*; OD]. Small, costellate, subcircular to subovate; concave cardinal plate supported for all or part of its length by median septum; loop unknown (198). *M.Dev.*, N.Am.——FIG. 610,4. **R. plicata*, composite figure; *4a,b*, brach.v. view, brach.v. int., ×2 (167).

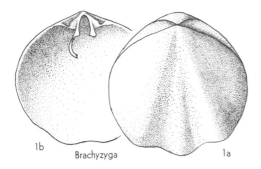

FIG. 611. Mutationellidae (Brachyzyginae)
(p. *H750*).

Septothyris COOPER & WILLIAMS, 1935, p. 849 [**S. septata*; OD]. Small, smooth, subcircular; concave cardinal plate supported throughout its length by median septum or medially sessile so that it appears divided; loop unknown (198). *M.Dev.*, N.Am.——FIG. 610,5. **S. septata,* composite figure; *5a,b,* brach.v. view, brach.v. int., ×4.3, ×4 (198).

Subfamily GLOBITHYRIDINAE Cloud, 1942
[*nom. correct.* STEHLI, herein (*pro* Globothyrinae CLOUD, 1942, p. 81)]

Large, subglobular, entirely costate or costellate; pedicle foramen hypothyridid; hinge plates discrete or united with crural plates to form septalium; dental plates present. *L.Dev.*

Globithyris CLOUD, 1942, p. 82 [**Rensselaeria callida* CLARKE, 1907, p. 241; OD]. Large, subglobose; entirely costate; dental plates thin, short to moderately long; pedicle collar present; hinge plates united in septalium. *L.Dev.,* N.Am.——FIG. 610,1. **G. callida* (CLARKE), Moose River Ss., USA (Maine); brach.v. int., ×1.3 (167).

Prorensselaeria RAYMOND, 1923, p. 468 [**P. nylanderi*; OD]. Large, moderately convex, costellate; subcircular in outline; dental plates short and thick; hinge plates discrete and supported by discrete crural plates; median septum absent (658). *L.Dev.,* N.Am.——FIG. 610,3. **P. nylanderi,* Chapman Ss., USA(Maine); brach.v. int., ×0.85 (167).

Rhenorensselaeria KEGEL, 1913, p. 126 [**Terebratula strigiceps* ROEMER, 1844, p. 58; SD SCHUCHERT & LEVENE, 1929, p. 107]. Moderate-sized to large, strongly convex, elongate in adults, subcircular in juveniles, entirely costate or costellate; dental places short; ventral myophragm prominent; hinge plates united in septalium; well-developed cardinal process present (669). *L.Dev.,* Eu.——FIG. 610,2. **R. strigiceps* (ROEMER), Ger.; *2a,* lat. view, ×1.05; *2b,* brach.v. int., ×1.3 (167).

Family MUTATIONELLIDAE Cloud, 1942
[*nom. transl.* STEHLI, herein (*ex* Mutationellinae CLOUD, 1942, p. 114)]

Generally small and considerably variable archaic Centronellidina with or without cardinal plate; without crural plates; loop highly variable, generally centronelliform but in some genera highly variable. *L.Dev.-Perm.*

Subfamily BRACHYZYGINAE Cloud, 1942
[Brachyzyginae CLOUD, 1942, p. 113]

Small, smooth shells with dorsal sulcus and ventral fold; dental plates present; hinge plates discrete, apparently unsupported by crural plates; loop imperfectly known but apparently short. *L.Dev.*

Brachyzyga KOZLOWSKI, 1929, p. 243 [**B. pentameroides*; OD]. Shell with general pentameroid aspect, anterior commissure intraplicate (487). *L.Dev.,* Eu.——FIG. 611,1. **B. pentameroides,* Borszczów Stage, Pol.; *1a,b,* brach.v. view, brach. v. int., ×2.9, ×1.3 (487).

Subfamily MUTATIONELLINAE Cloud, 1942
[Mutationellinae CLOUD, 1942, p. 114]

Small to moderate-sized, entirely or partially radially ornamented; deltidial plates discrete or conjunct; pedicle foramen mesothyridid to submesothyridid; dental plates short to obsolete; hinge plates normally joined anteriorly forming perforate cardinal plate but discrete or imperforate in some shells; cardinal process may be present but crural plates absent; loop extremely variable but apparently primitively centronelliform. *L.Dev.-M.Dev.*

Mutationella KOZLOWSKI, 1929, p. 236 [**Waldheimia podolica* SIEMIRADZKI, 1906, p. 177; OD]. Small, subcircular, biconvex to almost planoconvex, entirely and simply costate; hinge plates discrete or united to form perforate cardinal plate; loop extremely variable, ranging from typical rather long centronelliform condition to one approaching short terebratuliform (745). *L.Dev.,* Eu.-N.Am.-S.Am.-N.Z.-Antarctica-Afr. —— FIG. 612,5. **M. podolica* (SIEMIRADZKI), Czortków Stage, Eu.(Pol.); *5a,b,* brach.v. and lat. views, ×3.4, ×3.1; *5c,* brach.v. int., ×3.5; *5d,* loop, ×3 (*5a,b,* 167; *5c,* Stehli, n; *5d,* 487).

Cloudella BOUCOT & JOHNSON, 1963, p. 113 [*nom. subst. pro* Pleurothyris CLOUD, 1942, p. 123 (*non* LOWE, 1843; *nec* SCHRAMMEN, 1912] [**Rensselaeria stewarti* CLARKE, 1907, p. 239; OD]. Large for subfamily, entirely costellate, subglobose; deltidial plates discrete; dental plates obsolescent to obsolete; cardinal plate free and

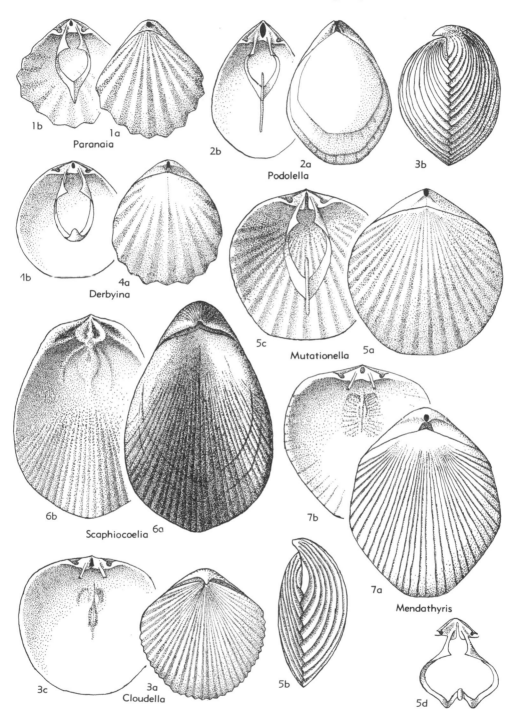

Fig. 612. Mutationellidae (Mutationellinae) (p. *H*750, *H*752).

perforate or medially divided; ventral umbones thickened; loop unknown (162). *L.Dev.,* N.Am.
——Fig. 612,3. **C. stewarti* (Clarke), Dalhousie Sh., Can.(N.B.); *3a,b,* brach.v. and lat. views, ×1.4; *3c,* brach.v. int., ×1.7 (167).

Derbyina Clarke, 1913, p. 210 [**Notothyris? smithi* Derby, 1890, p. 81; SD Clarke, 1913, p. 212] [*non Derbyina* Grabau, 1931]. Small, simply costate; anterior commissure feebly uniplicate; subcircular to slightly elongate; subequally biconvex, pedicle valve deeper than brachial valve; dental plates present; cardinal plate free and perforate; loop moderately long, basically centronelliform but with anterior end turned backward. *M.Dev.,* S.Am.(Brazil-Bol.).——Fig. 612,4. **D. smithi* (Derby), Brazil (Matto Grosso); *4a,b,* brach.v. view, brach.v. int., ×4.6, ×5 (250).

Mendathyris Cloud, 1942, p. 125 [**Rensselaeria mainensis* Williams, 1900, p. 80; OD]. Large, subglobular, entirely costellate; anterior commissure rectimarginate; pedicle foramen permesothyridid; palintrope conspicuous; dental plates obsolescent to obsolete; ventral umbone thickened; cardinal plate perforate, thickened and posteriorly sessile; loop unknown (881). *L.Dev.,* N.Am.——Fig. 612,7. **M. mainensis* (Williams), Chapman Ss., USA(Maine); *7a,b,* brach.v. view, brach. v. int., ×1.9 (167).

Paranaia Clarke, 1913, pl. 21, fig. 7, 8 [**Centronella? margarida* Derby, 1890, p. 84; OD (M)] [=*Brasilia* Clarke, 1913, p. 216 (*non* Buckman, 1898); *Brasilica* Greger, 1920, p. 70; *Chapadella* Greger, 1920, p. 70; *Brasilina* Clarke, 1921, p. 138; *Oliveirella* Oliveira, 1934, p. 167]. Like *Derbyina* except crura shorter and loop with anterior end extended forward (250). *L.Dev.,* S. Am.——Fig. 612,1. **P. margarida* (Derby), Maecuru Gr., Brazil; *1a,b,* brach.v. view, brach.v. int., ×3.4, ×3.5 (250).

Pleurothyrella Boucot, Caster, Ives, & Talent, 1963, p. 89 [**Scaphiocoelia? africana* Reed, 1906, p. 306; OD]. Large, costellate shells resembling *Cloudella*; anterior commissure rectimarginate; biconvex, pedicle valve deeper than brachial valve; cardinal plate bulbous and imperforate in adults; cardinal process present or not; loop unknown (659). *L.Dev.,* Antarctica-N.Z.-Afr.-S.Am.

Podolella Kozlowski, 1929, p. 232 [**P. rensselaeroides*; OD]. Small, terebratuliform; ornamentation restricted to anterior portion of shell; deltidial plates discrete; dental plates present, cardinal plate perforate; with or without crural plates; loop centronelliform, with vertical plate (487). *L.Dev.,* Eu.——Fig. 612,2. **P. rensselaeroides,* Borszczów Stage, Pol.; *2a,b,* brach.v. view, brach.v. int., ×3.3, ×3.6 (2a, 167; 2b, Stehli, n).

Scaphiocoelia Whitfield, 1891, p. 105 [**S. boliviensis;* OD (M)]. Large, elongate, simply costate shells; brachial valve gently concave and bearing sulcus, pedicle valve strongly convex and bearing fold; internally similar to *Mendathyris* but some species exhibiting cardinal process; loop unknown (864). *L.Dev.,* S.Am.-S.Afr.——Fig. 612,6. **S. boliviensis,* S.Am.(Bol.); *6a,b,* brach.v. view, brach.v. int., ×0.8 (Stehli, n).

Subfamily CIMICINELLINAE Stehli, n. subfam.

Moderate-sized, terebratuliform early probable derivatives of Mutationellidae; with crural plates and complex loop. *L.Dev.*

Cimicinella Schmidt, 1943 [**Terebratula cimex* Richter & Richter, 1918, p. 156; OD]. Moderate-sized, smooth; elongate, biconvex; anterior commissure rectimarginate; dental plates present; cardinal plate perforate, supported by inclined crural plates; loop long, recurved bands developing but united with main bands (667). *L.Dev.,* Eu.(Ger.).——Fig. 613,4. **C. cimex* (Richter & Richter); brach.v. int., ×2.2 (719).

Subfamily CRYPTACANTHIINAE Stehli, n. subfam.

Moderate-sized to small, with dorsal sulcus and ventral fold; anterior commissure more or less sulcate; pedicle foramen minute, mesothyridid; dental plates present; cardinal plate extending between socket plates, supported by median septum or not; crural plates absent; loop primitively centronelliform but tending to become long, with recurved branches and somewhat cryptonelliform save for hoodlike transverse band. *?L.Dev., Miss.-U.Perm.*

Cryptacanthia White & St. John, 1867, p. 119 [**Waldheimia? compacta* White & St. John, 1867, p. 119; OD]. Small, strongly biconvex with pronounced ventral fold and dorsal sulcus; externally similar to *Glossothyropsis*; loop essentially cryptonelliform but with main bands closely approaching or actually joining for some distance near mid-length before separating farther forward; anterior extremities of loop spinose; dorsal median septum absent; cardinal plate perforate (860). *M.Penn.-L.Perm.,* N.Am.-Eu.-Asia.——Fig. 613,5. **C. compacta* (White & St. John), composite figure; brach.v. int., ×8 (Stehli, n).

Gacina Stehli, 1961, p. 458 [**G. moorefieldensis;* OD (M)]. Medium-sized; elongately subpentagonal in outline; ventral fold broad at front, narrower and more pronounced near mid-length where margins of valve tend to be flattened; dorsal sulcus extending almost to beak; loop modified centronelliform, main bands uniting near mid-length; vertical plate arising anteriorly and anteroventrally split to form incipient recurved bands; dorsal median septum absent; cardinal plate perforate (779). *?Up.L.Dev., U.Miss.,* N. Am.-Eu.——Fig. 613,3. **G. moorefieldensis,* Meramec.(Moorefield F.), USA(Okla.); *3a,b,* ped. v. ext., brach.v. int., ×1.2, ×4.2 (Stehli, n).

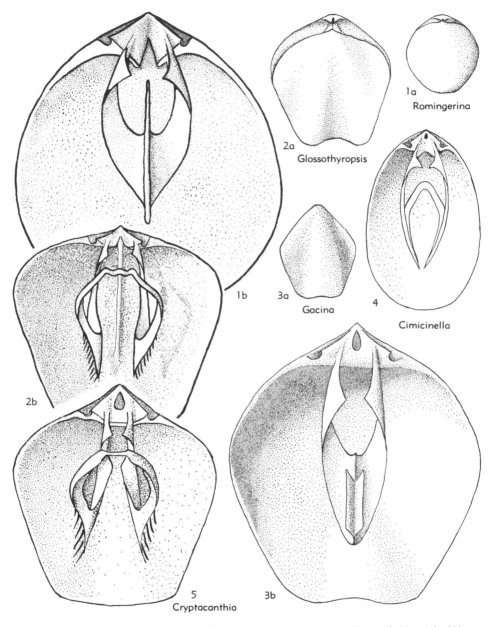

FIG. 613. Mutationellidae (Cimicinellinae) *(4)*, (Cryptacanthiinae) *(2-3,5)*; Family Uncertain *(1)*
(p. *H752-H754*).

Glossothyropsis GIRTY, 1934, p. 251 [*Cryptacanthia? robusta* GIRTY, 1934, p. 251; OD]. Small to moderate-sized; ventral fold pronounced; dorsal sulcus pronounced or not; cardinal plate more or less massive, imperforate, supported by median septum; loop long, more or less cryptonelliform, main bands widely separated and anteriorly

spinose (350). *L.Perm.-U.Perm.,* N.Am.-Eu.-Asia-Australia.——FIG. 613,2. **G. robusta* (GIRTY), composite figure; *2a,b,* brach.v. view, brach.v. int., ×3.5, ×6.8 (Stehli, n).

Family UNCERTAIN

Elmaria NALIVKIN, 1947 [**E. glabra*; OD]. Small,

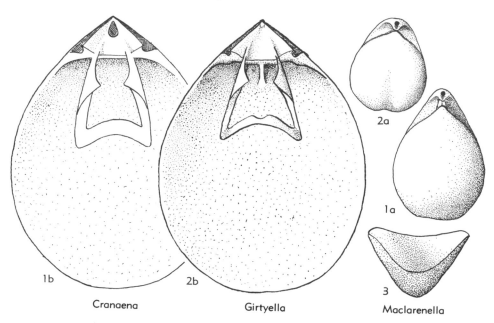

FIG. 614. Cranaenidae (Cranaeninae) *(1,3)*, (Girtyellinae) *(2)* (p. H754-H755).

rounded to elongate oval, smooth or faintly wrinkled, laterally keel-like, beak sharp; dental plates lacking. *M.Dev.(Givet.)*, USSR.

Romingerina HALL & CLARKE, 1894, p. 272 [*Centronella julia* WINCHELL, 1862, p. 405; OD] [=*Harttina* HALL & CLARKE, 1894, p. 292]. Small, smooth and almost circular in outline; subequally convex, pedicle valve more so than brachial valve; ventral beak short, deltidial plates conjunct; pedicle foramen submesothyridid; small dental plates present or absent; cardinal plate sessile; apparently divided; crural plates seemingly absent; loop long, centronelliform with large vertical plate (890). *L.Miss.-U.Miss.*, N.Am.——FIG. 613,*1. *R. julia* (WINCHELL), composite figure; *1a,b*, brach.v. view, brach.v. int., ×2, ×8.7 (Stehli, n).

Suborder TEREBRATULIDINA Waagen, 1883

A diagnosis of this assemblage is given in the section on Terebratulida—Main Groups (see p. H730).

Superfamily DIELASMATACEA Schuchert, 1913

[*nom. transl.* STEHLI, herein (*ex* Dielasmatinae SCHUCHERT, 1913, p. 402)]

Advanced derivatives of early radiation of Centronellidina which primitively possess short loop, though specialized descendants may have highly complicated loops; gen-

erally without true crural plates; pedicle foramen permesothyridid and labiate; pedicle collar present. *L.Dev.-U.Trias., ?L.Jur. (Lias.).* [Post-Paleozoic forms included in section by MUIR-WOOD, p. H762.]

Family CRANAENIDAE Cloud, 1942

[*nom. transl.* STEHLI, herein (*ex* Cranaeninae CLOUD, 1942, p. 131)]

Primitive Dielasmatacea, probably derived from Mutationellidae, possessing cardinal plate extending between socket ridges without support from crural plates and typically perforate but imperforate when plate is apically sessile or median septum is present. *L.Dev.-Miss.*

Subfamily CRANAENINAE Cloud, 1942

[Cranaeninae CLOUD, 1942, p. 131]

Moderate-sized, with terebratuliform loop, cardinal plate extending as apically perforate plate between socket ridges free of valve floor, or as imperforate plate apically united with valve floor; dental plates present. *L. Dev.-U.Miss.*

Cranaena HALL & CLARKE, 1893, p. 297 [*Terebratula romingeri* HALL, 1863, p. 48; OD] [=*Eunella* HALL & CLARKE, 1893, p. 290; *Cranaenella* FENTON & FENTON, 1924, p. 129]. Small to moderate-sized; anterior commissure rectimarginate to slightly uniplicate; both valves transversely con-

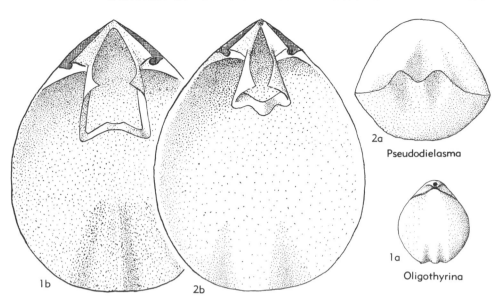

FIG. 615. Labaiidae (p. *H755*).

vex; cardinal plate free and perforate (389). *L. Dev.-U.Miss.,* cosmop.——FIG. 614,*1*. *C. romingeri* (HALL), composite figure; *1a,b,* brach.v. view, brach.v. int., ✕1.35, ✕3.8 (167).

Hamburgia WELLER, 1911, p. 445 [*H. typa*; OD (M)] [=*Stuartella* BELANSKI, 1929, p. 24]. Externally homeomorphous with *Cranaena*; internally like *Cranaena* except cardinal plate apically sessile and imperforate (857). *U.Dev.-U.Miss.,* N.Am.

Maclarenella STEHLI, 1955, p. 868 [*M. maculosa*; OD]. Moderate-sized; anterior commissure strongly uniplicate; dorsal valve of triangular cross section; ventral valve transversely concave; cardinal plate free and perforate (774). *U.Dev.,* N.Am. (Can.).——FIG. 614,*3*. *M. maculosa,* Waterways F., Alberta; ant. view, ✕1.35 (Stehli, n).

Subfamily GIRTYELLINAE Stehli, n.subfam.

Folded or unfolded shells with terebratuliform loop and imperforate cardinal plate supported by median septum; with or without dental plates. *Miss.*

Girtyella WELLER, 1914, p. 442 [*Harttina indianensis* GIRTY, 1908, p. 293; OD]. Small to moderate-sized; anterior commissure rectimarginate or modified by rounded plications; dental plates present (344). *Miss.,* N.Am.-Eu.——FIG. 614,*2*. *G. indianensis* (GIRTY), composite figure; *2a,b,* brach.v. view, brach.v. int., ✕2, ✕7.5 (Stehli, n).

Harttella BELL, 1929, p. 149 [*H. parva*; OD]. Small, similar to *Girtyella* except in being folded and lacking dental plates (64). *U.Miss.(Meramec.),* N.Am.

Family LABAIIDAE Likharev, 1960

[*nom. correct.* STEHLI, herein (*pro* Labaidae LIKHAREV, 1960, p. 293)]

Small, terebratuliform looped shells with tendency toward anterior folding; cardinal plate obsolete; crura arising from margins of socket ridges; dental plates absent. *M. Penn.-U.Perm.*

Labaia LIKHAREV, 1956, p. 65 [*L. Muir-Woodae*; OD]. Small, smooth, elongate, unfolded shell suboval and subrhomboidal; pedicle valve with pronounced shoulders in the umbonal region; pedicle interior without dental plates, pedicle collar probably present; brachial interior without cardinal plate; crura arising directly from the socket ridges and giving rise to a short loop. *U.Perm.,* USSR(N.Caucasus).

Oligothyrina COOPER, 1956, p. 525 [*O. alleni*; OD]. Small, with a weakly to strongly intraplicate anterior commissure; folds arising anterior to midlength; transverse band not projecting anteriorly (188). *M.Penn.-U.Perm.,* N.Am.——FIG. 615,*1*. *O. alleni,* composite figure; *1a,b,* brach.v. view, brach.v. int., ✕4.2, ✕19 (188).

Pseudodielasma BRILL, 1940, p. 317 [*P. perplexa*; OD]. Small, with weakly to strongly sulciplicate anterior commissure; folds arising near the front; loop with medial portion of transverse band projected anteriorly (118). *U.Perm.,* N.Am.-Australia. ——FIG. 615,*2*. *P. perplexa,* composite figure; *2a,b,* ant. view, brach.v. int., ✕7, ✕19 (*2a,* Stehli, n; *2b,* 118).

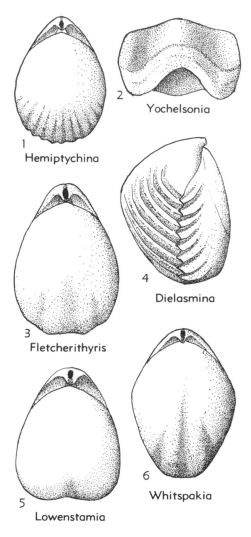

1 Hemiptychina

2 Yochelsonia

3 Fletcherithyris

4 Dielasmina

5 Lowenstamia

6 Whitspakia

Fig. 616. Dielasmatidae (Dielasmatinae) (p. H756, H758).

Family DIELASMATIDAE Schuchert, 1913

[nom. transl. SCHUCHERT & LEVENE, 1929, p. 23 (ex Dielasmatinae SCHUCHERT, 1913, p. 402)]

Smooth to plicate, folded or unfolded shells; pedicle valve with or without dental plates; brachial valve with terebratuliform loop and modified cardinal plate which is either divided or supported by septum. L. Carb.-U.Trias., ?L.Jur.(Lias.). [Post-Paleozoic forms included in section by MUIR-WOOD, p. H762.]

Subfamily DIELASMATINAE Schuchert, 1913

[Dielasmatinae SCHUCHERT, 1913, p. 402]

Dental plates present or absent; pedicle collar complete; pedicle beak not elongated. L.Carb.-U.Trias.

Dielasma KING, 1859, p. 7 [*Terebratulites elongatus VON SCHLOTHEIM, 1816, p. 27; OD] [=Dielasmoides WELLER, 1914, p. 253]. Small to large; normally with dorsal fold, ventral sulcus and uniplicate anterior commissure but in few species with folds anteriorly resulting in sulciplicate commissure; halves of cardinal plate separate or jointed near union with floor of valve; dental plates present (716). U.Miss.-Perm., cosmop.——FIG. 617,1. *D. elongatum (SCHLOTHEIM), composite figure; 1a, brach.v. int., ×2.4; 1b, lat. view, ×2.4; 1c, apical int., ×5.5 (Stehli, n).

Balanoconcha CAMPBELL, 1957, p. 86 [*B. elliptica; OD]. Medium-sized external homeomorph of Dielasma; anterior commissure rectimarginate to slightly uniplicate; cardinal plate as in Dielasma; dental plates absent (140). L.Carb.(Tournais.), Australia.——FIG. 617,4. *B. elliptica, composite figure; brach.v. int., ×3 (Stehli, n).

Dielasmina WAAGEN, 1882, p. 335 [*D. plicata; OD]. Moderate-sized to large, anteriorly ornamented by numerous low plications; anterior commissure rectimarginate; brachial valve geniculated sharply near mid-length; cardinal plate supported by median septum; dental plates present (845). Perm., Pakistan (Salt Range).——FIG. 616,4; 617,7. *D. plicata, composite figure; 616,4, lat. view, ×1.4; 617,7, brach.v. int., ×3 (Stehli, n).

Fletcherithyris CAMPBELL, 1965 [nom. subst. pro Fletcherina STEHLI, 1961, p. 452 (non LANG, SMITH, & THOMAS, 1955, p. 261)] [*Terebratula amygdala DANA, 1847, p. 152; OD]. Moderate-sized to large, folded or unfolded, when folded, brachial valve with median sulcus flanked by folds; cardinal plate supported by median septum; dental plates present (223). L.Perm., Australia.——FIG. 616,3; 617,6. *F. amygdala (DANA), composite figure; 616,3, brach.v. view, ×1.5 (Stehli, n); 617,6, brach.v. int., ×3.5 (779).

Hemiptychina WAAGEN, 1882, p. 335 [*Terebratula himalayensis DAVIDSON, 1862, p. 27; OD] [=Morrisina GRABAU, 1931, p. 97]. Moderate-sized, biconvex to subglobular; brachial valve and some pedicle valves geniculate anteriorly; anterior commissure rectimarginate; abundantly plicated anteriorly; halves of cardinal plate separate; dental plates absent (233). Perm., Asia(E.Tethyan area). ——FIG. 616,1; 617,5. *H. himalayensis (DAVIDSON), composite figure; 616,1, brach.v. view, ×1.7; 617,5, brach.v. int., ×4 (Stehli, n).

Lowenstamia STEHLI, 1961, p. 460 [*L. texana; OD]. Small, inflated external homeomorphs of Dielasma; halves of cardinal plate separate and

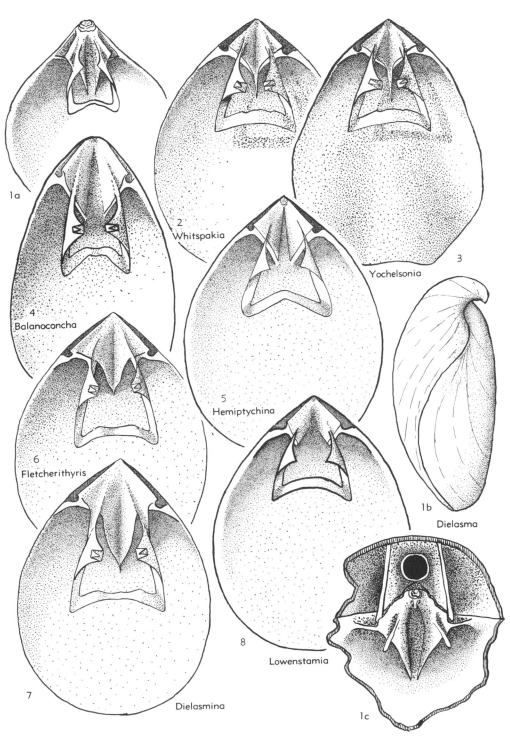

Fig. 617. Dielasmatidae (Dielasmatinae) (p. *H756, H758*).

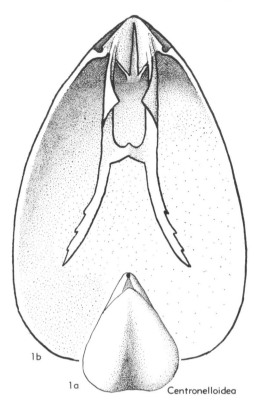

1b

1a   Centronelloidea

Fig. 618. Dielasmatidae (Centronelloideinae)
(p. H758).

becoming free of valve floor anteriorly; dental plates absent (779). *L.Perm.,* N.Am.——Fig. 616, *5;* 617,*8.* **L. texana,* Coleman Jct. Ls., USA (Tex.); 616,*5,* brach.v. view, ×3.4; 617,*8,* brach. v. int., ×7.5 ((616,*5,* Stehli, n; 617,*8,* 61).

Whitspakia STEHLI, 1964, p. 610 [**Dielasma biplex* WAAGEN, 1882, p. 249] [=*Pakistania* STEHLI, 1961, p. 462 (*non* EAMES, 1952)]. Medium-sized to large, subpentagonal to oval in outline; anterior commissure sulciplicate; front and sides not geniculate; cardinal plate like *Dielasma;* dental plates present (845). *Perm.,* N.Am.-Eu.-Asia.—— Fig. 616,*6;* 617,*2.* **W. biplex* (WAAGEN), composite figure; 616,*6,* brach.v. view, ×1.5; 617,*2,* brach.v. int., ×3 (Stehli, n).

Yochelsonia STEHLI, 1961, p. 454 [**Y. thomasi;* OD]. Small, subtriangular to pentagonal; brachial valve longitudinally flattened, bearing pronounced median sulcus; pedicle valve with high median fold bordered by sulci; front and sides of both valves geniculate; cardinal plate as in *Dielasma;* dental plates present. *U.Perm.,* W.Australia.—— Fig. 616,*2;* 617,*3.* **Y. thomasi;* composite figure; 616,*2,* ant. view, ×3.2; 617,*3,* brach.v. int., ×3.5 (Stehli, n).

Subfamily CENTRONELLOIDEINAE Stehli, n.subfam.

Small, somewhat aberrant dielasmatids with sulcate anterior commissure and elongate ventral beak; pedicle foramen permesothyridid but not telate; loop terebratuliform but modified by spinose anterior projections of main bands beyond transverse band; dental plates and partial pedicle collar present. *U.Miss.*

Centronelloidea WELLER, 1914, p. 246 [**Terebratula rowleyi* WORTHEN, 1884, p. 23; OD (M)]. Small, with sulcate anterior commissure; pedicle valve with rounded fold, brachial valve with more pronounced sulcus; ventral beak elongated; cardinal plate medially sessile (894). *U.Miss.,* N. Am.——Fig. 618,*1.* **C. rowleyi* (WORTHEN), composite figure; *1a,b,* brach.v. view, brach.v. int., ×3.6, ×14.5 (858).

Family NOTOTHYRIDIDAE Likharev, 1960

[*nom. transl. et correct.* STEHLI, herein (*ex* Notothyrinae LIKHAREV, 1960, p. 288)]

Folded or unfolded shells with apically perforate cardinal plate extending unsupported between socket plates; loop characteristically centronelliform but exhibiting stages in transformation from terebratuliform to quasicryptonelliform; dental plates absent. *U.Miss.-U.Perm.*

Notothyris WAAGEN, 1882, p. 375 [**Terebratula subvesicularis* DAVIDSON, 1862, p. 27; SD HALL & CLARKE, 1893, p. 275]. Small to moderate-sized with numerous plications toward front; anterior commissure rectimarginate to faintly sulcate; interior as in *Rostranteris* (233). *Perm.,* Eu.-Asia.——Fig. 619,*2.* **N. subvesicularis* (DAVID-SON), composite figure; *2a,b,* brach.v. view, brach.v. int., ×1.35, ×4.4 (845).

Alwynia STEHLI, 1961, p. 464 [**D. vesiculare* DE-KONINCK, 1887, p. 30; OD (M)]. Small, with antiplicate anterior commissure; loop basically terebratuliform but modified by close approach of main bands anteriorly and small transverse band (779). *L.Carb.,* Eu.——Fig. 619,*1.* **A. vesicularis* (DEKONINCK), Visean, Eng.(Isle of Man); *1a,b,* ped.v. ext. and ant. views, ×2.4; *1c,* brach.v. int., ×5.3 (Stehli, n).

Gefonia LIKHAREV, 1936, p. 264 [**G. cubanica;* OD]. Small, subpentagonal; anterior commissure basically sulcate but modified by folds into antiplicate condition; loop centronelliform but modified by union of main bands through transverse band anterior to mid-length and their subsequent separation with rise of diverging recurving bands which end without uniting (515). *U.Perm.,* USSR (Caucasus).——Fig. 619,*5.* **G. cubanica; 5a,b,*

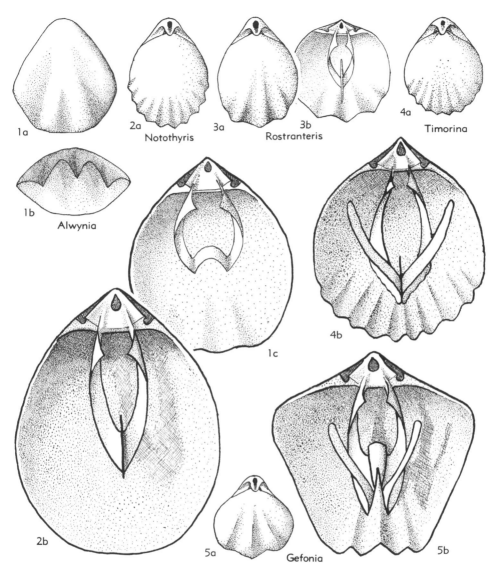

1a

1b Alwynia

2a Notothyris

3a

3b Rostranteris

4a Timorina

1c

4b

2b

5a Gefonia

5b

Fig. 619. Notothyrididae (p. *H758-H760*).

brach.v. view, brach.v. int., ×1.7, ×5.7 (Stehli, n).

Rostranteris GEMMELLARO, 1898 (1899), p. 306 [*D. adrianense* GEMMELLARO, 1894, p. 5; OD] [=*Mongolina* GRABAU, 1931, p. 105]. Small to moderate-sized, typically with intraplicate anterior commissure, more rarely sulcate; when intraplicate, major folds of pedicle valve may be flanked by one weak fold; loop centronelliform with high median plate extended anteriorly and posteriorly beyond union of main bands (330). *Perm.*, N.Am.-Eu.-Asia.

R. (Rostanteris). Distinguished by delicate cardi-

nalia. *Perm.*, N.Am.-Eu.-Asia.——FIG. 619,3. *R. (R.) adrianensis* (GEMMELLARO), composite figure; *3a,b,* brach.v. view, brach.v. int., ×1.7, ×2.2 (Stehli, n).

R. (Notothyrina) LIKHAREV, 1936 [*Notothyris (N.) pontica*; OD]. Very small pedicle valve with 2 strong folds each bordered by weak lateral fold; internal structures except for loop thickened with secondary shell material and identical to *Rostranteris* (515). *U.Perm.*, USSR (Caucasus).

Timorina STEHLI, 1961, p. 465 [*Notothyris minuta* BROILI, 116 (*non* WAAGEN, 1882) =*Timorina*

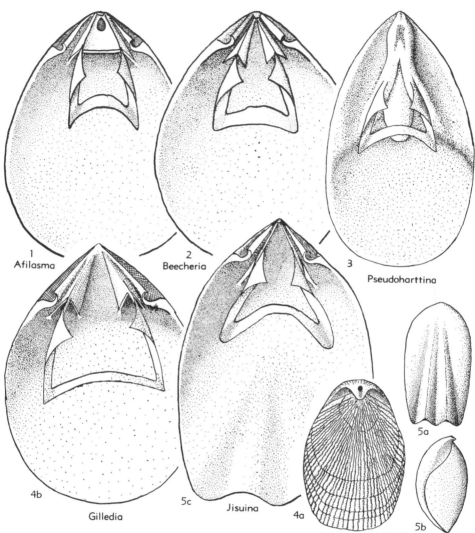

Fig. 620. Heterelasminidae *(1-2,4-5)*; Family Uncertain *(3)* (p. *H760-H762*).

broili *(recte broilii)* STEHLI, 1961, p. 465; OD]. Small, externally resembling *Notothyris* but with median 2, 3, or more plications on pedicle valve raised into slight fold; loop basically centronelliform but modified by origin from median plate of diverging recurved bands which end without uniting (120). *U.Perm.,* Timor.——FIG. 619,4. *T. broilii* STEHLI; *4a,b,* brach.v. view, brach.v. int., ×2.5, ×7 (Stehli, n; 799).

Family HETERELASMINIDAE
Likharev, 1956

[Heterelasminidae LIKHAREV, 1956, p. 64]

Shells with terebratuliform loop and cardinal plate which is supported by crural

plates but obsolete between them and socket ridges, primitively perforate apically and free of valve floor; advanced genera with cardinal plate medially sessile and divided or obsolete; dental plates present or absent. *U.Dev.-U.Perm.*

Jisuina GRABAU, 1931, p. 105 [*J. elegantula*; OD] [*=Heterelasmina* LIKHAREV, 1934, p. 212]. Small to moderate-sized, elongate and straight-sided; anterior commissure truncate to emarginate, primitively uniplicate but usually showing more complex folding; cardinal plate obsolete, crura arising from crural plates; dental plates absent (360). *Perm.,* Eu.-Asia.——FIG. 620,5. *J. elegantula,* composite figure; *5a,b,* ped.v. and lat. views,

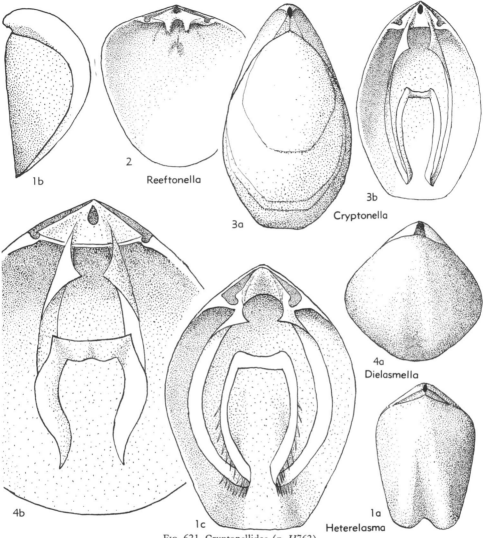

1b
2 Reeftonella
3a
3b Cryptonella
4a Dielasmella
4b
1c
1a Heterelasma

FIG. 621. Cryptonellidae (p. *H762*).

×1.25, ×1; *5c,* brach.v. int., ×3.9 (Stehli, n).
Afilasma STEHLI, 1961, p. 460 [**A. beecheri;* OD].
Moderate-sized, unfolded, thin; cardinal plate
apically perforate, extending free of valve floor
between crural plates, obsolete between crural
plates and socket ridges; dental plates present;
loop unknown but probably terebratuliform. *U.
Dev.,* N.Am.——FIG. 620,1. **A. beecheri,* Chemung, USA(N.Y.); brach.v. int., ×2.6 (Stehli, n).
Beecheria HALL & CLARKE, 1893, p. 300 [**B. davidsoni;* OD]. Unfolded to uniplicate, small to
large; cardinal plate imperforate, medially sessile
and divided into 2 plates each extending from
valve floor to crural plate and bearing crus; dental plates present. *L.Miss.-U.Perm.,* cosmop.——

FIG. 620,2. **B. davidsoni,* composite figure; brach.
v. int., ×2.6 (Stehli, n).
Gilledia STEHLI, 1961, p. 451 [**Terebratula cymbaeformis* MORRIS, 1845, p. 278; OD]. Large,
uniplicate shell ornamented with wavy radial
carinae; cardinal plate medially sessile forming
2 plates extending from floor of valve to top of
crural plates; internal structures greatly thickened
by secondary shell material; dental plates present
but massively united wtih sides of valve by secondary shell material (571). *L.Perm.,* Australia-
Tasmania.——FIG. 620,4. **G. cymbaeformis*
(MORRIS), Up. Marine Ser., New S. Wales; *4a,b,*
brach.v. view, brach.v. int., ×1.25, ×2.5 (Stehli,
n).

Family UNCERTAIN

Pseudoharttina LIKHAREV, 1934, p. 2112 [**P. ovalis*; OD]. Small to moderate-sized; anterior commissure rectimarginate; convexity of valves variable; cardinal plate obsolete; crura arising from socket ridges; dorsal median septum present; all internal structures except loop much thickened; dental plates present but ankylosed to wall of valve (515). *Perm.*, Asia-N.Am.——FIG. 620,*3*. **P. ovalis*, composite figure; brach.v. int., ×4 (Stehli, n).

Suborder TEREBRATELLIDINA Muir-Wood, 1955

A diagnosis of this assemblage is given in the section on Terebratulida—Main Groups (see p. *H730*).

Superfamily CRYPTONELLACEA Thomson, 1926

[*nom. transl.* STEHLI, herein (*ex* Cryptonellinae THOMPSON, 1926, p. 529)]

Generally smooth but rarely costate or costellate anteriorly, folded or unfolded; pedicle foramen mesothyridid to submesothyridid; dental plates present; pedicle collar absent; cardinal plate perforate or not, generally unsupported between socket plates but in few forms supported by small median septum; loop cryptonelliform. *L.Dev.-Perm.*

Family CRYPTONELLIDAE Thomson, 1926

[*nom. transl.* STEHLI, herein (*ex* Cryptonellinae THOMSON, 1926, p. 529)]

Characters of superfamily. *L.Dev.-Perm.*

Cryptonella HALL, 1861, p. 101 [**Terebratula rectirostra* HALL, 1860, p. 88; SD HALL & CLARKE, 1894, p. 861]. Small to moderate-sized; smooth or anteriorly faintly plicate, folded or not; anterior commissure rectimarginate to sulciplicate; pedicle foramen submesothyridid; cardinal plate perforate or imperforate and extending unsupported between socket plates; dental plates present (386). *L.Dev.-Perm.*, Eu.-N.Am.-S.Am.——FIG. 621,*3*. *C. planirostra* HALL, composite figure; *3a,b*, brach.v. view, brach.v. int., ×4.6, ×5.6 (167).

Dielasmella WELLER, 1911, p. 446 [**Eunella compressa* WELLER, 1906, p. 442; OD]. Small, subcircular to pentagonal in outline; both valves shallow; anterior commissure rectimarginate; pedicle foramen mesothyridid; delthyrium incompletely closed below foramen; perforate cardinal plate extending unsupported between socket plates; dental plates present (856). *Miss.*, N.Am.——FIG. 621,*4*. **D. compressa* (WELLER), composite figure; *4a,b*, brach.v. view, brach.v. int., ×4.3, ×11 (858; Stehli, n).

Heterelasma GIRTY, 1908, p. 337 [**H. shumardianum*; OD]. Small to moderate-sized; smooth; uniplicate to sulciplicate; pedicle valve moderately to highly convex longitudinally, brachial valve longitudinally concave to slightly convex; dental plates present; cardinal plate generally imperforate and extending unsupported between socket plates but in some shells supported apically by small medial septum (345). *Perm.*, N.Am.——FIG. 621, *1*. **H. shumardianum*, composite figure; *1a,b*, brach.v. and lat. views, ×3, ×3.4; *1c*, brach.v. int., ×7.5 (*1a*, 345; *1b,c*, Stehli, n).

?Reeftonella BOUCOT, 1959, p. 768 [**Meganteris neozelanica* ALLAN, 1935, p. 23; OD]. Moderate-sized; subequally convex, pedicle valve slightly more convex than brachial valve; outline subcircular to shield-shaped; smooth or ornamented with growth lines; anterior commissure rectimarginate; pedicle foramen submesothyridid; dental plates present but becoming obsolescent in adults; cardinal plate perforate, sessile; crural plates absent; loop unknown (18). [Systematic position quite uncertain but possibly belongs to Cryptonellidae.] *L.Dev.*, N.Z.——FIG. 621,*2*. **R. neozelanica* (ALLAN), composite figure; brach.v. int., ×1.15 (Stehli, n).

MESOZOIC AND CENOZOIC TEREBRATULIDINA

By HELEN M. MUIR-WOOD

INTRODUCTION

The present contribution deals with all known genera of the suborder Terebratulidina of Triassic to Recent age and family-group taxa to which they are assigned. Only one of the family assemblages (Dielasmatidae, L.Carb.-U.Trias., ?L.Jur.) includes pre-Mesozoic members. Among the remaining eight recognized families, five (Orthotomidae, L.Jur.; Cheniothyrididae, M.Jur.; Dictyothyrididae, M.Jur.-U.Jur.; Tegulithyrididae, U.Jur.; Pygopidae, ?L.Jur., M.Jur.-L. Cret.) are confined to Mesozoic deposits, and the remaining three (Terebratulidae, U. Trias.-Rec.; Cancellothyrididae, ?L.Jur.-?M. Jur., U.Jur.-Rec.; Dyscoliidae, ?U.Jur., U. Cret.-Rec.) include Mesozoic and Cenozoic genera. The world-wide distribution of the terebratuloid genera in post-Paleozoic formations is little known, mainly owing to lack of requisite information on the internal structures of many species. A majority of short-looped terebratuloids are still referred to as "*Terebratula*," long-looped species being designated as "*Waldheimia*" or "*Zeilleria*," which belong among the terebratelloids. Internal characters have been described by authors generally only when suitable weathered or silicified specimens were available.

Among Tertiary Terebratulacea, the internal characters are little known and relationships of the numerous species inferred from external characters is uncertain. Dissections, where possible, or serial sections will have to be prepared before any attempt can be made to classify these forms or work out their evolution. There are obviously a number of distinct stocks in addition to species of *Terebratula* (s.s.) and fossil species of *Gryphus, Liothyrella, Dallithyris,* and *Abyssothyris.*

The Dyscoliidae, like the Cancellothyrididae, may persist from Upper Jurassic, but most of the Jurassic and Cretaceous Terebratulidae do not survive after the end of the Mesozoic. A few Recent genera, such as *Cnismatocentrum* and *Agulhasia,* have not yet been found as fossils, but most of these genera range back into the Miocene or Pliocene. Some Recent genera are still imperfectly known in regard to their lophophore and its development stages.

PREVIOUS STUDIES

Of the very large number of authors who have described Mesozoic and Cenozoic terebratuloid species or genera it is only possible to mention a few. E. EUDES-DESLONGCHAMPS (1862-85) described and figured many Jurassic species mainly of France, and in 1884 erected the new genera *Flabellothyris, Fimbriothyris, Microthyris, Epicyra,* and *Disculina,* based on external characters.

QUENSTEDT (1868-71) illustrated the internal structure of many species whenever suitably preserved material was available but did not describe many new genera.

DOUVILLÉ (1879) proposed the genera *Dictyothyris, Glossothyris, Coenothyris, Plesiothyris,* and *Aulacothyris,* with reference to internal characters.

DAVIDSON's Mesozoic volumes (1851-55, 1874-82) portray mainly exteriors, though he illustrated some loops, and also some interiors in his "Introduction" to volume 1 (1853) but did not embark on any generic classification.

ROTHPLETZ (1886) was probably one of the first to employ transverse sections in his descriptions of Lower and Middle Jurassic

rhynchonellids of the Alps region, and he published longitudinal sections of terebratuloid species. KITCHIN (1900) also gave a few sections of Jurassic species from Pakistan (Cutch). BITTNER (1890, 1892) described a large number of Triassic species and some new genera.

S. S. BUCKMAN (1918) endeavored to classify Burmese and European (mainly British) species by means of the patterns of dorsal adductor muscle scars studied on internal molds, in addition to beak characters, folding of the shell, and surface ornament, and he described a number of new genera. Mostly internal characters of these brachiopods were not studied by BUCKMAN. The difficulty of preparing suitable internal molds and correctly interpreting the adductor scars shown by them has prevented the development of an acceptable basis of classification by this means.

ROLLIER (1915-1918) redescribed many Jurassic species and gave useful bibliographic references, but did little to advance generic or family classification.

SAHNI (1925, 1929) dissected out the loops of British Upper Cretaceous terebratulids and proposed a number of new genera based on the nature of the loop, form of the cardinal process and adductor muscle scars, and the presence or absence of inner hinge plates. He also pointed out the difference in length of these Upper Cretaceous loops from those of some of BUCKMAN's Middle Jurassic genera.

THOMSON's (1927) publication on Tertiary and Recent brachiopods summarized some of BUCKMAN's work and added much valuable information on fossil and Recent forms which gave a strong impetus to research.

MUIR-WOOD (1934-36) pointed out the importance of serial transverse sections in the identification and classification of genera and species, illustrating this mainly in relation to Jurassic and Cretaceous terebratulacean and zeilleriacean genera; she erected new genera based on external characters, as well as internal ones.

Serial sections of Mesozoic terebratuloids have been used by DAGIS (1958-63) in studying Upper Triassic forms from the Crimea; MIDDLEMISS (1959) in work on British Lower Cretaceous terebratulids, and also TOKU-

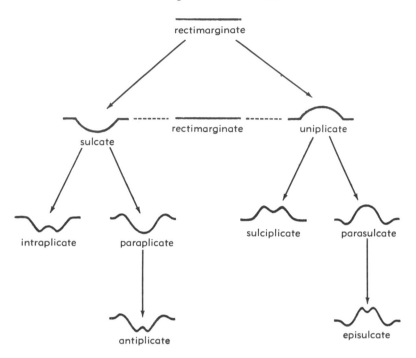

FIG. 622. Diagram showing anterior commissures of Terebratulida (after 810).

YAMA (1958a,b), PROSOROVSKAYA (1962), KYANSEP (1959, 1961) and MAKRIDIN (1960) among others, with the erection of many new genera, most of which require further investigation and research.

A considerable volume of literature relates to Cenozoic terebratuloids; some outstanding publications listed by THOMSON (1927) are works by R. S. ALLAN, C. E. BEECHER, F. BLOCHMANN, W. H. DALL, T. DAVIDSON, E. EUDES-DESLONGCHAMPS, K. HATAI, J. G. HELMCKE, J. W. JACKSON, J. G. JEFFREYS, E. S. MORSE, F. SACCO, C. SCHUCHERT, G. SEGUENZA, and J. A. THOMSON. Further lists of references were given by MUIR-WOOD in 1955 and 1959.

EXTERNAL MORPHOLOGY

The two valves—pedicle (or ventral) and brachial (or dorsal)—may be convex in all growth stages, or the brachial valve may be plane, or concave, or sulcate posteriorly only. The umbonal region is posterior. In *Dictyothyris* a deep ventral median sulcus is bounded by prominent ridges, and a low dorsal median fold is bordered by shallow sulci (**pliciligate** stage).

In the Terebratulacea the anterior commissure may be straight and not deflected either dorsally or ventrally, and is then known as plane or **rectimarginate**. It may be **everted** or dorsally deflected in a single **uniplication**. This may be medially sulcate, giving a **sulciplicate** stage, or a sulcus may be developed on each side of the uniplica in the **parasulcate** stage. A sulcus may develop medially in a parasulcation giving an **episulcate** stage. Further development of folds results in the **quadriplicate** stage in *Epithyris*. In some Mesozoic forms the uniplicate stage may be omitted when **biplication** develops directly from a rectimarginate commissure.

The reverse of everted is the **inverted type of folding** when the anterior commissure is deflected ventrally. The opposite of uniplicate is known as **sulcate,** the opposite of parasulcate is **paraplicate,** the opposite of sulciplicate is called **intraplicate; antiplicate** is the reverse of episulcate (Fig. 622).

Multiplication may be opposite in the two valves, but more commonly is alternate and may be superimposed on a uniplicate or sulcate stage. It may arise directly from a smooth stage or be the result of bifurcation of a few existing costae or the intercalation of new costae. In the **semiplicate** stage costae or costellae are developed on the anterior half or third of the shell.

The ventral umbo is erect, suberect, or incurved, massive, tapering, short, or produced. It is normally truncated by the foramen, except in the Orthotomidae, where the umbonal apex is intact and tapering and the delthyrium housing the pedicle lies anterior to the umbo (e.g., *Orthotoma,* Fig. 634,*1e*).

The foramen varies in size from a pinhole in the Upper Cretaceous terebratuloid genus *Gibbithyris* to large and commonly marginate or partly infilled with secondary deposits, or labiate, with a liplike development on the dorsal side of the foramen. Various terms have been applied by authors to the angle of incurvature of the umbo (Fig. 623).

The dorsal umbo is not prominent and

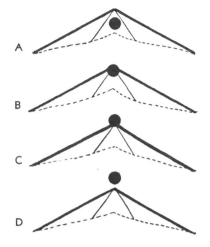

Fig. 624. Diagram showing position of foramen (after 810).——*A.* Hypothyridid.——*B.* Mesothyridid.——*C.* Permesothyridid.——*D.* Epithyridid. [Heavy lines represent beak ridges.]

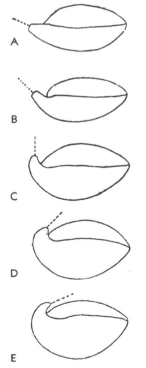

Fig. 623. Diagram showing incurvature of umbo (after 810).——*A.* Nearly straight.——*B.* Suberect.——*C.* Erect.——*D.* Slightly incurved.——*E.* Strongly incurved.

may be concealed by incurvature of the ventral umbo.

Beak ridges are more or less angular, curving, linear elevations of the shell extending from each side of the ventral umbo and commonly defining a palintrope. In the Terebratulacea they tend to be short and ill-defined. When the pedicle opening is on the dorsal side of the beak ridges it is known as **hypothyridid** (e.g. *Orthotoma*). It is termed **mesothyridid** when the foramen lies equally on each side of the beak ridges and is partly in the interarea and partly in the umbo; **permesothyridid** when the foramen is mostly within the ventral umbo, and **epithyridid** when the pedicle opening lies wholly within the ventral umbo and beak ridges are on the dorsal side of the umbo (Fig. 624). The beak ridges may project into the foramen or delthyrium as small points or **telae (telate)** or these may be worn away when the condition is known as **attrite.**

The deltidial plates in the Terebratulacea may be fused and form a single plate known as the **symphytium** without trace of median line of junction, whereas in zeilleriacean terebratelloids the deltidial plates may be **conjunct** or fused, or **disjunct** or discrete and not completely fused, when the foramen is referred to as **incomplete.**

The external sculpture (rather misleadingly known as ornament) of most of the terebratuloids consists rarely of radial ridges known as **costae** or **costellae,** or finer radial ornament comprised of **capillae.** When fewer than 15 radial ridges occur in a space of 10 mm., they are referred to as costae; if 15 to 25 are counted in 10 mm., they are referred to as costellae; if more than 25 are present in 10 mm., they are named capillae. [Attention is called to somewhat different definitions of these terms given in the glossary (p. *H*139.—Ed.]

Some genera have nodes and spines (e.g., *Dictyothyris,* or concentric rugae or lamellae (e.g., *Cheniothyris, Ornatothyris*), but most Mesozoic and also Cenozoic genera are smooth, with more or less prominent growth lines. Cenozoic Terebratulacea have a smooth shell or one that may be partly or wholly capillate. Most Recent shells are white or cream-colored, lacking the bright colors of many Terebratellacea. *Cnismatocentrum* has a brown shell, while *Cancellothyris* has concentric brown bands.

The Pygopidae (Jur.-L.Cret.) differ from all other Terebratulidina in having a biconvex early stage, then becoming sulco-convex, with the lateral slopes continuing to grow so as finally to converge and fuse, enclosing a median perforation. Some species do not develop the median perforation, but all stages of development of the lateral slopes and their convergence and possible complete fusion can be observed in other species (e.g., *Pygope*, Fig. 678,*1c*). An additional fold in the dorsal sulcus and a sulcus in the ventral fold are characteristic of the Neocomian genus *Pygites* (Fig. 678, *2a,d*).

Specimens are described as small when they are less than 0.75 in. or 20 mm. wide or long; medium-sized when they are 0.75 to 2 in. or 20 to 50 mm. wide or long; and large when they are more than 2 in. or 50 mm. in length or width.

INTERNAL MORPHOLOGY

The two valves articulate by means of **hinge teeth** in the pedicle valve which fit into **sockets** in the brachial valve. In addition accessory articulation is effected by means of **denticula** or toothlike terminations of the palintrope which fit into accessory sockets in the **outer socket ridges,** and also by means of the **denticular cavity** on the outer lateral side of the hinge teeth which articulates with a projection from the outer

socket ridges. The **inner socket ridges** may articulate with a depression on the inner face of the hinge teeth, as in many Cenozoic genera. The hinge sockets and teeth may be crenulated. Considerable variation in size and form of the teeth has been observed in different genera, but it is not known how far this can be used as a distinguishing character, and how much variation may occur in subsequent growth stages.

Articulation is also effected by means of the adductor or closing muscles and diductor or opening muscles. The **adductor muscles** leave four scars of attachment on the brachial valve, two placed farther forward and nearer the mid-line of the shell, being known as **anterior adductor scars** and two located behind the others and more laterally, being known as **posterior adductor scars.** In the brachial valve the **diductor muscles** are attached to the hinge plates, or to the cardinal process when this is developed. In the pedicle valve two adductor scars are visible between the broader **diductor scars.** The **pedicle muscles** of attachment are obscure in Mesozoic Terebratulidina, as a rule. Two scars may be detected on the outer lateral side of the diductor scars in the pedicle valve and rarely a single scar more umbonally and centrally placed, as in Recent forms.

In the Cancellothyrididae there are no hinge plates and pedicle muscles are attached to the floor of the dorsal valve.

Mantle canals are marked by furrows on interior surfaces of both valves, or by ridges on internal molds. They represent extensions of the coelom or body cavity into the dorsal or ventral mantles. The four main trunks in the pedicle and brachial valve observed in zeilleriacean terebratelloids are rarely observed in most of the Terebratulacea. In *Ornatothyris,* from the English Cenomanian, the mantle canals bifurcate (Fig. 666,*2f*), whereas in *Gibbithyris* two main nonbifurcating trunks are seen. Bifurcating mantle canals are frequently observed on internal molds of pygopids (Fig. 679,*3a*). In the Terebratulacea a more or less prominent cardinal process is developed. In the Rectithyridinae it is a low, medianly depressed plate, but in the Carneithyridinae it is commonly large and bulbous (Fig. 668,*1d*). In *Plectoidothyris* the cardinal process is prominent and bilobed (Fig. 694).

Hinge plates may be fused, resulting in development of a median hinge trough or septalium, as in most Zeilleriacea, or they may be free, as in most Terebratulacea. In most Mesozoic Terebratulacea only outer hinge plates are found, but in *Neoliothyrina* (Fig. 664) and some Tertiary forms inner hinge plates are present, being separated from the outer hinge plates by the crural bases.

In the Terebratulacea a low ridge may separate the adductor scars; this usually is referred to as the myophragm.

The brachial **loop** in most Jurassic and Cretaceous genera is attached to the hinge plates by the **crural bases**, which may be given off on the dorsal or ventral side of the hinge plates, as shown in serial sections. In the Cancellothyrididae the crural bases are fused with the inner socket ridges and there are no hinge plates. The portions of the loop posterior to the **crural processes** or crural points in the Terebratulidina are known as the **crura**. In front of the crural processes the **descending branches** are usually very short in terebratulids and they unite with the more or less arched **transverse band**.

In Triassic forms the loop is short, usually without crural processes, and centronellid or dielasmatid in form. The descending branches in the centronellid type are united by a median vertical plate which varies in length and position in different genera.

In the Terebratulacea the loops of most genera are imperfectly known, but in the Middle Jurassic loops of two distinct lengths occur, one about half or less of the length of the brachial valve, the other two-thirds of the length of the brachial valve, as in *Plectoidothyris*. In the pygopids the loop is very short, with a slightly arched transverse band. The Upper Cretaceous Carneithyridinae have loops about one-third of the length of the brachial valve. The precise implication of this is unknown. In Cenozoic genera the loop is usually about ¼ or ⅓ of the length of the brachial valve. The lophophore was probably plectolophous in most genera, but may have been schizolophous, ptycholophous or spirolophous in some forms.

The internal morphology as seen in serial transverse sections may be recorded graphically and recorded in generic diagnoses.

The form of hinge plates and inner socket ridges in section is found to be of diagnostic importance and certain terms additional to those proposed by MIDDLEMISS (1959) are needed. The hinge plates may be horizontal or deflected dorsally or ventrally, and may be ventrally convex, or ventrally concave. In some genera they may be rounded U-shaped or sharply V-shaped. When the crural bases are straight, more or less vertical, and at an angle to the hinge plates, they are here called **virgate**. The crural bases may be given off on the dorsal or ventral side of the hinge plates. The hinge plates and inner socket ridges are often indistinguishable in section, or they may be separated by a shallow sulcus. A keel may be developed dorsally below the hinge plates. The hinge plates in section may be thickened or clubbed, may taper medially, or be enlarged only at the tip or piped, or they may be bladelike (Fig. 697). The septalial plates seen in some Triassic genera (e.g., *Rhaetina*) extend from the hinge plates and converge and unite medially with the median septum, if present (Fig. 629). A pedicle collar, or continuation of the deltidial plates on the inner side of the umbo, may be developed in some Terebratulacea.

Additional terms relating to internal and external morphology of terebratuloids have been defined by BUCKMAN (1918), THOMSON (1927), and MUIR-WOOD (1934, 1936). Internal morphology is dealt with more fully in introductory chapters of the brachiopod volume.

HOMEOMORPHY

Homeomorphy occurs repeatedly among the Mesozoic and Cenozoic terebratuloids and constitutes one of the major problems in their identification and classification. It is frequently impossible to identify Mesozoic forms without examining the internal structure and to distinguish between representatives of the Terebratulacea (Terebratulidina) and of the superfamilies Zeilleriacea and Terebratellacea (Terebratellidina). For example, *Sphaeroidothyris* (terebratulacean) is almost identical externally to *Rugitela* (zeilleriacean); also nearly indistinguishable in outer appearance are four Upper Jurassic shells (new terebratulacean

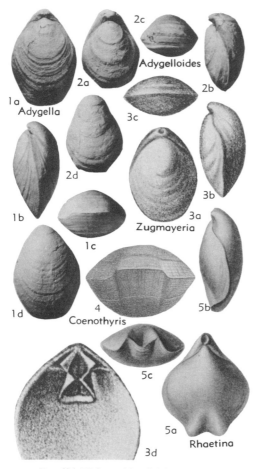

FIG. 625. Dielasmatidae (Dielasmatinae)
(p. *H768-H769*).

genus with short loop, *Cheirothyris* of the Zeilleriacea, and *Trigonellina* and *Ismenia* of the Terebratellacea) and *Tetractinella,* an Upper Triassic spiriferoid. Homeomorphs abound in the Upper Triassic, and lacking information about internal characters it is almost impossible to distinguish between spiriferoids, rhynchonelloids, zeilleriaceans, and terebratulaceans, all of which may have smooth shells and more or less sulcate brachial valves. The Cenozoic genera, *Dallithyris* (terebratuloid) and *Dallina* (long-looped dallinid), are close homeomorphs in external form.

Suborder TEREBRATULIDINA
Waagen, 1883

A diagnosis of this suborder is given in the section on Main Divisions of the Terebratulida (p. *H730*).

Superfamily DIELASMATACEA
Schuchert, 1913

[As defined by STEHLI in Paleozoic section, p. *H754*]

Family DIELASMATIDAE Schuchert, 1913

[As defined by STEHLI in Paleozoic section, p. *H756*]

Subfamily DIELASMATINAE Schuchert, 1913

[Dielasmatinae SCHUCHERT, 1913, p. 402] [=Zugmayeridae DAGIS, 1963, p. 171]

Small to medium-sized smooth forms having centronellid loop in early growth stages but later becoming short terebratuliform, with crural processes; pedicle collar developed; beak ridges rounded. Septalial plates uniting with hinge plates and bearing crural bases and cardinal process; dorsal median septum present; with or without dental plates. *L.Carb.-U.Trias.* [Pre-Mesozoic forms included in section by STEHLI, p. *H756*.]

Adygella DAGIS, 1959, p. 25 [**A. cubanica*; OD]. Shell small, valves biconvex, rounded-pentagonal, anterior commissure plane to incipiently uniplicate; umbo short, curved, foramen small, beak ridges obscure, permesothyridid. Loop about 0.3 length of valve, with crural processes and slightly arched transverse band; hinge plates fused; deep septalium supported by short septum; inner socket ridges scarcely distinguishable in section from horizontal hinge plates; no cardinal process; dental plates short, slightly diverging. *?M.Trias., U. Trias.,* Eu.(E.Alps-Caucasus)-?N.Z.——FIG. 625,*1*; 626,*2*. **A. cubanica,* Nor., Causasus; 625,*1a-d,* brach.v., lat. ant., ped.v. views, ×1 (Muir-Wood, n); 626,*2a-v,* transv. secs., ×1.5 (210).

Adygelloides DAGIS, 1959, p. 28 [**A. labensis*; OD]. Resembling *Adygella* externally and in short loop, but differs in more tapering and incurved umbo, and internally in longer dental plates, septalial plates fused posteriorly only, and becoming suspended free in dorsal umbonal cavity, dorsal septum lacking or very short, hinge plates in section not fused, slightly concave ventrally and distinguishable from inner socket ridges. *U.Trias.(Nor.),* Eu.(Caucasus).——FIG. 625,*2*; 626,*1*. **A. labensis*; 625,*2a-d,* brach.v., lat., ant., ped.v. views, ×1 (Muir-Wood, n); 626,*1a-t,* ser. transv. secs., ×1 (210).

Coenothyris DOUVILLÉ, 1879, p. 270 [**Terebratulites vulgaris* VON SCHLOTHEIM, 1820, p. 275; OD]. Medium-sized, sulco- to biconvex, with prominent dorsal fold, anterior commissure uniplicate; umbo erect to slightly incurved, beak ridges angular, telate, permesothyridid, symphytium exposed, pedicle collar developed; shell surface commonly

with radial color bands and rare capillae. Loop
terebratulid 0.3 length of valve, given off ventrally,
with long crural processes; cardinal process short,
bilobate; hinge plates ventrally concave in section,
supported by strong dorsal septum less than 0.5

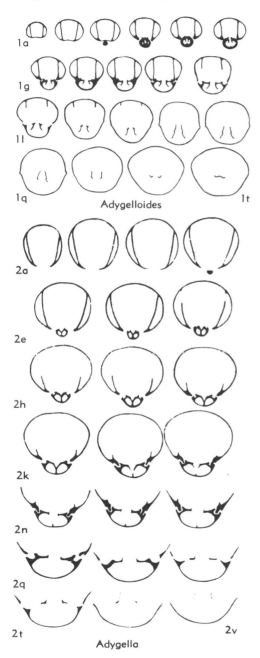

FIG. 626. Dielasmatidae (Dielasmatinae)
(p. *H768*).

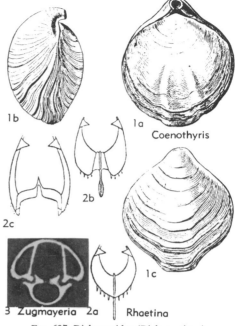

FIG. 627. Dielasmatidae (Dielasmatinae)
(p. *H768-H769*).

of valve length; crural bases prominent, demarcating rounded septalium; no dental plates. *M.Trias.
(Muschelkalk)*, Eu.-Asia.——FIG. 625,4; 627,1;
628,1. **C. vulgaris* (VON SCHLOTHEIM), Ger.; 625,
4, ant. view, ×1 (651); 627,1a-c, brach.v., lat.,
ped.v. views, ×1 (718); 628,1a-q, ser. transv.
secs., ×1.3 (651).

Rhaetina WAAGEN, 1882, p. 334 [**Terebratula
gregaria* SUESS, 1854, p. 14; OD]. Small to medium-sized, subpentagonal, biplicate, anterior
commissure sulciplicate; umbo suberect, deltidial
plates exposed, epithyridid. Loop centronellid in
early growth stages, later becoming terebratulid;
septal plates developed, dorsal septum low or absent; dental plates absent. *U.Trias.(Rhaet.)*, Eu.
(Austria - E. Alps-USSR-Caucasus).——FIG. 625,
5a-c. **R. gregaria* (SUESS), E.Alps; 5a-c, brach.v.,
lat., ant. views, ×1 (791).——FIG. 627,2; 629,1.
R. sp., USSR(Caucasus); 627,2a-c, immature, advanced immature, and adult loops, ×2 (791);
629,1a-u, ser. transv. secs., of immature form,
×3 (210).

Zugmayeria WAAGEN, 1882, p. 334 [**Terebratula
rhaetica* ZUGMAYER, 1880, p. 13; OD]. Small,
biconvex, elongate, anterior commissure plane or
incipiently uniplicate; growth lines prominent;
umbo tapering, suberect, beak ridges obscure. Loop
short, terebratulid, about 0.3 length of brachial
valve; crural processes developed; no dorsal septum; dental plates present. *U.Trias.(Rhaet.)*, Eu.
(E. Alps). —— FIG. 625,3; 627,3. **Z. rhaetica*
(ZUGMAYER); 625,3a-c, brach.v., lat., ant. views,
×2; 625,3d, loop, ×3; 627,3, transv. sec., ×3
(all 904).

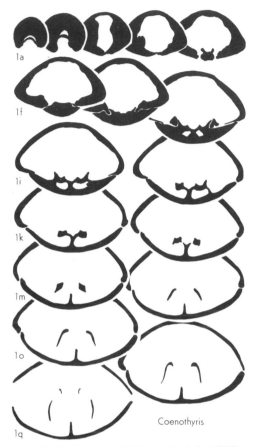

FIG. 628. Dielasmatidae (Dielasmatinae) (p. *H769*).

Subfamily NUCLEATULINAE Muir-Wood, n.subfam.

Loop centronellid or ringlike, crural processes, dental plates and dorsal median septum present or absent; brachial valve deep-ly sulcate. *U.Trias., ?L.Jur.(Lias.).*

Nucleatula BITTNER, 1888, p. 126 [**Rhynchonella retrocita* SUESS, 1855, p. 29; SD HALL & CLARKE, 1894, p. 858]. Small, concavo-convex, anterior commissure sulcate; umbo acute, incurved, beak ridges ill-defined. Loop barely more than 0.5 length of valve. Free vertical longitudinally ridged and fimbriated median plate projecting beyond loop; crural processes developed; no dorsal septum or ?dental plates; ?punctate in external shell layers only. *U.Trias.(Nor.),* Eu.(Austria-Alps).——FIG. 630,2. **N. retrocita* (SUESS); *2a-d,* brach.v., lat., ant., ped.v. views, ×2 (76); *2e,f,* loop, ×1.5 (75).

Dinarella BITTNER, 1892, p. 24 [**D. haueri*; OD]. Small, valves slightly convex, brachial valve with anterior sulcus, pedicle valve with corresponding fold and linguiform extension, anterior commissure sulcate; umbo acute, foramen small, beak ridges angular. Loop short, centronellid, descending branching uniting with median plate free of valve floor, dorsal median septum short, free of loop; dental plates weak; ?punctate in external shell layers only. *U.Trias.(Nor.),* Eu.(Bosnia-E. Alps).——FIG. 630,3. **D. haueri, 3a-d,* brach.v., lat., ant., ped.v. views, ×2; *3e,* loop, ×2 (77).

Propygope BITTNER, 1890, p. 210 [**Terebratula (Propygope) hagar*; OD]. Small, aulacothyridid, brachial valve with broad sulcus and long tapering linguiform extension, anterior commissure sulcate; umbo suberect, foramen small, beak ridges ill-defined. Loop almost ringlike, about 0.3 valve length; dorsal septum strong, less than 0.5 valve length; dental plates lacking. *U.Trias. (Carn.-Nor.),* Eu.(E.Alps-Austria-Yugosl.); *?Lias.,* Eu.——FIG. 630,*1.* **P. hagar* (BITTNER), E.Alps; *1a-d,* brach.v., lat., ant., ped.v. views, ×2; *1e,* loop, ×2 (76).

Subfamily JUVAVELLINAE Bittner, 1896

[*nom. correct.* MUIR-WOOD, herein (*pro* Juvavellinen BITTNER, 1896, p. 132)]

Shell biconvex, smooth, loop centronellid or ringlike, no crural processes, dental plates

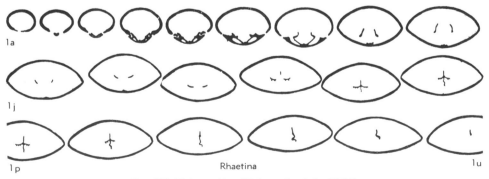

FIG. 629. Dielasmatidae (Dielasmatinae) (p. *H769*).

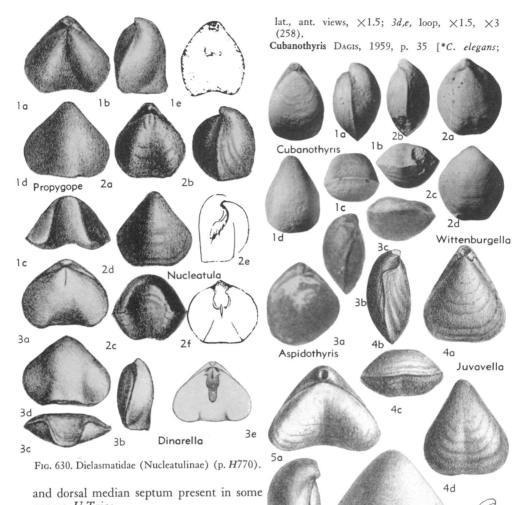

FIG. 630. Dielasmatidae (Nucleatulinae) (p. *H770*).

and dorsal median septum present in some genera. *U.Trias.*

Juvavella BITTNER, 1888, p. 127 [**J. suessi*; OD]. Small, valves biconvex, subtrigonal, shallow ventral sulcus, anterior commissure normally plane or incipiently uniplicate. Loop centronellid, about 0.25 valve length, with short median plate projecting posteriorly, no crural processes; dorsal septum and dental plates lacking. *U.Trias.(Nor.)*, C.Eu.——FIG. 631,4. **J. suessi*; *4a-d,* brach.v., lat., ant., ped.v. views, ×1.5 (76); *4e,f,* brach. loop, ×1.5 (75).

Aspidothyris DIENER, 1908, p. 58 [**A. krafft i*; OD]. Small, valves moderately convex, anterior commissure plane or incipiently uniplicate; umbo strongly incurved, almost concealing deltidial plates, other beak characters unknown. Loop centronelliform, with long median plate extending dorsally, dorsal septum about 0.3 length of valve, not supporting loop; septal plates developed, dental plates strong. *U.Trias.(Carn.)*, Asia(Himalayas).——FIG. 631,3. **A. kraffti*; *3a-c,* brach.v.,

lat., ant. views, ×1.5; *3d,e,* loop, ×1.5, ×3 (258).

Cubanothyris DAGIS, 1959, p. 35 [**C. elegans*;

FIG. 631. Dielasmatidae (Juvavellinae) *(1-4)*, (Subfamily Uncertain) *(5)* (p. *H771-H772*).

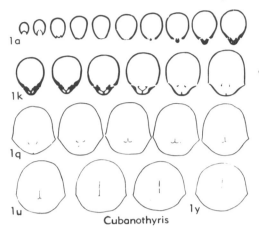

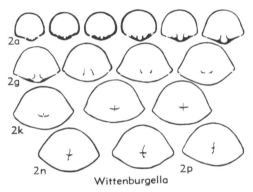

Fig. 632. Dielasmatidae (Juvavellinae)
(p. *H771-H772*).

OD]. Biconvex, valves without median sulci, anterior commissure plane. Loop with median centronellid plate extending ventrally; dorsal median septum present, deep septalium; no dental plates. *U.Trias.(Nor.)*, Eu.(NW.Caucasus).——Fig. 631,*1*; 632,*1*. *C. elegans*; 631,*1a-d*, brach.v., lat., ant., ped.v. views of holotype, ×1; 632,*1a-y*, ser. transv. secs., ×1 (210).

Juvavellina BITTNER, 1896, p. 132 [*P. kittli*; OD (M)]. Differs from *Juvavella* in greater length of loop, which equals half length of valve, and in incipient sulcation of anterior commissure. *U. Trias.(Nor.)*, Eu.(E.Alps).

Wittenburgella DAGIS, 1959, p. 30 [*W. minuta*; OD]. Small, valves biconvex, subpentagonal, anterior commissure incipiently sulcate; umbo short, curved, in contact with brachial valve, mesothyridid. Loop centronelliform, half of valve length, with high median plate extending dorsoventrally; no dorsal septum or dental plates, hinge plates divided, concave ventrally, becoming U-shaped. *U.Trias.(Nor.)*, Eu.(NW.Caucasus).——Fig. 631,*2*; 632,*2*. *W. minuta*; 631,*2a-d*,

holotype, brach.v., lat., ant., ped.v. views of holotype, ×2; 632,*2a-p*, ser. secs., ×2 (210).

Subfamily UNCERTAIN

Cruratula BITTNER, 1890, p. 66 [*Waldheimia eudora* LAUBE, 1866, p. 8; OD]. Medium-sized, with broad dorsal median sulcus, anterior commissure gently sulcate; umbo produced, slightly incurved, beak ridges obscure, permesothyridid. Loop imperfectly known, ?short; dorsal septum strong, 0.5 of valve length; dental plates absent or fused with thickened shell wall. Shell surface ?papillate or perforate. *M.Trias.-U.Trias.*, Eu.(E.Alps)-Asia (Himalayas).——Fig. 631,*5*. *C. eudora* (LAUBE), E.Alps; *5a-d*, brach.v. ext., lat., ant., ped.v. views, ×1; *5e*, incomplete loop, ×1 (76).

Pseudokingena BÖSE & SCHLOSSER, 1900, p. 177 [*Terebratulina deslongchampsi* DAVIDSON, 1850, p. 450; OD]. Small, rounded or quadrate, valves unequally convex, some brachial valves with shallow sulcus, anterior commissure plane or slightly waved; umbo short, palintrope well defined, beak ridges hypothyridid. Deltidial plates narrow, disjunct, pedicle collar present; shell surface granular, with 2 sizes of tubercles; inner shell surface capillate, especially around margin; loop short, centronellid, given off from socket ridges, about half of valve length, with short median plate, and crural processes; dorsal septum low, short; hinge plates fused, wide, gently concave ventrally, with median elevation; dental plates absent. *L.Jur.(M.Lias.-U.Lias.)*, Eu.(Eng.-Fr.-Italy). —— Fig. 633,*1*. *P. deslongchampsi* (DAVIDSON), M.Lias., Fr.; *1a*, brach.v. ext., ×4; *1b,c*, brach.v. lat., ant. views, ×2; *1d*, ped.v.

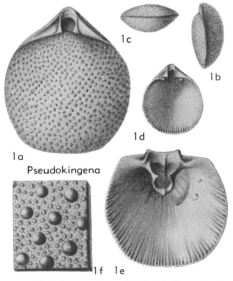

Fig. 633. Dielasmatidae (Subfamily Uncertain)
(p. *H772-H773*).

int., ×2; *1e*, brach.v. int. and loop, ×4; *1f*, ornament, enlarged (227).

Superfamily TEREBRATULACEA Gray, 1840

[*nom. transl.* SCHUCHERT & LEVENE, 1929, p. 22 (*ex* Terebratulidae GRAY, 1840, p. 143) (*non* WAAGEN, 1883, as suborder)]

Cardinal process and outer hinge plates commonly developed, inner hinge plates in some genera, or hinge plates absent; dental and septalial plates rarely developed except in early forms; adult lophophore trocholophous, schizolophous, spirolophous, subplectolophous or plectolophous. *U.Trias.-Rec.*

Family ORTHOTOMIDAE Muir-Wood, 1936

[Orthotomidae MUIR-WOOD, 1936, p. 224]

Small shells having short terebratulid loop, dental plates absent; adult shells hypothyridid, with triangular delthyrium bordered by jugate deltidial plates below tapering, acute umbo; shell rarely capillate. *L.Jur. (M.Lias.).*

Orthotoma QUENSTEDT, 1869, p. 315 [**Terebratula heyseana* QUENSTEDT, 1869, p. 315 (*non* DUNKER, 1847) =*Orthotoma spinati* RAU, 1905, p. 54; SD S. S. BUCKMAN, 1918, p. 96] [=*Orthoidea* FRIREN, 1876, p. 1 (type, *O. liasina*)]. Small, valves biconvex, becoming sulco-convex, anterior commissure rectimarginate to sulcate; umbo suberect to incurved, beak ridges angular, defining palintrope. Loop 0.3 length of valve, with low arched transverse band; cardinal process minute, projecting vertically as 2 small ears; hinge plates in transverse section, ventrally convex, dorsally inclined, tapering, not differentiated from inner socket ridges; adductor scars trigonal. *L.Jur.(M. Lias.)*, Eu. (Fr.-Ger.).——FIG. 634,*1a-d.* **O. spinati*, RAU, Lias., Ger. (Württemberg); *1a-c*, brach.v., lat., ant. views, ×4; *1d*, loop, ×2.5 (578).——FIG. 634,*1e-q. O. quenstedti*, M.Lias., Ger.; *1e*, umbonal region, ×2.5; *1f*, internal mold with dorsal adductor scars, ×2.5; *1g-q*, ser. transv. secs. at 0.1-0.3 mm. intervals, ×5 (578).

Family TEREBRATULIDAE Gray, 1840

[Terebratulidae GRAY, 1840, p. 143]

Valves smooth or with growth lamellae, semiplicate or part capillate; loop terebratulid, crural processes not united to form ringlike loop, outer hinge plates present, and inner hinge plates also in some genera; dorsal median septum and dental plates absent; lophophore plectolophous and filament spicules present in some Recent genera. *U.Trias.-Rec.*

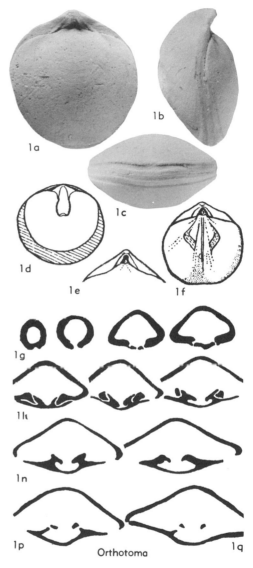

FIG. 634. Orthotomidae (p. H773).

Subfamily TEREBRATULINAE Gray, 1840

[*nom. transl.* WAAGEN, 1883, p. 330 (*ex* Terebratulidae GRAY, 1840, p. 143)] [=Gryphinae SAHNI, 1929, p. 8]

Small to large biconvex shells, or with brachial valve flat or concave, or rarely more convex than pedicle valve; smooth or partly capillate; anterior commissure normally plane, uniplicate or biplicate, rarely sulcate; beak ridges usually obscure, foramen mesothyridid to epithyridid. [Classification under review.] *U.Trias.-Rec.*

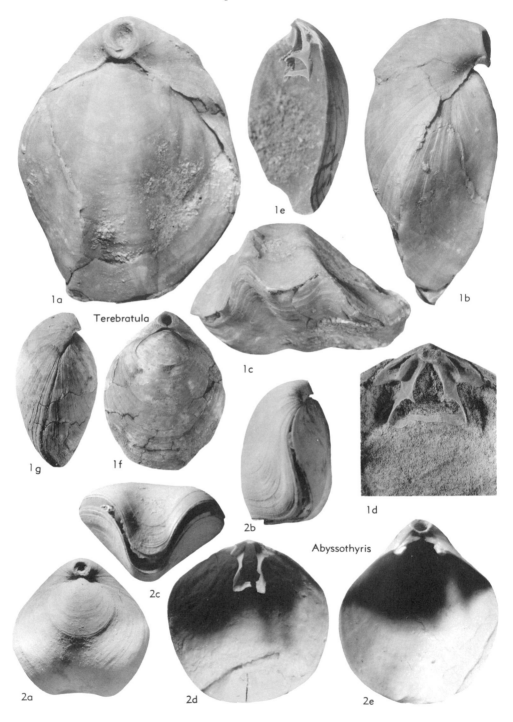

Fig. 635. Terebratulidae (Terebratulinae) (p. *H775-H777*).

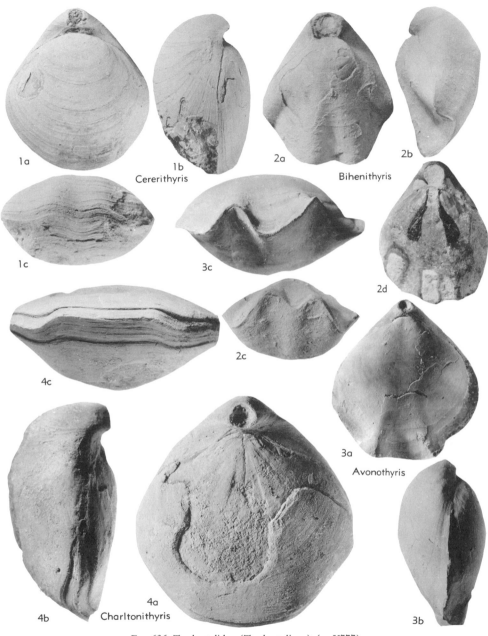

FIG. 636. Terebratulidae (Terebratulinae) (p. *H777*).

Terebratula MÜLLER, 1776, p. 249 [**Anomia terebratula* LINNÉ, 1758, p. 703; SD LAMARCK, 1799, p. 89]. Medium-sized to large, valves biconvex, anteriorly biplicate, anterior commissure uniplicate to sulciplicate; umbo short, massive, suberect to incurved, foramen mesothyridid to permesothyridid, symphytium narrow, commonly concealed, pedicle collar developed; shell smooth but growth lines prominent. Loop broadly triangular, about 0.25 to 0.33 of valve length, with narrow-ribboned, arched, and medially flattened transverse band, crural bases extending along edge of outer hinge plates, no inner hinge plates, crural processes long, tapering, cardinal process rounded, pos-

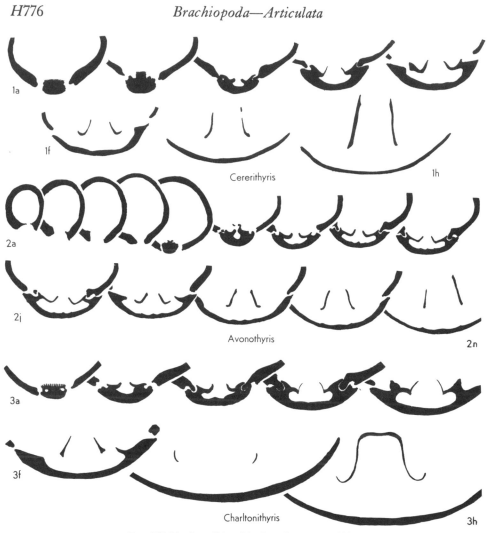

FIG. 637. Terebratulidae (Terebratulinae) (p. *H777*).

teriorly flattened boss, hinge plates concave, separated from prominent socket ridges by deep sulcus, rare short median septum; hinge teeth with swollen bases and posteriorly sulcate. [*Anomia terebratula* LINNÉ was not one of the species listed by MÜLLER in 1776, but was subsequently designated as type-species by LAMARCK in 1799. This case, as interpreted by BUCKMAN (1907) should be put to the ICZN for ratification.] *Mio.-Plio.,* Eu.——FIG. 635,*1a-c.* **T. terebratula* (LINNÉ), Plio. (Asti.), Italy(Rome); *1a-c,* brach. v., lat., ant. views, ×1.5 (696).——FIG. 635, *1d-g. T. ampulla* BROCCHI, Plio., Italy; *1d,e,* brach.v. int., lat. view of loop, ×0.9, ×1.5; *1f,g,* brach.v. and lat. views, ×0.8 (696).

Abyssothyris THOMSON, 1927, p. 190, *emend.* MUIR-WOOD, 1960, p. 521 [**Terebratula wyvillei* DAVIDSON, 1878, p. 436; OD]. Small, thin, trilobate,

dorsal valve anteriorly sulcate, pedicle valve carinate, anterior commissure ventrally uniplicate (sulcate); shell smooth except for growth lines; umbo short, slightly incurved, epithyridid, symphytium narrow, pedicle collar developed. Loop terebratulid, crura subparallel, crural processes short, blunt, transverse band broad, slightly arched, cardinal process transversely elongate, ridged, outer hinge plates depressed, bounded by elevated socket ridges, no inner hinge plates; lophophore plectolophous with small median coil. [THOMSON (1927) confused the terebratulid genus *Abyssothyris* with the rhynchonellid genus *Neorhynchia* in his original diagnosis. Subsequently *Abyssothyris* was redefined and two homeomorphs disentangled by MUIR-WOOD (1960).] *U.Mio.* or *L.Plio.,* Fiji I.; *Plio.,* Eu.(Italy); *Rec.,* Pac.O. (off S.Australia-N. Guinea-Galapagos I.-Chile).——FIG. 635,2. **A.*

wyvillei (DAVIDSON), Rec., off S. Australia; *2a-c*, brach.v., lat., ant. views of lectotype, ×3.1; *2d,e*, Rec., off Chile, brach.v. int. with loop, ped.v. int., ×3.1 (585).

Avonothyris S. S. BUCKMAN, 1918, p. 102 [*A. plicatina*; OD]. Small to medium-sized, biconvex, some sulcocarinate posteriorly, anterior commissure rectimarginate to episulcate; shell surface rarely capillate, growth lines numerous; umbo short, suberect, epithyridid; symphytium narrow, pedicle collar developed. Cardinal process low, lobate with or without posterior umbonal cavity, hinge plates in section dorsally deflected and ventrally concave, tapering, becoming U-shaped; adductor scars narrow, subparallel. *M.Jur.(Bathon.),* Eu. (Eng.-Fr.).——FIG. 636,*3*; 637,*2*. *A. plicatina*, Bradford, Eng. (Wilts.); 636,*3a,b*, brach.v., lat., ×1.25, *3c*, ant. view of holotype, ×1.5; 637, *2a-n*, transv. secs., ×1.8 (136).

Bihenithyris MUIR-WOOD, 1935, p. 110 [*B. barringtoni*; OD]. Medium-sized, biconvex, anterior commissure sulciplicate to episulcate; umbo massive, suberect to incurved, concealing symphytium, epithyridid, pedicle collar developed. Loop short, less than half of valve length; cardinal process short, broad, medianly depressed; no posterior umbonal cavity; hinge plates and inner socket ridges in section posteriorly thickened, gently concave, becoming U-shaped, tapering; dorsal adductor scars posteriorly threadlike, rapidly expanding and diverging. *U.Jur.(Callov.),* Afr. (Somaliland)-Asia(Arabia).——FIG. 636,*2*; 638,*1*. *B. barringtoni*, Somaliland (Madashon); 636,*2a-c*, brach.v., lat., ant. views of holotype, ×1.25; 636, *2d*, int. mold of paratype with adductor scars, ×1.25; 638,*1a-r*, ser. transv. secs., ×1.25 (577).

Cererithyris S. S. BUCKMAN, 1918, p. 109 [*Terebratula intermedia* J. SOWERBY, 1813, p. 48; OD] [=*Cererithyris* BUCKMAN, 1914, p. 2 (*nom. nud.*)]. Medium-sized, sulco- to plano- to biconvex, anterior commissure rectimarginate to uniplicate or sulciplicate; umbo short, stout, foramen marginate to labiate, epithyridid, symphytium well exposed. Loop half of valve length, transverse band with high arch; cardinal process low, short; no posterior umbonal cavity; hinge plates in transverse section not well demarcated from long inner socket ridges, ventrally concave, tapering, becoming V-shaped to U-shaped; adductors long, widely divergent. *M.Jur.(Bathon.),* Eu.—— FIG. 636,*1*; 637,*1*. *C. intermedia* (J. SOWERBY), Eng.; 636,*1a-c*, brach.v., lat., ant. views of lectotype, ×1.25; 637,*1a-h*, ser. transv. secs., ×1.8 (136).

Charltonithyris S. S. BUCKMAN, 1918, p. 106 [*Terebratula uptoni* S. S. BUCKMAN, 1895, p. 455; OD] [=*Charltonithyris* BUCKMAN, 1915, p. 78 (*nom. nud.*)]. Medium-sized to large, rounded, plano- to moderately biconvex, anterior commissure rectimarginate to uniplicate, rarely sulciplicate; umbo incurved, slightly carinate, foramen

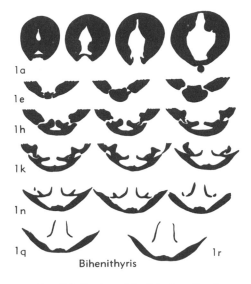

FIG. 638. Terebratulidae (Terebratulinae) (p. *H777*).

large, beak ridges strong, laterally extended, permesothyridid, symphytium exposed. Loop with highly arched, medianly horizontal, transverse band; cardinal process low, short; hinge plates in section dorsally deflected, distinguishable from inner socket ridges, tapering, V-shaped, crural bases virgate, keeled; adductor scars diverging, tapering posteriorly. *M.Jur.(L. M. Inferior Oolite),* Eu. (Eng.).——FIG. 636,*4*; 637,*3*. *C. uptoni* (BUCKMAN), *M.Inf.Ool.,* Eng.(Glos.); 636,*4a-c*, brach.v., lat., ant. views of paratype, ×1.25; 637,*3a-h*, ser. transv. secs., ×1.8 (127).

Cnismatocentrum DALL, 1920 (1921), p. 321 [*Terebratula (Liothyris) sakhalinensis* DALL, 1908, p. 28; OD]. Medium-sized, stout, biconvex, anterior commissure uniplicate, umbo stout, slightly incurved, symphytium exposed, foramen entire, epithyridid, pedicle collar with short septum; surface smooth or anteriorly capillate, with prominent growth lines. Loop very wide, slender, almost flattened, transverse band in same plane as loop, attached to wall of valve for some distance; crural processes short; cardinal process small, prominent; outer hinge plates narrow; dorsal median septum low; lophophore plectolophous. *Rec.,* Asia(Sakhalin I.-Okhotsk Sea)-N.Am.(Alaska).——FIG. 639,*1*. *C. sakhalinensis* (DALL), off Sakhalin; *1a-d*, brach.v., lat., ant., post. views of holotype, ×1 (427).

Dallithyris MUIR-WOOD, 1959, p. 302 [*D. murrayi*; OD]. Medium-sized to large, subtrigonal to subpentagonal, pedicle valve more convex than brachial; no median fold or sulcus, anterior commissure plane to uniplicate, lateral commissure dorsally convex; umbo short massive, foramen

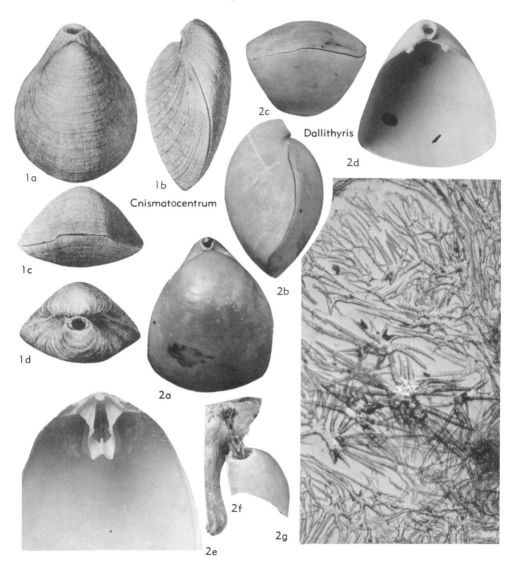

FIG. 639. Terebratulidae (Terebratulinae) (p. *H777-H778*).

epithyridid, symphytium short; pedicle collar short; shell surface smooth or with irregularly developed striations. Loop narrow, transverse band with broad ribbon having sharp median plication; crural bases extending along inner margins of concave outer hinge plates; cardinal process small transverse plate; inner socket ridges narrow, prominent, well demarcated from hinge plates; mantle canals much branched, adductor scars dendritic. *?U.Eoc.,* Eu.(Italy); *Mio.,* S.Eu.; *Rec.,* Carib.-E.Atl.O.-Medit.-Ind.O.(off Maldive I.-Mauritius)-E. Pac. O.(off Japan).——FIG. 639,2. **D. murrayi,* Rec., Ind. O. (Maldive Is.); *2a-c,*

brach.v., lat., ant. views of holotype, ×1; *2d,e,* ped.v. int., brach.v. int., ×1, ×2; *2f,* attachment by pedicle composed of separate strands, ×3; *2g,* spicules of mantle, ×25 (584).

Epithyris PHILLIPS, 1841, p. 55 [**Terebratula maxillata* J. DE C. SOWERBY, 1823, p. 52; SD BUCKMAN, 1906, p. 321] [*non Epithyris* KING, 1850, p. 146 (=*Dielasma* KING, 1859)]. Medium-sized to large, valves plano- to biconvex, anterior commissure plane to quadriplicate; umbo produced, becoming incurved, beak ridges subangular, epithyridid in adult, symphytium short, pedicle collar present. Loop about half length of valve with high-arched

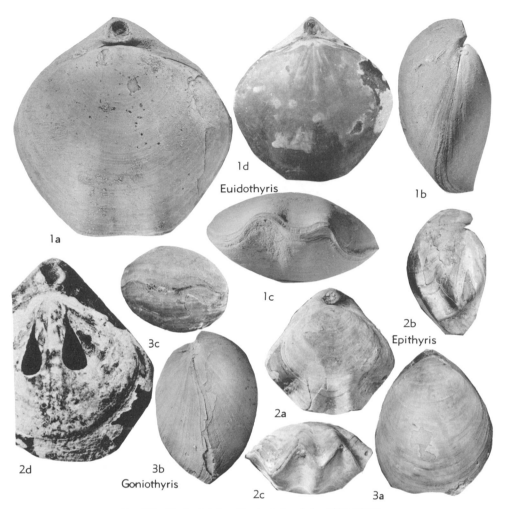

1d
Euidothyris
1b
1a
1c
3c
2b
Epithyris
3a
2a
2d
3b
Goniothyris
2c
3a

FIG. 640. Terebratulidae (Terebratulinae) (p. *H778-H780*).

transverse band; cardinal process small, bilobed; hinge plates in section scarcely demarcated from inner socket ridges, ventrally convex, with slight dorsal deflection, keeled; adductor scars elongate, pear-shaped. *M.Jur.(Bathon.)*, Eu.(Eng.-Fr.).——FIG. 640,2a-c; 641,2. *E. maxillata* (J. DE C. SOWERBY), Fullers Earth Rock, Eng. (Somerset); 640,2a-c, brach.v., lat., ant. views of holotype, ×1.2; 641,2a-j, ser. transv. secs., ×1.25 (579). ——FIG. 640,2d. E. oxonica ARKELL, Gt. Ool., Eng.; brach.v. int. mold showing adductor scars, ×1.2 (579).

Euidothyris S. S. BUCKMAN, 1918, p. 101 [*Terebratula euides* (broad form) BUCKMAN, 1886, p. 218 (=*E. extensa* BUCKMAN, 1918, p. 101); OD] [=*Euidothyris* BUCKMAN, 1915, p. 78 *(nom. nud.)*]. Medium-sized, valves sulcocarinate posteriorly, becoming biconvex, anterior commissure uniplicate to sulciplicate, umbo produced, laterally constricted, beak ridges long, conspicuous, epithyridid in adult. Loop about half of valve length; cardinal process trilobed, short; umbonal cavity present; hinge plates in transverse section well demarcated from inner socket ridges, V-shaped, tapering, crural bases virgate; adductor scars divergent. *M. Jur. (L. Inferior Oolite)*, Eu. (Eng.-Fr.).——FIG. 640, 1; 641,1. *E. extensa* BUCKMAN, Eng.; 640,1a-c, brach.v., lat., ant. views of holotype, ×1.5, ×1.2, ×1.2; 640,1d, brach.v. int. mold of paratype showing adductor scars, ×1.2; 641,1a-g, ser. transv. secs., ×1.25 (136).

Goniothyris S. S. BUCKMAN, 1918, p. 117 [*Terebratula gravida* SZAJNOCHA, 1881, p. 74; OD] [=*Goniothyris* BUCKMAN, 1914, p. 2 *(nom. nud.)*]. Medium-sized to large, trigonal, brachial valve highly convex, pedicle valve flat to convex

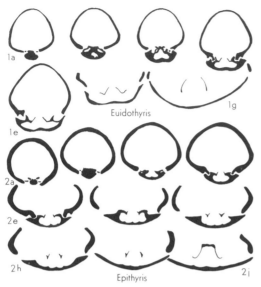

FIG. 641. Terebratulidae (Terebratulinae)
(p. *H778-H779*).

or carinate, anterior commissure plane, lateral
commissure dorsally curved; umbo very short,
foramen apical, epithyridid, beak ridges obscure,
symphytium narrow. Loop unknown; cardinal
process very small; hinge plates in section slightly
convex ventrally and deflected dorsally, keeled;
adductor scars almost parallel. *M.Jur.(M. Inferior
Oolite)*, Eu. (Eng.-Czech.-Aus.-Hung.) —— FIG.
640,3. *G. dorsetensis* (ROLLIER) (=*G. gravida*
DAVIDSON, 1884, and S. S. BUCKMAN, 1918, *non*
SZAJNOCHA, 1881), Blagdeni Zone, Eng.(Dorset);
3a-c, brach.v., lat., ant. views, ×1.2 (136).

Gryphus MEGERLE VON MUHLFELD, 1811, p. 64
[*Anomia vitrea* BORN, 1778, p. 104; OD (M)]
[not preocc. *Gryphus* BRISSON, 1760, not gen.]
[=*Liothyris* DOUVILLÉ, 1879, p. 265 (*non* CON-
RAD, 1875)*; Liothyrina* OEHLERT in FISCHER, 1887,
p. 1316]. Small to medium-sized, circular to subpentagonal, biconvex; anterior commissure plane
to incipiently uniplicate, lateral commissure vertical; surface smooth, rare fine capillation on flanks
and numerous growth lines; umbo short, suberect
to incurved, epithyridid, symphytium almost concealed, pedicle collar developed. Loop about 0.25
of valve length, descending branches slightly diverging, transverse band broad-ribboned, ventrally arched; crura very short, crural bases extending along inner margins of slightly concave
outer hinge plates, no inner hinge plates; cardinal
process small transverse plate, myophragm rare;
hinge teeth excavated by posteriorly placed socket,
spicules widely distributed, main mantle canals
almost straight, branching anteriorly, *?Eoc., Oligo.,*
USA; *Mio.-Plio.,* Sicily; *Rec.,* Medit.-Atl.O.——
FIG. 643,4. *G. vitreus* (BORN), Rec., Medit.; *4a-c,*

brach.v., lat., ant. views, ×1.2; *4d,* brach.v. int.,
×1.2 (810).

Heimia HAAS, 1890, p. 87 [*Terebratula mayeri*
CHOFFAT in HAAS, 1883, p. 254; OD]. Small to
large, valves plano- to incipiently sulco-convex or
carinate, anterior commissure sulcate or paraplicate; umbo short, stout, slightly incurved, permesothyridid, symphytium concealed, shell with
numerous growth lines or lamellae, and anteriorly
thickened. Loop unknown; cardinal process short,
prominent; umbonal cavity variably developed;
hinge plates in section well differentiated from
inner socket ridges, slightly concave ventrally,
clubbed, becoming thin, beveled, and rarely V-
shaped. *M.Jur.(Bajoc., Inferior Oolite)*, Eu.(Eng.-
Fr.-Switz.).——FIG. 642,1. *H. mayeri* (CHOF-
FAT), Switz.; *1a,b,* brach.v., ant. views, ×1; *1c-e,*
brach.v., lat., ant. views, ×1 (376).

Holcothyris S. S. BUCKMAN, 1918, p. 125 [*H.
angulata*; OD] [=*Holcothyris* BUCKMAN, 1915,
p. 78 (*nom. nud.*)]. Small to medium-sized, subpentagonal, valves moderately biconvex, with
continuous median dorsal furrow and ventral
fold or carination, anterior commissure sulcate to
paraplicate; umbo massive, short, symphytium
usually concealed, epithyridid; shell fully capillate.
Cardinal process short, bilobate, medianly depressed; myophragm long; hinge plates in section
slightly demarcated from inner socket ridges,
slightly concave ventrally, clubbed; adductor scars
narrow, tapering posteriorly, widely divergent.
M. Jur. (Bathon.), Asia (Burma). —— FIG. 642,2.
H. angulata, Namyau F.; *2a-c,* brach.v., lat.,
ant. views of holotype, showing capillation, ×1;
2d, brach.v. int. mold showing adductor scars,
×1 (94).

Jaisalmeria SAHNI & BHATNAGAR, 1958, p. 421 [*J.
taylori*; OD] [=*Jaisalmeria* SAHNI, 1955, p. 187
(*nom. nud.*)]. Small to medium-sized, valves
moderately biconvex, anterior commissure plane
to uniplicate, to incipiently biplicate; umbo thin,
erect, foramen small, incomplete, beak ridges angular, submesothyridid, deltidial plates disjunct;
ornament of fine capillae with intercalations. Loop
unknown; fine dorsal median septum; adductor
scars narrowly pear-shaped. *U.Jur.(?Portland.)*,
Asia(India-Pak.).——FIG. 642,3a-d. *J. taylori,*
India; *3a-c,* brach.v., ant., lat. views, ×1; *3d,*
ped.v. ext., ×2 (700).——FIG. 642,3e,f. *J. depressa* SAHNI & BHATNAGAR, Pak. (Cutch); *3e,f,*
brach.v., ped.v. views showing capillae, ×1
(700).

Juralina KYANSEP, 1961, p. 28 [*J. procerus*; OD].
Medium-sized, pedicle valve much more convex than brachial valve, anterior commissure
plane to uniplicate; umbo massive, produced and
projecting over brachial valve, beak ridges obscure,
mesothyridid, symphytium high, pedicle collar
present. Loop about 0.3 of valve length, with
arched transverse band; cardinal process prominent, medianly depressed; posterior umbonal cav-

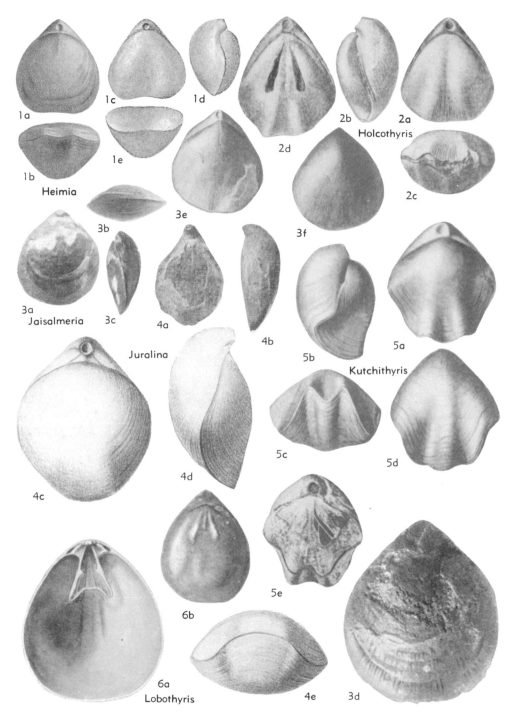

Fig. 642. Terebratulidae (Terebratulinae) (p. *H780, H783-H784*).

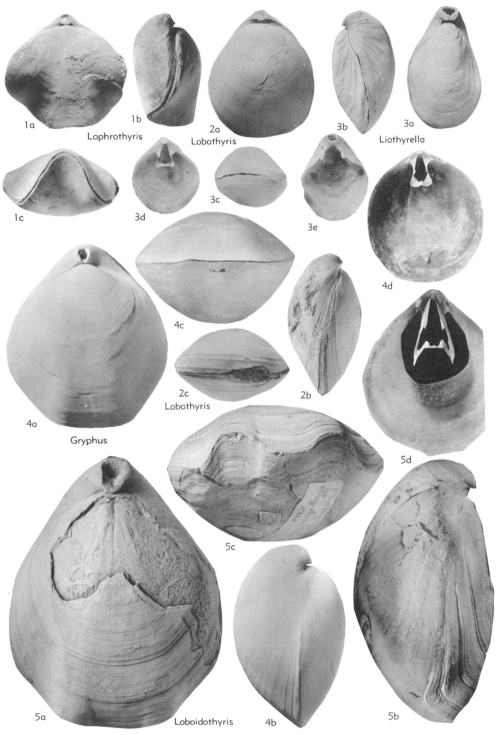

1a 1b Lophrothyris 2a Lobothyris 3b 3a Liothyrella 1c 3d 3c 3e 4d 4c 2c Lobothyris 2b 5d 4a Gryphus 5c 5a Loboidothyris 4b 5b

FIG. 643. Terebratulidae (Terebratulinae) (p. *H780, H783-H785*).

ity present; myophragm low; hinge plates in section not well differentiated from inner socket ridges, ventrally concave, becoming V-shaped, tapering, commonly keeled; adductor scars rounded-trigonal, threadlike posteriorly. *U.Jur. (Oxford.-Kimmeridg.),* W. Eu. - E. Eu. (Caucasus-Crimea).——Fɪɢ. 642,*4a,b*; 644,*1. *J. procerus,* L. Kimmeridg., Crimea; 642,*4a,b,* brach.v., lat. views, ×1; 644,*1a-o,* ser. transv. secs., ×1 (496).——Fɪɢ. 642,*4c-e. J. cotteaui* (Douvɪʟʟᴇ́), U.Jur.(U. Oxford.), Fr.; *4c-e,* brach.v., lat., ant. views, ×1 (264).

Kutchithyris S. S. Bᴜᴄᴋᴍᴀɴ, 1918, p. 113 [**Terebratula acutiplicata* Kɪᴛᴄʜɪɴ, 1897, p. 9; OD] [*=Kutchithyris* Bᴜᴄᴋᴍᴀɴ, 1915, p. 78 *(nom. nud.)*]. Medium-sized, valves unequally biconvex, biplicate, anterior commissure uniplicate to sulciplicate; umbo short, incurved, obliquely truncate, foramen large, circular, permesothyridid?, sym-

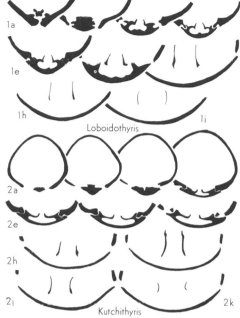

Fɪɢ. 645. Terebratulidae (Terebratulinae) (p. *H783-H784*).

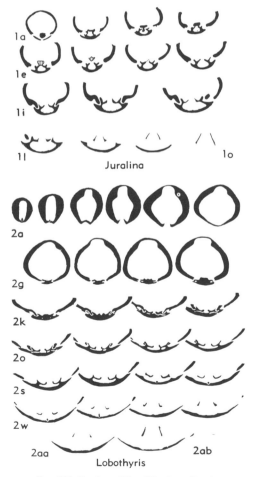

Fɪɢ. 644. Terebratulidae (Terebratulinae) (p. *H780, H783-H784*).

phytium rarely seen; growth lines numerous, shell rarely capillate. Loop about 0.3 of valve length, crural processes long; cardinal process low, short; no posterior umbonal cavity; hinge plates in section not well demarcated from long inner socket ridges, dorsally inclined, gently concave ventrally, clubbed to bladelike; adductor scars sharply divergent, posteriorly threadlike. *U.Jur.(Callov.),* Asia(India-Pak.). —— Fɪɢ. 642,*5*; 645,*2. *K. acutiplicata* (Kɪᴛᴄʜɪɴ); 642,*5a-d,* brach.v., lat., ant., ped.v. views of lectotype (Kɪᴛᴄʜɪɴ, 1897, pl. 1, fig. 1a-d) (herein designated), ×1; 642,*5e,* brach.v. view showing adductor scars, ×1; 645, *2a-k,* ser. transv. secs., ×1.3 (136).

Liothyrella Tʜᴏᴍsᴏɴ, 1916, p. 44 [**Terebratula uva* Bʀᴏᴅᴇʀɪᴘ, 1883, p. 124; OD]. Small to large, ovate, biconvex; anterior commissure plano- to sulciplicate; umbo short, foramen epithyridid, symphytium narrow, pedicle collar short; shell surface smooth or finely capillate, growth lines prominent. Loop short, broadly triangular, 0.3 length of valve, crural bases in contact with triangular crural processes without any crura; descending branches diverging, transverse band narrow-ribboned, slightly arched ventrally, crural bases extending along margin of concave outer hinge plates; cardinal process low, laterally extended plate; hinge teeth narrow with shallow sulcus. *Mio.-Plio.,* Italy-Alg.-N.Z.-Tasmania-S.Am.; *Rec.,* Arct.-E. Atl.-W. Atl.-S. Ind. O.-Pac.O.(off C.

FIG. 646. Terebratulidae (Terebratulinae) (p. *H785-H787*).

Am.-S.Am.)-Antarctic.——FIG. 643,*3*. **L. uva* (BRODERIP), Rec., off Mexico; *3a-c*, brach. v., lat., ant. views of holotype, ×1.2; *3d,e*, Rec., Falkland I.; brach.v. int. with loop, ped.v. int., ×1.2 (810).

Loboidothyris S. S. BUCKMAN, 1918, p. 112 [**L. latovalis*; OD] [=*Loboidothyris* BUCKMAN, 1914, p. 2 *(nom. nud.)*]. Medium-sized to large, biconvex, anterior commissure uniplicate to sulciplicate; umbo massive, short, foramen large, commonly labiate, epithyridid, beak ridges inconspicuous, symphytium hidden; valves rarely capillate anteriorly. Loop less than 0.5 of valve length; cardinal process low, short, lobate; no posterior umbonal cavity; hinge plates in section almost horizontal, well differentiated from socket ridges, slightly concave ventrally, keeled; crural bases virgate; adductor scars broad, tapering posteriorly, slightly divergent. *M.Jur.(Bajoc.)*, Eu.(Eng.-Fr.). ——FIG. 643,*5a-c*. **L. latovalis*, Eng.(Dorset); *5a-c*, brach.v., lat., ant. views of holotype, ×1.2

(136).——FIG. 643,*5d*; 645,*1*. *L. perovalis* (J. DE C. SOWERBY, Fr.(Normandy); 643,*5d*, loop, ×1.2; 645,*1a-i*, ser. transv. secs., ×1.3 (136).

Lobothyris S. S. BUCKMAN, 1918, p. 107 [**Terebratula punctata* J. SOWERBY, 1813, p. 46; OD] [=*Lobothyris* BUCKMAN, 1914, p. 2 *(nom. nud.)*]. Medium-sized, moderately biconvex, anterior commissure uniplicate, rarely sulciplicate; umbo suberect to incurved, epithyridid, pedicle collar with short septum, symphytium short. Loop about 0.3 length of valve; cardinal process small, trilobate; no posterior umbonal cavity; hinge plates and inner socket ridges in section slightly concave ventrally, clubbed, and gently inclined dorsally; adductor scars short, narrow, spatulate, diverging. *L. Jur. (Lias.)-M. Jur. (Bajoc.)*, Eu.(Eng.-Fr.-Ger.-Spain).——FIG. 642,*6*; 643,*2*; 644,*2*. **L. punctata* (J. SOWERBY), L.Jur.(M.Lias.), Eng.; 642,*6a*, loop (reconstr.), ×1.5; 642,*6b*, dorsal adductor muscle scars, ×1; 643,*2a-c*, brach.v., lat., ant. views of holotype, ×1.2; 644,*2a-ab*, ser. transv. secs., ×1.25 (576).

Lophrothyris S. S. Buckman, 1918, p. 114 [*L. lophus*; OD] [=*Lophrothyris* Buckman, 1914, p. 2 *(nom. nud.)*]. Small to medium-sized, biconvex, trilobate, anterior commissure markedly uniplicate, rarely sulciplicate; umbo short, symphytium narrow, beak ridges angular, epithyridid, Loop ?with median plication in transverse band; cardinal process low, short; no umbonal cavity posteriorly; hinge plates in transverse section well demarcated from inner socket ridges, gently concave ventrally, fine, tapering, developing slight flange at late stage; adductor scars slightly divergent. *M.Jur.(Bajoc.)*, Eu.(Eng.-Fr.).——Fig. 643,*1*. *L. lophus*, Eng.(Somerset); *1a-c*, brach.v., lat. ant. views of holotype, ×1.2 (136).

Naradanithyris Tokuyama, 1958, p. 2 [*N. kuratai*; OD]. Small to medium-sized, ovate to pentagonal, biconvex, anterior commissure angularly biplicate; umbo short, massive, suberect to incurved, in contact with brachial valve, foramen large, rounded, symphytium short, usually concealed, beak ridges obscure, ?mesothyridid or permesothyridid, pedicle collar absent; shell rarely capillate. Internal characters imperfectly known, owing to distortion of shell; loop less than 0.5 of valve length; no median septum; cardinal process short, wide; hinge plates in transverse section dorsally inclined, thickened hammer-shaped, separated from inner socket ridges by shallow sulcus; adductor scars subparallel, long. *M.Jur.(Bajoc.-Bathon.)*, Asia(Japan); *U.Jur.*, Asia.——Fig. 646,*1*. *N. kuratai*, M.Jur., Japan; *1a-c*, brach.v., lat., ant. views of holotype, ×1 (813).

Neumayrithyris Tokuyama, 1958, p. 120 [*N. torinosuensis*; OD]. Medium-sized, biconvex, anterior commissure incipiently uniplicate; umbo short, massive, suberect to moderately incurved, beak ridges rounded, permesothyridid. Cardinal process short, medianly depressed; commonly with posterior umbonal cavity; hinge plates in section almost horizontal, scarcely distinguishable from inner socket ridges, fine, tapering, crural bases given off ventrally at angle to hinge plates; no median septum. *M.Jur.*, (Eu.); *U.Jur.(L.Malm.)*, Eu.(Crimea)-Asia(Japan).——Fig. 646,*5*; 647,*1*. *N. torinosuensis*, U.Jur., Japan; 646,*5a-c*, brach.v., lat., ant. views of holotype, ×1.5; 647,*1a-k*, ser. transv. secs., ×2.4 (814).

Oleneothyris Cooper, 1942, p. 233 [*Terebratula harlani* Morton, 1828, p. 73; OD]. Large, elongate-oval, almost parallel-sided, biconvex; brachial valve less convex than pedicle valve; anteriorly biplicate, anterior commissure uniplicate to sulciplicate; umbo incurved, almost concealing narrow symphytium, foramen large, mesothyridid to permesothyridid; surface smooth except growth lines and rare capillation. Loop about 0.3 of valve length, with inverted V-shaped transverse band ventrally directed; crural processes massive, triangular, crural bases extending along margin of concave outer hinge plates forming prominent

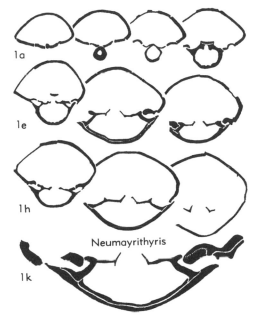

Fig. 647. Terebratulidae (Terebratulinae) (p. H785).

ridge, no crura; hinge plates extended anteriorly and joining descending branches of loop; cardinal process large, inner socket ridges elevated. *Eoc.*, USA (N.J.-Ala.-S. Car.-Del.).——Fig. 648,*1*. *O. harlani* (Morton), N.J.; *1a-c*, brach.v., lat., ant. views, ×1.3; *1d,e*, brach.v. int. with loop and lat. of same, ×0.7; *1f,g*, brach.v. int. with cardinalia, ped.v. int. with massive grooved teeth, ×0.7 (863).

Parathyridina Schuchert & Levene, 1929, p. 121 [*pro Parathyris* Douvillé, 1916, p. 35 (*non* Hübner, 1816)] [*Parathyris plicatoides* Douvillé, 1916, p. 36; OD]. Small to medium-sized, valves biconvex to spherical without prominent fold or sulcus; anterior commissure uniplicate with superimposed alternating costation, costae few, broad, near anterior margin only; umbo short, beak ridges obscure, ?permesothyridid, symphytium narrow. No dorsal median septum or dental plates, loop presumed to be terebratulid, other internal characters unknown. *M.Jur.(Bajoc.)*, Afr.(Egypt). ——Fig. 646,*4*. *P. plicatoides* (Douvillé); *4a-c*, brach.v., ant., views, ×1 (265).

Plectoconcha Cooper, 1942, p. 233 [*Rhynchonella aequiplicata* Gabb, 1864, p. 35; OD]. Small to medium-sized, subglobose, anterior commissure incipiently uniplicate, with superimposed alternating multiplication; costae regular, rounded on anterior half of shell; umbo large, incurved, foramen labiate, permesothyridid, deltidial plates not exposed, pedicle collar present. Loop with widely divergent descending lamellae, transverse band slightly arched, crural bases short, crural processes at anterior end of hinge plates, hinge plates divided. *U.Trias.* W.N.Am.(Nev.-Calif.). —— FIG. 646,2. **P. aequiplicata* (GABB), Nev.; *2a-c,* brach. v., lat., ant. views, ×1 (177).

Pseudoglossothyris S. S. BUCKMAN, 1901, p. 240 [**Terebratula curvifrons* DAVIDSON, 1878, p. 153 (*non* OPPEL, 1858, p. 423) (*=Aulacothyris leckhamptonensis* ROLLIER, 1919, p. 347); SD MUIR-WOOD, herein]. Medium-sized, valves plano- be-

FIG. 648. Terebratulidae (Terebratulinae) (p. *H785-H787*).

coming increasingly sulco-convex, anterior commissure sulcate; umbo slightly incurved, foramen telate in some, permesothyridid, symphytium exposed, lamellose anteriorly and rarely striated. Loop about 0.5 of shell length; cardinal process large, prominent, lobate, medianly excavate; no posterior umbonal cavity; hinge plates in transverse section not demarcated from inner socket ridges, almost horizontal, slightly concave ventrally, clubbed, becoming V-shaped; adductor scars long, divergent. *M.Jur.(Bajoc.)*, Eu.(Eng.-Fr.).——FIG. 648,2; 649,2. **P. leckhamptonensis* (ROLLIER), Glos.; 648,2a-c, brach.v., lat., ant. views, ×1.3; 648,2d, brach.v. int. mold showing adductor scars, ×1.3; 649,2a-i, ser. transv. secs., ×1.2 (136).

Ptyctothyris S. S. BUCKMAN, 1918, p. 101 [**Terebratula stephani* DAVIDSON, 1877, p. 12; OD] [=*Ptyctothyris* BUCKMAN, 1914, p. 2 (*nom. nud.*)]. Medium-sized, biconvex, biplicate, with prominent median ventral fold, anterior commissure rectimarginate to sulciplicate, umbo stout, incurved, foramen large, epithyridid, symphytium exposed. Loop about 0.3 of valve length; cardinal process low; no umbonal cavity; hinge plates and inner socket ridges in transverse section dorsally inclined, ventrally concave, clubbed, becoming broad U-shaped, tapering; adductor scars broad, spatulate. *M.Jur.(Bajoc.-Bathon.)*, Eu.(Eng.-Fr.). FIG. 648,3. **P. stephani* (DAVIDSON), Bajoc., Eng. (Dorset); 3a-c, brach.v., lat., ant. views, ×1.3 (136).

Rouillieria MAKRIDIN, 1960, p. 295 [**Terebratula michalkowii* FAHRENKOHL, 1856, p. 228; OD]. Large, subcircular to elongate-oval, biconvex, anterior commissure uniplicate to sulciplicate; umbo short, in contact with brachial valve, foramen large, beak ridges obscure. Cardinal process broad, lamellar, dorsal septum fine, half valve length; adductor scars elongate pear-shaped, somewhat divergent. Other internal characters unknown. *U.Jur. (Volg.)-L.Cret.(L.Valangin.)*, Eu.——FIG. 646,3. **R. michalkowii* (FAHRENKOHL), L.Volg., USSR; 3a,b, brach.v., lat. views, ×1 (294).

Rugithyris S. S. BUCKMAN, 1918, p. 127 [**Terebratula subomalogaster* BUCKMAN, 1901, p. 259; OD] [=*Rugithyris* BUCKMAN, 1915, p. 79 (*nom. nud.*)]. Medium-sized, valves sulco- to planoconvex, anterior commissure rectimarginate, uniplicate or incipiently sulciplicate; umbo produced, slightly incurved, beak ridges prominent, permesothyridid, symphytium exposed; growth lamellae numerous, commonly squamose and overlapping. Loop length unknown; cardinal process short, lobate; no posterior umbonal cavity; hinge plates in transverse section mallet-shaped, slightly convex ventrally, rarely keeled; adductor scars short, broad with inner margins parallel. *M.Jur. (Bajoc.)*, Eu.(Eng.).——FIG. 648,4; 649,1. **R. subomalogaster* (BUCKMAN), Eng.(Dorset); 648, 4a-c, brach.v., lat., ant., views, ×2; 649,1a-g, ser. transv. secs., ×1.2 (136).

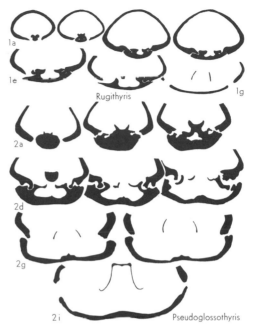

FIG. 649. Terebratulidae (Terebratulinae) (p. H786-H787).

Somalithyris MUIR-WOOD, 1935, p. 124 [**S. macfadyeni*; OD]. Medium-sized, biconvex, anterior commissure uniplicate to sulciplicate; umbo short, suberect, symphytium concealed, foramen small, beak ridges obscure, epithyridid, pedicle collar absent; shell rarely capillate. Loop with high-arched transverse band; cardinal process short, bilobed, with shallow umbonal cavity; hinge plates in transverse section wide, not well demarcated from inner socket ridges, concave ventrally and slightly deflected dorsally, tapering; adductors narrow, lenticular, diverging. *U.Jur.(Oxford.)*, Afr.(Somaliland).——FIG. 650,1; 651,1. **S. macfadyeni*; 650,1a-c, brach.v., lat., ant. views of holotype, ×1.3; 650,1d, adductor scars, ×1.3; 651,1a-n, ser. transv. secs., ×1 (577).

Sphaeroidothyris S. S. BUCKMAN, 1918, p. 115 [**S. globisphaeroidalis*; OD] [=*Sphaeroidothyris* BUCKMAN, 1914, p. 2 (*nom. nud.*)]. Small to medium-sized, plano- to biconvex to sphaeroidal, anterior commissure plane, slightly waved or uniplicate; umbo short, incurved, foramen small, beak ridges obscure, epithyridid, symphytium not exposed; growth lines few, prominent. Loop greater than 0.5 length of valve; low septum present; cardinal process prominent, lobate, medianly depressed; hinge plates in transverse section differentiated from inner socket ridges, dorsally inclined, ventrally concave with short U-shaped kink (spoon-shaped); adductor scars spatulate, posteriorly threadlike, not divergent. *M.Jur.(Bajoc.-*

1c

2a

2b

Sphaeroidothyris

1b

2c

2d

1a   Somalithyris   1d

FIG. 650. Terebratulidae (Terebratulinae) (p. *H787-H788*).

Bathon.), Eu.(Eng.-Fr.-Ger.), ?*U.Jur.*——FIG. 650, 2; 651,2. **S. globisphaeroidalis,* U.Inferior Oolite, Eng.(Somerset); 650,2a-c, brach.v., lat., ant. views, ×1.3; 650,2d, dorsal view showing adductor scars of holotype (Dorset), ×2; 651,2a-q, ser. transv. secs., ×1 (579).

Stiphrothyris S. S. BUCKMAN, 1918, p. 109 [**Terebratula globata* var. *tumida* DAVIDSON, 1878, p. 149; OD] [=*Stiphrothyris* BUCKMAN, 1915, p. 78 (*nom. nud.*)]. Medium-sized, concavo-convex, becoming biconvex to sphaeroidal, biplicate, plications strong, anterior commissure uniplicate to sulciplicate; umbo short, obliquely truncated, incurved in later stages, symphytium usually exposed, beak ridges obscure, permesothyridid to epithyridid, foramen small. Cardinal process prominent, bilobed with median depression; hinge plates thickened, clubbed, slightly concave, scarcely demarcated from inner socket ridges, becoming

U- or V-shaped; crural processes short, converging, virgate; adductor scars subparallel. *M.Jur.* (*Bajoc.*), Eu.(Eng.).——FIG. 652,3; 653,1; 657,2. **S. tumida* (DAVIDSON), Glos.; 652,3a-c, brach.v., lat., ant. views, ×1.2; 653,1a-h, ser. transv. secs., ×1.5; 657, 2, artificial internal mold showing adductor scars, ×1.2 (229).

Stroudithyris BUCKMAN, 1918, p. 111 [**Terebratula pisolithica* BUCKMAN, 1886, p. 41; OD] [=*Stroudithyris* BUCKMAN, 1915, p. 78 (*nom. nud.*)]. Small to medium-sized, valves moderately biconvex, biplicate anteriorly, anterior commissure uniplicate to sulciplicate; umbo short, incurved, beak ridges obscure, epithyridid, symphytium short. Loop less than half of valve length, transverse band highly arched, medianly rounded; cardinal process low; no umbonal cavity; hinge plates in transverse section slightly demarcated from inner socket ridges, concave, clubbed becoming U-shaped, tapering, slightly keeled; adductor scars narrow, tapering posteriorly, slightly divergent. *M.Jur.(Bajoc.),* Eu. (Eng.).——Fig. 652,*1*; 654,*2.* **S. pisolithica* (Buckman), Glos.; 652,*1a-c,* brach.v., lat., ant. views of paratype, ×1.2; 654, *2,* adductor muscle scars of paratype (after Buckman); ×1 (136).

Taurothyris Kyansep, 1961, p. 27 [**T. avundaensis*; OD]. Large, moderately biconvex, elongate-oval in outline, anterior commissure plane to incipiently uniplicate; umbo small, tapering, suberect, symphytium well exposed, beak ridges obscure; shell capillate when decorticated. Loop short, triangular; crural bases given off ventrally; cardinal process massive, bilobate, prominent; hinge plates and inner socket ridges in section slightly deflected ventrally and very slightly concave. *U.Jur.(Ox-*

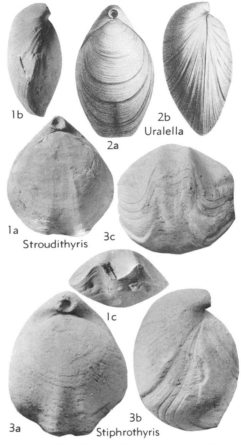

Fig. 652. Terebratulidae (Terebratulinae) (p. *H788-H789, H792*).

ford.), Eu.(Crimea).——Fig. 654,*7*; 655,*2.* **T. avundaensis*; 654,*7a-c,* brach.v., lat., ped.v. views, ×1; 655,*2a-h,* ser. transv. secs., ×1.3 (496).

Triadithyris (*see* p. *H905*).

Trichothyris S. S. Buckman, 1918, p. 125 [**Dictyothyris compressa* Kitchin, 1897, p. 28; OD] [=*Trichothyris* Buckman, 1915, p. 78 (*nom. nud.*)]. Small, valves biconvex, anterior commis-

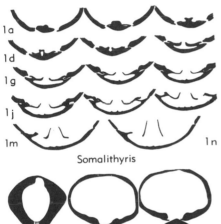

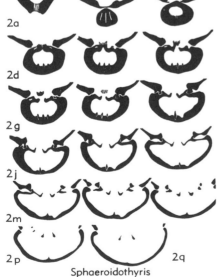

Fig. 651. Terebratulidae (Terebratulinae) (p. *H787-H788*).

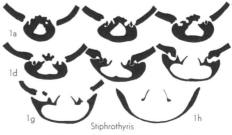

Fig. 653. Terebratulidae (Terebratulinae) (p. *H788*).

sure uniplicate to parasulcate; umbo short, stout, obliquely truncate, foramen large, epithyridid, deltidial plates discrete; shell finely capillate.

Loop and internal characters largely unknown; adductor scars broad, inner edges of scars parallel. *U.Jur.(Callov.)*, Asia(Pak.).——Fig. 654,*3*. *T.*

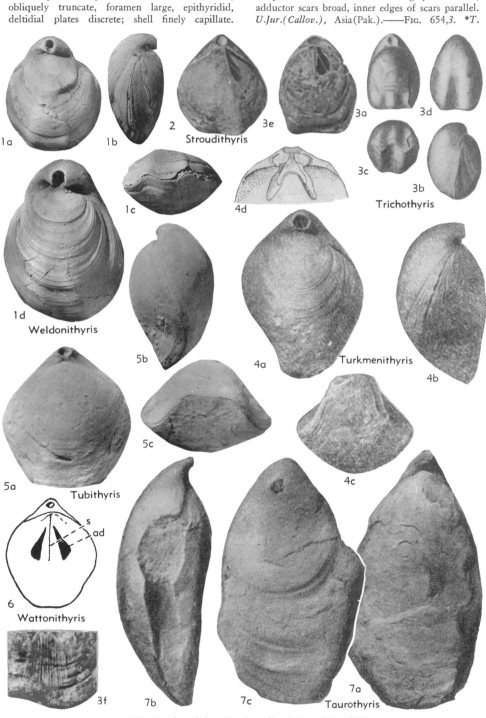

Fig. 654. Terebratulidae (Terebratulinae) (p. *H788-H793*).

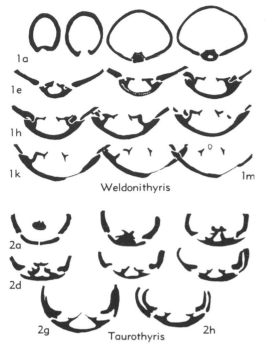

Fig. 655. Terebratulidae (Terebratulinae)
(p. *H789, H792-H793*).

compressa (KITCHIN); *3a-d*, brach.v., lat., ant., ped.v. views, ×1; *3e*, brach.v. int. mold with adductor scars, ×1.5; *3f*, ornament, ×4.5 (136).

Tubithyris S. S. BUCKMAN, 1918, p. 115 [**Terebratula wrighti* DAVIDSON, 1855, p. 20] [*Tubithyris* BUCKMAN, 1915, p. 78 *(nom. nud.)*]. Small to medium-sized, biconvex to spheroidal, anterior commissure uniplicate to sulciplicate; umbo massive, incurved, foramen tubular, telate in young, permesothyridid, beak ridges subangular, symphytium narrow, well exposed; growth lines prominent. Loop about 0.5 of valve length; cardinal process bilobed, prominent, medially depressed; umbonal cavity present; hinge plates in transverse section slightly demarcated from socket ridges, inclined dorsally, gently concave to markedly V-shaped or broad U-shaped; adductor scars elongate-oval slightly divergent. *M.Jur.(Bajoc.-Bathon.), Eu.(Eng.-Fr.).*——FIG. 654,5; 656,2. **T. wrighti* (DAVIDSON), Eng.(Glos.); *654,5a-c*, brach.v., lat., ant. views, ×2; *656,2a-s*, ser. transv. secs., ×1.4 (579).

Turkmenithyris PROZOROVSKAIA, 1926, p. 108 [**T. krimholzi*; OD]. Medium-sized, biconvex, pentagonal, tapering strongly anteriorly, brachial valve with median fold, pedicle valve with sulcus, anterior commissure markedly uniplicate, lateral commissure deflected ventrally; umbo incurved, usually concealing symphytium, mesothyridid,

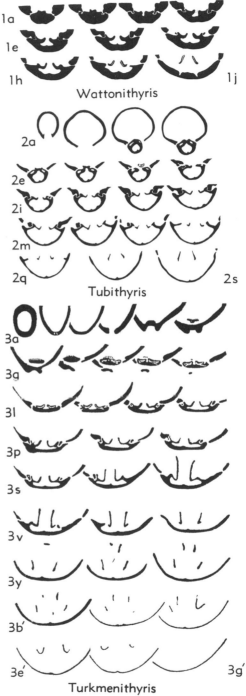

Fig. 656. Terebratulidae (Terebratulinae)
(p. *H791-H792*).

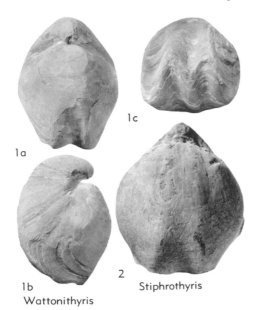

1c

1a

1b   2   Stiphrothyris
Wattonithyris

Fig. 657. Terebratulidae (Terebratulinae)
(p. H788, H792).

pedicle collar present; shell smooth with conspicuous growth lines. Loop about 0.3 length of valve,
ribbon broad anteriorly; cardinal process broad,
bilobed; hinge plates in section tapering, concave
ventrally, becoming U-shaped to acutely V-shaped,
crural bases virgate. *U.Jur.(Callov.),* Asia (W.
Turkoman).——Fig. 654,4; 656,3. **T. krimholzi;*
654,4a-c, brach.v., lat., ant. views, ×1; 654,4d,
loop, ×1; 656,3a-g'; ser. transv. secs., ×0.85
(649).

Uralella Makridin, 1960, p. 295 [**Terebratula
strogonofii* d'Orbigny, 1845, p. 483; OD]. Large,
elongate-oval, biconvex, anterior commissure plane;
umbo massive, foramen large, marginate, permesothyridid. Cardinal process large, almost
quadrate in cross section, dorsal septum strong;
dorsal adductor scars narrow, elongate-triangular,
strongly converging; other internal characters unknown. *U.Jur.(L.Volg.),* Eu.(USSR-S.Urals)-Asia
(E.Siberia)-Arctic.——Fig. 652,2. **U. strogonofii*
(d'Orbigny), USSR; 2a,b, brach.v., lat., ×0.6
(624).

Wattonithyris Muir-Wood, 1936, p. 91 [**W. wattonensis;* OD]. Small to medium-sized; valves
sulco-convex to biconvex to biplicate, anterior commissure uniplicate to sulciplicate, or episulcate;
umbo rounded, suberect to incurved, beak ridges
rounded, permesothyridid, symphytium short,
pedicle collar not observed. Loop equal to or
greater than half length of brachial valve; cardinal process small, with shallow median sulcus;
normally no posterior umbonal cavity; hinge plates
in section ventrally concave, clubbed, not de-

marcated from inner socket ridges; adductors sub-
crescentic, diverging. *M.Jur.(Bathon.),* Eu.(Eng.-
Fr.-Switz.).——Fig. 654,6; 656,1; 657,1. **W.
wattonensis,* Eng.(Dorset); 654,6, brach.v. showing dorsal adductor scars *(ad),* septum *(s),* ×1;
656,1a-j, ser. transv. secs., ×1.1; 657,1a-c, brach.
v., lat., ant. views, ×1.3 (579).

Weldonithyris Muir-Wood, 1952, p. 130 [**W.
weldonensis;* OD]. Small to medium-sized, elongate-oval, valves biconvex, anterior commissure
uniplicate to sulciplicate; umbo incurved in adult,
beak ridges indistinct, epithyridid, symphytium
commonly concealed by labiate foramen; growth

1c
Musculina

1a   1b

3c

2c

3b

2a

2b
Sellithyris

3a   Platythyris

Fig. 658. Terebratulidae (Sellithyridinae)
(p. H793, H795).

lines numerous, few prominent lamellae. Loop about 0.3 length of valve, high-arched, medially horizontal transverse band; cardinal process low, short, medially depressed; posterior umbonal cavity present; hinge plates and inner socket ridges in section slightly deflected dorsally, gently concave, keeled; adductor scars slightly divergent, increasing in width. *M.Jur.(Bajoc.)*, Eu.(Eng.).——Fig. 654,*1*; 655,*1*. **W. weldonensis*; 654,*1a-c*, brach.v., lat., ant. views of holotype, ×1; 654,*1d*, dors. view of paratype showing lamellae, ×1.5; 655,*1a-m*, ser. transv. secs., ×1.3 (582).

Subfamily SELLITHYRIDINAE Muir-Wood, n.subfam.

Pentagonal, smooth or partly capillate, biplicate shells with low bilobate cardinal process, loop short, broad with low arched transverse band. *L. Cret.-U. Cret.(Cenoman.)*.

Sellithyris Middlemiss, 1959, p. 113 [**Terebratula sella* J. de C. Sowerby, 1823, p. 53; OD]. Small to medium-sized, anterior commissure episulcate, lateral commissure with angular deflection ventrally; shell partly capillate; umbo erect to suberect, foramen marginate, mesothyridid to permesothyridid; hinge plates in section ventrally concave to broad U-shaped, tapering, virgate and keeled in some. *L.Cret.(Neocom.)-U.Cret.(Cenoman.)*, Eu.(Eng.-Fr.-Belg.-Ger.-Switz.).——Fig. 658,*2*; 659,*1*. **S. sella* (J. de C. Sowerby), L.Cret. (Apt.), Eng.; 658,*2a-c*, brach.v., lat., ant. views, ×1.5; 659,*1a-j*, ser. transv. secs., ×1.5 (558).

Musculina Schuchert & LeVene, 1929, p. 120 [pro *Musculus* Quenstedt, 1869, p. 4 (*non* Rafinesque, 1818)] [**Terebratula acuta* Quenstedt, 1869, p. 384; SD S. Buckman, 1907, p. 530 (*non T. acuta* J. Sowerby, 1816, p. 115) (=**M. biennensis* Muir-Wood, n.sp., herein, syntypes figured by Quenstedt, 1871 (651), pl. 48, figs. 70-74, Neocom., Bielersee, Switz., as *Terebratula acuta*)]. Resembling *Sellithyris* externally but less pentagonal and having smaller dimensions, biplication developed at earlier stage, brachial valve more convex posteriorly and flattening anteriorly, lateral commissure without angular deflection; shell surface less capillate; sharp beak ridges mesothyridid, larger symphytium bordered by ridges; hinge plates in section rounded V-shaped, slightly clubbed, crural bases virgate. *L.Cret.(Neocom.)*, Eu.(Switz.-Spain).——Fig. 658,*1*; 659,*2*. **M. biennensis* Muir-Wood, Neocom., Switz.; 658,*1a-c*, brach.v., lat., ant. views, ×2; 659,*2a-i*, ser. transv. secs., ×1.5 (651).

?Platythyris Middlemiss, 1959, p. 109 [**P. comptonensis* (=*Terebratula moutoniana* auctt., *partim*); OD]. Medium-sized, elongate-oval, tapering anteriorly, anterior commissure rectimarginate to uniplicate, rarely sulciplicate; beak short, suberect, mesothyridid to permesothyridid, symphy-

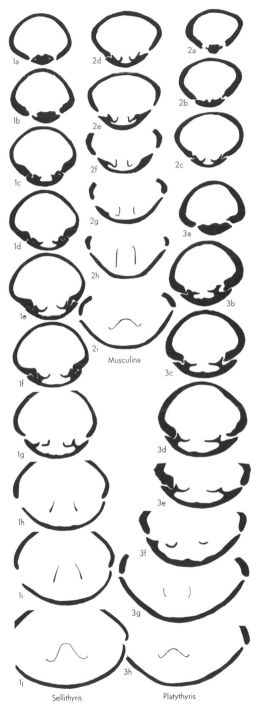

Fig. 659. Terebratulidae (Sellithyridinae) (p. *H793, H795*).

Rhombothyris

Neoliothyrina

Rectithyris

FIG. 660. Terebratulidae (Rectithyridinae) (p. *H795-H797*).

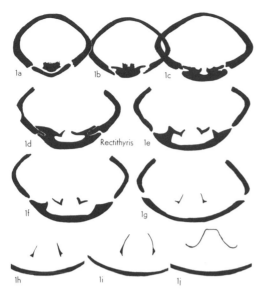

FIG. 661. Terebratulidae (Rectithyridinae)
(p. *H795*).

tium very short. Cardinal process oval, low, hinge
plates in section horizontal, tapering, well demarcated from socket ridges, loop short narrow,
about 0.3 length of valve, with low-arched transverse band; myophragm well developed. *L.Cret.
(Apt.),* Eu.(Eng.).——FIG. 658,*3;* 659,*3.* *P.
comptonensis; 658,*3a-c,* brach.v., lat., ant. views,
×1.5; 659,*3a-h,* ser. transv. secs., ×1.5 (558).

Subfamily RECTITHYRIDINAE Muir-Wood, n.subfam.

Medium-sized to large, smooth, biconvex,
with numerous growth lines. Cardinal process low oval plate, crural bases extending
near inner margins of hinge plates, or narrow inner hinge plates present, loop about
0.3 of valve length, broadly triangular, with
high arched transverse band. *L.Cret.-U.Cret.*

Rectithyris SAHNI, 1929, p. 9 [**Terebratula depressa* VALENCIENNES in LAMARCK, 1819, p. 249
(=*T. nerviensis* D'ARCHIAC, 1847, p. 313); OD].
Large, anterior commissure rectimarginate to
uniplicate, rarely sulciplicate; umbo erect or
curved, foramen large, mesothyridid, symphytium
high; hinge plates in section dorsally deflected,
slightly concave ventrally, becoming V-shaped,
keeled, demarcated from long inner socket ridges
by shallow sulcus, transverse band of loop medianly horizontal. *U.Cret.(Cenoman.),* Eu.(Eng.-
Belg.-Ger.).——FIG. 660,*2;* 661,*1.* *R. *depressa*
(VALENCIENNES), Ger.; 660,*2a-c,* brach.v., lat.,
ant. views, ×1; 660,*2d,* loop, ×3; 661,*1a-j,* ser.
transv. secs., ×1.5 (697).

Cyrtothyris MIDDLEMISS, 1959, p. 123 [**Terebratula
depressa* var. *cyrta* WALKER, 1868, p. 404; OD].
Medium-sized to large, subcircular to subpentagonal in outline, valves convex, with dorsal fold
anteriorly, anterior commissure rectimarginate to
markedly uniplicate, rarely sulciplicate; umbo
short, suberect, foramen large, circular to labiate,

FIG. 662. Terebratulidae (Rectithyridinae)
(p. *H795-H796*).

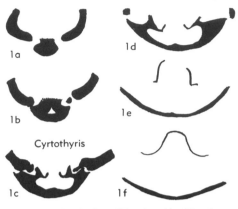

FIG. 663. Terebratulidae (Rectithyridinae)
(p. H795-H796).

mesothyridid, symphytium narrow, exposed. Hinge plates in section dorsally deflected, ventrally concave, slightly clubbed, keeled. *L.Cret.(Apt.)*, Eu. (Eng.-Ger.-Switz.).——Fig. 662,1. *C. uniplicata* (WALKER), *1a-c*, brach.v., lat., ant. views of holotype, ×1 (558).——FIG. 663,1. *C. cyrta* (WALKER), Eng.; *1a-f*, ser. transv. secs., ×1.5 (558).
Neoliothyrina SAHNI, 1925, p. 375 [*Terebratula obesa* DAVIDSON, 1852, p. 53 (*non* J. DE C. SOWERBY, 1823, p. 54=*N. obesa* SAHNI, 1925, p. 375); OD]. Large, anterior commissure plane to biplicate (episulcate); valves with prominent growth

lines and rare capillae; umbo massive, short, foramen large, marginate, permesothyridid, symphytium narrow, commonly concealed. Hinge plates in section, well differentiated from socket ridges, asymmetrically developed in some specimens, dorsally deflected, U-shaped, keeled, rarely with slightly concave inner hinge plates. *U.Cret.(Senon.)*, Eu.(Eng.).——FIG. 660,3; 664,1,2. *N. obesa* SAHNI; 660,3a-c, brach.v., lat., ant. views, ×1; 660,3d, loop, ×3; 664,1a-k, ser. transv. secs., ×1.5 (695); 664,2, transv. sec. showing inner hinge plate, ×1.5 (75).
Praelongithyris MIDDLEMISS, 1959, p. 134 [*P. praelongiforma* (=*Terebratula praelonga* AUCTT., *partim*); OD]. Medium-sized to large, valves elongate-oval with anterior margin truncate, anterior commissure rectimarginate, sulciplicate only at late stage; umbo massive, erect to suberect, foramen large, circular to slightly labiate, permesothyridid; hinge plates in section ventrally concave, becoming markedly V-shaped, clubbed, and anteriorly keeled. *L.Cret.(Apt.)*, Eu.(Eng.-?Ger.).——FIG. 662,2. *P. praelongiforma*, Eng.; *2a-c*, holotype, brach.v., lat., ant. views, ×1 (558).
Rhombothyris MIDDLEMISS, 1959, p. 99 [*Terebratula extensa* MEYER, 1864, p. 252; OD]. Medium-sized, elongate-oval, anterior commissure rectimarginate to uniplicate, rarely biplicate; umbo

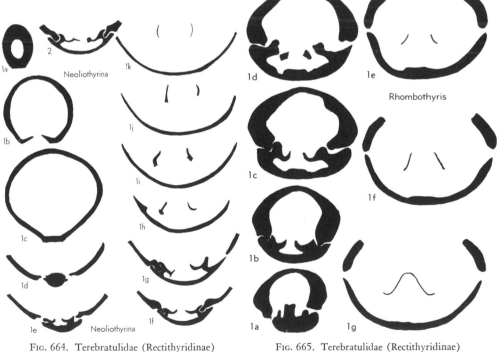

FIG. 664. Terebratulidae (Rectithyridinae)
(p. H796).

FIG. 665. Terebratulidae (Rectithyridinae)
(p. H796-H797).

very short, suberect, truncated by circular or transversely oval foramen, mesothyridid. Cardinal process oval plate, prominent in some, hinge plates concave in section, clubbed, well differentiated from socket ridges. *L.Cret.(Apt.),* Eu. (Eng.).——FIG. 660,*1*; 665,*1*. *R. extensa* (MEYER); 660,*1a-d*, brach.v., lat., ant., ped.v. views, ×1.5; 665,*1a-g*, ser. transv. secs., ×1.5 (558).

Subfamily GIBBITHYRIDINAE Muir-Wood, n.subfam.

Shell smooth or with prominent growth

lamellae or rugae, and rare capillae. Teeth small, sockets narrow, concealed beneath fused socket ridges and hinge plates, cardinal process low, transversely oval plate; loop about 0.3 of valve length, given off dorsally. *U.Cret.(Cenoman.-Senon.).*

Gibbithyris SAHNI, 1925, p. 372 [*G. gibba*] [=*Kestonithyris* SAHNI, 1925, p. 363 (type, *K. inflata*)]. Medium-sized, biconvex, anterior commissure plane to biplicate; umbo incurved, symphytium usually concealed, foramen minute, epithyridid, beak ridges strong; hinge plates in section ventrally convex or horizontal, with pendent dorsally directed crural bases. *U.Cret.(Cenoman.-Turon.-Senon.),* Eu.——FIG. 666,*1a-e.* G.

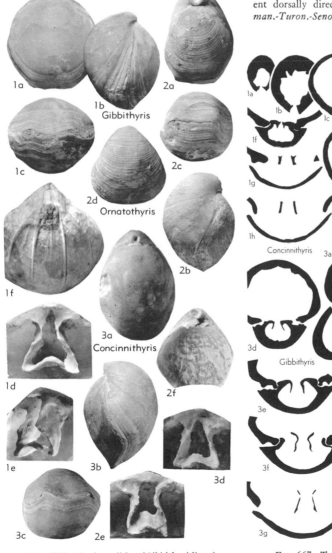

FIG. 666. Terebratulidae (Gibbithyridinae) (p. *H797-H799*).

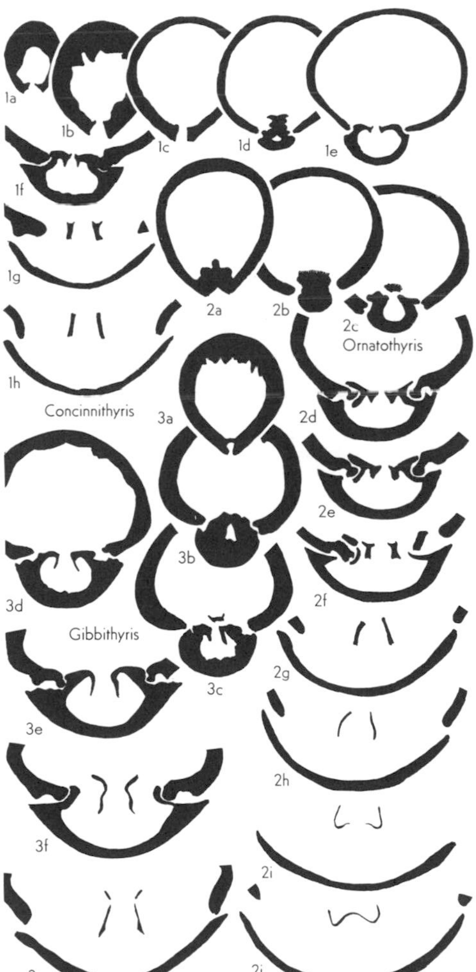

FIG. 667. Terebratulidae (Gibbithyridinae) (p. *H797-H799*).

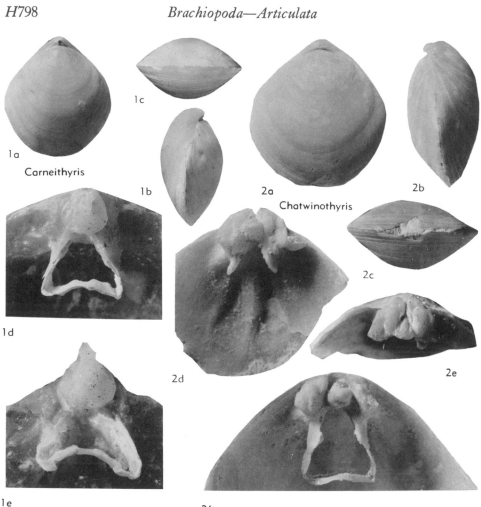

FIG. 668. Terebratulidae (Carneithyridinae) (p. *H799*).

semiglobosa (J. SOWERBY), Cenoman., Eng.; *1a-c,* brach.v., lat., ant. views of holotype, ×0.9; *1d,e,* loop and cardinalia, front and lat. views, ×1.8 (695).——FIG. 666,*1f. G. pyramidalis* SAHNI, Senon., Eng.; ped.v. flint cast with mantle canals, ×0.9 (695).——FIG. 667,*3. G. subrotunda* (J. SOWERBY), Turon., Eng.; *3a-g,* ser. transv. secs., ×1.5 (695).

Concinnithyris SAHNI, 1929, p. 11 [**Terebratula obesa* J. DE C. SOWERBY, 1823, p. 54]. Medium-sized, biconvex, anterior commissure plane to uniplicate or incipiently biplicate (sulciplicate); umbo much incurved, symphytium concealed, foramen usually large, epithyridid, beak ridges indistinct; crural bases given off dorsally, hinge plates in section ventrally convex, and dorsally deflected, keeled. *U.Cret.(Cenoman.-Turon.),* Eu.——FIG. 666,*3a-c. *C. obesa* (J. DE C. SOWERBY),

Cenoman., Eng.; *3a-c,* brach.v., lat., ant. views of holotype, ×0.6 (697).——FIG. 666,*3d;* 667,*1. C. subundata* (J. SOWERBY), Turon., Eng.; 666,*3d,* loop, ×1.8; 667,*1a-h,* ser. transv. secs., ×1.5 (697).

Ornatothyris SAHNI, 1929, p. 45 [**Terebratula sulcifera* MORRIS, 1847, p. 254]. Small to medium-sized, plano-convex to biconvex, anterior commissure plane to uniplicate, rarely sulciplicate; ornament of transverse rugae and rare capillae; umbo massive, short, foramen large, circular to labiate, symphytium exposed, beak ridges obscure; mantle canals bifurcating anteriorly, hinge plates in section not demarcated from long socket ridges, dorsally deflected, keeled, crural bases given off dorsally. *U.Cret.(Cenoman.-?Senon.),* Eu.(Eng.). ——FIG. 666,*2a-c;* 667,*2. *O. sulcifera* (MORRIS), Cenoman., Eng.; 666,*2a-c,* brach.v., lat., ant. views of holotype, ×0.6; *667,2a-j,* ser. transv. secs., ×1.5 (695).——Fig. 666,*2d-f.* *O. latissima* Sahni, Cenoman., Eng.; *2d,* ped.v. view, ×0.6; *2e,* loop, ×1.8; *2f,* ped.v. int. with bifurcated mantle canal markings, ×0.6 (695).

Subfamily CARNEITHYRIDINAE Muir-Wood, n.subfam.

Smooth shells, broad, deep sockets, socket ridges, crural bases and hinge plates tending to fuse in more or less prominent ridge, cardinal process bulbous, lobate or reduced, loop one-third length of valve, given off ventrally. *U.Cret.(Senon.-Dan.).*

Carneithyris Sahni, 1925, p. 364 [**C. subpentagonalis;* OD] [=*Ellipsothyris* Sahni, 1925, p. 371 (type, *E. similis*); *Magnithyris* Sahni, 1925, p. 367 (type, *M. magna*); *Ornithothyris* Sahni, 1925, p. 374 (type, *O. carinata*); *Piarothyris* Sahni, 1925, p. 370 (type, *P. rotunda*); *Pulchrithyris* Sahni, 1925, p. 361 (type, *P. gracilis*)]. [These genera are considered to be variants of *Carneithyris* and not distinct genera.] Small to medium-sized, biconvex, anterior commissure rectimarginate to incipiently uniplicate; umbo incurved, foramen variable, commonly pinhole, beak ridges angular, mesothyridid to permesothyridid. Cardinal process large, bulbous, with 2 lateral knobs and median ridge, hinge plates in section inseparable from socket ridges, dorsally deflected, ven-

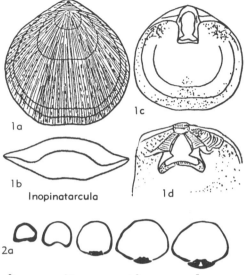

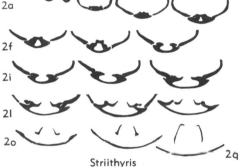

Fig. 670. Terebratulidae (Inopinatarculinae) *(1),* (Subfamily Uncertain) *(2)* (p. H800).

trally curved, thickened, crural bases given off ventrally. *U.Cret.(Senon.).*——Fig. 668,*1a-c;* 669, *1. C. carnea* (J. Sowerby), Senon., 668,*1a-c,* brach. v., lat., ant. views, ×1; 669,*1a-l,* ser. transv. secs., ×1.5 (695).——Fig. 668,*1d,e. C. subovalis* Sahni, Senon., Eng.; *1d,* card. process and loop, ×2; *1e,* ant. view of loop, ×2 (695).

Chatwinothyris Sahni, 1925, p. 368 [**C. subcardinalis;* OD]. Externally like *Carneithyris,* but differs in having ill-defined beak ridges; socket ridges and hinge plates thickened, fused and forming prominent broad ridge with reduction in size of cardinal process; hinge plates and socket ridges in section ventrally convex, as in *Carneithyris,* but more thickened posteriorly and more deflected dorsally. *U.Cret.(Senon.-Dan.),* Eu.(Eng.-Denm.-Belg.-Ger.).——Fig. 668,*2a-e;* 669,*2. *C. subcardinalis,* Maastricht., Eng.; 668,*2a-c,* brach.v., lat., ant. views, ×1; 668,*2d,e,* brach.v. int. and post. view of thickened cardinalia, ×3; 669,*2a-g,* ser. transv. secs., ×1.5 (695).——Fig. 668,*2f. C. ciplyensis* Sahni, Belg.; loop, ×3 (695).

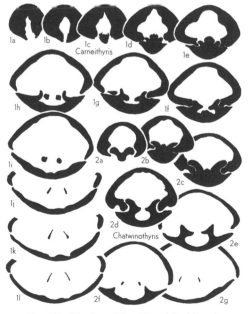

Fig. 669. Terebratulidae (Carneithyridinae) (p. H799).

Fig. 671. Terebratulidae (Subfamily Uncertain)
(p. *H*800).

Fig. 672. Cheniothyrididae (p. *H*800-*H*801).

Subfamily INOPINATARCULINAE Muir-Wood, n.subfam.

Capillate and spinose forms with short, wide, terebratulid loop; inner and outer socket ridges and cardinal process present, median septum and dental plates absent. *U.Cret.(Santon.).*

Inopinatarcula Elliott, 1952, p. 2 [*Trigonosemus acanthodes* R. Etheridge, Jr., 1913, p. 15; OD]. Medium-sized, thick, valves biconvex, with median fold and sulcus, anterior commissure uniplicate; ornament of capillae bearing fine spines, growth lines few, prominent; umbo short, suberect, foramen minute, permesothyridid, symphytium triangular, transversely rugose; loop about 0.3 length of valve. *U.Cret.(Santon.),* W. Australia.——Fig. 670,*1.* **I. acanthodes* (R. Etheridge); *1a,b,* brach.v., ant. views, ×2; *1c,d,* loop, young and adult, ×4, ×2.25 (281).

Subfamily UNCERTAIN

Striithyris Muir-Wood, 1935, p. 129 [**S. somaliensis*; OD]. Medium-sized, biconvex, biplicate anteriorly, anterior commissure rectimarginate to sulciplicate; umbo short, massive, suberect, beak ridges obscure, epithyridid, symphytium short, pedicle collar absent; whole shell finely costellate with numerous intercalations. Loop with broad ribbon and high-arched transverse band; cardinal

process low, medially depressed, with posterior umbonal cavity; hinge plates in section mallet-like, becoming ventrally concave, tapering, not well demarcated from inner socket ridges; adductor scars narrow spatulate, diverging. *U.Jur.(Oxford.),* Afr.(Somaliland).——Fig. 670,2; 671, *1.* **S. somaliensis*; 671,*1a-c,* brach.v., lat., ant. views, ×1.3; 671,*1d,* adductor scars, ×1.3; 670, *2a-q,* ser. transv. secs., ×1.25 (577).

Family CHENIOTHYRIDIDAE Muir-Wood, n. fam.

Short-looped forms without dental plates or dorsal septum; shell folding ligate to bilobate, each valve with median furrow, ornament of steplike squamose lamellae with numerous papillae. *M.Jur.(U. Inferior Oolite).*

Cheniothyris S. S. Buckman, 1918, p. 128 [*Terebratula morierei* E. Eudes-Deslongchamps in Davidson, 1852, p. 256; OD] [=*Cheniothyris* Buckman, 1915, p. 79 *(nom. nud.)*]. Small, subpentagonal to elongate, coarsely punctate, valves moderately biconvex, anterior commissure rectimarginate; umbo incurved, foramen large, beak

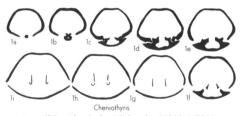

Fig. 673. Cheniothyrididae (p. *H*800-*H*801).

FIG. 674. Dictyothyrididae (p. *H*801).

ridges angular, permesothyridid, symphytium high. Loop about 0.3 length of valve; cardinal process low, short, hinge plates in section thick, squat, somewhat trigonal, ventrally directed and concave. *M.Jur.(U. Inferior Ooolite)*, Eu.(Eng.-Fr.).——FIG. 672,1; 673,1. *C. morierei* (EUDES-DESLONGCHAMPS), 672,1a-c, brach.v. lat., ant. views, ×2.7; 672,1d, Eng.(Dorset.); brach.v. int. mold, ×2.7; 673, 1a-i, ser. transv. secs., ×1.4 (230).

Family DICTYOTHYRIDIDAE Muir-Wood, n. fam.

Short looped forms without median septum or dental plates, shell folding pliciligate, with deep ventral sulcus bounded by strong folds, and low dorsal median fold

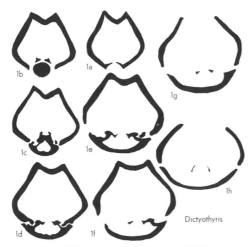

FIG. 675. Dictyothyrididae (p. *H*801).

bounded by shallow furrows; ornament longitudinal and transverse, reticulate, with low nodes or spinules at point of intersection. *M. Jur. (Bathon.) - U. Jur. (Oxford.-?Kimmeridg.)*.

Dictyothyris DOUVILLÉ, 1879, p. 267 [*Terebratulites coarctatus* PARKINSON, 1811, p. 229; OD]. Small to medium-sized, anterior commissure W-shaped; umbo short, becoming incurved, foramen large, permesothyridid, symphytium exposed in some shells. Cardinal process prominent, bilobed, medianly flattened; with posterior umbonal cavity; hinge plates in section scarcely demarcated from socket ridges, slightly convex ventrally, horizontal or slightly deflected dorsally, crural bases given off dorsally; adductor scars large, pear-shaped; *M. Jur. (Bathon.)-U. Jur. (Oxford.-?Kimmeridg.)*, Eu.——FIG. 674,1; 675,1. *D. coarctata* (PARKINSON), Bathon.(Bradford Cl.), Eng.; 674, 1a-c, brach.v., lat., ant. views, ×1.8; 675, 1a-h, ser. transv. secs., ×1.5 (263).

Family TEGULITHYRIDIDAE Muir-Wood, n.fam.

Shell folding ligate to pliciligate, with sulcus in each valve and additional dorsal median fold, antiplicate anterior commissure, shell surface smooth or capillate, with prominent growth lamellae. *U.Jur.(Callov.)*.

Tegulithyris S. S. BUCKMAN, 1918, p. 123 [*Terebratula bentleyi* DAVIDSON, 1851, p. 58; OD] [=*Tegulithyris* BUCKMAN, 1915, p. 78 (*nom. nud.*)]. Medium-sized, biconvex to sulco-convex, becoming sulcosulcate to pliciligate, dorsal sulcus becoming median fold, and angular folds separating deep ventral sulcus from concave flanks, anterior commissure antiplicate; umbo massive, foramen large, beak ridges long, angular, permesothyridid, symphytium well exposed; shell smooth or rarely capillate, with prominent growth lines. Loop with high-arched, medianly horizontal transverse band; cardinal process short, massive, trilobed; hinge plates in section short, keeled, nearly horizontal. *U.Jur.(Callov.)*, Eu.(Eng.-Fr.-Ger.). ——FIG. 676,1; 677,1. *T. bentleyi* (DAVIDSON), *U.Cornbrash*, Eng.(Bedfordsh.); 676,1a-c, brach. v., lat., ant., views, ×1.2; 677,1a-i, ser. transv. secs., ×1.5 (136).

Family PYGOPIDAE Muir-Wood, n. fam.

Valves smooth, medium-sized to large, convex, with dorsal median sulcus and ventral fold posteriorly, usually developing as 2 lateral lobes separate in young, but in contact and fusing in adult and enclosing median perforation, anterior commissure plane, lateral commissure vertical or sig-

FIG. 676. Tegulithyrididae (p. *H*801).

moid. Loop very short, with low arched transverse band; no median septum or dental plates; mantle canal markings well defined, with several bifurcations. ?*L.Jur. (Lias.), M.Jur.-L.Cret.(Neocom.).*

Pygope LINK, 1830, p. 451 [**Terebratula antinomia* CATULLO, 1827, p. 169 (pl. 5, fig. r) (=*T. deltoidea* VALENCIENNES in LAMARCK, 1819, p. 250); SD BUCKMAN, 1906, p. 445]. Shell rounded-trigonal in outline, lateral commissure almost vertical, perforation central, or valves imperforate. *U. Jur.(Kimmeridg.)-L. Cret.(Neocom.),* Eu.(Fr.-Switz.).——FIG. 678,*1.* **P. deltoidea* (VALENCIENNES), L.Tithon., Fr.; *1a,b,* brach.v., lat. views, ×0.6; *1c,* ped.v. view showing imperfectly fused lobes, ×0.6 (128).

Antinomia CATULLO, 1851, p. 74 [**Terebratula dilatata* CATULLO, 1851, p. 75; SD BUCKMAN, 1906, p. 435]. Differs from *Pygope* in its smaller and more posteriorly placed perforation, sigmoid lateral commissure, and lateral flattening of valves. *U. Jur. (Portland.)-L. Cret. (Neocom.),* Eu.-Arctic. ——FIG. 678,*3;* 679,*3a,b. A. catulloi* (PICTET), Portland., Italy(Tyrol); 678,*3,* ant. view, ×1.25; 679,*3a,b,* brach.v. int. mold showing mantle canal markings and post. median perforation, lat. view showing commissure, ×1 (642).——FIG. 679,*3c. A.* sp., Portland., Italy; long. sec. showing tube lining perforation, ×1 (642).

Linguithyris S. S. BUCKMAN, 1918, p. 99 [**Terebratula bifida* ROTHPLETZ, 1886, p. 114] [=*Linguithyris* BUCKMAN, 1914, p. 2 (*nom. nud.*)]. Small, cordate, posteriorly biconvex, becoming sulco-convex, anterior commissure deeply sulcate medianly, lateral commissure slightly curved; umbo incurved, beak ridges angular, epithyridid, symphytium narrow. Loop about 0.25 of valve length, slightly arched, rounded, transverse band; cardinal process low, short, lobate;

hinge plates in transverse section short, slightly concave ventrally, tapering, well demarcated from inner socket ridges; adductor scars short, divergent. *M. Jur. (Bajoc.),* Eu. (Eng.-Ger.-Aus.-Italy). FIG. 679,*4a-c;* 680,*3. L. umbonata* BUCKMAN, M. Inferior Oolite, Eng.; 679,*4a-c,* brach.v., lat., ant. views, ×2; 680,*3a-i,* ser. transv. secs., ×1.3 (Muir-Wood, n).——FIG. 679,*4d.* **L. bifida* (ROTHPLETZ), M. Inferior Oolite, Eng.(Dorset); loop, ×1 (136).

Nucleata QUENSTEDT, 1868, p. 25 [**N. collina* (=*Terebratulites nucleata* VON SCHLOTHEIM, 1820, p. 281); OD] [=*Glossothyris* DOUVILLÉ, 1879, p. 267 (obj.)]. Small, valves concavo-convex, anterior commissure sulcate; umbo short, truncated by large foramen, beak ridges obscure, epithyridid, no symphytium. Loop very short, rounded, transverse band not demarcated; cardinal process absent or very short; hinge plates in transverse section hardly demarcated from long inner socket ridges, slightly concave ventrally; adductor scars short, curved, converging. *U.Jur.(Kimmeridg.-Portland.),* C.Eu.-S.Eu.——FIG. 679,*1;* 680,*1.* **N. nucleata* (SCHLOTHEIM), U. Jur. (gamma), Ger. (Bavaria); 679,*1a-c,* brach.v., ped.v. with loop exposed, ant. views, ×2; 679,*1d,* int. mold showing dorsal adductor scars, ×4; 680,*1a-g,* ser. transv. secs., ×1.3 (136).

Pygites S. S. BUCKMAN, 1906, p. 449 [**Terebratula diphyoides* D'ORBIGNY, 1849, p. 87; OD]. Resembling *Pygope* but pedicle valve posteriorly with sulcus in median fold, brachial valve with median fold in sulcus, large central perforation in adult. Cardinal process low; posterior umbonal cavity developed; hinge plates in section dorsally deflected, slightly concave ventrally, tapering, not well demarcated from long inner socket ridges. *L. Cret. (L. Neocom.),* Eu. (Fr.-Switz.)-N. Afr.-

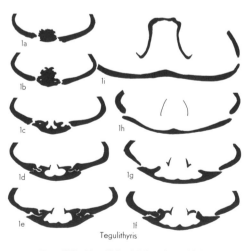

FIG. 677. Tegulithyrididae (p. *H*801).

Arctic.——Fig. 678,2; 679,2; 680,2. *P. diphyoides* (d'Orbigny), Berrias., Fr.(Ardèche); 678,2*a-d,* brach.v., lat., ant., ped.v. views, showing large median perforation, sulcus in ventral fold, and fold in dorsal sulcus, ×1.25; 678,2*e,f,* ped.v. and brach.v. int. showing short loop, ×1.25; 679,2*a,* ped.v. int. mold showing mantle canal markings, ×1; 679,2*b,* immature shell, ped.v. view, ×1; 679, 2*c,* long. sec. showing tube lining perforation, ×1; 680,2*a-i,* ser. transv. secs., ×1.3 (642).

Family DYSCOLIIDAE
Fischer & Oehlert, 1891

[Dyscoliidae Fischer & Oehlert, 1891, p. 23]

Brachial loop very short, with inconspicuous crural processes and ventrally convex transverse band; valves biconvex, with incurved lateral and anterior margin or flange in each valve; shell surface smooth or capillate; lophophore in Recent genus a short

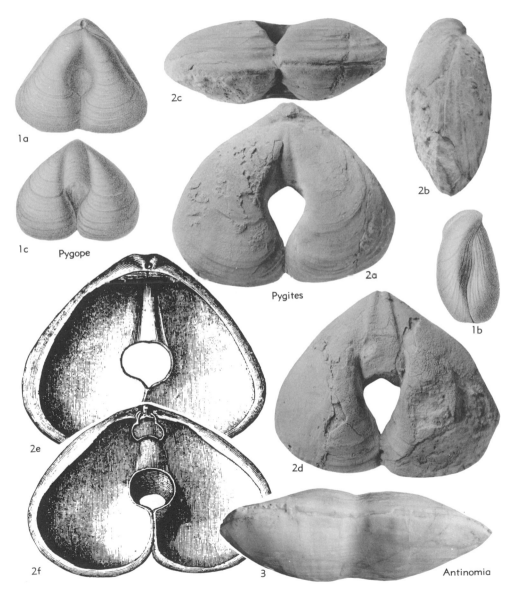

Fig. 678. Pygopidae (p. *H*802-*H*803).

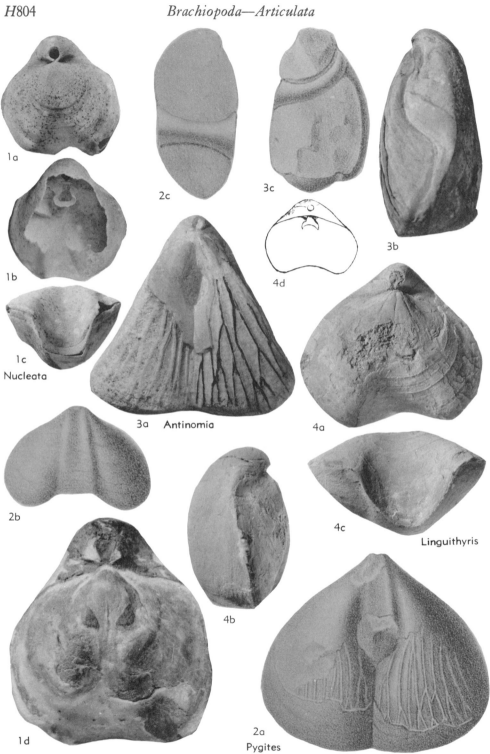

1a

1b

1c

Nucleata

2c

3c

3b

4d

3a   Antinomia

4a

2b

4c

Linguithyris

4b

1d

2a

Pygites

FIG. 679. Pygopidae (p. *H802-H803*).

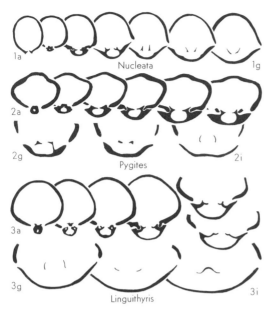

FIG. 680. Pygopidae (p. *H802-H803*).

subrectangular disc, concave ventrally, with long centrifugal filaments, spicules abundantly developed. *?U.Jur., U.Cret.(Cenoman.)-Rec.*

Dyscolia FISCHER & OEHLERT, 1890, p. 70 [**Terebratulina wyvillei* DAVIDSON, 1878, p. 436; OD]. Shell thickened, medium-sized to large, subtrigonal; anterior commissure plane to incipiently biplicate; surface smooth or capillate, growth lines prominent; umbo short, foramen nearly apical, epithyridid, symphytium almost concealed; pedicle collar present. Four main mantle canals in each valve, much branched; cardinal process small, transverse, myophragm rarely present; hinge plates divided, narrow. *Plio.*, Eu.(Italy); *Rec.*, Ind. O.-E.Atl.O.(off Afr.-Spain)-Carib.——FIG. 681,*1*; 682,*3a*. **D. wyvillei* (DAVIDSON), Rec., Carib.; 681, *1a-c*, brach.v., lat., ant. views of holotype, ×1; 681,*1d*, same, valves separated showing imperfect loop and mantle canals, ×1; 681,*1e*, brach.v. int. with loop post., ×1; 682,*3a*, brach.v. int. with lophophore, ×1 (642).——FIG. 682,*3b*. *D. johannisdavisi* (ALCOCK), Rec., Ind.O.; brach.v. int. with loop, ×1 (299).

Moraviaturia SAHNI, 1960, p. 19 [**Terebratula*

FIG. 681. Dyscoliidae (p. *H805*).

diphimorpha STOLICZKA, 1872, p. 25; OD]. Shell solid, medium-sized, subtrigonal, biconvex, with anterior sulcation in brachial valve; anterior commissure plane, with incurved lateral and anterior margin or flange in each valve; surface capillate, with prominent, steplike, growth lamellae; umbo massive, slightly incurved, symphytium narrow, foramen large, beak ridges obscure. Internal characters unknown, probably dyscoliid. *U.Cret.* *(Cenoman.)*, Asia(S.India).——FIG. 682,*1.* **M. diphimorpha* (STOLICZKA); *1a-d,* brach.v., lat., ped.v., ant. views of holotype, ×1 (784).

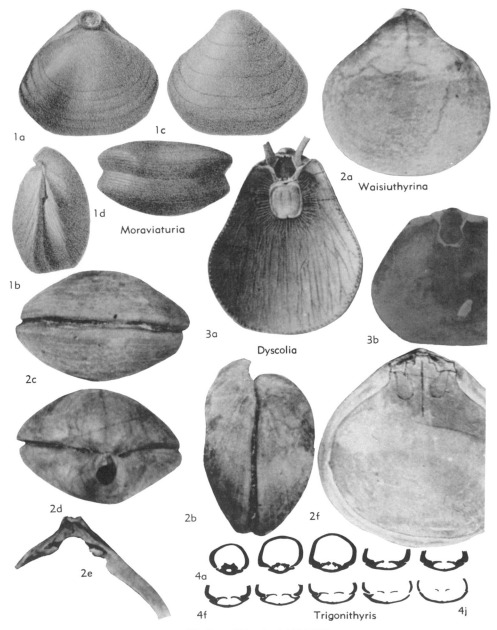

FIG. 682. Dyscoliidae (p. *H805-H807*).

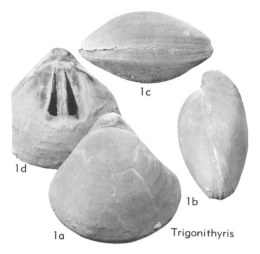

FIG. 683. Dyscoliidae (p. *H806-H807*).

?**Trigonithyris** MUIR-WOOD, 1935, p. 131 [*T. eruduwensis*; OD]. Medium-sized, trigonal, biconvex, anterior commissure plane; umbo short, suberect, epithyridid, symphytium short, shell surface smooth, pedicle collar absent. Loop very short, 0.2 length of valve, ribbon narrow, crural bases given off dorsally; cardinal process broad, medianly depressed; with posterior umbonal cavity; hinge plates in transverse section horizontal, becoming very slightly convex ventrally, not differentiated from inner socket ridges; adductor muscle scars subparallel, narrow; teeth crenulated. *U.Jur. (?Oxford.)*, Afr.(Brit. Somaliland).——FIG. 682, 4; 683,1. *T. eruduwensis*; 682,4a-j, ser. transv. secs., ×1.25; 683,1a-c, brach.v., lat., ant. views, ×1.3; 683,1d, dors. adductor scars, ×1.3 (577).

Waisiuthyrina BEETS, 1943, p. 341 [*W. margineplicata*; OD]. Shell thick, medium-sized to large, valves biconvex, with anterior and lateral flanges, subcircular; anterior commissure almost plane; surface smooth except for growth lines; umbo short, erect, foramen mesothyridid or epithyridid, symphytium narrow, pedicle collar absent. Loop unknown, probably dyscoliid; cardinal process transverse, bilobed, myophragm low; hinge teeth small, grooved and articulating with socket ridges; dorsal adductor scars, broad, spatulate, posteriorly placed. *U. Oligo. (?=Mio. - Plio.)*, Asia (E. Ind. - Celebes).——FIG. 682,2. *W. margineplicata*, Celebes; 2a-d, brach.v., lat., ant., post. views of holotype; 2e,f, ped.v. int., brach.v. int. showing flanges and adductor scars, all ×1 (291).

Family CANCELLOTHYRIDIDAE Thomson, 1926

[*nom. correct.* MUIR-WOOD, herein (*pro* Cancellothyridae, *nom. transl.* MUIR-WOOD, 1955, p. 93 (*ex* Cancellothyrinae THOMSON, 1926, p. 525)]

Valves capillate, brachial loop short, crural processes may unite to form ringlike loop, crural bases attached to socket ridges; hinge plates and median septum only exceptionally developed; adult lophophore spirolophous, plectolophous, or subplectolophous. *?L.Jur.-?M.Jur., U.Jur.-Rec.*

Subfamily CANCELLOTHYRIDINAE Thomson, 1926

[*nom. correct.* MUIR-WOOD, herein (*pro* Cancellothyrinae THOMSON, 1926, p. 525)] [=Terebratulinae DALL, 1870, p. 99 (*partim*)]

Valves biconvex, anterior commissure plane to incipiently sulciplicate; crural processes fused in adult forming complete ringlike loop; foramen hypothyridid, mesothyridid, or epithyridid; lophophore plectolophous. *U.Jur.-Rec.*

Cancellothyris THOMSON, 1926, p. 525 [*Terebratulina hedleyi* FINLAY, 1927 (March), p. 533 (=*Cancellothyris australis* THOMSON, 1927, p. 188), both *nom. subst. pro* *Terebratulina cancellata* KOCH in KÜSTER, 1844, p. 35; (*non* EICHWALD, 1829, p. 276); OD]. Small to medium-sized, ovate to subpentagonal, valves biconvex; anterior commissure uniplicate to sulciplicate; hinge margin terebratulid; surface finely capillate, growth lamellae prominent; umbo short, massive,

FIG. 684. Cancellothyrididae (Cancellothyridinae) (p. *H807-H809*).

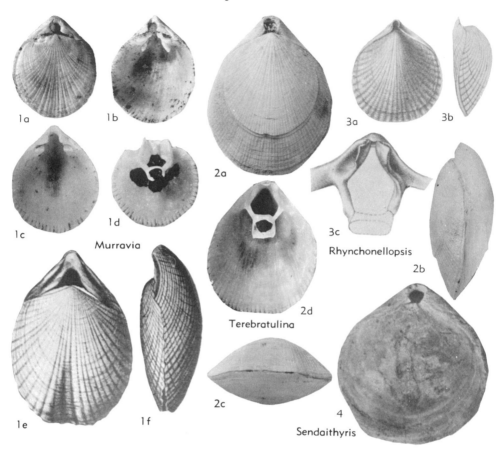

1a 1b 3a 3b

2a

1d 3c

1c

Murravia Rhynchonellopsis

2b

2d

Terebratulina

2c

4

1e 1f Sendaithyris

Fig. 685. Cancellothyrididae (Cancellothyridinae) (p. *H808-H810*).

suberect, foramen large, entire, epithyridid, labiate, symphytium narrow, pedicle collar developed. Loop terebratulinid, transverse band broad, slightly arched ventrally, cardinal process low, bilobed; hinge plates extending posteriorly beyond hinge margin; lophophore plectolophous, median coil long. *Mio.,* N.Z.; *Rec.,* off Australia.——Fig. 684, *1. *C. hedleyi* (Finlay), *Rec.,* S.Australia; *1a-c,* brach.v., lat., ant. views, ×1.2; *1d,* brach.v. int. with loop, ×1.2 (810).

Murravia Thomson, 1916, p. 45 [**Terebratulina catinuliformis* Tate, 1896, p. 130 (footnote), *nom. subst. pro *Terebratulina davidsoni* Ether-idge, 1876, p. 16 (*non* Boll, 1856, p. 37)]. Shell small, thick, ovate, with wide hinge line, pedicle valve slightly convex, brachial valve almost flat, rarely with shallow anterior sulcus; anterior commissure rectimarginate or incipiently sulcate; surface capillate, crossed by numerous growth lines; umbo slightly produced, deltidial plates disjunct, pedicle collar present, foramen below apex of umbo, hypothyridid. Adult loop annular, terebratulinid; hinge plates narrow, myophragm may

be present; cardinal process prominent elongate boss, socket ridges prominent; hinge teeth massive, with sulcus along their inner margin, internal margin of both valves crenulated. *Mio.,* S. Australia-Tasm.-N.Z.; *Rec.,* S. Australia.——Fig. 685,*1a-d.* *M. catinuliformis* (Tate), Mio., Australia (Victoria); *1a-d,* brach.v. view, ped.v. int. with ped. collar, ped.v. int. with hinge teeth, brach.v. int. with loop and septum, ×3 (810).
——Fig. 685,*1e,f.* M. exarata (Verco), Rec., S. Australia; *1e,f,* brach.v., lat. (holotype), ×6 (810).

Rhynchonellopsis Vincent, 1893, p. 50 [**Tere-bratulina nysti* Bosquet, 1862, p. 349; OD (M)] [*non Rhynchonellopsis* Böse, 1894; *nec* de Gregorio, 1930]. Small, thick-shelled, rounded, brachial valve highly convex, pedicle valve flat or slightly convex with median sulcus anteriorly; anterior commissure slightly waved dorsally; surface capillate, capillae bifurcating, enlarged by intersecting growth lines; umbo small, short, suberect, beak ridges obscure, foramen incomplete, deltidial plates disjunct. Hinge plates absent, socket ridges

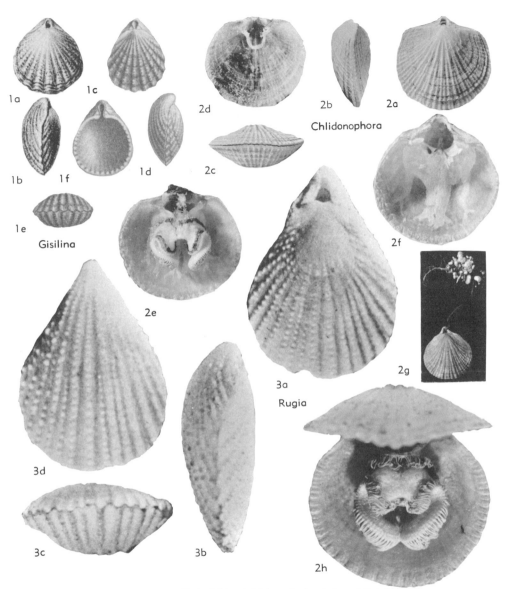

FIG. 686. Cancellothyrididae (Chlidonophorinae) (p. *H*810-*H*811).

prominent and united with crural bases, crura converging, loop probably terebratulinid; inner shell margin crenulated. *L.Oligo.*, Eu.(Belg.-Neth.-Ger.-USSR).——FIG. 685,*3*. *R. nysti* (BOSQUET), Belg.; *3a,b,* brach.v., lat. views, ✕3; *3c,* brach.v. int. with loop (restored), ✕9 (844).

Sendaithyris HATAI, 1940, p. 253 [*S. otutumiensis;* OD]. Medium-sized, circular, valves almost equally convex; surface smooth. Interior imperfectly known, cardinalia and loop said to be as in *Terebratulina,* but having short bifurcated septum in brachial valve. *Mio.,* Japan.——FIG. 685,*4.*

S. otutumiensis, Rikuzen; brach.v. view, ✕1 (399).

Surugathyris YABE & HATAI, 1934, p. 587 [*S. (Terebratulina) suragaensis;* OD]. Imperfectly known, may be immature form of some species of *Terebratulina. Rec.,* Japan.——FIG. 684,*2.* *S. surugaensis;* brach.v. view, ✕2.4 (399).

Terebratulina D'ORBIGNY, 1847, p. 249 [*Anomia caputserpentis* LINNÉ, 1767, p. 1153 (*non* LINNÉ, 1758, p. 703) =*A. retusa* LINNÉ, 1758, p. 701; OD]. Small to large, ovate to subpentagonal, slightly auriculate, valves biconvex; anterior commis-

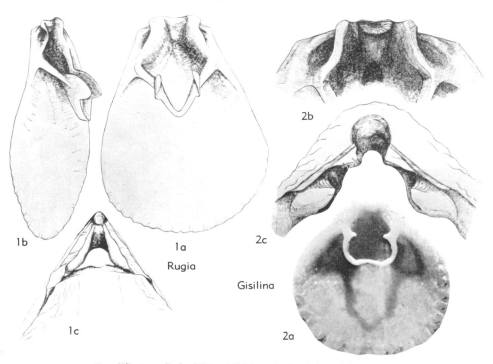

Fig. 687. Cancellothyrididae (Chlidonophorinae) (p. *H810-H811*).

sure rectimarginate to uniplicate; surface capillate, capillae may be enlarged or granular, with prominent nodules in young; umbo suberect, foramen incomplete, mesothyridid to permesothyridid, deltidial plates disjunct, pedicle collar present. Crura converging, crural processes uniting, forming ringlike loop, transverse band ventrally arched; median septum and hinge plates absent; socket ridges and crural bases fused, forming prominent ridge, hinge teeth without swollen bases, but with sulcus on inner face; lophophore plectolophous, median coil short, spicules abundant. [*Anomia caputserpentis* Linné, 1758, p. 703, is a smooth form, probably a terebratulid, whereas *Anomia caputserpentis* Linné, 1767, p. 1153 (=*A. retusa* Linné, 1758, p. 701) is a capillate terebratulinid. This case should be submitted to ICZN.] *U.Jur.*, Eu.; *Rec.*, cosmop.——Fig. 685,*2a-c*. **T. retusa* (Linné), Rec., off Nor.; *2a-c,* brach.v., lat., ant. views, ×2 (810).——Fig. 685,*2d*. *T. japonica* (G. B. Sowerby), Rec., Japan Sea; brach.v. int., ×3 (810).

Subfamily CHLIDONOPHORINAE Muir-Wood, 1959

[Chlidonophorinae Muir-Wood, 1959, p. 259]

Crural processes converging but not uniting to form ringlike loop, transverse band ventrally arched, lophophore subplectolophous. *U.Cret.-Rec.*

Chlidonophora Dall, 1903, p. 1538 [**Terebratulina? incerta* Davidson, 1878, p. 438; OD]. Small, semicircular, valves moderately biconvex, with shallow ventral sulcus; anterior commissure rectimarginate to incipiently uniplicate; hinge line straight, interarea narrow; umbo small, foramen incomplete, deltidial plates narrow, disjunct, pedicle collar developed. Socket ridges projecting above hinge line and united with transverse cardinal process, myophragm developed; pedicle long, composed of rosette of fibers. *?U.Cret.*, Eu.-N.Am.; *Eoc.*, N.Am.; *Rec.*, Ind.O.-Atl.O. (off Afr.)-Carib.——Fig. 686,*2a-f*. **C. incerta* (Davidson), Rec., N.Atl.O.; *2a-c,* brach.v., lat., ant. views of holotype, ×2.5; *2d,e,* brach.v. int. with loop and lophophore, ×2.5, ×3; *2f,* ped.v. int. showing teeth and mantle canals, ×3 (584).——Fig. 686,*2g,h*. *C. chuni* (Blochmann), Rec., Ind.O.(Maldive Is.); *2g,* brach.v. view with long divided pedicle attached to *Globigerina,* ×2; *2h,* valves opened, showing subplectolophous lophophore, ×10 (584).

Gisilina Steinich, 1963, p. 735 [**Terebratula gisii* Roemer, 1840, p. 40; OD]. Small, rounded, auriculate, valves biconvex, with narrow interarea; anterior commissure rectimarginate or incipiently uniplicate; capillae simple, prominent, smooth or enlarged at intersection with growth lines; umbo short, suberect or erect, mesothyridid, deltidial plates disjunct, pedicle collar developed. Loop

FIG. 688. Cancellothyrididae (Eucalathinae) (p. *H*811-*H*812).

chlidonophorid, ventrally directed, crural processes converging, hinge teeth with swollen bases; lophophore from spicule arrangement probably plectolophous. *U.Cret.,* Eu.(Denm.-Ger.-G.Brit.).——FIG. 686,*1*; 687,*2*. *G. gisii* (ROEMER), Rügen; 686,*1a,b*, brach.v., lat. views, ×3; 686,*1c-f,* brach. v., lat., ant., ped.v. int. views, ×4; 687,*2a,* brach. v. int. with loop, ×10; 687,*2b,c,* brach.v. int. with hinge, ped.v. int. post. hinge, both ×11 (584).

Rugia STEINICH, 1963, p 735 [**R. tenuicostata;* OD]. Resembles *Terebratulina* in hinge characters and probable plectolophous lophophore, but differs in small size of adult shells, in having plane commissure, more elongate umbo, long tapering deltidial plates, and long pedicle collar, hypothyridid foramen, more granular shell surface; and in

chlidonophorid loop having broad spoon-shaped crural processes ventrally directed and not fused. *U.Cret.(L.Maastricht.),* Eu.(Denm.).——FIG. 686, *3*; 687,*1.* **R. tenuicostata;* 686,*3a-d,* brach.v., lat., ant., ped.v. views, ×20; 687,*1a,b,* brach.v. int. showing loop and crural processes in vent. and lat. views, ×20; 687,*1c,* ped.v. umbo showing deltidial plates, ×20 (782).

Subfamily EUCALATHINAE Muir-Wood, n. subfam.

Crural processes converging or ventrally directed but not united, transverse band with median plication dorsally directed, hinge plates absent, capillae simple, prominent, foramen mesothyridid, lophophore spirolophous. ?*U.Cret., Rec.*

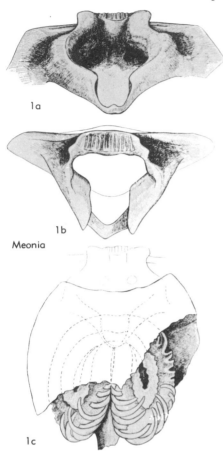

FIG. 689. Cancellothyrididae (Eucalathinae)
(p. *H*812).

Eucalathis FISCHER & OEHLERT, 1890, p. 72 [**Tere-bratulina? murrayi* DAVIDSON, 1878, p. 437; OD]. Minute, subtrigonal, auriculate, ventribiconvex, anterior commissure rectimarginate, or incipiently uniplicate, hinge straight; surface capillate with rare intercalations or granular, growth lines numerous; umbo short, slightly incurved, obliquely truncate, pedicle collar present, deltidial plates minute, disjunct. Loop chlidinophorid but transverse band dorsally directed, socket ridges as narrow plates uniting with cardinal process, and anteriorly with crural bases; lophophore with 2 single whorl spirals set at angle to plane of symmetry, long filaments. *Mio.,* Eu.(Italy); *Rec.,* Pac.O.-E.Atl.O.-Medit.-Carib. ——FIG. 688,*2a-e.* **E. murrayi* (DAVIDSON), Rec., Pac.O. (off Fiji Is.); *2a-c,* brach.v., lat., ant. views, ×16, ×9, ×9; *2d,e,* brach.v. int. showing lophophore, with loop, ×20 (298).——FIG. 688,*2f-j.* *E. ergastica* FISCHER & OEHLERT, Rec., N.Atl.O.;

2f-h, brach.v., lat., ant. views, ×4; *2i,* ped.v. int., ×4; *2j,* brach.v. int. with loop, ×4 (298).

Meonia STEINICH, 1963, p. 733 [**Terebratula semi-globularis* POSSELT, 1894, p. 35]. Small rounded shells with straight hinge line and narrow interarea, highly convex pedicle valve and flat or weakly sulcate brachial valve; anterior commissure plane or slightly sulcate; umbo short, foramen small, mesothyridid; capillae simple, granular, numerous growth lines. Loop short, chlidonophorid, ventrally projecting; crural processes scarcely developed, transverse band narrow uniting broader converging crura; cardinal process crenulated; hinge teeth with deep sockets on inner face articulating with socket ridges; lophophore probably spirolophous, with spirals set at angle to plane of symmetry. *U.Cret.,* Eu.(Denm.).——FIG. 688,*1;* 689,*1.* **M. semiglobularis* (POSSELT); 688, *1a-d,* brach.v., lat., ant., ped.v. int. views, ×3; 689,*1a,b,* brach.v. int., post view showing loop, same, more post. view, ×22; 689,*1c,* brach.v. int. with brachial spirals silicified and part of int. restored, ×21 (782).

Subfamily AGULHASIINAE Muir-Wood, n. subfam.

Crural processes short, converging but not uniting, transverse band of loop ventrally arched, cardinal process bilobed, supported by short septum, narrow hinge plates pres-

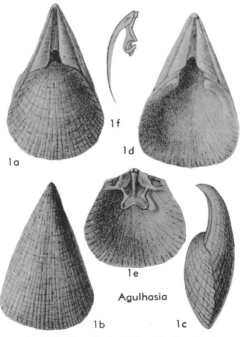

FIG. 690. Cancellothyrididae (Agulhasiinae)
(p. *H*813).

FIG. 691. Cancellothyrididae (Orthothyridinae) (p. H813).

ent, umbo much produced, tapering, with high interarea; shell surface capillate; lophophore subplectolophous. *Rec.*

Agulhasia KING, 1871, p. 109 [**A. davidsoni*; OD]. Minute, rounded trigonal, with produced umbo 0.3 of valve length; valves biconvex, with faint ventral median sulcus, anterior commissure plane to incipiently sulcate; pedicle collar long, deltidial plates disjunct, foramen anterior, hypothyridid. Hinge teeth longitudinally grooved and articulating with socket ridges, internal shell margin strongly crenulated. *Rec.*, Atl.(off S. Afr.).——FIG. 690,*1*. **A. davidsoni*; *1a-c*, brach.v., ped.v., lat. views of holotype, ×8; *1d-f*, ped.v. and brach.v. int., lat. of loop, ×8 (474).

Subfamily ORTHOTHYRIDINAE Muir-Wood, n. subfam.

Small, concavo-convex, costate or costellate, with short, ?terebratulinid loop, and broad, medianly depressed hinge plate. *U. Cret.*

Orthothyris COOPER, 1955, p. 64 [**O. radiata*; OD]. Small, valves with wide straight hinge, slightly biconvex, becoming concavo-convex, anterior commissure broadly sulcate; ornament of coarse simple costae medianly and anteriorly, and medio-laterally directed costellae posteriorly; delthyrium margined by elevated deltidial plates, interarea well developed, beak ridges strong, foramen small. Sockets deep, with erect socket ridges; loop projecting ventrally, attached to socket ridges; interior margin scalloped anteriorly. *U.Cret.*, W. Indies.——FIG. 691,*1*. **O. radiata*, Cuba; *1a-e*, brach.v., lat., ant., ped.v., post. views, ×8 (186).

Subfamily UNCERTAIN

Disculina E. EUDES-DESLONGCHAMPS, 1884, p. 147 [**Terebratula hemisphaerica* J. DE C. SOWERBY, 1826, p. 69; OD]. Small, rounded, valves concavo- to plano-convex, anterior commissure rectimarginate to incipiently sulcate; umbo incurved,

foramen large, incomplete, mesothyridid, deltidial plates disjunct, interarea in pedicle valve, pedicle collar present; valves finely capillate and nodose. Loop short, imperfectly known, median hinge trough wtih 2 diverging spinelike processes, possibly representing cardinal process lobes which project slightly above hinge; inner socket ridges prominent; dorsal median septum and dental plates absent. *M.Jur.(Gt. Oolite Ser.)-U.Jur.(U.Oxford.)*, Eu.(Eng.-Fr.).——FIG. 692,*6*. **D. hemisphaerica* (J. DE C. SOWERBY), Bathon., Fr.(Normandy); *6a-c*, brach.v., lat., ant. views, ×4; *6d*, ped.v. view showing 2 diverging processes (possibly cardinal process), and part of right-hand crus, ×4; *6e,f*, brach.v. and ped.v. int., ×2 (254).

Dzirulina NUTSUBIDZE, 1945, p. 188 [**Terebratula dzirulensis* ANTHULA, 1899, p. 70; OD]. Small, subpentagonal, valves unequally biconvex, anterior commissure plane; shell smooth, or capillate near shell periphery; umbo incurved, concealing symphytium, foramen rounded, beak ridges short, angular, permesothyridid. Internal characters imperfectly known, dorsal median septum ?supporting short loop. *L.Cret.(Apt.-Alb.)*, USSR(Caucasus).——FIG. 692,*1*. **D. dzirulensis* (ANTHULA), Alb.; *1a-d*, brach.v., lat., ant., ped.v. views, ×1 (39).

Hesperithyris DUBAR, 1942, p. 78 [**Terebratula renierii* CATULLO, 1827, p. 167, var. *sinuosa* DUBAR, 1942, p. 83; OD]. Small to large, biconvex, anterior commissure uniplicate, umbo massive, incurved, foramen large, symphytium short, beak ridges ?permesothyridid; ornament of few broad subangular costae from umbo, alternating on opposite valves, normally 2 on fold, 1 in sulcus and 2 bounding sulcus. Loop short, imperfectly known; cardinal process large, with short myophragm; adductor scars posteriorly parallel, threadlike, becoming slightly diverging, spoon-shaped. *L.Jur.(Pliensbach.)*, Eu.(Fr.-Spain-Port.-Alps-Italy-Hung.)-N.Afr.; *L.Jur.(Domer.)*, N.Afr.(Morocco)-Asia(Timor).——FIG. 692,*3*. **H. sinuosa* (DUBAR), Morocco; *3a-e*, brach.v., lat., ped.v., ant., post. views, ×1; *3f*, brach.v. int. mold with adductor scars, ×1 (267).

Phymatothyris COOPER & MUIR-WOOD, 1951, p. 195 [*pro Pallasiella* RENZ, 1932, p. 40 (*non* SARS, 1895)] [**Pallasiella kerkyraea* RENZ, 1932, p. 41; OD]. Small to medium-sized, externally resembling athyridid shells, valves concavo-convex, anterior commissure sulcate, umbones swollen, much incurved, that of pedicle valve in contact with brachial valve usually concealing foramen, beak ridges obscure. Loop presumably short but all internal characters unknown. *L.Jur.(M.Lias.-U. Lias.)*, Eu.(Italy-Albania-Corfu-Alps). —— FIG. 692,*2*. **P. kerkyraea* (RENZ), U.Lias., Greece; *2a-d*, brach.v., lat., ant., ped.v. views, ×1 (666).

Plectoidothyris S. S. BUCKMAN, 1918, p. 122 [**Terebratula polyplecta* S. S. BUCKMAN, 1901, p. 242; OD] [=*Plectoidothyris* BUCKMAN, 1914, p. 2

(*nom. nud.*)]. Medium-sized to large, biconvex, brachial valve posteriorly sulcate, anterior commissure plane to uniplicate, becoming multipli-cate; umbo short, obliquely truncate, foramen sub-apical, epithyridid, symphytium short, pedicle collar developed. Loop 0.7 length of valve, with

Fig. 692. Cancellothyrididae (Subfamily Uncertain) (p. *H813-H816*).

Plectothyris

1a

1b

2a

2c

2b

Plectoidothyris

Fig. 693. Cancellothyrididae (Subfamily Uncertain)
(p. *H813-H816*).

wide ribbon anteriorly; cardinal process prominent, bilobed with deep umbonal cavity; hinge
plates and inner socket ridges in section V-shaped,
tapering, slightly keeled; crural bases converging;
dorsal adductor scars long, narrow, slightly diverging. *L.Jur.(Bajoc.)*, Eu.(Eng.-Fr.).——Fig.
692,5; 693,2; 694,1. *P. polyplecta* (Buckman),
Glos.; 692,5a, dorsal adductor scars, ×1; 692,5b,
loop (reconstr.), ×1.25; 693,2a-c, ant., brach.v.,
lat. views of holotype, ×1; 694, *1a-w*, ser. transv.
secs., ×1 (576).

Plectothyris S. S. Buckman, 1918, p. 121 [*Terebratula fimbria* J. Sowerby, 1822, p. 27; OD]
[=*Plectothyris* Buckman, 1914, p. 2 (nom.
nud.)]. Medium-sized, plano- to biconvex, anterior commissure plane, multiplicate on anterior
0.3 or 0.5 of both valves; umbo short, stout,

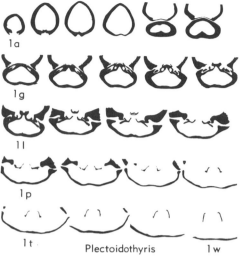

1a

1g

1l

1p

1t

Plectoidothyris   1w

Fig. 694. Cancellothyrididae (Subfamily Uncertain)
(p. *H813-H815*).

obliquely truncate or incurved, permesothyridid,
symphytium short. Loop with high arched medianly horizontal transverse band; cardinal process lobate; posterior umbonal cavity present; hinge
plates in section differentiated from socket ridges,
very slightly concave ventrally, dorsally inclined,
becoming V-shaped; crural bases virgate, converging; adductor scars long, widely divergent. *M.
Jur.(Bajoc.)*, Eng.——Fig. 693,1; 695,2. *P. fim-*

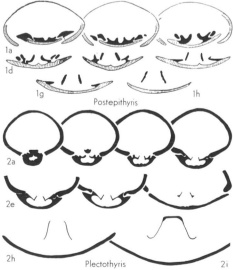

1a
1d

1g   1h
Postepithyris

2a

2e

2h
Plectothyris   2i

Fig. 695. Cancellothyrididae (Subfamily Uncertain)
(p. *H815-H816*).

bria (J. SOWERBY), Glos.; 693,*1a-c*, ant., brach.v., lat. views, ✕1; 695,*2a-i*, ser. transv. secs., ✕1.2 (136).

Postepithyris MAKRIDIN, 1960, p. 294 [**Terebratula cincta* COTTEAU, 1857, p. 137]. Medium-sized, subcircular, moderately biconvex, anterior commissure plane to sulciplicate; umbo massive, suberect to incurved, beak ridges inconspicuous, permesothyridid, symphytium narrow, exposed in some. Loop about 0.5 length of valve, descending branches divergent; cardinal process short, medianly flattened; posterior umbonal cavity variably developed; hinge plates in section ventrally concave, clubbed, becoming V-shaped; adductor muscle scars subparallel, diverging only at extremities. *U. Jur.(U. Oxford.-L.Kimmeridg.),* Eu.(Fr.-Switz.-USSR).——FIG. 692,*4*; 695,*1*. **P. cincta* (COTTEAU); U.Oxford., Fr.; 692,*4a-c*, brach.v., lat., ant. views, ✕1· 695,*1a-h*, ser. transv. secs., ✕0.6 (264).

Terebratularius DUMÉRIL, 1806, p. 170 [no type-species] [=*Terebratulier* DUMÉRIL, 1806, p. 171 (vernacular)]. Group name for terebratuloids, considered invalid.

[**Family Uncertain—Magharithyris,** *see* p. H904.]

MESOZOIC AND CENOZOIC TEREBRATELLIDINA

By H. M. MUIR-WOOD, G. F. ELLIOTT, and KOTORA HATAI

Terebratuloids included in the Mesozoic and Cenozoic Terebratellidina are divided among two superfamilies, the Zeilleriacea (U.Trias.-L.Cret.) and the Terebratellacea (U.Trias.-Rec.). The general form of the shell of both these assemblages is closely similar to that of the Terebratulidina, chief distinctions being found in the nature of internal features, especially characters of the loop.

ZEILLERIACEA

[Discussion by H. M. MUIR-WOOD]

The external surface of zeilleriacean shells is commonly smooth. In the superfamily both valves may have opposite sulci anteriorly (**ligate**) or longitudinal folds may also be developed (**strangulate**). Folds and sulci are equally developed in the bilobate stage. Prominent anterior carinae occur in *Digonella* and *Obovothyris,* or opposite lateral folds may be developed with no median sulci (**ornithellid**). In *Cheirothyris* four prominent opposite angular ridges are seen

in each valve. Multiplication of folds and sulci commonly is alternate (e.g., *Eudesia*) and may be superimposed on a uniplicate or sulcate stage.

The foramen varies in size from a minute pinhole (e.g., *Cincta*) to a large opening, as in several terebratuloids. Beak ridges tend to be angular, well defined, and relatively long, being commonly more evident than in the terebratuloids. The deltidial plates may be conjunct or fused, or disjunct and incompletely fused, the foramen being then referred to as incomplete.

A cardinal process usually is lacking in the Zeilleriacea, but it is present in the genus *Zeillerina* and a complicated process occurs in the Eudesiidae (Fig. 714,*1h*). A septalium of varying form and depth is broadly U-shaped in *Zeilleria* (Fig. 701,*1f*), but deeply V-shaped in *Modestella* (Fig. 709, *3e*). The septalial plates unite with the median septum, which may be a prominent platelike structure.

The descending branches of the loop in Zeilleriacea extend almost to the anterior margin of the shell in some genera and are recurved in the form of ascending branches which are united by a transverse band (Fig. 696). The developmental stages of the loop in zeilleriaceans, if any, still are unknown, and whether the loop is attached to the septum in juvenile growth stages is undetermined.

Various features of internal morphology are important for identification and classification of terebratellid shells, as in other terebratulids, rhynchonellids, and diverse brachiopod groups. Commonly, the only feasible way to determine them is to prepare somewhat closely spaced serial sections, especially of the beak region, and from these such features as hinge plates, socket ridges, median septa, and brachidial loops can be reconstructed (Fig. 697, 698). Serial sections of *Aulacothyropsis reflexa,* an Upper Triassic terebratellacean species, are illustrated here (Fig. 699) for comparison with those of zeilleriaceans. These sections of a Triassic shell are remarkably similar to sections of the Cretaceous terebratellacean *Kingena,* a genus of the Dallinidae.

Dental plates are developed in the pedicle valve of Zeilleriacea. They are usually short and either subparallel or divergent. In serial sections they are often seen to be angled (e.g., *Zeilleria, Cincta*) and one part of each plate is set at an angle to the other, suggesting that they may be composed of two plates, like the adminicula of some spirifer-oids. The dental plates in Zeilleriacea tend to split apart, one portion remaining in contact with the teeth and the other attached to the shell wall.

A pedicle collar or continuation of the deltidial plates on the inner side of the

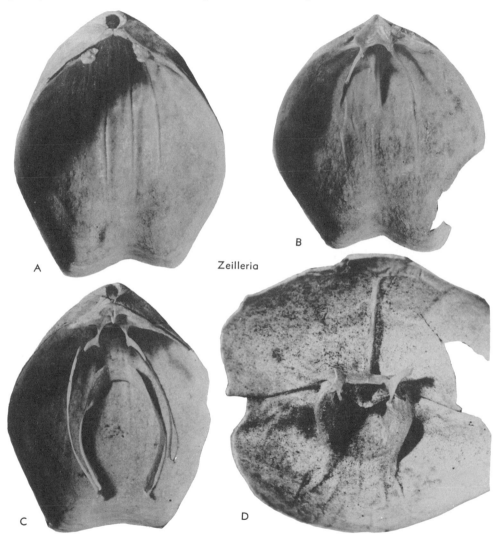

Zeilleria

Fig. 696. Internal morphological features of terebratellaceans illustrated by interior of brachial and pedicle valves of *Zeilleria cornuta* (J. DE C. SOWERBY), from the Jurassic (M.Lias.) of northern France (Muir-Wood, n).——*A.* Interior of pedicle valve showing angular beak ridges, conjunct deltidial plates, hinge teeth, and 4 main vascular trunks appearing as deep incisions, ×2.——*B.* Interior of brachial valve showing one hinge socket, inner and outer socket ridges, fused hinge plates and septalium, median septum, crural processes foreshortened, and descending branches of loop; the massive rounded adductor scars are adjacent to 4 vascular trunks, ×2.——*C.* Loop seen in reverse from dorsal side, showing broad transverse band, and fine ribbon of ascending and descending branches, ×2.——*D.* Two valves in contact along hinge, showing teeth embedded in hinge sockets and supported by dental plates; hinge plates appear between dental plates and are supported by strong septum, ×2.5.

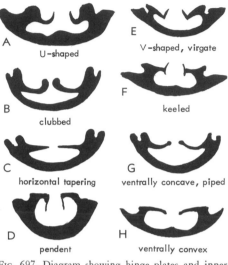

FIG. 697. Diagram showing hinge plates and inner socket ridges in transverse sections (Muir-Wood, n).

umbo may be developed in some Zeilleriacea, and in *Digonella and Obovothyris* it may be supported by a septum.

MESOZOIC TEREBRATELLACEA

[Discussion by G. F. ELLIOTT]

The Mesozoic Terebratellacea form a minority among other contemporaneous long-looped brachiopods, except in the Upper Cretaceous. Current views on the origin of the superfamily are those of MUIR-WOOD (1955), STEHLI (1956), and ELLIOTT (1957); also, I have discussed relationships among the five modern component families. Of these, the Kraussinidae and Platidiidae are not known from the Mesozoic, and the Cretaceous members of the Megathyrididae are mentioned in this volume in the section dealing with the Tertiary and Recent Tere-

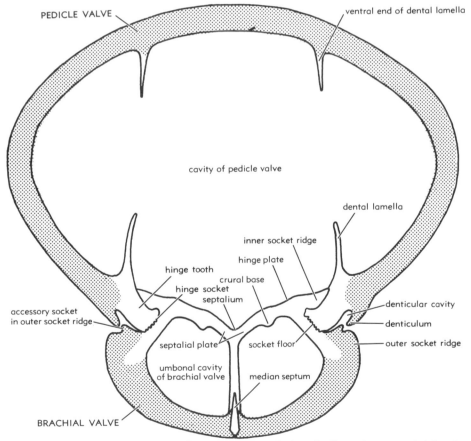

FIG. 698. Diagrammatic representation of transverse section of *Digonella digona* (J. SOWERBY) (after 576).

bratellacea. All of those dealt with in this Mesozoic section are referred to the Dallinidae or Terebratellidae, the former ranging upward from the Upper Triassic, the latter from the Lower Cretaceous. The Dallinidae show hinge teeth supported by dental plates, weak cardinalia with small cardinal processes, and loops with growth stages which commonly are spinous and which show early septal hoods. The Terebratellidae do not have dental plates; many of them possess strong cardinalia with large cardinal processes and their nonspinous loop growth stages show early septal rings. The brachial growth stages of both series are listed in the glossary; a detailed account has been given by ELLIOTT (1953). Reference of the fossil genera to a family is often dependent on suitably preserved growth stages and adult material for dissection. Future work will undoubtedly add to the number of known Mesozoic Terebratellacea and perhaps clarify understanding of their relationships to other brachiopods.

Knowledge of terebratellacean brachial development is largely based on that of Recent species, but occasional good evidence exists of similar development in Mesozoic forms, particularly in the Dallinidae (ELLIOTT, 1947, 1953).

The earliest brachial structure to appear in the Dallinidae is a small median septal pillar, rising from the valve floor. The descending branches develop from the cardinalia, free of the valve floor, and grow anteriorly to meet and fuse with the septal pillar. On the posterior sloping edge of the latter a small cone (hood), open above, closed below, develops. This developmental stage is known as **precampagiform.**

Enlargement of the hood, resorption of its lower (dorsal) closed end, widening of the attachments of the descending branches to the septum, and their fusion with the lower portion of the cowl-like modified hood, leads to the **campagiform stage.**

With continued enlargement of this structure, and alteration in the proportions of its component parts, lacunae or gaps appear by resorption in the sides of the loop, thus individualizing 2 pairs of lateral loop ribbons; this is the **frenuliniform stage.**

With continued enlargement, further resorption removes the posterior lateral rib-

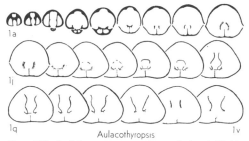

FIG. 699. Serial transverse sections of the dallinid species *Aulacothyropsis reflexa* (BITTNER) (Upper Triassic) belonging to Terebratellacea and resembling the Cretaceous genus *Kingena,* for comparison with sections of zeilleriaceans, ×0.6. [Figures indicate distance in mm. from preceding section; pedicle valve at top.] (Muir-Wood, n).

bons, and the early ascending branches and transverse band appear; this is the **terebrataliiform stage.**

Finally, resorption of the remaining connections from descending branches to the septum leads to the **dalliniform stage,** in which the long loop is wholly free of the valve floor.

The earliest brachial structure to appear in the Terebratellidae is a small high median septum, on which a tiny ring appears at the posterior upper (ventral) angle. Each descending branch, free of the valve floor, grows both from cardinalia and septum, and these portions meet and unite. This developmental step is the **premagadiniform stage** up to completion of the descending branches, when the **magadiniform stage** is attained.

With enlargement of ring and descending branches, proportional changes lead to union of their attachments on the septum, left and right; this is the **magelliform stage.**

With further development, these unions recede from the septum, each being still joined to the latter by connecting bands; this is the **terebratelliform stage,** parallel to the terebrataliiform stage of the Dallinidae.

Finally, resorption of these connecting bands leads to the **magellaniiform stage,** parallel to the dalliniform stage of the Dallinidae.

CENOZOIC TEREBRATELLACEA

[Discussion by KOTORA HATAI]

The surface of brachiopods belonging to Cenozoic Terebratellacea may be marked by concentric growth lines, capillae plications or costae, and in some shells by both concentric and radial sculpture. Other sculpture may be the corrugation or foliation developing from gerontism or by stunted growth. Shells having capillae (e.g., *Terebratalia gouldi*) are thought to have been formed by rows of closely spaced setae on the surface of the mantle. In forms with plications or costae (e.g., *Coptothyris grayi*) the mantle is crenulated or wrinkled along the anterior margin, these corrugations being reflected on the exterior of the shell by radially disposed ridges separated by troughs. The ridges and troughs serve to lock the shells together at the anterior margins of the valves. This kind of sculpture may consist of simple or bifurcated ridges, or of smaller ones alternating with larger ones.

When concentric growth lines are the only markings on the shell surface, they may be quite uniform in size and distribution, or they may be unevenly spaced, with some much stronger than others, and they may show foliation or corrugation at their margins. When both concentric and radial sculpture cover the surface, the former appear as wavy lines crossing the radials, and do not make reticulate sculpture in the Terebratellidae, although reticulation can be observed in forms of other superfamilies.

Terebratellaceans found in Cenozoic deposits, including numerous genera represented in faunas of the present day, are differentiated almost entirely on the basis of their internal characters. Chiefly important are characters of the calcareous loop serving as support of the lophophore. Supplementing discussion given by Elliott in his section devoted to Mesozoic forms, it may be noted that two primary lines of loop development have been recognized by Thomson (810) in the Terebratellidae and Dallinidae and described by Elliott (275). These are defined as terebratellid and dallinid. The trends of the dallinid type are precampagiform (or preismeniform) to campagiform (or ismeniform) to frenuliniform, terebrataliiform, and dalliniform, then with further modification to laqueiform, and finally to pictothyridiform. That of the terebratellid type is premagadiniform to magadiniform, to magelliform, to terebratelliform, and fi-

nally to magellaniiform. No adult brachiopod is known to have either the premagadiniform or precampagiform (preismeniform) type of loop, although the stages are passed. The campagiform stage is represented by the adult loop of *Campages* and *Jolonica,* that of the frenuliniform stage by the adult loop of *Frenulina,* that of the terebrataliiform stage by the adult loop of *Terebratalia, Dallinella,* and *Japanithyris,* and the dalliniform stage by the adult loop of *Dallina, Coptothyris,* and *Macandrevia.* A modified stage of the dalliniform or laqueiform stage is represented by *Laqueus,* while the more advanced type of loop is represented by the pictothyridiform stage of *Pictothyris.*

Distribution of Cenozoic terebratellaceans partly depends on their mode of life and reproduction. In this group of brachiopods the sperm and ova are discharged into the sea water around the parents, except in *Argyrotheca,* in which the sperms enter with the inhalant water current and early larval development takes place in the brood pouch (263). The fertilized ova develop into ciliated larvae with a feebly free-swimming life of at most a few days before settling and metamorphosis into a tiny fixed brachiopod. The fact that terebratelloids usually occur in patches or clusters may signify that the larvae do not disperse widely, as judged from their absence from large areas of adjacent possible anchorage. Their possession of a pedicle provides them with better chance for survival in waters unaffected by currents or by mud drained in large quantities from the land. Since the migration of terebratelloids is much limited by their very short larval period, the occurrence of identical or closely similar forms in remote areas is significant. Their occurrence in remote areas is related to old migration and a long geological history.

In general, terebratellacean brachiopods can be classed in three broad groups according to their geographical distribution: a world-wide group, a Northern Hemisphere group, and a Southern Hemisphere group. The first group is represented by the families Megathyrididae, Platidiidae and Kraussinidae; the second is represented by the Dallinidae; and the third group comprises the Terebratellidae. The members of the first group are the most primitive in structure and development. The second group comprises geologically oldest members with certain primitive characters when compared with the Terebratellidae.

Cenozoic Terebratellacea are classified in six families and ten subfamilies, as follows: Megathyrididae (U.Cret.-Rec.); Platidiidae (Eoc.-Rec.); Kraussinidae (Mio.-Rec.); Dallinidae (U.Trias.-Rec.) [Dallininae (?L.-Cret., Eoc.-Rec.), Frenulininae (Mio.-Rec.), Nipponithyridinae (Mio.-Rec.)]; Laqueidae (Mio.-Rec.) [Laqueinae (Mio.-Rec.), Pictothyridinae (Plio.-Rec.), Kurakithyridinae (Plio.)]; Terebratellidae (L.Cret.-Rec.) [Terebratellinae (Oligo.-Rec.), Bouchardiinae (U.Cret.-Rec.), Magadinae (U.Cret.-Rec.), Neothyridinae (Oligo.-Rec.)]. The nature of the lophophore, including its spiculation, and the presence or absence of dental plates in general of the six families are indicated in the following tabulation (614):

Lophophore and Dental Plate Characters in Cenozoic Terebratellacean Families

Families	Lophophore in Highest Genera	Spiculation	Dental Plates
Megathyrididae	plectolophous	weak or absent	absent
Platidiidae	plectolophous	strong	present in some
Kraussinidae	plectolophous	strong	absent
Dallinidae	plectolophous	weak or absent (rarely strong)	typically present
Laqueidae	plectolophous	moderate	present
Terebratellidae	plectolophous	absent	absent

The subfamilies are distinguished chiefly by characteristics of the cardinalia, crural bases, median septum and foramen. Of the cardinalia, shape of the hinge platform and presence or absence of a hinge trough are important. Similarly the presence or absence of a cardinal process, degree of development of the dental plates, and strength and type of median septum are important. The loop, important in Recent forms, is rarely seen in the fossils.

Suborder TEREBRATELLIDINA Muir-Wood, 1955

Diagnosis of this assemblage is given in the section on Terebratulida—Main Groups (p. *H730*).

Superfamily ZEILLERIACEA Allan, 1940

[*nom. transl.* KYANSEP, 1961, p. 80 (*ex* Zeilleriidae ALLAN, 1940, p. 269)] [Materials for this superfamily prepared by H. M. MUIR-WOOD]

Loop long, descending branches spinose, not attached to dorsal median septum in adult, possibly attached in early growth stages, crural bases and loop given off dorsally or ventrally; cardinal process rarely developed; hinge plates fused, shallow septalium, or hinge trough, composed of septalial plates which unite with median septum; dental plates present; shell attached throughout life by pedicle emerging through delthyrium. *Trias.-L.Cret.*

Family ZEILLERIIDAE Allan, 1940

[*nom. correct.* ALLAN, 1940, p. 269 (*pro* Zeilléridés ROLLIER, 1915, p. 14); authorship and date of this family would be ROLLIER, 1915, if generally accepted by paleontologists (Code, Art. II, c, iii) but ALLAN, 1940, has come to be recognized instead]

Valves normally ligate, strangulate, bilobate or quadrilobate, or both valves convex, or brachial valve flat or concave; anterior commissure plane, rarely uniplicate or sulcate; valves normally smooth; deltidial plates conjunct or disjunct, beak ridges commonly angular and persistent, mesothyridid or permesothyridid. *Trias.-L.Cret.*

Zeilleria BAYLE, 1878, expl. pl. 9 (no page number) [**Terebratula cornuta* J. DE C. SOWERBY, 1824, p. 66; SD DOUVILLÉ, 1879, p. 275]. Small to large, biconvex, with no posterior dorsal sulcation, becoming strangulate to bilobate or quadrilobate, anterior commissure plane; umbo suberect to much incurved, beak ridges angular, persistent, demarcating interarea, permesothyridid, pedicle collar not observed. Loop given off dorsally, dorsal septum about 0.3 of valve length; hinge plates and socket ridges in section ventrally deflected and ventrally convex, septalium broad and medianly horizontal, U-shaped; adductor scars subcircular. *?U.Trias., L.Jur., ?M.Jur.,* Eu.(or cosmop.).——FIG. 700,*3a-f*; 701,*1*. **Z. cornuta* (J. DE C. SOWERBY), L.Jur.(M.Lias.), Eng.(Somerset.);700,*3a-c*, brach.v., lat., ant. views, ×1.3; 700,*3d*, dors. adductor scars and vascular trunks, ×1.3; 700,*3e*, brach.v. int. showing cardinalia and median septum, ×2.4; 701,*1a-j*, ser. transv. secs., ×1.2 (263).——FIG. 700,*3f*. *Z. quadrifida* (VALENCIENNES in LAMARCK), M.Lias., Fr.; ext. showing quadrilobation, ×0.7 (263).

Antiptychina ZITTEL, 1880, p. 704 [**Terebratula bivallata* E. EUDES-DESLONGCHAMPS, 1859, p. 200

(p. 7 of sep.); SD Eudes-Deslongchamps, 1884, p. 268].. Small, smooth, pedicle valve carinate with median sulcus, brachial valve flatter, sulcate, with median fold, anterior commissure antiplicate; umbo fine, tapering, beak ridges long, curving, angular, ?epithyridid. Loop long, both ascending and descending branches spinose, broad transverse band with 2 lateral posteriorly projecting carinae,

separated from median lobe by deep concavity. *M. Jur. (Bajoc.) - U. Jur. (Oxford.),* Eu. (Fr.-Ger.-Czech.-Aus.) [Cret. record relates to unnamed terebratulid homeomorph].——Fig. 702,*1.* *A. bivallata* (Eudes-Deslongchamps), U.Bajoc., Fr.; *1a-d,* brach.v., lat., ant., ped.v. views, ×2; *1e,f,* loop, ×3, ×2.5 (901).

Aulacothyris Douvillé, 1879, p. 277 [*Terebratula

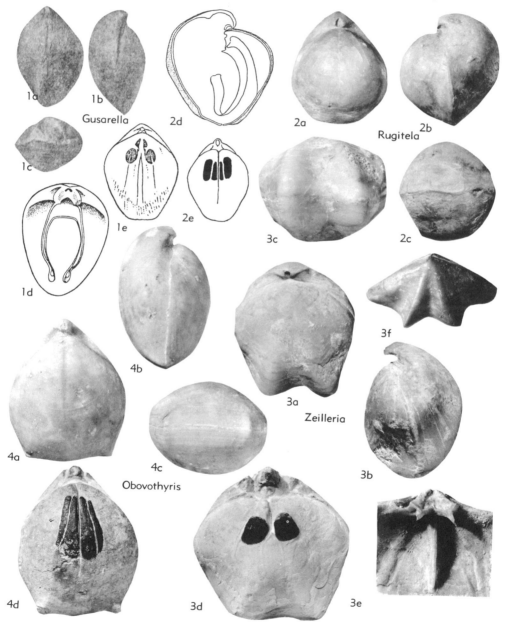

Fig. 700. Zeilleriidae (p. *H821, H825-H828*).

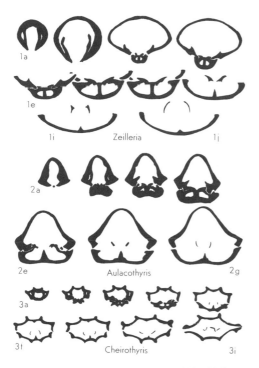

FIG. 701. Zeilleriidae (p. *H821, H822-H824*).

resupinata J. SOWERBY, 1816, p. 116; OD]. Small to medium-sized, valves concavo-convex or concavo-carinate to plano-convex, anterior commissure normally sulcate; umbo flattened, incurved, with angular, permesothyridid beak ridges, strongly demarcating interarea, pedicle collar not observed. Interior commonly with much callus thickening, crural bases ventrally directed; median septum about half of valve length; hinge plates ventrally curved, not well demarcated from inner socket ridges, septalium shallow, rounded, V-shaped, adductor scars elongate-oval, anterior scars about half size of posterior scars; dental plates short, angled. ?Trias., L.Jur.(L.Lias.), ?U.Jur., Eu. ?cosmop.——FIG. 701,2; 703,2. *A. resupinata* (J. SOWERBY), L.Jur.(M.Lias.), Eng.; 701,2a-g, ser. transv. secs., ×1.2; 703,2a-c, brach.v., lat., ant. views, ×1.3; 703,2d, brach.v. int. mold showing adductor scars, ×1.3 (263).

Camerothyris BITTNER, 1890, p. 318 [*Terebratula ramsaueri* SUESS, 1855, p. 25; SD HALL & CLARKE, 1894, p. 887]. Small, valves strangulate, with deep median sulcus; anterior commissure sulcate; umbo tapering, produced, suberect to incurved, deltidial plates fused, beak ridges rounded. Loop zeilleriiform, dorsal median septum and septalium present; dental plates converging and in some shells uniting with ventral septum. *U.Trias.,* Eu. (E. Alps). —— FIG. 705,7. *C. ramsaueri* (SUESS); 7a-d, brach.v., lat., ant., ped.v. views,

×1;7e-g, long. secs. showing loop and transv. band, and transv. sec., ×1 (792).

Cheirothyris ROLLIER, 1919, p. 338 [*Terebratula fleuriausa* D'ORBIGNY, 1850, p. 25; OD] [=Trigonella QUENSTEDT, 1868, p. 25 (*non* DA COSTA, 1778; *nec* CONRAD, 1837; *nec* HEHL, 1842); Neotrigonella COSSMAN, 1910, p. 74]. Small to medium-sized, valves slightly biconvex, pentangular, with 4 prominent carinae, anterior commissure plane; umbo short, broad, foramen large, incomplete, mesothyridid, deltidial plates disjunct. Loop zeilleriiform, given off ventrally with low-arched transverse band, dorsal septum about half of valve length, hinge plates and inner socket ridges in section convex and slightly deflected ventrally, septalium shallow, broad; dental plates short, divergent. [Homeomorph with short loop, no septum or dental plates, as well as 2 terebratelloid homeomorphs *(Ismenia, Trigonellina)* exist in U.Jur.] *U.Jur.(M.-U.Kimmeridg.) (White Jura ε, ζ,* Eu. (Fr.-Switz.-Ger.).——FIG. 701,3; 705,2. *C. fleuriausa* (D'ORBIGNY), Ger.; 701,3a-i,

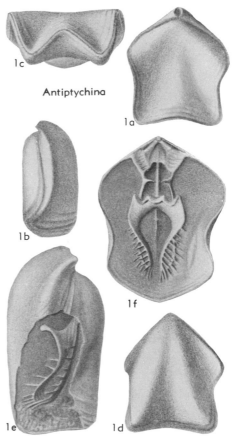

Antiptychina

FIG. 702. Zeilleriidae (p. *H821-H822*).

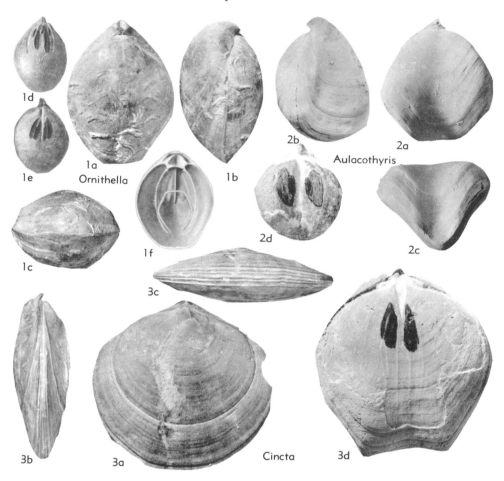

1d
1e
1a
Ornithella
1c
1f
3c
3b
3a
2b
2a
Aulacothyris
2d
2c
1b
3d
Cincta

FIG. 703. Zeilleriidae (p. *H822-H824, H827*).

ser. transv. secs., ×1.2; 705,*2a-c,* brach.v., lat., ant. views, ×2 (672).

Cincta QUENSTEDT, 1868, p. 25 [**Terebratula numismalis* VALENCIENNES in LAMARCK, 1819, p. 249; SD DALL, 1877, p. 20]. Small to medium-sized, subcircular to pentagonal in outline, valves slightly convex, ligate to strangulate, anterior commissure plane; umbo acute, suberect, foramen minute, beak ridges short, angular, curved, mesothyridid; growth lines prominent. Loop given off dorsally, dorsal septum about 0.3 of valve length, hinge plates and inner socket ridges in section ventrally inclined, septalium shallow, rounded V-shaped, becoming deep U-shaped; adductor scars elongate-oval, tapering posteriorly; dental plates angled, convergent, commonly embedded in callus. *L.Jur.(Lias.)-M.Jur.(L.Bajoc.),* Eu.(?cosmop.).
——FIG. 703,*3a-c. *C. numismalis* (VALENCIENNES), L.Lias.(Pliensbach.), Fr.; *3a-c,* brach.v., lat., ant. views, ×1.3 (130).——FIG. 703,*3d;*

708,*1. C. pernumismalis* S. S. BUCKMAN, Eng.; 703,*3d,* brach.v. int. mold showing adductor scars and vascular trunks, ×1.3; 708,*1a-g,* ser. transv. secs., ×1.2 (130).

Digonella MUIR-WOOD, 1934, p. 550 [**Terebratula digona* J. SOWERBY, 1815, p. 217]. Small to medium-sized, concavo-carinate posteriorly, becoming biconvex, elongate-oval to sac-shaped in outline, greatest width anteriorly with development of angular carinae; umbo flattened, suberect, beak ridges short, angular, mesothyridid or permesothyridid, telate, pedicle collar with septum. Loop with numerous long spines, given off dorsally, transverse band with posteriorly projecting carinae; dorsal median septum high, plate-like, slightly greater than half length of valve, hinge plates in section demarcated from inner socket ridges, slightly concave ventrally; dorsal adductor scars linear, adjacent to septum; dental plates nearly parallel. *M.Jur.(Bathon.),* Eu.(Eng.-

Fr.).——Fig. 705,5; 706,1. **D. digona* (J. Sower-by), Gt. Oolite Ser., Eng.; 705,5*a-c*, brach.v., lat., ant. views, ×2; 705,5*d*, dorsal adductor scars, ×1; 705,5*e*, loop (reconstr.), ×1.2; 706,1*a-m'*, ser. transv. secs., ×1.25 (576).

Epicyrta E. Eudes-Deslongchamps, 1884, p. 275 [**Terebratula eugenii* von Buch in Davidson, 1850, p. 72; OD]. Medium-sized, pedicle valve depressed-convex with deep median sulcus, brachial valve highly convex, anterior commissure dorsally arched; umbo short, erect, flattened, foramen small, apical, mesothyridid, beak ridges angular, persistent, delimiting interarea; shell with rarely preserved fine capillation. Loop zeilleriid, given off ventrally, spines not observed, inner socket ridges ventrally directed at high angle, septalium deep, broad, U-shaped, hinge plates keeled, dorsal median septum less than 0.5 of valve length; dental plates short, subparallel. *L.Jur.(M.Lias.),* Eu.(Fr.).——Fig. 707, 2; 708,2. **E. eugenii* (von Buch); 707,2*a-e*, brach. v., lat., ant., ped.v., post. views, ×1; 707,2*f*, umbo, enl., ×2; 708,2*a-h*, ser. transv. secs., ×1.2 (253).

Fimbriothyris E. Eudes-Deslongchamps, 1884, p. 273 [**Terebratula guerangeri* Eudes-Deslong-champs, 1856, p. 304; OD]. Small to medium-sized, subpentagonal, laterally compressed, anteriorly truncated, no median fold or sulcus, anterior commissure plane; umbo short, suberect, beak ridges subangular, long, defining narrow interarea, foramen small, permesothyridid, telate, symphytium narrow; costate medially on anterior half of shell, costae simple subparallel, rare on lateral slopes. Loop zeilleriid, given off dorsally; dorsal median septum about 0.3 of valve length; hinge plates and inner socket ridges in section convex ventrally, septalium deep, becoming broad, shallow U-shaped, septalial plates incompletely fused, leaving small cavity below septalium; dental plates long, slightly divergent. *L.Jur.(M.Lias.),* Eu. (Fr.)-?Afr. (Morocco).—— Fig. 707,1; 709,1. **F. guerangeri* (Eudes-Deslongchamps), Fr. (Sarthe); 707,1*a-e*, brach.v., lat., ant., ped.v., post. views, ×1; 709,1*a-l*, ser. transv. secs., ×1.2 (252).

Flabellothyris E. Eudes-Deslongchamps, 1884, p. 293 [**Terebratula flabellum* Defrance, 1828, p. 160; OD]. Small, valves slightly convex, flabellate, anterior commissure plane to incipiently uniplicate, multiplicate, ill-defined fold and sulcus; umbo massive, short, foramen large, beak ridges angular, mesothyridid, deltidial plates disjunct to conjunct, commonly concealed, pedicle collar developed. Crural bases given off ventrally; cardinal process present; median septum thin plate supporting hinge plates posteriorly only; hinge plates in section demarcated from inner socket ridges, ventrally directed, slightly concave; dental plates short. [Specimens from L.Jur. and U.Cret. are homeomorphs.] *M.Jur.(Bathon.),* Eu.(Eng.-Fr.).

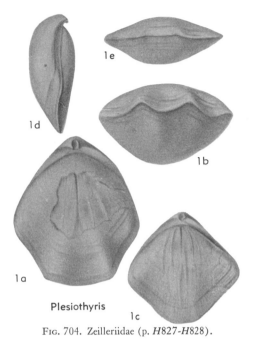

Plesiothyris

Fig. 704. Zeilleriidae (p. *H827-H828*).

——Fig. 705,8; 709,2. **F. flabellum* (Defrance), Fr.; 705,8*a-c*, brach.v., lat., ant., views, ×4; 709, 2*a-h*, ser. transv. secs., ×1.2 (253).

Gusarella Prozorovskaya, 1962, p. 111 [**Zeilleria gusarensis* Moisseev, 1944, p. 58; OD]. Medium-sized, biconvex, or pedicle valve carinate, elongate-pentagonal in outline, anterior commissure incipiently uniplicate; umbo moderately incurved, beak ridges angular, ?permesothyridid, pedicle collar absent. Loop long, delicate, transverse band rather posterior; hinge plates W-shaped, resembling those of *Rugitela,* but lacking median septum; dorsal adductor scars large, oval; dental plates short. *U.Jur.(Callov.),* Asia(W.Turkoman near Caspian Sea).——Fig. 700,1. **G. gusarensis* (Moisseev); 1*a-c*, brach.v., lat., ant. views, ×0.7; 1*d*, loop, ×0.7; 1*e*, brach.v. int. mold, ×0.7 (649).

Modestella E. F. Owen in Casey, 1961, p. 573 [**M. modesta*; OD]. Small, biconvex, cinctiform shells, ligate or strangulate anteriorly, anterior commissure plane; umbo produced, suberect, beak ridges angular, mesothyridid, foramen large, pedicle collar not observed, deltidial plates disjunct to conjunct. Crural bases given off ventrally, dorsal septum half of valve length; hinge plates not distinguishable in section from inner socket ridges, ventrally inclined, septalium exceptionally deep, V-shaped; dental plates short, angled. *L. Cret.(Alb.),* Eu.(S.Eng.).——Fig. 705,10; 709,3. **M. modesta;* 705,10*a-c*, brach.v., lat., ant. views, ×3; 709,3*a-h*, ser. transv. secs., ×1.2 (127).

Obovothyris S. S. Buckman, 1927, p. 32 [**O.*

magnobovata; OD]. Small to medium-sized, sulcocarinate, becoming biconvex, subpentagonal,
with angular anterolateral carinae; umbo suberect to incurved, beak ridges short, subangular,

permesothyridid, pedicle collar with stout septum.
Crural bases given off dorsally; dorsal median
septum high, slightly greater than half of valve
length and supporting hinge plates; septalium very

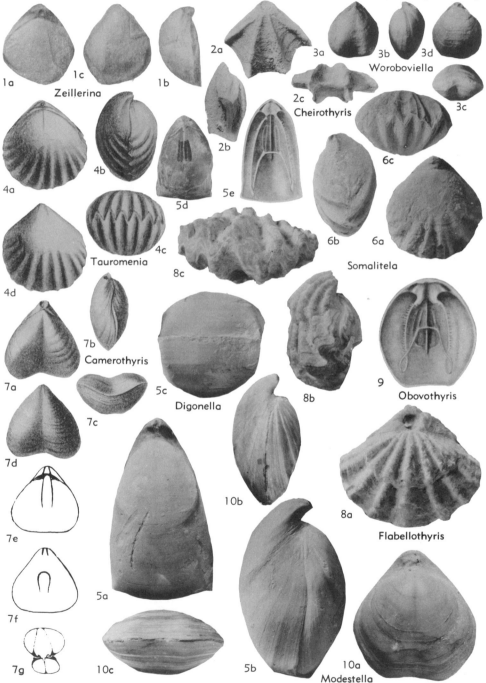

Fig. 705. Zeilleriidae (p. *H823-H829*).

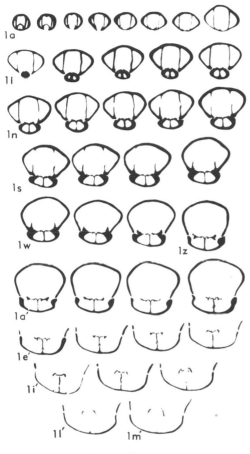

Digonella

FIG. 706. Zeilleriidae (p. *H824-H825*).

shallow, rounded, hinge plates not demarcated from inner socket ridges, slightly convex ventrally; dorsal adductor scars elongate-oval tapering posteriorly; dental plates short, curved. *M.Jur.* (*Bathon.*), Eu.(Eng.-Fr.).——FIG. 700,4; 705,9; 710,1. *O. magnobovata*, L.Cornbrash, Eng.; 700, *4a-c*, brach.v., lat., ant. views, ×1.3; 700,*4d*, dorsal adductor scars, ×1.3; 705,*9*, loop (reconstr.), ×1; 710,*1a-z*, ser. transv. secs., ×1.2 (576).

Ornithella E. EUDES-DESLONGCHAMPS, 1884, p. 273 [*Terebratula ornithocephala* J. SOWERBY, 1815, p. 227; OD] [=Microthyris E. EUDES-DESLONG-CHAMPS, 1884, p. 274 (*non* LEDERER, 1863)= Microthyridina SCHUCHERT & LEVENE, 1929, p. 120 (type, *Terebratulites lagenalis* VON SCHLOTH-EIM, 1820, p. 284)]. Small to medium-sized, biconvex, elongate-oval to pentagonal, ligate or strangulate anteriorly, anterior commissure plane; umbo suberect to incurved, beak ridges short, rounded, permesothyridid, pedicle collar rarely

observed. Crural bases given off ventrally; median septum less than half of valve length, supporting hinge plates, which in section are slightly deflected ventrally, becoming gently undulating, with shallow V-shaped septalium commonly filled with callus knob; adductor scars lens-shaped, set at slight angle to septum; dental plates short, curved, converging. *M.Jur.(Bajoc.-Bathon.)-U.Jur.(Callov.), ?Cret.*, Eu.——FIG. 703,1; 710,2. *O. ornitho-cephala* (J. SOWERBY), Bathon. (L.Cornbrash), Eng.; 703,*1a-c*, brach.v., lat., ant. views of holotype, ×1.3; 703,*1d*, dorsal adductor scars, ×0.7; 703,*1e*, ventral muscle scars, ×0.7; 703,*1f*, loop (reconstr.), ×0.8; 710,*2a-z, 2a'-b'*, ser. transv. secs., ×1 (576).

Plesiothyris DOUVILLÉ, 1879, p. 275 [*Terebratula (Waldheimia) verneuili* E. EUDES-DESLONG-CHAMPS, 1864, p. 268 (sep. publ. 1863); OD]. Medium-sized, pentagonal, moderately biconvex, anteriorly bilobate or quadrilobate, anterior commissure plane or sulciplicate; umbo suberect to incurved, beak ridges angular strong, demarcating interarea, symphytium short. Loop presumed to be

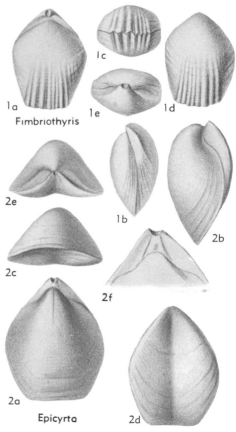

FIG. 707. Zeilleriidae (p. *H825*).

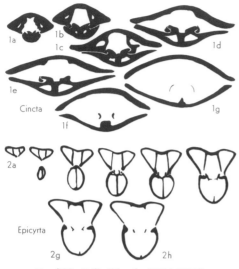

FIG. 708. Zeilleriidae (p. *H824-H825*).

zeilleriiform; dorsal septum about half of valve length; dental plates present. L.Jur., Eu.(Spain-Fr.).——FIG. 704,1. *P. verneuili* (EUDES-DESLONGCHAMPS), M.Lias., Spain; *1a-b*, brach.v., ant., views of adult specimens (lectotype, herein), ×1; *1c-e*, immature specimen, brach.v., lat., ant. views, ×1 (253).

Rugitela MUIR-WOOD, 1936, p. 121 [*Terebratula bullata* J. DE C. SOWERBY, 1823, p. 49; OD]. Medium-sized, elongate-oval, sulcocarinate in early stages, becoming biconvex, commonly globose, anteriorly ligate or bilobate; umbo suberect to incurved, beak ridges short, subangular, mesothyridid, or permesothyridid, telate, pedicle col-

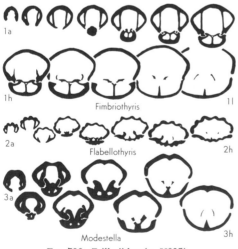

FIG. 709. Zeilleriidae (p. *H825*).

lar rarely observed; shell surface with concentric rugae. Crural bases given off dorsally, loop possibly without spines; median septum long, about 0.7 length of valve, supporting hinge plates posteriorly, septalium shallow, replaced by callus ridge anteriorly; hinge plates and inner socket ridges and median callosity form W-shaped structure; adductor scars elongate-oval; dental plates short and slightly divergent. M.Jur.(Bathon.), ?L.Cret.(Neocom.), Eu.(Eng.-Fr.).——FIG. 700,2a-d. *R. bullata* (J. DE C. SOWERBY), M.Jur.(Fullers Earth Rock), Eng.; *700,2a-c*, brach.v., lat., ant. views, ×1.3; *700,2d*, long. sec. showing loop, ×1.3 (579).——FIG. 700,2e; 711,2. *R. cadomensis* (EUDES-DESLONGCHAMPS); *700,2e*, brach.v. int. mold showing median septum and adductor scars, ×0.7; *711,2a-k*, ser. transv. secs., ×1.2 (579).

Somalitela MUIR-WOOD, 1935, p. 140 [*S. ambalensis*; OD]. Small, valves biconvex, anterior commissure incipiently uniplicate; umbo flattened, suberect, permesothyridid, telate, deltidial plates conjunct; anterior half of valves with prominent, angular plications. Loop given off dorsally; dorsal septum about half of valve length, supporting hinge plates posteriorly; inner socket ridges not separable from hinge plates, slightly concave ventrally, median callosity replacing septalium and whole structure with flattened W-shaped section; dental plates short. U.Jur.(?Kimmeridg.), Afr.(Somaliland).——FIG. 705,6; 711,1. *S. ambalensis*; *705,6a-c*, brach.v., lat., ant. views, ×2; *711, 1a-h*, ser. transv. secs., ×1.2 (577).

Tauromenia SEGUENZA, 1885, p. 253, footnote [*T. polymorpha*; OD] [*non Tauromenia* FUCINI, 1931]. Small, circular to elongate-oval or pentagonal, biconvex, without definite fold or sulcus, anterior commissure plane, multiplicate; umbo small, short, beak ridges angular, impersistent, permesothyridid, deltidial plates conjunct, short; anterior half of valves prominently costate. Loop given off ventrally; dorsal median septum less than half of valve length, supporting hinge plates; inner socket ridges not differentiated from hinge plates, in section slightly convex ventrally with shallow septalium; dental plates short. [Probably same as *Fimbriothyris*.] ?U.Trias.(Rhaet.), Eu.(Italy); L.Jur.(L.Lias.), Eu.(Italy-Port.-Spain)-N.Afr.——FIG. 705,4. *T. polymorpha*, L.Lias., Sicily; *4a-d*, brach.v., lat., ant., ped.v. views, ×1 (771).

Walkeria HAAS, 1890, p. 102, footnote (*nom. nud.*) (*non Walkeria* FLEMING, 1828). Proposed for forms of *Zeilleria* with spines on both ascending and descending branches of loop. No type-species designated.

Woroboviella DAGIS, 1959, p. 33 [*W. caucasica*; OD]. Small, valves biconvex, with shallow dorsal median sulcus, anterior commissure slightly sulcate; umbo short, curved, mesothyridid. Loop long, with narrow descending branches, broad ascending branches, and narrow convex transverse band; hinge plates and inner socket ridges dorsally inclined, septalium shallow; median sep-

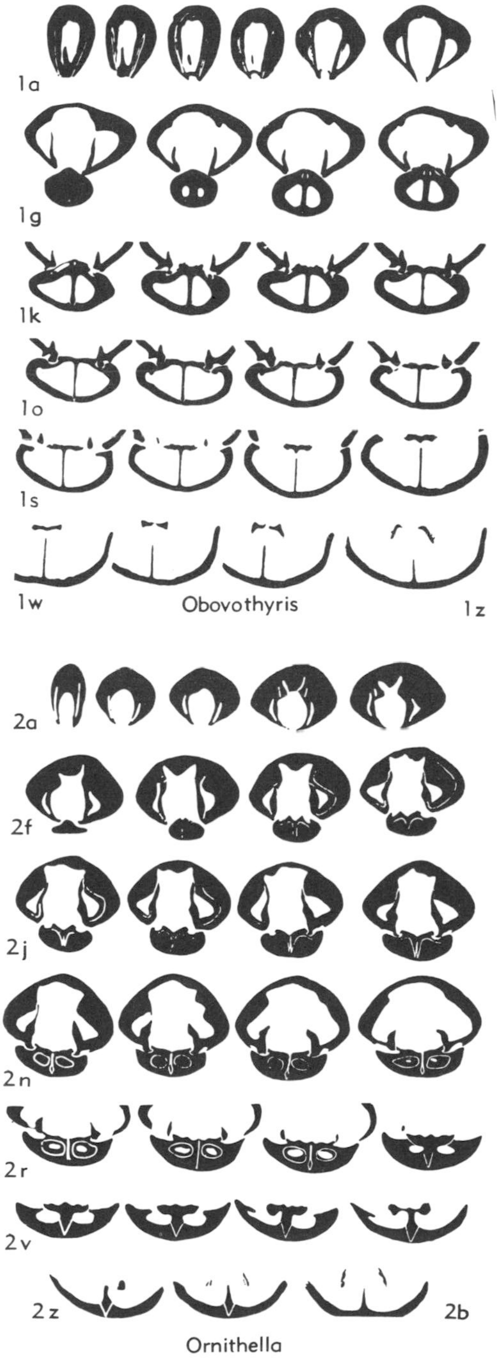

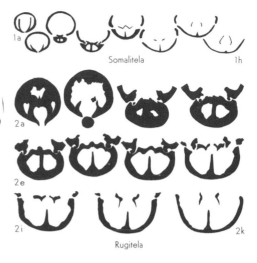

FIG. 710. Zeilleriidae (p. *H825-H827*).

FIG. 711. Zeilleriidae (p. *H828*).

tum 0.3 length of valve, not attached to loop; dental plates short, slightly divergent. *U.Trias.* (*Nor.*), Eu.(NW.Caucasus).——FIG. 705,3; 712, 1. *W. caucasica,* 705,3a-d, brach.v., lat., ant., ped.v. views of holotype, ×1; 712,1a-u, ser. transv. secs., ×1 (210).

Zeillerina KYANSEP, 1959, p. 119 [*Zeilleria belbekensis* MOISSEEV, 1934, p. 149; OD]. Differs from *Zeilleria* in its depressed convex, oval-pentagonal valves, anterior commissure incipiently uniplicate; more produced slightly incurved umbo, shorter and less angular beak ridges, development of pedicle collar, cardinal process and no septalium, crural bases given off ventrally; median septum 0.5 to 0.75 of valve length; hinge plates and inner socket ridges in section not differentiated and dorsally deflected; straight dental plates. *U.Jur.* (*Oxford.-Kimmeridg.*), Eu. (Ger.-Fr.-Switz.-USSR).——FIG. 705,1; 712,2. *Z. belbekensis* (MOISSEEV), Kimmeridg., Crimea; 705,1a-c, brach. v., lat., ped.v. views, ×1; 712,2a-u, ser. transv. secs., ×1 (495).

Family EUDESIIDAE Muir-Wood, n. fam.

Loop zeilleriid, given off dorsally; adult cardinal process complicated in structure, hollow, with 2 small cavities, trilobed, prominent, and elevated above fused thickened hinge plates, which commonly are pierced by 3 small cavities; median dorsal septum and dental plates present; shell, biconvex, fully costate or costellate. *M.Jur.(Bathon.).*

Eudesia KING, 1850, p. 144 [*Terebratula orbicularis* J. DE C. SOWERBY, 1826, p. 68 (=*T. cardium* VALENCIENNES in LAMARCK, 1819, p. 255); OD]. Small to medium-sized, elongate-oval, anterior

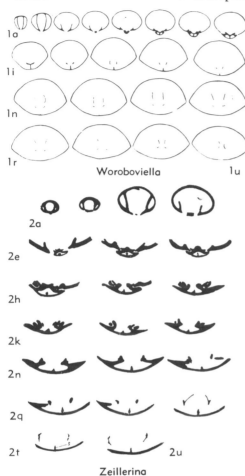

Fig. 712. Zeilleriidae (p. *H828-H829*).

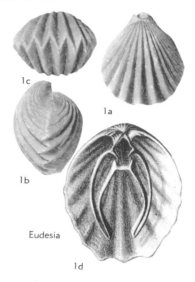

Fig. 713. Eudesiidae (p. *H829-H830*).

lophophore up to schizolophous stage, loop in advanced genera developing both from cardinalia and median septum but ultimately becoming free from septum which is partly or entirely resorbed (614). *U.Trias.-Rec.*

Family MEGATHYRIDIDAE Dall, 1870

[*nom. correct.* HATAI & ELLIOTT, herein (*pro* Megathyridae ALLAN, 1940, p. 269, *nom. transl. ex* Megathyrinae DALL, 1870, p. 100)] [Materials for this family prepared by KOTORA HATAI and G. F. ELLIOTT, with addition by R. C. MOORE as indicated, diagnosis of family by HATAI]

commissure plane, multiplicate; umbo short, massive, suberect or incurved, concealing deltidial plates, foramen large, beak ridges obscure, ?mesothyridid, pedicle collar present. Hinge plates in section not distinguishable from inner socket ridges, slightly convex ventrally, keeled, median septum about 0.5 of valve length, supporting hinge plates in some species; dental plates subparallel, short. *M.Jur.(Bathon.)*, Eu.-Asia-Afr.——FIG. 713, *1*; 714,*1*. **E. cardium* (VALENCIENNES), Fr.; 713, *1a-c*, brach.v., lat., ant. views, ×1; *1d*, loop, ?×1.5; 714,*1a-p*, ser. secs., ×1.2 (253).

Superfamily TEREBRATELLACEA King, 1850

[*nom. transl.* ALLAN, 1940, p. 269 (*ex* Terebratellidae KING, 1850, p. 245)] [Materials of this superfamily prepared by G. F. ELLIOTT and KOTORA HATAI as indicated]

Brachial loop long, undergoing development in association with median septum,

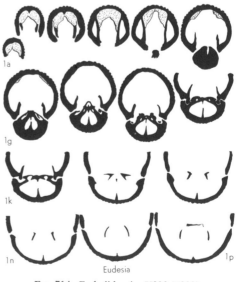

Fig. 714. Eudesiidae (p. *H829-H830*).

Loop composed of descending branches only, passing in most advanced genus through stages correlative with *Gwynia, Argyrotheca,* and *Megathiris,* lower genera not completing the series; lophophore trocholophous to ptycholophous; posterior margin strophic or nearly so; spicules absent or very feebly developed; dental plates absent (810). *U.Cret.-Rec.*

Megathiris D'ORBIGNY, 1847, p. 269 [*pro Argiope* EUDES-DESLONGCHAMPS, 1842, p. ix (*non* AUDOUIN in SAVIGNY, 1827)] [**Anomia detruncata* GMELIN, 1792; OD] [=*Megathyris* BRONN, 1848, p. 244 (*nom. van.*); *Argyope* DAVIDSON, 1850, p. 65 (*non* SAVIGNY, 1826)]. Biconvex to multiplicate opposite with 8 to 14 rounded opposite plicae; foramen submesothyridid, deltidial plates disjunct. Hinge teeth small, no dental plates; cardinalia with low hinge platform uniting 2 prominent socket ridges; cardinal process small; 2 lateral septa reaching to near middle of brachial valve; crura short; loop of 2 descending branches free only near crura, attached to sides of median septum; lophophore attached to dorsal mantle, ptycholophous; adductor muscles attached to pedicle valve in front of diductors; ventral pedicle muscles attached to hinge platform (810). *U.Cret.,* Eng.-Fr.; *Eoc.,* Italy; *Oligo.,* Ger.; *Mio.,* Italy; *Plio.,* Eu.; *Rec.,* Medit.-E.Atl. (Guernsey to Madeira and Aegean Sea, 32-260 m.).——FIG. 715,3. **M. detruncata* (GMELIN), Rec., Medit.; *3a,* brach. v. int.; *3b,c,* brach.v., ped.v. views, all ×5.4 (244). [HATAI.] [Several species are recorded from Upper Cretaceous strata of Europe (e.g., *M. davidsoni* BOSQUET).—ELLIOTT.]

Argyrotheca DALL, 1900, p. 44 [**Terebratula cuneata* RISSO, 1826, p. 388; OD] [=*Cistella* GRAY, 1853, p. 114 (*non* GISTL, 1848)]. Biconvex to strangulate to oppositely multiplicate, smooth or more commonly multiplicate, punctae rather coarse; beak fairly short, subtruncate; foramen large, submesothyridid but almost hypothyridid; deltidial plates small; pedicle collar well developed, supported by median septum. Cardinal process forming transversely elongate, subrectangular boss that projects slightly behind posterior margin, buttressed by median septum; crura widely separate, short; loop relatively long, formed of 2 descending branches, anteriorly converging to join end of median septum; lophophore large, schizolophous (810). *U.Cret.,* Eu.-N.Am.; *Eoc.,* Eu.-N.Am.-S.Am.-W.Ind.; *Oligo.,* Eu.-Mex.; *Mio.,* Eu.(USSR)-N.Am.-W.Ind.-N.Z.; *Plio.,* Eu.(Eng.-Italy); *Rec.,* Atl.(60-1280 m.)-Pac. (160 m.)-Medit.(60-400 m.).——FIG. 715,4. **A. cuneata* (RISSO), Rec., Medit.; *4a,b,* brach.v. int., ped.v. int., ×11 (244). [HATAI.] [Several species from Upper Cretaceous rocks are known from Europe (e.g., *A. megatremoides* BOSQUET) and North America.—ELLIOTT.]

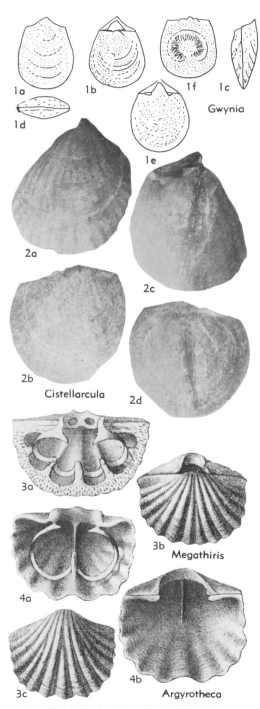

FIG. 715. Megathyrididae (p. *H831-H832*).

FIG. 716. Megathyrididae (p. *H832*).

Cistellarcula ELLIOTT, 1954, p. 726 [**C. wrigleyi*; OD]. Elongate, sulcate, low triangular interarea, triangular pedicle opening, pedicle collar supported by median septum. Cardinalia with median trough, high socket ridges and high median septum (284). *Eoc., Fr.*——FIG. 715,2. **C. wrigleyi*; *2a,b,* ped.v. and brach.v. ext., ×5; *2c,d,* ped.v. and brach.v. int., ×5 (284). [HATAI.]
Gwynia KING, 1859, p. 258 [**Terebratula capsula* JEFFREYS, 1859, p. 43; OD]. Pouch-shaped, almost linguloid, minute, biconvex rectimarginate, smooth, thin, punctae rather large and remote; rostrum apicate, foramen delthyridid, deltidial

plates rudimentary. Hinge teeth without dental plates, no hinge plates; cardinalia weak, cardinal process minute, lophophore trocholophous, attached to dorsal mantle, no median septum, traces of loop with its lower sides cemented to valve (810). *Pleist.(postglacial),* Norway; *Rec.,* E.Atl. (16-4,400 m.)-Fr.-Neth.——FIG. 715,1. **G. capsula* (JEFFREYS); Rec., E.Atl.; *1a-d,* ped.v., brach. v., lat., ant. views, ×9; *1e,f,* ped.v. int., brach.v. int., ×9 (299). [HATAI.]
Phragmothyris COOPER, 1955, p. 65 [**P. cubensis*; OD]. Small, width ranging to 15 mm., moderately to strongly biconvex, pedicle valve deeper than brachial, anterior commissure rectimarginate to broadly sulcate; surface multicostellate; large submegathyridid foramen, symphytium rarely complete. Hinge teeth large, not supported by dental plates, median ridge extending from beak nearly to front margin; brachial valve with wide, deep sockets bounded by elevated socket ridges; adductor muscle scars on elevated platform, with median septum rising well above it; loop consisting of broad ribbon which extends around muscle platform and unites with floor of valve beneath it. *Eoc.-Oligo.,* Cuba.——FIG. 716,1. **P. cubensis*; Eoc., Camaguey Prov.; *1a,* brach.v. view (holotype), ×4; *1b-e,* lat., ped.v., ant., post. views, ×3; *1f,g,* ped.v. int., brach.v. int., ×3 (187). [MOORE.]

Family PLATIDIIDAE Thomson, 1927

[*nom. transl.* ALLAN, 1940, p. 269 (*ex* Platidiinae THOMSON, 1927, p. 215)] [Materials for this family prepared by KOTORA HATAI and G. F. ELLIOTT as indicated, diagnosis of family by HATAI]

Plano-convex, amphithyridid, spiculate forms with loop in most advanced genera composed of descending and ascending branches separately attached to median septum, lophophore plectolophous (810). *Eoc.-Rec.*

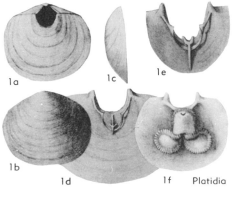

FIG. 717. Platidiidae (p. *H833*).

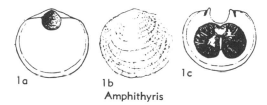

Amphithyris

Fig. 718. Platidiidae (p. *H833*).

Platidia Costa, 1852 (Jan.), p. 47 [*Orthis ano-mioides* Scacchi & Philippi, 1844, p. 69; OD]. [=*Morrisia* Davidson, 1852, p. 371 (May)]. Smooth or with radiating lines or spinules on pedicle valve, shell thin and penetrated by minute caeca; deltidial plates narrow, pedicle collar short, sessile. Hinge teeth with feeble dental plates; crura long, converging, crural processes short; descending branches of loop converging. Dorsal pedicle muscles attached to inner side of cardinalia; mantles, body wall, brachial membrane, and bases of filaments strongly spiculate (810). *Eoc.,* N.Am.; *Oligo.,* Ger.-Italy; *Mio.,* Italy-Pol.; *Plio.,* Italy; *Pleist.,* N.Am. *Rec.,* cosmop.(E. Atl., 50-1340 m.; W.Atl.-Carib., 170-1290 m.; E. Pac., 100-400 m.; W.Pac., 130 m.).——Fig. 717, *1.* **P. anomioides* (Scacchi & Philippi); Rec., E. Atl.; *1a-c,* brach.v., ped.v., lat. views, ×1.2; *1d,e,* brach.v. int., ×1.8, ×1.2; *1f,* brach.v. int. showing lophophore, ×1.2 (810). [Hatai.]

Amphithyris Thomson, 1918 [**A. buckmani*; OD]. Broadly suborbicular, capillate, punctae fine and dense; hinge line nearly straight. Beak apicate, triangular delthyrium in pedicle valve and semicircular notch in cardinal margin of brachial valve, dorsal umbo being resorbed. Hinge teeth without dental plates or swollen bases; cardinalia with socket ridges only, no loop, median septum fairly high, lophophore schizolophous, supported by spicules (810). *Rec.,* N.Z.-Medit.——Fig. 718,*1.* **A. buckmani*, N.Z.; *1a,b,* brach.v., ped.v. views of whole shell, ×3.5; *1c,* brach.v. int., ×3.5 (299). [Hatai.]

Family KRAUSSINIDAE Dall, 1870

[*nom. transl.* Allan, 1940 (*ex* Kraussininae Dall, 1870)] [=Mühlfeldtiinae Oehlert, 1887] [Materials for this family prepared by Kotora Hatai]

Spiculate, without dental plates and with zygolophous to plectolophous lophophores, loop in most advanced genera composed of ascending lamellae attached to low median septum and descending lamellae attached to sides of ring formed by ascending lamellae; in more primitive genera ring of ascending lamellae not completed ventrally and descending lamellae not developed at all or only incipiently (810). *Mio.-Rec.*

Kraussina Davidson in Suess, 1859, p. 210 [*pro* **Kraussia** Davidson, 1852, p. 369 (*Anomia rubra Pallas, 1776, p. 182; SD Davidson, 1853, p. 69) (*non Kraussia* Dana, 1852)]. Biconvex to sulcate, smooth or multicostate, punctae conspicuous; hinge line broad, beak subtruncate, foramen submesothyridid, deltidial plates disjunct; pedicle collar fused to floor in umbonal cavity. Hinge teeth without dental plates; cardinalia with socket ridges projecting behind hinge line, cardinal process small, prominent; dorsal adductor impressions strong, separated by median septum behind brachidium, which consists of 2 stout di-

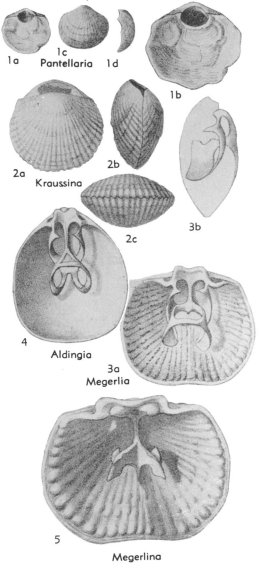

Fig. 719. Kraussinidae (p. *H833-H834*).

la

lb

Pumilus

lc

ld

FIG. 720. Kraussinidae (p. *H834*).

vergent processes extending lateroventrally from median septum; mantle canals of 2 large trunks commencing close to cardinalia, terminating anteriorly close to median line, each canal giving off 6 or 7 branches which bifurcate as they reach shell margin; spicules very small (810). *Rec.,* Ind.O.(20-300 m.).——FIG. 719,2. **K. rubra* (PALLAS), S.Afr.; *2a-c,* brach.v., lat., ant. views, ×1 (244).

Aldingia THOMSON, 1916, p. 501 [**Terebratella furculifera* TATE, 1880, p. 161; OD]. Incipiently sulcate, punctae fine and dense; beak suberect, foramen submesothyridid, deltidial plates almost united. Hinge teeth with swollen bases, grooved for reception of socket ridges; hinge trough moderately large but with median ridge extending anteriorly halfway from umbo. Brachial valve thickened posteriorly to form platform, buttressed by short median septum, crural bases fused with socket ridges; cardinal process low, loop long, reflected, descending branches attached by connecting bands to sides of septum, ascending branches united to top of septal pillar by connecting bands. Ventral muscular impressions strong (810). *Mio.(Janjukian),* Australia; *Rec.* (220-240 m.), Australia.——FIG. 719,4. *A. willemoesi* (DAVIDSON), Rec., Australia; brach.v. int., ×3 (244).

Megerlia KING, 1850, p. 145 [**Anomia truncata* LINNÉ, 1767, p. 1152; OD] [not preoccupied by *Megerlea* ROBINEAU-DESVOIDY, 1830] [=*Mühl-*

feldtia BAYLE, 1880, p. 240]. Incipiently narrowly sulcate, capillate, striae slightly nodulose, interior of valves radially tuberculate; foramen submesothyridid, deltidial plates disjunct, pedicle collar free anteriorly. Pedicle valve with small median septum extending up under pedicle collar but not supporting it; cardinalia consisting of widely divergent socket ridges, excavate below, to inner sides of which crural bases are attached, no cardinal process, loop forming complete ring, on median septum, descending branches extending from cardinalia to ring; filaments long, slender, and numerous. Brachial valve with 2 long central mantle canals and 2 much larger many-branched lateral canals; spicules present both in mantle and arms (810). *Mio.-Plio.,* Italy-Fr.; *Plio.-Rec.,* Gibraltar-Algeria; *Rec.,* Medit.-E.Atl.(60-600 m.)-Ind. O.-Pac.O., Persian Gulf.——FIG. 719,3. **M. truncata* (LINNÉ), Rec., Medit.; *3a,* brach.v. int. showing loop, ×2; *3b,* shell int. lat. showing loop, ×2 (237).

Megerlina EUDES-DESLONGCHAMPS, 1884, p. 243 [**Kraussia lamarckiana* DAVIDSON, 1852, p. 80; OD]. Sulcate, with fine alternate multiplication developing directly upon smooth stage, punctae fine, interior tuberculate; beak suberect, foramen submesothyridid but almost hypothyridid, deltidial plates discrete; pedicle collar anteriorly excavate, embayed in middle. Hinge teeth without dental plates; socket ridges rather stout, giving off on their inner anterior corners 2 spurs which approach septum without joining it and thus enclosing an imperfect hinge trough; cardinal process feeble; median septum with 2 diverging Y-shaped lamellae slightly expanded at their extremities in front and concave toward each other. Spicules stouter than in *Kraussina* (810). *Neog.,* Tasm.; *Rec.,* Australia.——FIG. 719,5. **M. lamarckiana* (DAVIDSON), Rec., Australia (New S. Wales); brach.v. int., ×4.5 (244).

Pantellaria DALL, 1919, p. 251 [**Terebratula monstruosa* SCACCHI, 1836, p. 8; OD]. Differs from *Megerlia* only in possessing amphithyridid instead of submesothyridid foramen, in flattening of brachial valve, and in absence of radial sculpture (221). *Mio.-Pleist.,* Italy; *Rec.,* Medit.-Adriatic-Aegean-Ind.O., E.Atl.(460-2780 m.)-W.Atl.—— FIG. 719,1. **P. monstruosa* (SCACCHI), Rec., Medit.; *1a,b,* brach.v. view, ×1, ×2; *1c,d,* ped.v., lat. views, ×1 (244).

Pumilus ATKINS, 1958, p. 560 [**P. antiquatus*; OD]. Adult lophophore schizolophous. Beak suberect; foramen incomplete; deltidial plates, narrow, disjunct; pedicle collar well developed. Hinge teeth without dental plates; no cardinal process or loop, median septum of brachial valve low, terminating anteriorly in small protuberance. Spicules present (42). *Rec.,* N.Z.——FIG. 720,1. **P. antiquatus*; *1a,b,* brach.v. views, ×9, ×10; *1c,d,* ped.v. views, ×10, ×9 (42).

Family DALLINIDAE Beecher, 1893

[*nom. transl.* ALLAN, 1940, p. 270 (*ex* Dallininae BEECHER, 1893, p. 391)] [Materials for this family prepared by G. F. ELLIOTT and KOTORA HATAI as indicated, diagnosis of family by ELLIOTT & HATAI]

Loop passing through all or part of precampagiform, campagiform, frenuliniform, terebrataliiform, and dalliniform growth stages, or modifications of these; spicules present in some forms but never abundant; dental plates generally present (810). *U.-Trias.-Rec.*

Subfamily DALLININAE Beecher, 1893

[Diagnosis of subfamily by G. F. ELLIOTT]

Folding sulcate to intraplicate; cardinal process small, loops campagiform, frenuliniform, terebrataliiform, or dalliniform. *L. Cret.-Rec.*

Dallina BEECHER, 1893, p. 377, 382 [*Terebratula septigera* LOVÉN, 1846, p. 29; OD]. Broadly sulcate to intraplicate, with tendency to quadriplication. Beak erect, no beak ridges; foramen large, mesothyridid, attrite; symphytium concave. Hinge teeth supported in young by dental plates which become thin with age and absent in adult; cardinalia with excavate hinge plates supported by median septum, separated into inner and outer parts by crural bases, cardinal process absent or rudimentary, shelflike; loop dalliniform (810). *?Eoc.,* Japan; *Mio.,* Italy-Japan; *Plio.,* Italy-Japan; *Pleist.,* Italy-Japan-Norway; *Rec.,* Atl.O.(40-1560 m.)-Pac.O.(90-210m.)-Medit.——FIG. 721,3. *D. septigera* (LOVÉN), Rec., NE.Atl.O.; *3a-d,* brach.v., lat., ped.v., ant. views, ×1; *3e,* brach.v. int., oblique lat. view showing loop, ×1.5 (244). [HATAI.]

Campages HEDLEY, 1905, p. 43 [*C. furcifera*; OD]. Rectimarginate to intraplicate, punctae fine; beak short, suberect to erect; foramen marginate, permesothyridid, symphytium narrow, pedicle-collar short. Hinge teeth strong, without dental plates or swollen bases; cardinalia with excavate hinge plates resting on median septum, cardinal process small, shelflike, loop early terebrataliiform, funnel-like in appearance (810). *Mio.-Rec.,* Japan; *Tert.,* N.Z.; *Rec.,* W.Pac.(140-1400 m.)-Ind.O.——FIG. 721,5. *C. furcifera,* Rec., Australia; *5a,* ped.v. cxt., lat. view, ×1.8; *5b,* brach.v. int. showing loop (reconstr.), ×2.5; *5c,* slightly oblique lat. view of loop, ×2.5 (810). [HATAI.]

Chathamithyris ALLAN, 1932, p. 15 [*C. traversi*; OD]. Incipiently sulcate, punctae small and dense. Beak prominent, suberect; foramen mesothyridid, attrite; symphytium rather low. Cardinalia weak, loop terebrataliiform or early dalliniform (16). *Tert.,* N.Z.——FIG. 721,7. *C. traversi,* Tert., N.Z.; *7a,b,* brach.v. and lat. views, ×4.6 (16). [HATAI.]

Coptothyris JACKSON, 1918, p. 479 [*Terebratula grayi* DAVIDSON, 1852, p. 76 (=?*Magasella adamsi* DAVIDSON, 1871, p. 54); OD] [=*Thomsonia* JACKSON, 1916, p. 22 (*non* SIGNORET, 1879; *nec* KONOW, 1884); *Pereudesia* DALL, 1920, p. 360 (*nom. subst. pro Thomsonia*); *Cacata* STRAND, 1928, p. 38 (*pro Thomsonia*)]. Widely oval to transverse, hinge line long and little curved, sulcate, test coarsely and irregularly multiplicate, punctate; beak obtuse, foramen large, permesothyridid, deltidial plates conjunct, commonly broken, pedicle collar short, sessile. Hinge teeth with strong dental plates; cardinalia as in *Terebratalia,* median septum reduced, loop dalliniform (399). *Mio.-Rec.,* Japan-Korea.——FIG. 721,4. *C. grayi* (DAVIDSON), Rec., Korea Str.; *4a,b,* brach.v. and lat. views, ×1; *4c,* brach.v. int. showing loop (reconstr.), ×1 (244). [HATAI.]

Dallinella THOMSON, 1915, p. 75 [*Terebratalia obsoleta* BEECHER, 1893, p. 382, 392 (=*Terebratella occidentalis obsoleta* DALL, 1891, p. 186); OD]. Differs from *Terebratalia* in having permesothyridid foramen and narrowly intraplicate folding (810). *Mio.-Rec.*(100-220 m.), NW.Am.——FIG. 721,6. *D. obsoleta* (BEECHER), Rec., brach.v. view, ×1 (220). [HATAI.]

Diestothyris THOMSON, 1916, p. 503 [*Terebratula frontalis* MIDDENDORFF, 1849, p. 518; OD]. Narrowly sulcate, thick, smooth, punctae rather large and widely spaced; beak large, obtuse; foramen submesothyridid, attrite; deltidial plates rudimentary; pedicle collar sessile, long, striate. Hinge teeth strong, with ventrally recessive dental plates; cardinalia strong, consisting of strong socket ridges separated by callous deposit; cardinal process low; median septum in front of valve; loop terebrataliiform. Dorsal adductor impressions strong, anterior and posterior muscle impressions lying side by side (810). *Mio.,* N.Am.; *Plio.,* N.Am.-E.Asia; *Pleist.,* N.Am.-Eu.; *Rec.,* N.Pac. [HATAI.]

D. (Diestothyris). Small to medium-sized, distinctly sulcate, foramen moderately large. *Mio.,* N.Am.; *Plio.,* N.Am.-Kamchatka-Japan; *Pleist.,* N.Am.; *Rec.,* N.Pac.-Okhotsk Sea-Japan Sea.——FIG. 722,2. *D. (D.) frontalis* (MIDDENDORFF), Rec., N.Pac.; brach.v. int. showing loop (reconstr.), ×2 (244). [HATAI.]

D. (Tisimania) HATAI, 1938, p. 97 [*Diestothyris tisimana* NOMURA & HATAI, 1936, p. 132; OD]. Differs from *D. (Diestothyris)* in large size, rectimarginate to incipient sulcate folding, smaller foramen, cardinalia of *Terebratalia* type, median septum intermediate between *Terebratalia* and *D. (Diestothyris),* strong muscle impressions separated by septum in brachial valve and lying anterior to septal ridge in pedicle valve (399). *Rec.,* NW.Pac. [HATAI.]

Fallax ATKINS, 1960, p. 72 [*F. dalliniformis*; OD]. Hinge teeth supported by dental plates, pedicle collar sessile, folding sulcate to intraplicate, beak erect, beak ridges rounded. Cardinalia with platform and thick septum, hinge plates excavated, no

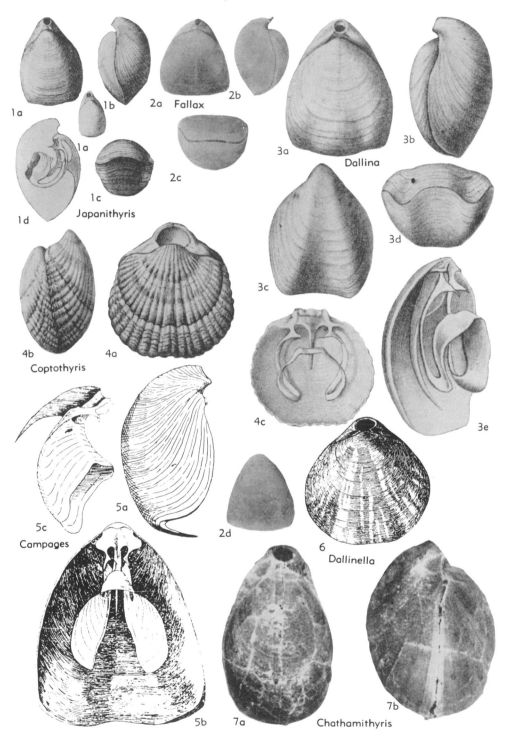

FIG. 721. Dallinidae (Dallininae) (p. *H*835, *H*837).

crural bases, no cardinal process, loop campagiform, adult lophophore plectolophous. Spicules abundant, occurring in lophophore, body wall and mantle canals (43). *Rec.*, E.Atl.——Fɪɢ. 721,2. **F. dalliniformis*; *2a-d*, brach.v., lat. ant., ped.v. views, ×1 (43). [Hatai.]

Glaciarcula Elliott, 1956, p. 285 [**Terebratella spitzbergensis* Davidson, 1852, p. 78; OD]. Small, elongate-pyriform, biconvex, rectimarginate; test thin; ventral beak prominent, incurved, abraded; pedicle-opening elongate-triangular; deltidial plates long, narrow, disjunct; cardinal process minute, crural bases and inner socket-ridges fused, septum passing posteriorly into deep hinge-trough; loop terebrataliiform. *Pleist.*, Scand.; *Rec.*, N.Atl.-?Japan.——Fɪɢ. 723,*1*. **G. spitzbergensis* (Davidson), Rec., N.Atl.; *1a,b*, brach.v., lat. views, ×3 (244); *1c*, loop, ×7 (917). [Elliott.]

Japanithyris Thomson, 1927, p. 251 [**Terebratella Mariae* A. Adams, 1860, p. 412; OD]. Like *Dallina* in shape, folding, beak characters, cardinalia, and absence of dental plates, differing only in not having passed the terebrataliiform loop stage and in much smaller adult size; loop less advanced than in *Campages* (810). *Plio.*, Italy; *Pleist.*, E. China Sea (Ryukyu Is.); *Rec.*, Japan.——Fɪɢ. 721, *1*. **J. mariae* (A. Adams), Rec., Japan; *1a,a'*, brach. view, ×2, ×1; *1b,c*, lat., ant. views, ×2; *1d*, slightly oblique lat. view of loop, ×2 (244). [Hatai.]

Macandrevia King, 1859, p 261 [**Terebratula cranium* Müller, 1776, p. 249; OD] [not preoccupied by *Macandrewia* Gray, 1860] [=*Frenula* Dall, 1871, p. 55 (type, *F. jeffreysi*); *Waldheimiathyris* Helmcke, 1939, p. 331]. Biconvex, rectimarginate to sulcate, thin, smooth or with fine radial sculpture, punctae minute and rather distant; beak obtuse, suberect to erect; beak ridges ill-defined; foramen ?permesothyridid, attrite; deltidial plates weak. Hinge teeth strong, supported by dental plates, which are united by callus deposit closely applied to floor of valve; crural bases fused with socket ridges, from which 2 hinge plates steeply descend to valve floor, not supported by septum but excavated anteriorly at their sides; loop dalliniform. Diductor muscles attached to small transverse impression over dorsal umbo (810). *Mio.*, Japan; *Plio.*, Italy; *Pleist.*, Norway-Sweden - Italy, *Rec.*, Atl.(10-2,900 m.) - Pac.(240-4,400 m.)-Antarctic(400-2,800 m.).——Fɪɢ. 722, *1*. **W. cranium* (Müller), Rec., Atl.; *1a-d*, brach. v., lat., ped.v., ant. views, ×1.2 (*1a,b*, with protruding pedicle); *1e,f*, foramen, ×3 (*1f* showing hinge teeth); *1g*, loop (reconstr.), ×3 (299). [Hatai.]

Pacifithyris Hatai, 1938, p. 98 [**Terebratalia xanthica* Dall, 1920, p. 346; OD]. Differs from *Terebratalia* in having no collar and cardinal process, widely divided crura, crural stems being united to concave platform which is continuous with posterior end of median septum dividing

space beneath platform into 2 cavities, lack of septum or mesial ridge between muscle scars (643). *Rec.*, Japan(170 m.)-USSR(Sakhalin Is., 125 m.). [Hatai.]

Pegmathyris Hatai, 1938, p. 225 [**Dallina miyatokoensis* Hatai, 1936, p. 315; OD]. Differs from

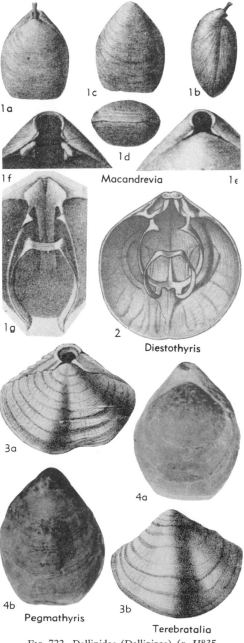

1a　1b　1c

1d

1f　Macandrevia　1ε

1g　2　Diestothyris

3a

4a

3b

4b　Pegmathyris　Terebratalia

Fɪɢ. 722. Dallinidae (Dallininae) (p. *H835*, *H837-H838*).

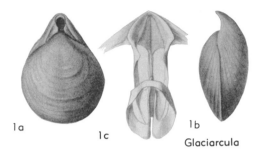

1a
1c
1b
Glaciarcula

FIG. 723. Dallinidae (Dallininae) (p. *H837*).

Dallina in much thicker test, rectimarginate folding, high symphytium, straighter and stronger beak, stronger median septum, inner hinge plates horizontal instead of inclined, stronger muscle impressions; cardinal process strong (399). *Mio.,* Japan.——FIG. 722,4. **P. miyatokoensis* (HATAI); *4a,b,* brach.v. and ped.v. views, ×1 (399). [HATAI.]

Terebratalia BEECHER, 1893, p. 377 [**Terebratula transversa* G. B. SOWERBY, 1846, p. 94 (=*Magasella radiata* DALL, 1877); OD]. Rectimarginate to sulcate, smooth or radial ribs, punctae moderately developed; beak generally suberect, beak ridges usually sharp; foramen mesothyridid, attrite, incomplete or uncommonly complete; pedicle collar short, sessile. Hinge teeth with ventrally recessive dental plates which may be obsolete; cardinalia strong, with callus between socket ridges in umbonal region joined to septum; cardinal process variable, fused with callus; median septum generally stout; loop terebrataliiform. Muscle impressions may be very strong (810). *Oligo.,* W. N.Am.-?Japan; *Mio.-Pleist.,* W.N.Am.-Japan; *Rec.,* N.Pac.(10-1,750 m.).——FIG. 722,3. **T. transversa* (SOWERBY), Rec., Can.(Vancouver Is.); *3a,b,* brach.v. and ped.v. views, ×1 (244). [HATAI.]

Terebrataliopsis SMIRNOVA, 1962, p. 98 [**T. quadrata;* OD]. Circular to rounded pentagonal, pedicle valve strongly convex, carinate; brachial valve slightly convex, with faint sulcus, commissure rectimarginate to slightly sulcate; umbo overhanging brachial valve, foramen small, beak ridges sharp; hinge line terebratulid. Teeth broad, dental plates divergent; ventral septum fourth of valve length; cardinal process not observed, hinge plate supported by median septum; loop passing through massive campagiform and frenuliniform stages to adult terebrataliiform, with wide, irregular spinous bands. *L.Cret.,* USSR.——FIG. 724,2. **T. quadrata; 2a,b,* brach.v. int. (reconstr.), ×1.25 (748). [ELLIOTT.]

Subfamily GEMMARCULINAE Elliott, 1947

[Gemmarculinae ELLIOTT, 1947, p. 145] [Materials for this subfamily prepared by G. F. ELLIOTT]

Cardinal process large, fused with cardinalia; accessory structures present on all growth stages of terebrataliiform loop. *L.Cret.-U.Cret.*

Gemmarcula ELLIOTT, 1947, p. 145 [**G. aurea* (=*Terebratula truncata* J. DE C. SOWERBY, 1826, p. 71, *non Anomia truncata* LINNÉ, 1767, p. 1152); OD]. Small to medium-sized, biconvex, ovate-quadrilateral, coarsely striate, rectimarginate to parasulcate; umbo short, suberect, markedly truncate; foramen large, vertically ovate, submesothyridid; prominent symphytium, area high; hinge line submegathyridid; pedicle collar present. Pedicle valve with strong median ridge; cardinalia strong, buttressed by median septum, cardinal process transverse and countersunk in cardinal platform, inner and outer socket ridges present;

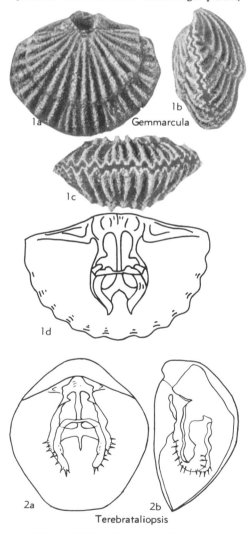

1b
1a
Gemmarcula
1c
1d
2a
2b
Terebrataliopsis

FIG. 724. Dallinidae (Dallininae) *(2),* (Gemmarculinae) *(1)* (p. *H838-H839*).

loop terebrataliiform, with accessory ears, passing through precampagiform to frenuliniform growth stages. *L.Cret.-U.Cret.,* Eu.; *U.Cret.,* N.Am.——Fig. 724,*1a-c. G. arizonensis* COOPER, U.Cret., USA(Ariz.); *1a-c,* brach.v., lat. ant. views, ×3.3 (187).——Fig. 724,*1d. *G. aurea,* L.Cret., Eng., brach.v. int., ×6.5 (276).

Subfamily KINGENINAE Elliott, 1948

[Kingeninae ELLIOTT, 1948, p. 311] [Materials for this subfamily prepared by G. F. ELLIOTT]

Loop development after campagiform passing to kingeniform, modified kingeniform, or belothyridiform. *U.Jur.-U.Cret.*

Kingena DAVIDSON, 1852, p. 40 [**Terebratula lima* DEFRANCE, 1828, p. 156; OD] [*=Kingia*

SCHLOENBACH, 1865, p. 296 *(nom. null.)* (*non* THEOBALD, 1910; *nec* MALLOCH, 1921)]. Medium-sized, biconvex, rounded-pentagonal, rectimarginate to slightly ligate; test thin, with tiny external asperities and color-traces; umbo short, suberect, truncated, foramen permesothyridid, deltidial plates disjunct and obscured by beak ridges. Pedicle collar sessile, muscle marks and pallial sinus grooves faint. Cardinalia with inner socket ridges prominent, cardinal process small, transverse, concave hinge trough supported by thin low median septum, loop with very broad hood-like transverse band doubly attached to septum above attachments from descending branches, passing through precampagiform and campagiform growth stages before diverging to kingeniform.

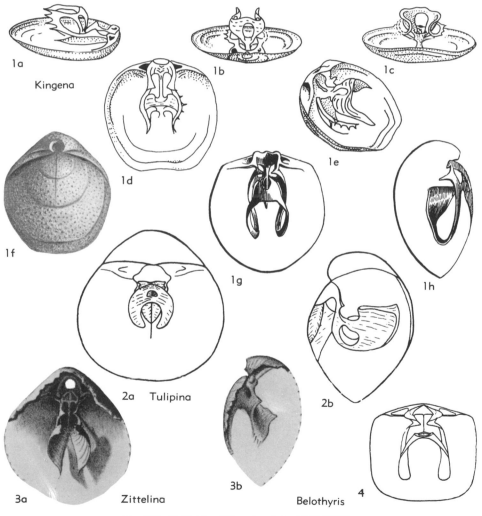

1a

Kingena

1b

1c

1d

1e

1f

1g

1h

2a Tulipina

2b

3a Zittelina

3b

Belothyris 4

FIG. 725. Dallinidae (Kingeninae) (p. *H839-H841*).

L. *Cret.* - *U. Cret.*, Eu.-Asia-Afr.-Australia-N. Am.
——FIG. 725,*1a-e. K. mesembrina* (ETHERIDGE),
U.Cret., W.Australia; *1a-e,* lat., post., ant., dorsal,
oblique of loop, ×4 (281).——FIG. 725,*1f-h.* **K.*

lima (DEFRANCE), U.Cret., Eng.; *1f-h,* brach.v.
view, brach.v. int., lat. view of loop, ×2 (229).
Belothyris SMIRNOVA, 1960, p. 117 [**B. plana*; OD].
Small, terebratuliform, pentagonal to elongate in

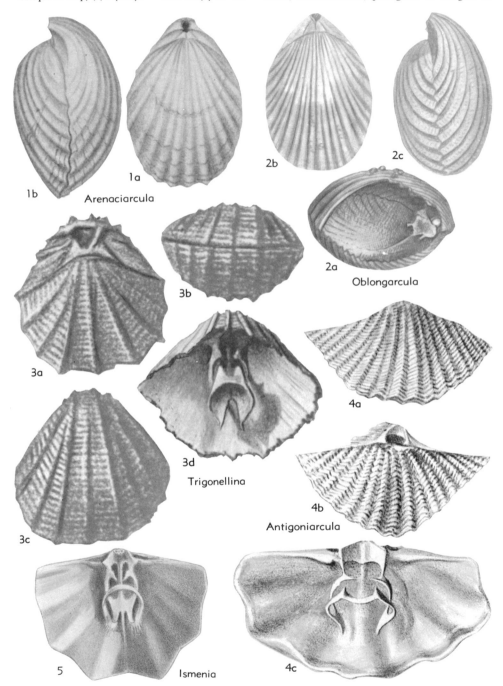

FIG. 726. Dallinidae (**Trigonellininae**) *(3-5)*, (Subfamily Uncertain) *(1-2)* (p. *H*841-*H*842, *H*844-*H*845).

outline, biconvex, smooth, commissure rectimarginate, folding ligate to strangulate, beak characters similar to those of *Kingena*; test thin, dental plates well developed, curved, teeth massive. Dorsal valve with inner socket ridges and crural bases fused and overhanging concave median hinge trough; thin median septum rising steeply to its apex at about 0.3 of valve length. Loop fairly stout; with conspicuous triangular crura, subparallel descending branches widening at points of recurvature, continuing as broad ascending branches which curve posteriorly and inwardly forming dorsoventrally curved arc, from which 2 short connecting bands are individualized out to apex of median septum; accessory loop structures present. [Thus no connecting bands from septum to descending branches occur, or between ascending and descending branches, and although connection between septum and ascending branches is somewhat reminiscent of that in *Kingena,* proportions of ascending branch structure are quite different and there is no hood.] *L.Cret.,* USSR.——Fig. 725,4. **B. plana;* brach.v. int. (reconstr.), ×1.3 (747).

Tulipina Smirnova, 1962, p. 102 [**Zeilleria koutaisensis* De Loriol, 1896, p. 145; OD]. Small, globose, terebratuliform, with shallow dorsal sulcus. Beak incurved, beak ridges rounded, pinhole foramen. Anterior commissure rectimarginate to sulcate. Teeth wide, dental plates close to shell walls, low rounded ventral septum to half valve length. Cardinal process not seen, crural bases massive, septalium deep and cuplike, dorsal median septum formed of 1 median and 2 lateral plates. Loop developing from campagiform hood by frenuliniform resorption during growth to adult loop somewhat like that of *Kingena,* but with different proportions and with transverse hoodlike structure developed dorsally and not vertically or ventrally. *L.Cret.,* USSR.——Fig. 725,2. **T. koutaisensis* (De Loriol); *2a,b,* brach.v. int. (reconstr.), ×4.5 (748).

Zittelina Rollier, 1919, p. 368 [**Terebratula orbis* Quenstedt, 1858, p. 639; OD]. Small, smooth, kingeniform. Pedicle collar present. Cardinalia with small central concave hinge platform, septum thin, loop campagiform, with short descending branches fringed with long spines, and large hood with annular flutings, laterally angled and anteriorly produced into gracefully curved projections. *U.Jur.,* Eu.-W.Asia; *?L.Cret.,* Eu.——Fig. 725,3. **Z. orbis* (Quenstedt), U.Jur., Ger.; *3a,b,* dorsal and lat. views of loop, ×4 (900).

Subfamily TRIGONELLININAE Elliott, n. subfam.

[Materials for this subfamily prepared by G. F. Elliott]

Shell transverse, with costate, scaly surface; cardinalia small; inner socket ridges conspicuous; loops angular, spinose, campagiform to dalliniform. *L.Jur.-U.Jur.*

Trigonellina Buckman, 1907, p. 342 [**Terebratu-*

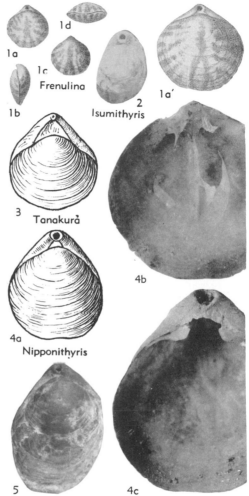

Fig. 727. Dallinidae (Frenulininae) *(1)*, Nipponithyridinae *(2-5)* (p. H842-H843).

lites pectunculus von Schlotheim, 1820, p. 272; OD]. Small, transverse terebratuliform, test thick, scaly, ornamented by several well-spaced, rounded, opposite carinae, foramen with conjunct deltidial plates, lateral areas delimited by beak ridges, hinge line nearly strophic; small median ridge in ventral umbo. Cardinalia with inner socket ridges prominent, cardinal process small, small cardinal platform buttressed by median septum extending to middle of valve, loop small, campagiform. *U.Jur.,* Eu.——Fig. 726,3. **T. pectunculus* (von Schlotheim), Ger.; *3a-c,* brach.v., ant., ped.v. views, ×4; *3d,* loop, ×4 (900).

Antigoniarcula Elliott, 1959, p. 146 [**Argiope perrieri* Eudes-Deslongchamps, 1853, p. 5; OD]. Small, transverse, alate; test costate and scaly, with shallow median sulcus in brachial valve opposed by opposite fold in pedicle valve; foramen large, with narrow deltidial plates and pedicle collar, lateral areas wide and low, hinge line strophic. Brachial valve interior with small hinge plates, delimited by inner socket ridges, supported anteriorly by very short thin median septum, cardinal process small, crura thin and delicate, loop angular dalliniform, anteriorly produced into sharp points. *L.Jur.,* W.Eu.——FIG. 726,4. **A. perrieri* (DESLONGCHAMPS), Fr.; *4a,b,* ped.v. and brach.v. views, ×3; *4c,* brach.v. int. showing loop (reconstr.), ×4.5 (256).

Ismenia KING, 1850, p. 81, 142 [**Terebratulites pectunculoides* VON SCHLOTHEIM, 1820, p. 271; OD]. Small, transverse, biconvex, with about 5 prominent alternate rounded carinae on each valve; test thick, scaly; beak blunt, lateral areas low, foramen large, rounded, deltidial plates small; hinge line nearly strophic. Brachial interior with small central cardinalia, prominent inner socket ridges, cardinal platform small, buttressed by median septum, cardinal process transverse, loop terebrataliiform, anteriorly spinous. *U.Jur.,* Eu. ——FIG. 726,5. **I. pectunculoides* (VON SCHLOTHEIM), Ger.; brach.v. int. showing loop (reconstr.), ×3 (227).

Subfamily FRENULININAE Hatai, 1938

[Materials for this subfamily prepared by KOTORA HATAI]

Biconvex, rectimarginate to sulcate, deltidial plates conjunct to disjunct, foramen submesothyridid. Cardinal process slender; dental plates present; cardinalia widely divergent; spiculation absent; loop passing through preismeniform stage and attaining frenuliniform stage in adult (399). *Mio.-Rec.*

Frenulina DALL, 1894, p. 724 [**Anomia sanguinolenta* GMELIN, 1792, p. 3347; OD]. Sulcate, thin, smooth, punctae coarse and dense; beak suberect; foramen submesothyridid, attrite; deltidial plates disjunct, appearing conjunct; pedicle collar closely applied to floor of valve. Hinge teeth with dental plates; socket ridges rather strong, united on their inner sides to crural bases; cardinalia divergent clear to apex; cardinal process small, striated over umbo; loop frenuliniform. Muscle impressions feeble (81). *Rec.,* Australia to Ryukyu Is.(E.China Sea).——FIG. 727,1. **F. sanguinolenta* (GMELIN), Rec., Hawaiian Is.; *1a-d,* brach.v., lat., ped.v., ant. views, ×1 (*1a',* ×2) (244).

Jolonica DALL, 1920, p. 366 [**Campages (Jolonica) hedleyi* DALL; OD]. Rectimarginate to weakly sulcate, punctae fine and rather dense; beak rather short; foramen complete; pedicle collar feeble. Hinge teeth with dental plates, leaving narrow area between it and shell wall; pedicle valve with short septal ridge; cardinalia with deep sockets, divergent clear to apex and headed by cardinal process; median septum prominent. Muscle im-

pressions weak (643). *Pleist.,* E. China Sea (Ryukyu Is.); *Rec.,* S. Japan-Philip. Is.(to 640 m.).

Kamoica HATAI, 1936, p. 313 [**Jolonica (Kamoica) iduensis* HATAI, 1936; OD]. Biconvex, rectimarginate, thick, punctae oval, dense; beak suberect; foramen incomplete, ?submesothyridid; deltidial plates disjunct. Hinge teeth strong, with slightly recessive dental plates; cardinalia divergent clear to apex, cardinal process prominent; crural bases united to inner part of divergent socket ridges but separated from them by groove, crura obliquely vertical, slender; median septum fused with weak callus. Muscle impressions weak (399). *Mio.-Plio.,* Japan.

Subfamily NIPPONITHYRIDINAE Hatai, 1938

[*nom. correct.* HATAI, herein (*pro* Nipponithyrinae HATAI, 1938)] [Materials for this subfamily prepared by KOTORA HATAI]

Rectimarginate, sulcate to intraplicate; beak nearly straight to erect, foramen complete, symphytium straight to concave. Hinge teeth strong, pedicle valve with septal ridge; loop not passing terebrataliiform stage in Recent genera; cardinalia with inner hinge plate troughlike or partially filled; cardinal process present in all forms; median septum much thickened. Spicules unknown (399). *Mio.-Rec.*

Nipponithyris YABE & HATAI, 1934, p. 588 [**N. nipponensis;* OD]. Punctae minute and dense. Foramen complete, symphytium solid, concave or nearly straight. Hinge teeth stout, with swollen bases and grooves for reception of socket ridges; pedicle valve with strong septal ridge separating muscle impressions; inner hinge plates deeply sunken, partially excavated beneath, generally calloused; cardinal process small; loop terebrataliiform; crura attached to lower part of socket ridges, short and slender, posterior edge of ascending branches with square notch. Dorsal muscle impressions fairly strong (399). *Mio.-Rec.,* Japan. ——FIG. 727,4. **N. nipponensis,* Rec.; *4a,* brach. v. view, ×1; *4b,c,* brach.v. int., ped.v. int., ×2 (399).

Isumithyris HATAI, 1948, p. 498 [**I. kazusaensis;* OD]. Differs from *Nipponithyris* in smaller size, erect beak, mesothyridid foramen, intraplicate folding, shorter symphytium without any mesial ridge, and lack of septal ridge on floor of pedicle valve. Punctae fine and rather dense (401). *Plio.,* Japan.——FIG. 727,2. **I. kazusaensis;* brach.v. view, ×2 (280).

Miyakothyris HATAI, 1938, p. 237 [**Nipponithyris subovata* HATAI, 1936; OD]. Differs from *Nipponithyris* in its rectimarginate folding, incipient truncation, very solid and thick test, higher but less incurved beak, suberect position of foramen, stronger cardinalia, and stronger median septum, which is less swollen at bases (399). *Mio.,* Japan.

Tanakura HATAI, 1936, p. 322 [**Magasella fibula*

HAYASAKA, 1921, p. 1 (*non* REEVE, 1861, p. 180) (=*Tanakura tanakura* HATAI, 1936, *nom. nov.*)]. Rectimarginate, punctae fine; beak erect, nearly epithyridid; symphytium concave; pedicle collar sessile. Hinge teeth strong, bases swollen and grooved for reception of socket ridges; septal ridge short; cardinalia with callus between socket ridges uniting with septum; cardinal process trefoil on top; crural bases closely applied to socket ridges. Muscle impressions rather strong (399). *Mio., Japan.*——FIG. 727,3. **T. tanakura* HATAI; brach. v. view, ×2 (244).

Yabeithyris HATAI, 1948, p. 498 [**Y. notoensis*; OD]. Rectimarginate, smooth, rather thick, punc-

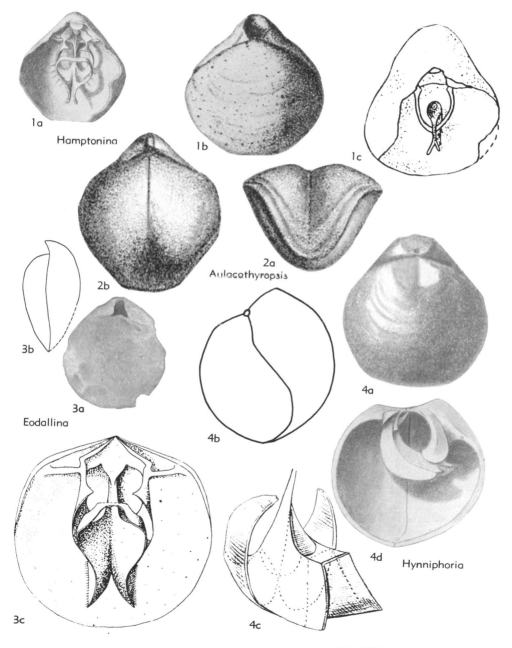

FIG. 728. Dallinidae (Subfamily Uncertain) (p. *H*844-*H*845).

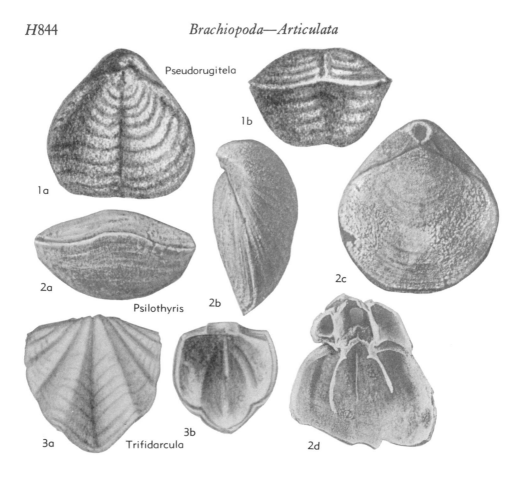

FIG. 729. Dallinidae (Subfamily Uncertain) (p. *H*845).

tae minute, elongate-oval and dense; beak erect; foramen complete; symphytium short, with median raised ridge; pedicle collar indistinct. Hinge teeth with ventrally recessive dental plates; cardinalia with deep trough posteriorly, anteriorly margined by divergent crural bases and prominent median septum; no cardinal process, only small shelflike process over umbo (401). *Mio.,* Japan.——FIG. 727,5. *Y. notoensis;* brach.v. view, ×2 (401).

Subfamily UNCERTAIN

[Materials for this assemblage prepared by G. F. ELLIOTT]

Arenaciarcula ELLIOTT, 1959, p. 147 [*Terebratella fittoni* MEYER, 1864, p. 250; OD]. Like *Oblongarcula* but smaller, with fewer and coarser costae, thicker test, commissure uniplicate to parasulcate, dental plates fused to sides of valve, and cardinal platform with well-marked separate raised inner socket ridges, crural bases, and median septal ridge. Loop believed terebrataliiform. *L.Cret.,* W. Eu.——FIG. 726,1. *A. fittoni* (MEYER), Eng.; *1a,b,* brach.v. and lat. views, ×3 (557).

Aulacothyropsis DAGIS, 1959, p. 99 [*Waldheimia* (*Aulacothyris*) *reflexa* BITTNER, 1890, p. 258;

OD]. Terebratuliform, plano-convex, like *Aulacothyris* externally. Umbo short, foramen small, mesothyridid, beak ridges well developed. Pedicle valve with short dental plates united by callus posteriorly; teeth thick, wedge-shaped, not denticulate; brachial valve with hinge plate showing inner socket ridges and crural bases, median septum very long and thin; loop long, descending branches united posteriorly to septum, then proceeding anteriorly with each branch apparently united along its length to corresponding ascending branch; loop set dorsally with thick short spines; possibly an early dallinid (terebrataliiform) pattern. ?*M.Trias., U.Trias.,* C.Eu.-USSR.——FIG. 728,2. *A. reflexa* (BITTNER), U.Trias., C.Eu.; *2a,b,* ant., brach.v. views, ×3 (76). [See also Fig. 699.]

Eodallina ELLIOTT, 1959, p. 146 [*E. peruviana;* OD]. Small, terebratuliform, shallow dorsal sulcus only; beak straight, entire, pedicle opening triangular, without deltidial plates. Pedicle valve interior with medianly divided muscle field; cardinalia with inner socket ridges enclosing concave hinge plates supported by low median septum, which extends anteriorly past mid-length, supporting spinous modified campagiform loop with short crural points, descending branches broadly attached and hood thinning to narrow transverse band. *U.Trias.*, S.Am.——FIG. 728,3. **E. peruviana*; *3a,b,* brach.v., lat. views, ×4; *3c,* brach.v. (reconstr.), ×8 (776).

Hamptonina ROLLIER, 1919, p. 360 [**Terebratella buckmani* MOORE, 1860, p. 441; OD]. Small, terebratuliform, rectimarginate; beak suberect, foramen rounded, small disjunct triangular deltidial plates. Brachial valve interior with inner socket ridges enclosing concave hinge plate, cardinal process small, transverse; septum low posteriorly, rising steeply anteriorly, supporting spinous modified campagiform loop with moderately long crural points, broadly attached descending branches, long anterior spurs and ascending branches modified to backwardly directed ring. Precampagiform growth stages with high septal pillar and nearly vertical hood known; abnormal individuals with short brachial structure free of valve floor known. *M.Jur.*, Eng.——FIG. 728,1. **H. buckmani* (MOORE); *1a,* brach.v. int. showing loop (reconstr.), ×4; *1b,* brach.v. view, ×4 (569); *1c,* brach.v. int. of juvenile shell (reconstr.), ×20 (279).

Hynniphoria SUESS, 1859, p. 44 [**H. globularis*; OD]. Small, terebratuliform, globose, with both valves markedly inflated, length about equal to width; test smooth, pedicle valve with broad shallow median sulcus occasioning wide boxlike anterior uniplication, lateral commissures falciform; umbonal areas flattened, foramen very small, deltidial structures small and obscure, beak ridges round and indistinct, hinge line terebratuliform. Pedicle valve interior with dental plates and stout, apparently composite, median septal structure which projects free anteriorly as curved and thickening blade- or scimitar-like structure, extending dorsally into shell cavity. Brachial valve with divided cardinalia, stout inner socket ridges; small, short posterior median septum diminishing rapidly anteriorly; descending branches of loop broad and bladelike, posteriorly attached to septum, then twisting to apparent squarish boxlike campagiform hood with divergent anterior corners. *U.Jur.*, C.Eu.——FIG. 728,4. **H. globularis*; *4a,b,* brach.v. and lat. views, ×6; *4c,* oblique view of loop (reconstr.), ×10; *4d,* lat. view of shell int. (reconstr.), ×6 (794).

Oblongarcula ELLIOTT, 1959, p. 147 [**T. oblonga* J. DE C. SOWERBY, 1826, p. 68; OD]. Medium-sized, elongate-ovoid, biconvex, commissure rectimarginate to sulcate, test thin, ornamented by fine regular costae bifurcating only near umbo; beak suberect, foramen mesothyridid, deltidial plates conjunct, areas delimited by beak ridges, hinge line terebratulid. Cardinalia thin, platelike, cardinal process raised, transverse, sockets narrow, inner socket ridges enclosing wide thin hinge plate, medium septum thin, supporting hinge plate beneath and extending anteriorly to half of valve length, crura delicate, loop believed terebrateliiform. *L.Cret.*, Eu.——FIG. 726,2. **O. oblonga* (SOWERBY), Eng.; *2a-c,* shell int. from side, brach. v. and lat. views, ×2 (761).

Pseudorugitela DAGIS, 1959, p. 100 [**Waldheimia (Aulacothyris) pulchella* BITTNER, 1890, p. 200; OD]. Terebratuliform, biconvex, tending to anterior ligation; valve surfaces with strongly developed concentric growth steps. Dental plates thin and parallel; brachial valve hinge plate divided by deep narrow V-shaped hinge trough to which median septum is joined; loop like that of *Aulacothyropsis* but not spinous. *U.Trias.*, C.Eu.-USSR.——FIG. 729,1. **P. pulchella* (BITTNER), C.Eu.; *1a,b,* brach.v. and ant. views, ×8 (76).

Psilothyris COOPER, 1955, p. 10 [**P. occidentalis*; OD]. Small to medium-sized, smooth, biconvex, ovate to subpentagonal, rectimarginate to uniplicate; umbo erect, foramen small to large, round, submesothyridid to mesothyridid, deltidial plates disjunct to conjunct, hinge line terebratulid. Cardinalia small, hinge plate undivided, medianly concave; median septum, short, slender, with or without buttressing cardinalia; loop dalliniform, with long crural processes and short crura, passing through precampagiform to terebratuliform growth stages. *L.Cret.*, Eu.; *L.Cret.-U.Cret.*, N.Am.——FIG. 729,2. **P. occidentalis*, L.Cret., USA (Ariz.); *2a-c,* ant., lat., brach.v. views, ×4; *2d,* post. part of shell int. showing crura and loop, ×8 (187).

Trifidarcula ELLIOTT, 1959, p. 147 [**Terebratella trifida* MEYER, 1864, p. 167; OD]. Small, transverse, test thick with 3 principal high-raised, rounded, straight-sided dorsal folds alternating with 2 ventral folds; foramen large, deltidial plates small, area sloping, hinge line megathyridid; pedicle collar present. Cardinal platform small, thick, elements fused, cardinal process small, septum thick, extending anteriorly for half valve length, loop believed terebrataliiform. *L.Cret.*, Eng.——FIG. 729,3. **T. trifida* (MEYER); *3a,b,* ped.v. ext., brach.v. int., ×4 (557).

Family LAQUEIDAE Hatai, n. fam.

[=Laqueinidae YABE & HATAI, 1941 (invalid because contains no genus providing stem Laquein-, name being erroneously derived from *Laqueus*)] [Materials for this family prepared by KOTORA HATAI]

Dental plates present, spiculation of mantle canals moderate, loops differing from terebrataliiform and dalliniform loop stages of the Dallinidae in incomplete separation of ascending and descending branches which remain united by interconnecting bands on each side (400). *Mio.-Rec.*

Subfamily LAQUEINAE Hatai, n. subfam.

Biconvex, rectimarginate to ligate or strangulate, smooth, with rather coarse punctae; beak fairly prominent, beak ridges

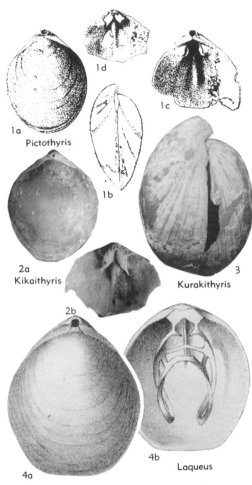

FIG. 730. Laqueidae (Laqueinae) *(4)*, (Pictothyridinae) *(1-2)*, (Kurakithyridinae) *(3)* (p. *H845-H847*).

sharp; foramen permesothyridid, slightly remigrant, telate; deltidial plates conjunct, concave. *Mio.-Rec.*

Laqueus DALL, 1870, p. 123 [**Terebratula californiana* KOCH, 1848, p. 38 (=*Laqueus californicus* CARPENTER, 1864) (*non* KOCH, 1848=*L. erythraeus* DALL, 1920); OD]. Hinge teeth with ventrally recessive dental plates; pedicle collar sessile. Cardinalia with inner and outer hinge plates separated by crural bases; inner hinge plates resting on median septum; no cardinal process; loop laqueiform. Muscle impressions not strong. Small spicules present over mantle canals but not extending to body wall or lophophore (810). *Mio.-Rec.*, N. Am.-Japan-N. Pac.O. (30-1,350 m.).—— FIG. 730,*4*. **L. californianus* (KOCH), Rec., USA (Calif.); *4a,b*, brach.v. view, brach.v. int. showing loop (reconstr.), ×1 (244).

Subfamily PICTOTHYRIDINAE Yabe & Hatai, 1941

[*nom. correct.* HATAI, herein (*pro* Pictothyrinae YABE & HATAI, 1941)]

Beak suberect; foramen permesothyridid, attrite. Hinge teeth with swollen bases; cardinalia divergent, no inner hinge plates; cardinal process prominent, trefoil on top by enfolding of wings; loop without connecting bands from descending branches to median septum, being more advanced than *Laqueus*; median septum stout (810). *Plio.-Rec.*

Pictothyris THOMSON, 1927, p. 260 [**Anomia picta* DILLWYN, 1817, p. 295; OD]. Biconvex, rectimarginate, smooth, punctate; beak suberect; foramen permesothyridid, attrite; deltidial plates conjunct. Hinge teeth strong, bases swollen, appearing soldered to sides, grooved for reception of socket ridges; cardinalia divergent; clear to apex, headed by bilobed cardinal process; median septum stout; descending branches of lophophore not united to median septum, ascending branches united with descending ones by interconnecting bands at corners of transverse band. Ventral muscle scars separated by septal ridge, dorsal muscle scars separated by median septum (810). *Plio.-Rec.*, Japan-Formosa-Ryukyu Is. (40-160 m.).—— FIG. 730,*1*. **P. picta* (DILLWYN), Rec., Japan; *1a,b*, brach.v. and lat. views, ×0.9; *1c,d*, ped.v. int., brach.v. int., ×0.9 (810).

Kikaithyris YABE & HATAI, 1941, p. 491 [**Pictothyris hanzawai* YABE, 1934; OD]. Resembles *Pictothyris* in shape, folding, and cardinalia, differing in very small foramen pierced in strongly incurved beak and by much shorter median septum in brachial valve (400). *Pleist.*, Japan-Ryukyu Is.-Formosa.——FIG. 730,*2*. **K. hanzawai* (YABE), Ryukyu Is.; *2a,b*, brach.v., brach.v. int., ×1 (after 897).

Subfamily KURAKITHYRIDINAE Hatai, 1948

[*nom. correct.* HATAI, herein (*pro Kurakithyrinae* HATAI, 1948)]

Rectimarginate to sulcate; pedicle collar obsolete. Cardinal process indistinct in adult, developing from very weak in young, ventrally recessive dental plates in adult developing from what appear to be swollen bases in young; cardinalia divided into inner and outer hinge plates, former appearing as deep trough bordered by latter, which appear as swollen crural bases, whole excavated beneath and supported by prominent median septum; descending branches of lophophore attached to median septum by short connecting bands in young but becoming free in adult (401). *Plio.*

Kurakithyris HATAI, 1946, p. 98 [**K. quantoensis*;

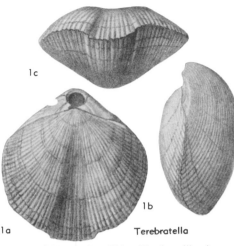

FIG. 731. Terebratellidae (Terebratellinae)
(p. *H847*).

OD]. Biconvex, rectimarginate to sulcate; foramen complete, ?permesothyridid; deltidial plates conjunct, median ridge low. Hinge teeth not strong, with ventrally recessive dental plates, no median ridge in pedicle valve; pedicle collar and muscle impressions indistinct; cardinalia weak, divided into inner and outer hinge plates by swollen bases of crural processes, inner hinge plate troughlike, supported by median septum, excavated beneath. Finely punctate (401). *Plio.,* Japan.——FIG. 730,3. **K. quantoensis*; lat. view, ×1.8 (401).

Family TEREBRATELLIDAE King, 1850

[Terebratellidae KING, 1850, p. 245] [Materials for this family prepared by G. F. ELLIOTT and KOTORA HATAI as indicated, diagnosis by ELLIOTT & HATAI]

Loop passing through all or part of premagadiniform, magadiniform, magelliform, terebratelliform, and magellaniiform growth stages or modifications of these; dental plates absent; animal nonspiculate (810). *U.Cret.-Rec.*

Subfamily TEREBRATELLINAE King, 1850

[*nom. transl.* DAVIDSON, 1866 (*ex* Terebratellidae KING, 1850, p. 245)] [=Magellaniinae BEECHER, 1893] [Materials for this subfamily prepared by KOTORA HATAI]

Cardinalia weak, lamellar, with excavate hinge plates meeting on septum; loop magelliform to magellaniiform (217). *Oligo.-Rec.*

Terebratella D'ORBIGNY, 1847, p. 229 [**Terebratula chilensis* BRODERIP, 1833, p. 141 (=**Anomia dorsata* GMELIN, 1792, p. 3348; *Terebratula flexuosa* KING, 1835, p. 337; *Terebratula patagonica* GOULD, 1850, p. 347); OD]. Sulcate, smooth or with irregular, somewhat wavy, multiplication developing directly on smooth stage, punctae coarse and dense; beak suberect to erect; foramen submesothyridid to mesothyridid; deltidial plates conjunct or almost conjunct. Hinge teeth without swollen bases; excavated hinge plates meeting on median septum, which is low; lophophore plectolophous, filaments long, slender and close. Muscle impressions not strong (810). *Oligo.-Mio., S.Am.-N.Z.; Mio.,* Australia-N.Z.; *Plio.,* N.Z.; *Rec.,* S.

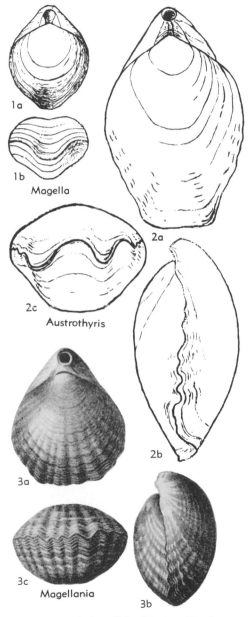

FIG. 732. Terebratellidae (Terebratellinae)
(p. *H848-H849*).

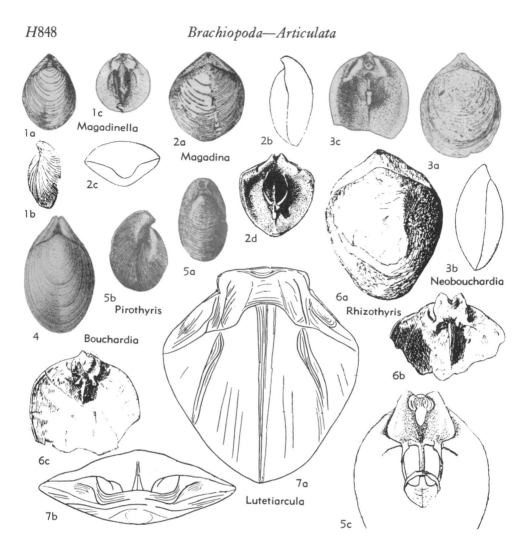

FIG. 733. Terebratellidae (Bouchardiinae) *(3-4)*, (Magadinae) *(1-2,5-7)* (p. H849, H851).

Atl. O. (10-240　m.)-S. Pac. (40-360　m.).——FIG. 731,*1*. **T. dorsata* (GMELIN), Rec., Magellan Str.; *1a-c,* brach.v., lat., ant. views, ×1 (244).

Aerothyris ALLAN, 1939, p. 245 [**Magellania macquariensis* THOMSON, 1918, p. 30; OD]. With internal characters of *Magellania* but differs in being completely smooth, in having discrete deltidial plates and much coarser punctation (24). *Rec.,* Antarctic-S.Pac.O.(600 m.).

Austrothyris ALLAN, 1939, p. 238 [**Waldheimia gambierensis* ETHERIDGE, JR., 1876, p. 19 (=**W. grandis* TENISON-WOODS, 1865)]. Smooth or marginally multiplicate, with intraplicate anterior commissures, differing from *Magellania* in possessing hinge plates adpressed to floor of valve and meeting low on sides of median septum which extends to base of cardinal process, which, while transverse, is supported from floor of hinge trough

(24). *Mio.,* N.Z.——FIG. 732,*2*. **A. grandis* (TENISON-WOODS); *2a-c,* brach.v., lat., ant. views, ×1 (24).

Magasella DALL, 1870, p. 134 [**Terebratella Evansii* DAVIDSON, 1852, p. 77 (=*Terebratula sanguinea* LEACH, 1814, p. 76); OD]. Multicostate. *Tert.-Rec.,* N.Z.

Magella THOMSON, 1915, p. 396 [**M. carinata* (=*Terebratella kakanuiensis* THOMSON, 1908, p. 102, *non* HUTTON, 1905, p. 479)]. Sulcate, smooth, thin; beak suberect; foramen submesothyridid, incomplete; deltidial plates discrete, triangular. Hinge teeth without swollen bases; cardinalia weak, with excavated hinge plates; median septum long; crura short; loop magelliform (810). *Oligo.-Mio.,* Antarctic; *Mio.,* N.Z.; *Pleist.,* Antarctic.——FIG. 732,*1*. **M. carinata* (THOMSON), Mio., N.Z.; *1a,b,* brach.v. and ant. views, ×3 (810).

Magellania BAYLE, 1880, p. 24 [*nom. subst. pro Waldheimia* KING, 1850 (*non* BRULLÉ, 1846)] [**Terebratula australis* QUOY & GAIMARD, 1834, p. 551 (=?*T. flavescens* LAMARCK, 1819, p. 246); OD]. Sulcate to intraplicate, smooth to multiplicate, plicae developing on smooth stage, punctae coarse and dense; beak suberect to erect; foramen mesothyridid, attrite; deltidial plates united in symphytium. Hinge teeth without swollen bases; cardinalia weak, hinge plates excavated, meeting on median septum; cardinal process transverse; crura short, crural processes prominent; loop magellaniform, lophophore plectolophous, filaments long, slender, numerous; setae numerous, short. Four main trunks of mantle canals in both ventral and dorsal mantles, all bearing genital glands except 2 inner trunks of dorsal mantle (810). *Oligo.-Mio.*, Australia-S.Am.; *Mio.*, Australia; *Rec.*, S.Pac.O.(12-600 m.)-Antarctic (600 m.).——FIG. 732,*3*. **M. australis* (QUOY & GAIMARD), *Rec., Australia; 3a-c,* brach.v., lat., ant. views, ×1 (244).

Waltonia DAVIDSON, 1850, p. 474 [**W. valenciennesi* (=*Terebratula inconspicua* G. B. SOWERBY, 1846, p. 339); OD]. Surface smooth. *Tert.-Rec.,* N.Z.

Subfamily BOUCHARDIINAE Allan, 1940

[Bouchardiinae ALLAN, 1940, p. 270] Materials for this subfamily prepared by KOTORA HATAI and G. F. ELLIOTT as indicated, diagnosis by HATAI & ELLIOTT]

Small thick-shelled forms with slightly concave palintrope anterior to sharp beak ridges, lacking grooves which usually mark outlines of delthyrium; cardinalia thick, fused, crural bases uniting in hinge platform; median septum nonbifurcate, no hinge trough; loop premagadiniform (217). *U.Cret.-Rec.*

Bouchardia DAVIDSON, 1850, p. 62 [*Anomia rosea MAWE, 1823, p. 65; OD] [=*Pachyrhynchus* KING, 1850, p. 70 (*non* WAGLER, 1822, *nec* GERMAR, 1824) (obj.)]. Sulcate, smooth and thick, punctae dense; beak obtuse, not incurved; foramen epithyridid; symphytium slightly concave. Hinge teeth strong, with swollen bases grooved for reception of socket ridges; inner high socket ridges enclosing hinge platform; cardinal process fused with platform; lophophore with no descending branches; pedicle valve with low septal ridge (810). *Oligo.-Mio.*, S.Am.; *Mio.*, Antarctic; S.Am. (25-120 m.).——FIG. 733,*4*. **B. rosea* (MAWE), *Rec.*, S.Am.; brach.v. view, ×1 (244). [HATAI.]

Bouchardiella DOELLO-JURADO, 1922, p. 200 [**Bouchardia patagonica* IHERING, 1903, p. 210; OD]. Small, biconvex (dorsal umbo flattened), elongate ovoid-pentagonal, test smooth, thick, commissure sulcate, beak short and nearly straight, beak ridges sharp, foramen epithyridid to permesothyridid, symphytium fused in concave area,

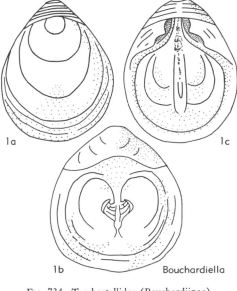

FIG. 734. Terebratellidae (Bouchardiinae) (p. H849).

hinge line slightly sloping. Interior of pedicle valve beak constricted, hinge teeth stout, with grooved swollen bases, muscle marks anterior, divided. Cardinalia with solid platform, prominent socket ridges, cardinal process a subquadrate muscle pit, septum low posteriorly and margined by small cavities in anterior face of platform, then rising steeply anteriorly, as in *Magas,* and bearing 2 curved posteriorly directed plates (retrograde premagadiniform loop). *U.Cret.,* Australia-S.Am. ——FIG. 734,*1*. *B. cretacea* (ETHERIDGE), W. Australia; *1a-c,* brach.v. view, brach.v. int., ped.v. int., ×7.5 (281). [ELLIOTT.]

Neobouchardia THOMSON, 1927, p. 270 [**Bouchardia minima* THOMSON, 1911, p. 260; OD]. Incipiently sulcate, smooth, punctae fine; beak suberect; symphytium with median groove. Cardinal process rounded anteriorly, with posterior tongue and converging lateral wings meeting near umbo; hinge platform pierced by 3 caves, including large and deep central one and 2 smaller lateral ones, separated by 2 small projections (810). *Mio.*, N.Z.-Australia.——FIG. 733,*3*. **N. minima* (THOMSON), *Mio.*, N.Z.; *3a,b,* brach.v. and lat. views, ×3; *3c,* brach.v. int., ×3 (810). [HATAI.]

Subfamily MAGADINAE Davidson, 1886

[*nom. correct.* ELLIOTT & HATAI, herein (*pro* Magasinae DAVIDSON, 1886, p. 4)] [Materials for this subfamily prepared by G. F. ELLIOTT and KOTORA HATAI as indicated, diagnosis by ELLIOTT & HATAI]

Brachial valve with stout socket ridges and crural bases, swollen bases to hinge teeth, posterior hinge trough, unbifurcated septum, loop magadiniform to magellanii-form; foramen permesothyridid (217). *L. Cret.-Rec.*

Magas J. SOWERBY, 1816, p. 39 [**M. pumilus*; OD]. Small, smooth, unequally biconvex or plano-convex, sulcate, beak entire, incurved, overhanging foramen margined by narrow to triangular deltid-

ial plates, areas delimited by sharp beak ridges, hinge line submegathyridid. Ventral interior with constricted beak area, hinge teeth with swollen bases, short, low, thick median ridge tapering anteriorly and posteriorly with deep muscle areas to left and right. Cardinalia occupying median two-thirds of hinge line, inner socket ridges thick, sunken median cardinal platform with cardinal

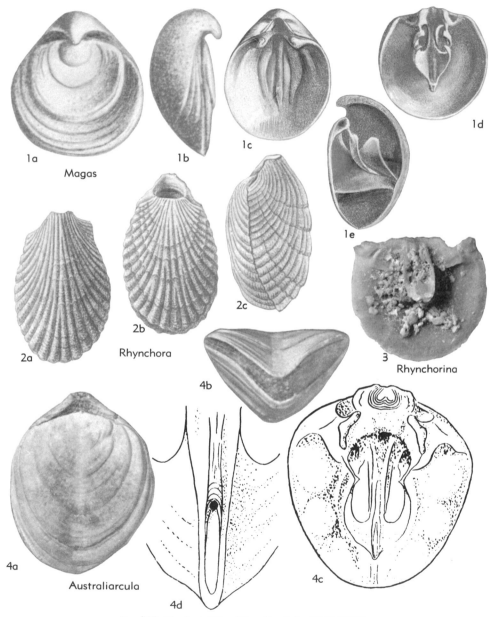

FIG. 735. Terebratellidae (Magadinae) (p. *H849-H851*).

process raised area on this; septum buttressing cardinalia and rising very high as anteriorly directed pillar, crura short, descending branches narrow, straight, broadly attached to septum beneath 2 posteriorly directed curved lamellae (modified magadiniform loop). *U.Cret.*, Eu.——Fig. 735,1. **M. pumilus*; *1a,b*, brach.v. and lat. views; *1c,d*, ped.v. int., brach.v. int. showing loop (reconstr.); *1e*, lat. view of shell int. (reconstr.); all ×4 (229). [Elliott.]

Australiarcula Elliott, 1960, p. 26 [**A. artesiana*; OD]. Small, ovoid, narrowing anteriorly, with median dorsal sulcus matched by ventral keel; test smooth, commissure sulcate. Beak erect, foramen permesothyridid, symphytium concave, beak ridges sharp, hinge line gently sloping. Hinge teeth stout, with grooved swollen bases. Sockets deep, socket ridges strong, cardinal platform solid with posterior muscle pit, septum high anteriorly, loop primitive magadiniform with well-developed descending branches and rudimentary hood. *L. Cret.*, S.Australia.——Fig. 735,4. **A. artesiana*; *4a,b*, brach.v. and ant. views, ×3; *4c*, brach.v. int. (reconstr.), ×4.5; *4d*, loop, ×18 (289). [Elliott.]

Lutetiarcula Elliott, 1954, p. 727 [**L. perplexa*; OD]. Biconvex, with solid strong brachial cardinalia, cardinal process low, median septum well developed, extending nearly to anterior margin, curved lateral brachial ridges on valve floor (284). *Eoc.*, Fr.——Fig. 733,7. **L. perplexa*; *7a,b*, brach. v. int., post. view of brach.v., ×15 (284). [Hatai.]

Magadina Thomson, 1915, p. 399 [**M. browni*; OD]. Sulcate to intraplicate, smooth, solid, punctae dense; beak erect; foramen permesothyridid; symphytium solid, concave; pedicle collar strong, sessile, uniting at sides with posterior part of bases of hinge teeth, which are strong, with swollen bases, and grooved for reception of socket ridges. Pedicle valve with septal ridge, situated anteriorly; crural bases united with socket ridges; hinge trough short, deep; descending and ascending branches of loop separately attached to septum; lophophore zygolophous, median lobe being uncoiled (810). *Oligo.-Mio.*, Australia; *Mio.*, Australia-N.Z.; *Rec.*, S.Australia(25-400 m.).——Fig. 733,2. **M. browni*, Mio., N.Z.; *2a-c*, brach.v., lat., ant. views, ×1; *2d*, brach.v. int., ×1 (810). [Hatai.]

Magadinella Thomson, 1915, p. 400 [**Magasella woodsiana* Tate, 1880, p. 163; OD]. Brachial valve thickened posteriorly and laterally; muscle impressions forming deep pit on each side of median septum; hinge trough moderately large, only partially filled by small, anteriorly directed cardinal process, which is trefoil on top; loop late magelliform or early terebratelliform (810). *Oligo-Mio.*, Australia.——Fig. 733,1. **M. woodsiana* (Tate), Mio.; *1a,b*, brach.v. and lat. views, ×1; *1c*, brach.v. int., ×1 (810). [Hatai.]

Pirothyris Thomson, 1927, p. 280 [*Magasella vercoi

Blochmann, 1910, p. 91; OD]. Uniplicate, punctae fine and dense; beak suberect; foramen apparently permesothyridid, attrite. Cardinalia strong, hinge trough deep, large; cardinal process occupying half length of hinge trough; loop magelliform. Ventral muscle impressions extending anteriorly beyond middle of valve, separated by median ridge; dorsal muscle impressions strong (810). *Rec.*, Australia(30-40 m.).——Fig. 733,5. **P. vercoi* (Blochmann); *5a,b*, brach.v. and lat. views, ×3; *5c*, brach.v. int. showing loop (reconstr.), ×6 (810). [Hatai.]

Rhizothyris Thomson, 1915, p. 399 [**Bouchardia rhizoida* Hutton, 1905; p. 480; OD]. Incipiently sulcate, punctae dense; beak erect, foramen permesothyridid. Hinge teeth strong, bases swollen, restricting umbonal cavity; ventral muscle impressions strong, separated by septal ridge; sockets large, socket ridges high, overhanging, projecting behind umbo, laterally fused with crural bases; hinge trough with large cardinal process; crura originating just above septum; median septum short, stout, uniting with cardinalia; loop magelliform (810). *Mio.*, N.Z.——Fig. 733,6. **R. rhizoida* (Hutton); *6a*, brach.v. view, ×1; *6b,c*, brach.v. int., ped.v., int., ×1 (810).

Rhynchora Dalman, 1828, p. 135 [**Terebratula costata* Nilsson, 1827, p. 37 (=**Anomites costatus* Wahlenberg, 1821, 1819, p. 62=*Anomia pectinata* Linné, 1758, p. 701)]. Large, thick, pedicle valve deeper than brachial, coarsely costate, sulcate; area nearly at right angles to plane of lateral commissures, foramen very large, vestigial deltidial plates at anterior corners. Hinge teeth large, widely separated, low median pedicle valve ridge fading anteriorly. Cardinalia thick, rounded, elements fused, socket ridges thick and fused with cardinal platform, cardinal process a large, slightly raised median surface area on platform, hollows under cardinal platform beneath crural processes; septum thin, buttressing cardinalia, high posteriorly and diminishing anteriorly to midpoint of valve; loop unknown, believed to be of terebratelliform series. *U.Cret.*, NW.Eu.——Fig. 735,2. **R. costata* (Nilsson); *2a-c*, ped.v., brach. v., lat. views, ×1.5 (920). [Elliott.]

Rhynchorina Oehlert, 1887, p. 1326 [**Anomites spathulatus* Wahlenberg, 1821, p. 62; OD]. Similar in size, outline, and area to *Rhynchora*, differing in smooth exterior with concentric growth lines; cardinalia with very wide concave outer hinge plates, marked crural bases with convex inner hinge plates arching over septum and meeting in median ridge which runs back to cardinal process, loop like that of *Magas*. *U.Cret.*, NW.Eu.——Fig. 735,3. **R. spathulata*; brach.v. int., ×5 (Elliott, n). [Elliott.]

Subfamily TRIGONOSEMINAE Elliott, n. subfam.

[Materials for this subfamily prepared by G. F. Elliott]

Shell striate to multicostate, sulcate; cardinalia strong, with massive or pillar-like

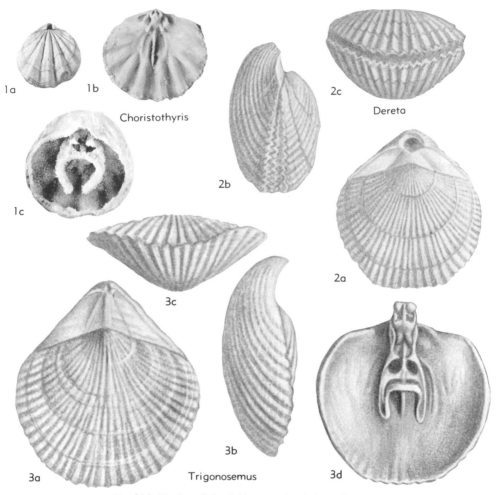

1a 1b

Choristothyris

1c

2c

Dereta

2b

2a

3c

3b

3a Trigonosemus 3d

FIG. 736. Terebratellidae (Trigonoseminae) (p. *H851-H854*).

cardinal process, and terebratelliform loop. *U.Cret.*

Trigonosemus KOENIG, 1825, p. 3 [*T. elegans*; OD] [=*Delthyridea* M'COY, 1844, p. 150; *Fissirostra* D'ORBIGNY, 1847, p. 269; *Fissurirostra* D'ORBIGNY, 1850, p. 132]. Medium-sized to large, unequally biconvex to plano-convex, terebratuliform, with very high overhanging beak, sulcate, test thick, striate, area very high, smooth, concave, with terminal pinhole foramen, symphytium high and narrow, hinge line submegathyridid. Pedicle valve interior posteriorly constricted, teeth heavy, muscle marks to left and right of median ridge posteriorly. Cardinalia dominated by massive cardinal process with swollen base, sockets to left and right, deep muscle marks to left and right of septum, which is thick buttress posteriorly and extends anteriorly to little more than half of valve length, loop narrow, rather small, tere-

bratelliform. *U.Cret.*, Eu.-W.Asia.——FIG. 736,3. *T. elegans*, W.Eu.; *3a-d*, brach.v., lat., ant., brach.v. int. views, ×8 (229).

Choristothyris COOPER, 1942, p. 233 [*Terebratula plicata* SAY, 1820, p. 43; OD]. Small, subcircular, multicostate to plicate, sulcate, test thick; hinge line subterebratulid, beak suberect to erect, foramen large, submesothyridid, deltidial plates small, disjunct. Hinge teeth large, with deep fossettes in supporting callus, ventral muscular area large, flabellate, divided by low stout median ridge. Cardinalia strong, inner socket ridges strong and high; hinge plate small, concave, cardinal process massive, trilobed; median septum high, thin, reaching to center of valve, dividing muscle marks; crural processes long, slender, loop terebratelliform. *U.Cret.*, N.Am.——FIG. 736,1. *C. plicata* (SAY), USA(N.J.); *1a*, brach.v. view, ×1; *1b,c*, brach.v. int., int. showing calcite encrusted loop, ×2 (178).

Dereta ELLIOTT, 1959, p. 147 [**Terebratula pectita* J. SOWERBY, 1816, p. 83; OD]. Medium-sized to large, transverse terebratuliform, strongly biconvex, test thick, striate, commissure sulcate; fora-

men round to oval, encroaching slightly on gently incurved beak, lateral areas smooth, bordering symphytium, beak ridges sharp, hinge line subterebratulid. Pedicle collar present, median ridge

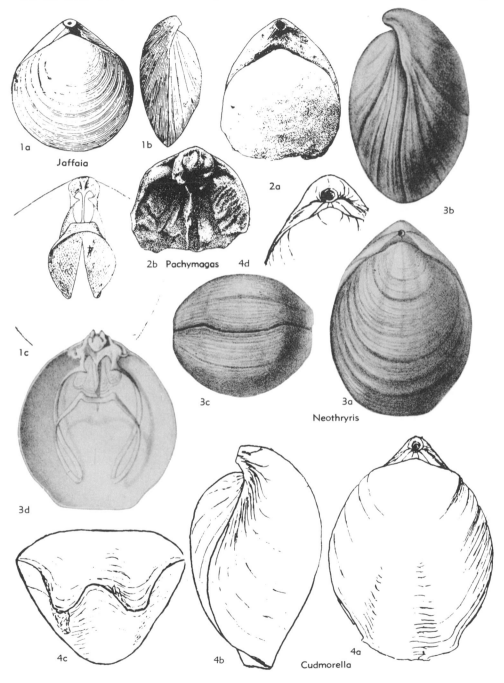

FIG. 737. Terebratellidae (Neothyridinae) (p. H854-H855).

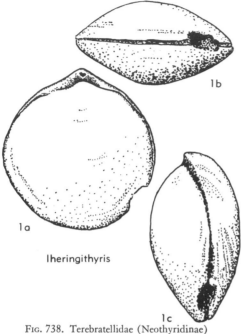

Iheringithyris

FIG. 738. Terebratellidae (Neothyridinae)
(p. H854).

in ventral umbonal area. Cardinal platform small, thick, elements fused, dominated by high narrow pillar-like cardinal process, septum thick, loop believed terebratelliform. *U.Cret.*, Eu.——FIG. 736, 2. **D. pectita* (SOWERBY), Eng.; *2a-c,* brach.v., lat., ant. views, ×2 (229).

Subfamily NEOTHYRIDINAE Allan, 1940

[Materials for this subfamily prepared by KOTORA HATAI]

Socket ridges, crural bases, and septum thick and solid; cardinalia characterized by distinct hinge trough, and bifurcated septum. Foramen ranging from hypothyridid to permesothyridid, loop from magadiniform to magellaniiform, and folding from sulcate to intraplicate (25). *Oligo.-Rec.*

Neothyris DOUVILLÉ, 1879, p. 274 [**Terebratula lenticularis* DESHAYES, 1839, p. 359; OD]. Sulcate, punctae large, close; beak erect to incurved; foramen mesothyridid, attrite; deltidial plates conjunct. Hinge teeth large, bases swollen. Cardinalia strong; crural bases fused with socket ridges; median septum short, high, bifurcating; hinge trough roomy, but almost completely filled by large cardinal process; loop magellaniiform; crura rather short; crural processes prominent; lophophore plectolophous. Muscle impressions strong (810). *Mio.-Rec.,* N.Z.(30-85 m.).——FIG. 737,3. **N. lenticularis* (DESHAYES), Rec., N.Z.; *3a-c,* brach.v., lat., ant., views, ×1; *3d,* brach.v., int. showing loop (reconstr.), enlarged (244).

Cudmorella ALLAN, 1939, p. 242 [**C. tatei*; OD]. Biconvex, intraplicate, punctae fine, dense; beak obtuse, suberect to erect; foramen permesothyridid, attrite; symphytium low, with median ridge. Hinge teeth strong with swollen bases; cardinalia as in *Pachymagas*; socket ridges descending obliquely to hinge trough; crural bases stout; cardinal process small; crura short, thick; median septum short, bladelike, forking to join anterior edge of socket ridges; hinge trough roomy; loop magelliform (24). *Mio.,* Australia.——FIG. 737,4. **C. tatei*; *4a-c,* brach.v., lat., ant. views, ×1; *4d,* foramen, enlarged (24).

Gyrothyris THOMSON, 1918, p. 23 [**Gyrothyris mawsoni*; OD]. Incipiently sulcate, obsoletely and rather finely multicostate; beak erect; foramen mesothyridid, attrite; deltidial plates conjunct, concave, almost hidden. Median septum narrowly bifurcating; hinge trough shallow; loop terebratelliform (810). *Mio.,* N.Z., *Rec.,* Antarctic.

Iheringithyris LEVY, 1961, p. 84 [**Magellania ameghinoi* IHERING, 1903, p. 326; OD]. Large, smooth, thick, biconvex shell; circular in outline, commissure rectimarginate, umbo suberect, foramen very small, mesothyridid, deltidial plates discrete, beak ridges sharp. Cardinalia thickened like those of *Pachymagas,* cardinal process large and excavate, septum thick, loop magellaniiform. *Mio.,* Patagonia.——FIG. 738,1. **I. ameghinoi* (IHERING); *1a-c,* brach.v., ant., lat. views, ×1 (506a).

Jaffaia THOMSON, 1927, p. 254 [**Magasella jaffaensis* BLOCHMANN, 1910, p. 92; OD]. Biconvex, incipiently sulcate, punctae moderately fine, dense; beak suberect; foramen submesothyridid (almost mesothyridid), attrite; symphytium solid. Cardinalia fairly strong; crural bases and socket ridges united; hinge trough wide; cardinal process triangular, confined to posterior part of hinge trough; crura and crural processes short; descending branches of loop extending beyond their union with septum, ascending branches united with descending branches throughout their entire length, transverse band narrow (810). *Rec.,* Australia (80-500 m.).——FIG. 737, 1. **J. jaffaensis* (BLOCHMANN); *1a,b,* brach.v. and lat. views, ×1.8; *1c,* brach.v. int., showing loop (reconstr.), ×3.6 (810).

Malleia THOMSON, 1927, p. 283 [**Terebratella portlandica* CHAPMAN, 1913, p. 187; OD]. Planoconvex; beak short; foramen hypothyridid, incomplete; deltidial plates rudimentary. Hinge teeth strong, transversely striated, with swollen bases deeply grooved for reception of socket ridges. Median septum bifurcating as 2 low ridges before reaching cardinalia, enclosed hinge trough with flatly inclined sides; cardinal process swollen above, with 2 lateral wings and median ridge; loop probably magadiniform. Ventral muscle impressions not strong, separated by septal ridge (810). *Mio.,* Australia.——FIG. 739,1. **M. port-*

landica (CHAPMAN); *1a,b,* brach.v. and lat. views, ×2; *1c,d,* ped.v. int., brach.v. int., ×1; *1e,* lat. view of loop, ×2 (810).

Pachymagas IHERING, 1903, p. 332 [*Terebratella (Pachymagas) tehuelcha*; OD]. Sulcate to exceptionally intraplicate, punctae large and close; beak suberect to erect, rarely incurved; foramen mesothyridid, attrite; deltidial plates conjunct. Hinge teeth strong, with swollen bases; cardinalia strong; median septum bifurcating rather widely; hinge trough roomy; cardinal process small; loop terebratelliform (810). *Oligo.-Mio.,* Antarctic; *Mio., N.Z.; Oligo.-Plio.,* S.Am.——FIG. 737,2. *P. tehuelcha,* Plio., S.Am.; *2a,b,* brach.v. view, brach.v. int., ×1 (810).

Stethothyris THOMSON, 1918, p. 23 [*S. uttleyi;* OD]. Beak suberect to incurved; symphytium with median ridge. Hinge teeth small, strong, with swollen bases, posterior thickening of valve restricting beak cavity; cardinalia strong; bifurcating narrow median septum uniting with swollen and rather flattened crural bases beyond points of origin of crura; hinge trough broad and shallow posteriorly, narrow in front; cardinal process slightly raised central boss with 2 lateral wings projecting on each side of umbo; loop magellaniiform. Ventral muscle impressions separated by triangular median ridge (810). *Oligo.-Mio.,* Australia; *Mio.,* N.Z.; *Rec.,* Antarctic(650 m.).—— FIG. 739,3. *S. uttleyi,* Mio., N.Z.; *3a-c,* brach.v., lat., ant. views, ×1; *3d,e,* ped.v. int., brach.v. int., ×1 (810).

Victorithyris ALLAN, 1940, p. 289 [*V. peterboroughensis;* OD]. Biconvex, sulcate to intraplicate, smooth, punctae fine and dense; beak suberect to strongly incurved; foramen permesothyridid, attrite. Cardinalia strong, socket ridges distinct or massive; crural bases swollen, massive, or not swollen, forked anteriorly; median septum short, thick, solid, or long and bladelike, thick at base; cardinal process small to large, with wings or bilobed; loop magellaniiform; hinge trough deep to almost completely filled (26). *Tert.,* Australia.—— FIG. 739,4. *V. peterboroughensis; 4a-c,* brach.v., lat., ant. views, ×1 (26).

Waiparia THOMSON, 1920, p. 380 [*Pachymagas abnormis* THOMSON, 1917, p. 412; OD]. Sulcate, smooth, punctae fine and dense; beak subapicate, erect; foramen submesothyridid (almost hypothyridid); deltidial plates conjunct. Hinge teeth strong, bases swollen; cardinalia strong; median septum short, bifurcating widely, fused with socket ridges and crural bases; hinge trough moderately large; cardinal process pyramidal, low; loop presumably terebratelliform (810). *Mio.,* N.Z.—— FIG. 739,2. *W. intermedia* (THOMSON); *2a-c,* brach. v., lat., ant. views, ×1 (810).

Family UNCERTAIN

[Materials for this assemblage prepared by H. M. MUIR-WOOD and G. F. ELLIOTT as indicated]

Leptothyrella MUIR-WOOD, herein [*nom. subst. pro*

Leptothyris MUIR-WOOD, 1959 (*non* CONRAD in KERR, 1875, p. 20)] [*Leptothyris ignota* MUIR-WOOD, 1959, p. 308; OD]. Small, elongate-oval, valves slightly convex, anterior commissure plane; surface smooth; hypothyridid; deltidial plates narrow, bordering open delthyrium. Loop with descending branches attached to high medially developed platelike septum, no ring or hood present (precampagiform stage); cardinal process small, uniting high inner socket ridges, which

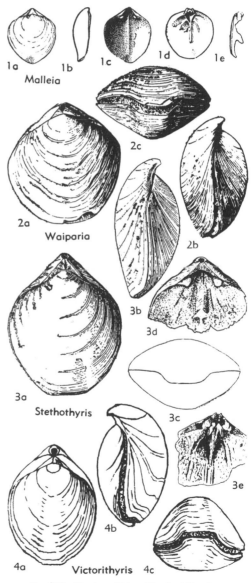

FIG. 739. Terebratellidae (Neothyridinae) (p. *H854-H855*).

are continuous with crural bases; hinge plates and dental plates absent; lophophore spirolophous, septum posterior to spirals. *Rec.,* Ind.O. (off Zanzibar)-Gulf of Aden.——Fig. 740,3. **L. ignota* (Muir-Wood), Gulf of Aden *(3a,b),* off Zanzibar *(3c,d); 3a,b,* brach.v. and lat. views, ×7; *3c,d,* ped.v. and brach.v. int., ×10 (584). [Muir-Wood.]

Terebrirostra d'Orbigny, 1847, p. 269 [**Terebratula lyra* J. Sowerby, 1816, p. 83; OD] [=*Lyra* Cum-

berland in J. Sowerby, 1816, p. 84 *(nom. nud.)*]. Biconvex, ornament of wavy radial costae, folding subintertext to ligate, dorsal valve elongate-oval to subtrigonal in outline, pedicle valve resembling brachial valve but with very long curving suberect umbo; anterior commissure rectimarginate or slightly sulcate. Beak ridges angular, deltidial plates fused, dental plates extending whole length of umbo anteriorly curved and uniting with lateral margin. Dorsal cardinalia

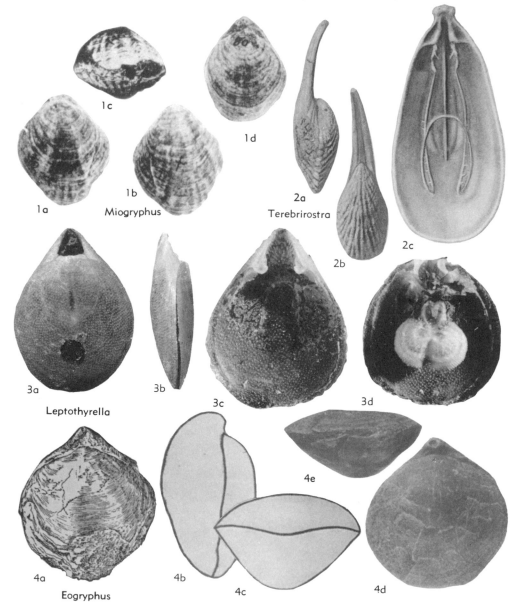

1a, 1b, 1c, 1d Miogryphus

2a, 2b, 2c Terebrirostra

3a, 3b, 3c, 3d Leptothyrella

4a, 4b, 4c, 4d, 4e Eogryphus

Fig. 740. Family Uncertain *(2,3);* Family Unknown *(1,4)* (p. *H855-H857*).

with large triangular sockets, deep central hinge trough, large conspicuous trilobed cardinal process; long thin median septum extending anteriorly from cardinalia; loop long, thin, spinose, not attached to septum in adult. *L.Cret., W.Eu.*——FIG. 740,2*a,b*. **T. lyra* (SOWERBY), Eng.; *2a,b*, lat., brach.v. views, ×1.5 (576).——FIG. 740,2*c*. *T. incurvirostrum* LAMPLUGH & WALKER, Eng.; brach.v. int., ×2.5 (576). [ELLIOTT.]

Zellania MOORE, 1855, p. 111 [**Z. davidsoni*; OD]. Minute, valves smooth or feebly striate, flattish, convex at umbo, outline triangular or shield-shaped, hinge line megathyridid or submegathyridid, pedicle-opening amphithyridid. Brachial valve interior with smooth or granulate margin bounded by inner ridges commencing anterior of sockets and reflexed anteriorly into posteriorly directed septum. *Jur., W.Eu.*——FIG. 741,*1*. **Z. davidsoni*, L.Jur., Eng.; *1a-c*, brach.v., ped.v., brach.v. int. views, ×30 (568). [ELLIOTT.]

Suborder, Superfamily, and Family
UNKNOWN

[Materials for this assemblage prepared by H. M. MUIR-WOOD]

Eogryphus HERTLEIN & GRANT, 1944, p. 88 [**E. tolmani*; OD]. Medium-sized, circular to ovate, valves moderately convex, brachial valve with shallow median sulcus; anterior commissure incipiently sulcate or rectimarginate; umbo short, slightly incurved, foramen small, permesothyridid; surface smooth. Dorsal septum present, mantle canal impressions straight and evenly spaced; other internal characters unknown. *Eoc.*, USA (Calif.).——FIG. 740,*4*. **E. tolmani*; *4u-c*, brach. v., lat., ant. views (holotype); *4d,e*, brach.v. and ant. views (paratype), all ×1 (427).

Miogryphus HERTLEIN & GRANT, 1944, p. 95 [**M. willetti*; OD] Medium-sized, ovate to subpentagonal, biconvex, with low dorsal median fold, with sulcus or flattening of pedicle valve in some specimens; anterior commissure uniplicate to sulciplicate; shell surface smooth or with few anterior radial plications in both valves and prominent growth lines; umbo small, erect, foramen small, ?mesothyridid, symphytium developed. Dorsal median septum present, other internal characters unknown. *Mio.*, USA(Calif.).——FIG. 740, *1*. **M. willetti*; *1a-c*, brach.v., ped.v., ant. views (holotype), ×1; *1d*, brach.v. view (paratype), ×1 (427).

ORDER UNCERTAIN—
THECIDEIDINA
By G. F. ELLIOTT
[Iraq Petroleum Company, Limited, London]

The Thecideidina are here classed tentatively as a suborder equal to the Strophomenidina. The case for regarding them as

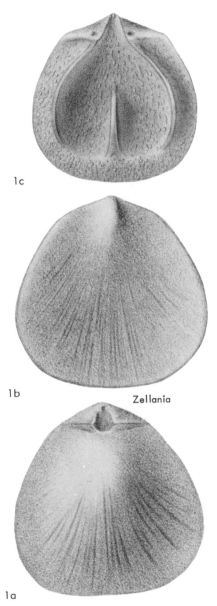

1c

1b Zellania

1a

FIG. 741. Family Uncertain (p. *H857*).

highly modified terebratuloids rests on their punctate shell structure and on belief in their profound modification due to neoteny. Pending a better knowledge of the varieties and functions of punctae, and a restudy of the larval development of the living *Lacazella,* it seems best to assign them as is done below.

Although members of the Thecideidina show a profusion of small varied detail, they are remarkably homogeneous, considered as a whole. In the present account the single superfamily is divided into two families, each without subfamilies. BACKHAUS (1959) in his detailed and profound study dealing with the Cretaceous thecideoids only, recognized but one family, with a single subfamily divided into two tribes, each with a single genus and three subgenera. These tribes are based on his interpretation of thecideoid phylogeny within the Cretaceous; they do not correspond to the major divisions defined here. Future studies may show how far this classification can be extrapolated to pre- and post-Cretaceous forms. For the present account all genera so far described are given: the reader is referred to BACKHAUS (44) for the different interpretation of the Cretaceous genera, among which *Parathecidea* BACKHAUS is described here from the type-species.

Suborder THECIDEIDINA Elliott, 1958

[*nom. correct.* ELLIOTT, herein (*pro* Thecideoidea ELLIOTT, 1958, p. 373)]

Small articulates, shell usually attached by cementation, without pedicle, rarely free, outline variable and irregular; valves hinged by teeth and sockets permitting wide opening, usually smooth externally, commonly granular internally; no mantle canal markings or obvious ovarian scars; test thick, fibrous with scattered punctae or densely punctate; high interarea with convex or flat pseudodeltidium; pedicle valve deep, with hemispondylium sessile or supported by septum, 2 diductor muscle-scars prominent; 2 median and 2 lateral adductor muscle scars inconspicuous; brachial valve lidlike, without area, with square cardinal process, inconspicuous median and lateral adductor scars, median septum simple or branched, extending from anterior margin to terminate posteriorly, with or without bridge, with or without brachial ridges. Mantle thin, without marginal setae, spicules present or absent in mantle; lophophore thin, centripetal, schizolophous or ptycholophous; muscles paired, not branching, diductors and median and lateral adductors, muscle scars smooth; marsupium present in some species. *Trias.-Rec.* (max. *U.Cret.*).

Superfamily THECIDEACEA Gray, 1840

[*nom. transl.* TERMIER & TERMIER, 1949 (*ex* Thecideidae GRAY, 1840)]

Characters of suborder. *Trias.-Rec.*

Family THECIDELLINIDAE Elliott, 1958

Small forms with bilobed brachial interiors, relatively simple dorsal septum, lophophore schizolophous. *Trias.-Rec.*

Thecidellina THOMSON, 1915 [**Thecidea barretti* DAVIDSON, 1864; OD] [=*Thecidellella* HAYASAKA, 1938 (type, *T. japonica*]. Similar to *Bifolium,* but test densely punctate, pseudodeltidium fused with interarea, brachial ridges secondarily elaborated by spiny processes which may roof over brachial cavities with transverse bars; spicules present; lophophore schizolophous, with long filaments. [*Thecidellella* resembles *Thecidellina,* but accessory shelly structure ("reversed spondylium" of HAYASAKA) between the bridge and posterior end of the median septum in the brachial valve. The distinction is slight and the structure occurs also in *Thecidellina* spp. (184); *Thecidellella* therefore is considered to be a synonym of *Thecidellina.*] *Tert.,* W.Indies; *Rec.,* W.Indies-Ind.O.-Pac.O.——FIG. 742,3. *T. blochmanni* (DALL), *Rec.,* Christmas Is., Ind.O.; *3a,b,* brach.v. int., ped.v. int., ×9 (Elliott, n).

Bifolium ELLIOTT, 1948 [**Thecidea faringdonensis* DAVIDSON, 1874; OD]. Similar to *Moorellina,* but with bridge always present, adult females with marsupial notch; dorsal septum more variable than in *Moorellina,* brachial ridges forming complete lateral subcircular features, spicules not seen. *Cret.,* Eu.-SW.Asia; *Tert.,* Eu.-N.Am.-E.Afr.-N.Z. ——FIG. 742,1. **B. faringdonense* (DAVIDSON), L.Cret., S.Eng.(Faringdon); *1a,b,* brach.v. int., ped.v. int., ×20 (Elliott, n).

Moorellina ELLIOTT, 1953 [**Thecidea duplicata* MOORE, 1855; OD]. Similar to *Thecidella,* test fibrous, with scattered punctae or clearly punctate, but with brachial ridges present, varying from pustulose strips to incomplete arcuate lines, and exceptionally in form of posteriorly directed or anteriorly branching ridges; bridge usually developed; spicules present in some. *U.Trias.(Rhaet.)-U.Jur.,* W.Eu.——FIG. 742,2. **M. duplicata* (MOORE), M.Jur., S.Eng.(Dundry); brach.v. int., ×16 (Elliott, n).

Thecidella OEHLERT, 1887 [**Thecidea (Thecidella) normaniana*; OD]. Very small, irregularly trigonal to transverse in outline, test apparently fibrous, with scattered punctae, triangular pseudodeltidium present, hemispondylium developed, dorsal median septum simple, wide, variable in outline with denticulate margins, bridge usually absent but exceptionally present, brachial ridges absent. *U. Trias.(Rhaet.)-L. Jur.,* W. Eu.——FIG. 742,4.

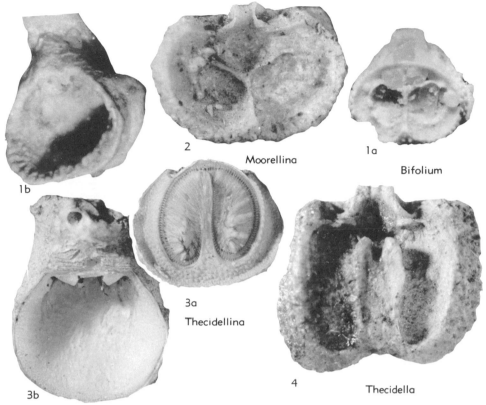

FIG. 742. Thecidellinidae (p. *H*858).

T. normaniana MUNIER-CHALMAS, L.Jur., Fr. (May); brach.v. int., ×16 (Elliott, n).

Family THECIDEIDAE Gray, 1840

Larger forms with much-divided brachial interiors, only bilobed in adult in single exceptional genus, possible not a thecideid; lophophore ptycholophous. *L.Jur.-Rec.*

Thecidea DEFRANCE, 1822 [**Thecidea radians* (=**Terebratulites papillata* VON SCHLOTHEIM, 1813); OD] [=*Thecidium* SOWERBY, 1823 *(nom. null.)*; *Thecidaea* KEFERSTEIN, 1829, p. 82 *(nom. null.)*; *Thecideum* FISCHER DE WALDHEIM, 1834, p. 279 *(nom. null.)*; *Thecedea* D'ORBIGNY, 1847, p. 249 *(nom. null.)*]. Small, free, symmetrical, beak entire, otherwise terebratuliform, radial ornament, test densely punctate, pseudodeltidium narrow and convex in high area; pedicle valve with internal median ridge, prominent lateral muscle platforms left and right of hemispondylium; brachial valve with very wide granular internal margin anterolaterally, median septum with 2 or 3 curved branches on either side, brachial ridges following course of septa; bridge present. *Cret.,* W.Eu.——FIG. 743,*3.* **T. papillata* (SCHLOTHEIM),

U.Cret., N.Fr.(Fréville); *3a-c,* brach.v. ext., ped. v. int., brach.v. int., ×6 (Elliott, n).

Davidsonella MUNIER-CHALMAS, 1881 [**Thecidea sinuata* E. EUDES-DESLONGCHAMPS, 1853; OD]. [*non Davidsonella* WAAGEN, 1885; *nec* FREDERIKS, 1926]. Small, elongate, pseudodeltidium present, test pseudopunctate; pedicle valve deep, anteriorly bilobed by median sulcus; brachial valve with bridge, long median septum terminating posteriorly in sharp point, 2 long deep brachial cavities thus formed roofed over or filled by coarse spicular growth, brachial ridges not present. [This may not be a true thecideid but is conveniently placed here at present.] *L.Jur.,* W. Eu.-N.Afr.——FIG. 743,*2.* **D. sinuata* (DESLONG-CHAMPS), L.Jur., Fr.(May); *2a,b,* brach.v. int., ped.v. int., ×4.5 (Elliott, n).

Eolacazella ELLIOTT, 1953 [**Thecidea affinis* BOS-QUET, 1860; OD]. Small, irregular or trigonal, high narrow pseudodeltidium, test punctate; internal median ventral ridge spinose; dorsal median septum with few branches, or several branches originating together anteriorly, brachial ridges following course of septal branches; bridge present; spicules present. *Cret.,* W.Eu.——FIG. 743,*1.* **E. affine* (BOSQUET), U.Cret.(Maastricht.),

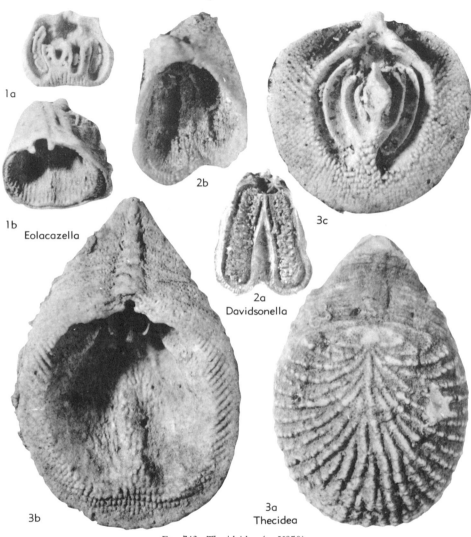

1a

1b
Eolacazella

2b

3c

2a
Davidsonella

3b

3a
Thecidea

FIG. 743. Thecideidae (p. *H859*).

Neth.; *1a,b,* brach.v. int., ped.v. int., ×12 (Elliott, n).

Eudesella MUNIER-CHALMAS, 1881 [**Thecidea mayalis* E. EUDES-DESLONGCHAMPS, 1853; OD]. Small, transverse, test apparently fibrous, with scattered punctae, pseudodeltidium present; pedicle valve with hemispondylium; interior of brachial valve lobed by varying number of septa, commonly 6 to 8, extending from valve margin to terminate near visceral cavity, center septum bifurcating in some and uncommonly attached to bridge, if present; brachial ridges not present. *L. Jur.,* W.Eu.——FIG. 744,*1.* **E. mayalis* (DESLONGCHAMPS), Fr.(May); *1a-c,* brach.v. ext., ped. v. int., brach.v. int., all ×6 (Elliott, n).

Lacazella MUNIER-CHALMAS, 1881 [**Thecidea*

mediterranea RISSO, 1826; OD]. Small, irregular or trigonal, pseudodeltidium triangular, test densely punctate; interior of pedicle valve granular to spinose except over muscle scars, hemispondylium projecting anteriorly as 2 spurs; brachial valve with trifurcating median septum, brachial ridges arcuate to left and right; mantle thin, coarsely spicular; lophophore ptycholophous, filaments long; paired diductor, median and lateral adductors; body very small, marsupium present in adult females, shells of both sexes similar except for notch in bridge. *Tert.,* Eu.-?Australia; *Rec.,* Medit.-W.Indies-Mauritius.——FIG. 744,*2.* **L. mediterranea* (RISSO), *Rec.,* Alg.(Bône); *2a,b,* brach.v. int., ped.v. int., ×9 (Elliott, n).

Parathecidea BACKHAUS, 1959 [**Thecidea hiero-*

glyphica GOLDFUSS, 1840; OD]. Small, square to elongate, pseudodeltidium merging into area, test thick, punctate; pedicle valve with several irregular longitudinal-radial internal ridges, deep muscle scars, hemispondylium supported by thin septum: brachial valve with bridge present, more numerous septa than in *Thecidiopsis*, extending inward both from valve margins and median septum, brachial ridges interdigitating with septa. *Cret.*, Eu.——FIG. 745,*1*. **P. hieroglyphica* (GOLD-FUSS), U.Cret.(Maastricht.), Neth.; *1a*, brach.v. int., ✕5; *1b*, ped.v. int., ✕4 (44).

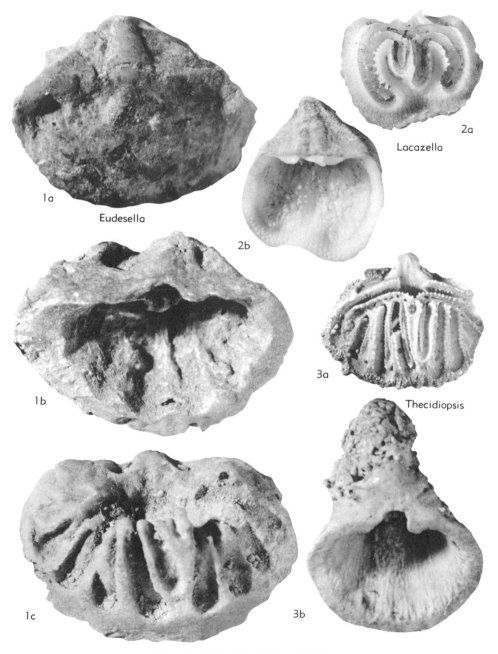

FIG. 744. Thecideidae (p. *H*860, *H*862).

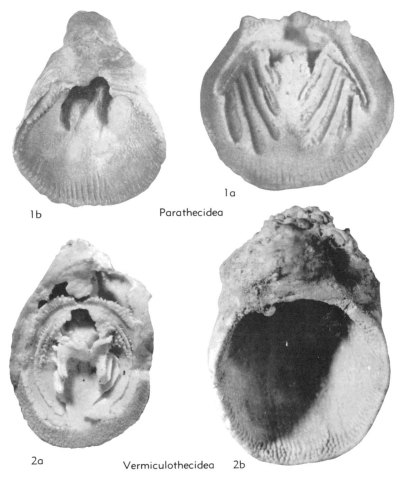

1b   1a   Parathecidea

2a   Vermiculothecidea   2b

FIG. 745. Thecideidae (p. *H860-H862*).

Thecidiopsis OEHLERT, 1887 [**Thecidea digitata* G. B. SOWERBY, 1823; OD]. Small, transverse, pseudodeltidium high and narrow, test very thick, punctate; pedicle valve with broad median internal rounded ridge, muscle scars prominent, median supporting septum of hemispondylium conspicuous; brachial valve with bridge, numerous septa (average about 10) extending inward from valve margin, septal terminations rounded; brachial ridges interdigitating with septa. *Cret.*, Eu.—— FIG. 744,*3*. **T. digitata* (G. B. SOWERBY), U.Cret. (Maastricht.), Neth.; *3a,b*, brach.v. int., ped.v. int., ×4.5 (Elliott, n).

Vermiculothecidea ELLIOTT, 1953 [**Terebratulites vermicularis* VON SCHLOTHEIM, 1813; OD]. Small, irregularly elongate, test thick, punctate, high narrow pseudodeltidium not sharply delimited; pedicle valve deep, granulation well developed except over ill-defined muscle scars; dorsal valve with high marginal ridge, median septum wide, low, anteriorly attached to valve floor, but posteriorly curving upward free of valve floor, with 4 or 5 anteriorly and upwardly directed branches on either side; branches in form of split tubes, within which corresponding brachial ridges occur, joining posteriorly; bridge present, spicules present. *U.Cret.*, W.Eu.——FIG. 745,*2*. **V. vermicularis* (VON SCHLOTHEIM), Maastricht, Neth.; *2a,b*, brach.v. int., ped.v. int., ×9, ×6 (Elliott, n).

Order, Suborder, and Family UNCERTAIN

Amblotrema RAFINESQUE, ?1831, p. 8 [=*Amblytrema* AGASSIZ, 1847, p. 45 *(nom. van.)*].

Arctitreta WHITFIELD, 1908, p. 57 [**A. pearyi*; OD]. Pedicle valve convex with high interarea, ramicostellate, ventral muscle scar large, suboval; brachial valve unknown. [Three pedicle valves are figured. The delthyrial cover of one is probably a pseudodeltidium comparable with that of the davidsoniaceans: the delthyrium of another is open but underlain by a delthyrial plate like that of the Spiriferidina.] *U.Carb.,* Canada (Cape Sheridan, Grant Land). [WILLIAMS.]

Australostrophia CASTER, 1939, p. 83 [**Leptostrophia??* mesembria CLARKE, 1913, p. 286; OD]. Semioval, gently plano-convex, finely costellate, interareas low, with well-defined chilidium and pseudodeltidium; teeth elongate ?unsupported, ventral muscle field large, subtriangular with low median ridge; cardinal process bilobed, socket ridges strong, curved. [A few nodes along the posterolateral parts of ventral internal molds suggest that the interarea was penetrated by canals in the manner of Chonetidina, but no septa are found in the brachial valve and at present it is impossible to describe the genus more closely than as belonging to the Strophomenida.] *M.Dev.,* Brazil-Falkland Is. [WILLIAMS.]

Biarea TORBAKOVA, 1959.

Brachiopus RAFINESQUE, ?1831, p. 7.

Branconia GAGEL, 1890, p. 62 [**B. borussica;* OD, M]. *Ord.,* Eu.

Bufocephalus LINNÉ, 1779, p. 49.

Bursula HERRMANNSEN, 1846, p. 148 [=*Bursula* KLEIN, 1753, non. binom.].

Comelicania FRECH, 1901, p. 551 [**Athyris megalotis* STACHE, 1878; SD].

Cornwallia WILSON, 1932, p. 388 [**C. minuta;* OD]. Genus poorly understood, known from single ?pedicle valve. Small, suboval outline, convex, ornament of fine, radiating striae. *U.Ord.,* N.Am.(Can.). [ROWELL.]

Diclipsites RAFINESQUE, ?1831, p. 8.

Diclisma RAFINESQUE, 1820, p. 232.

Didymospira SALOMON, 1895, p. 81.

Diphyites HERRMANNSEN, 1846, p. 390.

Diphytes SCHRÖTER, 1779, p. 411, non-binom.

?Dirinus M'COY, 1844, p. 44 [**D. bucklandi;* OD]. Described as gastropod, may possibly be craniid. Type material lost. [ROWELL.]

Euorthisina HAVLÍČEK, 1950.

Gamdaella MILORADOVICH, 1947.

Gasconsia NORTHROP, 1939, p. 161 [**G. schucherti;* OD]. Very large, approximately semicircular in outline, posterior margin straight. [When erected, provisionally included in the Trimerellidae.] *Sil.* (Gascon & Bouleaux F.), N. Am. (Gaspé). [ROWELL.]

Gaspesia CLARKE, 1907, p. 277 [**Orthis aurelia* BILLINGS, 1874, p. 34; OD]. Semi-elliptical, ?convexo-concave, ?lacking interareas, costate and imbricate; interiors unknown (may not be a brachiopod). *L.Dev.,* Can.(Que.). [WILLIAMS.]

Goniclis RAFINESQUE, ?1818, p. 107.

Hemisterias RAFINESQUE, 1832, p. 122.

?Ivanovia IVANOVA, 1949.

Lamanskya MOBERG & SEGERBERG, 1906, p. 71 [**L. splendens;* OD]. Concavo-convex, geniculate, lustrous but with subdued costellae (probably

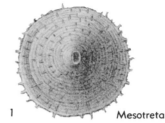

FIG. 746. Family Uncertain (p. *H863*).

Strophomenida). *L.Ord.,* Sweden. [WILLIAMS.]

Lampas ANON. (HUMPHREY), 1797, p. 45 [=*Lampus* SOWERBY, 1842, p. 169 *(nom. null.)*].

Larium DE GREGORIO, 1930, p. 25 [**L. inventum;* OD]. Biconvex, with subconical pedicle valve, exterior spinose (possibly Productidina, Scaccinellidae). *L.Perm.,* Italy(Sicily). [WILLIAMS.]

Liocoelia SCHUCHERT & COOPER, 1931, p. 248 [**Pentamerus proximus* BARRANDE, 1879; OD]. Similar to *Clorinda* externally but having rhynchonelloid-type of cardinalia suggestive of *Camarotoechia.* *Sil.,* Eu.(Czech.). [SCHUCHERT & COOPER.]

Martinigisis LEBEDEV, 1926.

Megarites RAFINESQUE, ?1820, p. 8.

Mesotreta KUTORGA, 1848, p. 271 [**Siphonotreta tentorium* KUTORGA, 1848; OD]. Genus poorly understood, known only from pedicle valve. Subcircular in outline, known only from pedicle valve. Subcircular in outline, depressed conical, with central apex perforated by foramen; without pseudointerarea. Ornament of concentric growth lines and scattered spines. *Ord.,* Eu.(Est.).
——FIG. 746,*1.* **M. tentorium* (KUTORGA); ped.v. ext., ×1.5 (396). [ROWELL.]

Minororthis IVANOV, 1950 [**M. malivkini;* OD]. Quadrate in outline, biconvex, costellate with strong dorsal median fold (Orthidina). *M.Ord.,* USSR. [WILLIAMS.]

Neogypidula LIKHAREV, 1934, p. 211.

Peridiolithus HUPSCH, 1768, p. 144.

Platilites RAFINESQUE, ?1820, p. 8.

Pleuranisis RAFINESQUE, ?1820, p. 8.

Plicoprothyris DAHMER, 1940.

Pomatospirella BITTNER, 1892, p. 26 [**Spirigera (Pomatospirella) thecidium;* OD]. *Trias.*

Priambonites FISCHER DE WALDHEIM, 1834, p. 289 [*?nom. null. pro Plectambonites* PANDER, 1830].

Reflexa ROTAI, 1931, p. 24 [**R. reflexa;* OD].

Rhynchoferella SPRIESTERSBACH, 1942 [*non* STRAND, 1915, p. 182].

Rictia DE GREGORIO, 1930, p. 27 [**R. simplex;* OD]. [May not be a brachiopod.] *L.Perm.,* Italy(Sicily). [WILLIAMS.]

Selenella HALL & CLARKE, 1894, p. 270 [**S. gracilis;* OD]. [Terebratulida, suborder, superfamily, and family uncertain]. *M.Dev.,* Can.(Ont.).

Socraticum DE GREGORIO, 1930, p. 25 [**S. firmum;* OD]. ?Plano-convex, with subconical pedicle valve, exterior ?spinose (probably Strophomenida). *L.Perm.,* Italy(Sicily). [WILLIAMS.]

Sphenorthis GRUBBS, 1939, p. 554 [*S. niagarensis*; OD]. Subtriangular biconvex, sulcate, hinge line short, curved, delthyrium open, costate, cardinal process absent (possibly Rhynchonellida or Orthida). *L.Sil.,* USA(Ill.). [WILLIAMS.]

Spondylobolus M'COY, 1851, p. 407 [*S. craniolaris*; SD DALL, 1870, p. 164] [=*Spondylobus* DALL, 1870, p. 154 *(nom. null.)*; *Spondilobolus* GORYANSKY, 1960, p. 174 *(nom. null.)*]. Genus poorly understood, possibly not a brachiopod. Shell calcareous. Larger valve subtriangular in outline, with 2 low bosses on either side of beak directed toward opposite valve; smaller valve with apex in posterior quarter of valve. *Sil.,* Eu. [ROWELL.]

Swantonia WALCOTT, 1905, p. 296 [*Camerella antiquata* BILLINGS, 1861, p. 10; OD]. Pedicle valve ovate, moderately convex, beak pointed, incurved; surface bearing 8 to 12 rounded ribs. *L. Cam.,* USA(Vt.). [WALCOTT.]

Syntrophoides SCHUCHERT & COOPER, 1931, p. 247 [*Billingsella harlanensis* WALCOTT, 1905, p. 236; OD]. Differs from *Billingsella* in being concentrically marked externally, instead of multicostellate, and in having different brachial-valve musculature. *M.Cam.,* USA(Tenn.). [SCHUCHERT & COOPER.]

Telistrophis RAFINESQUE, 1832, p. 142.

Thecospirella BITTNER, 1900, p. 46 [*T. loczyi*; OD].

Venezuelia WEISBORD, 1926.

Virbium DEGREGORIO, 1930, p. 25.

Wynnia WALCOTT, 1908, p. 142 [*Orthis warthi* WAAGEN, 1891, p. 102; OD]. Subcircular in outline, biconvex with dorsal median sulcus, delthyrium and notothyrium open; ventral interior with subflabellate muscle field and subparallel *vascula media*; dorsal interior with vaguely impressed adductor scars; other details and exterior unknown (?Orthida). *M.Cam.,* India. [WILLIAMS.]

Yeosinella REED, 1932, p. 193 [*Y. consignata*; OD]. Semioval and mucronate in outiine, unequally biconvex with dorsal median sulcus, costate; cardinal process bilobed, sockets widely divergent crenulated; other internal features unknown (probably Orthidina). *?U.Ord.,* Burma. [WILLIAMS.]

GENERIC NAMES ERRONEOUSLY ASCRIBED TO BRACHIOPODA

Arbusculites MURRAY, 1831, p. 147.

Aulacomerella VON HUENE, 1900, p. 209. Mollusk.

Badiotella BITTNER, 1890, p. 94. Mollusk.

Bagenovia RADUGHIN, 1937, p. 301 [*B. sajanica*; OD]. ?Mollusk. [ROWELL.]

Curvulites RAFINESQUE, 1831, p. 4. ?Mollusk.

Delgadella WALCOTT, 1912, p. 560 [*Lingulepis lusitanica* DELGADO, 1904, p. 365]. Trilobite (*Treatise,* p. O190). [ROWELL.]

Discinella HALL, 1871, p. 3. Hyolithid operculum (*Treatise,* p. W132). [ROWELL.]

Khmeria MANSAY, 1914, p. 53. Coelenterate (*see Treatise,* p. F477).

Macquartia ROULLIER & VORINSKY, 1848, p. 271 [*non* ROBINEAU-DESVOIDY, 1830, p. 204]. ?Mollusk.

Neoproductus NIKITIN, 1900, p. 385. Invalid hypothetical genus.

Orthonote CONRAD, 1841, p. 50. Mollusk.

Pectenoproductus LIKHAREV, 1930, p. 438 [*P. proprius*]. Insufficiently known, may be a lamellibranch. *L.Perm.,* Eu.(N.Caucasus). [MUIR-WOOD.]

Protobolella CHAPMAN, 1935, p. 117 [*P. jonesi*] [=*Fermoria* SAHNI, 1936; *Fermoria* CHAPMAN, 1935 *(nom. vet.)*; *?Vindhyanella* SAHNI, 1936]. Problematic, possibly algal (*Treatise,* p. W240). [ROWELL.]

NOMINA NUDA

Apleurotis RAFINESQUE, 1819, p. 427 *(?nom. nud.)*.

Brynella BANROFT, 1933, p. 3 *(nom. nud.)*. [WILLIAMS.]

Clipsilis RAFINESQUE, 1820, p. 5 *(nom. nud.)* [=*Clipsilia* DALL, 1877, p. 20 *(nom. null.)*].

Cranicella RAFINESQUE, 1815, p. 148 *(nom. nud.)*. [ROWELL.]

Delthyridaea M'COY, 1844, p. 150 *(nom. nud.)*. [ELLIOTT.]

Gonotrema RAFINESQUE, 1820, p. 232 *(?nom. nud.)*.

Hunanella GRABAU & TIEN, ?date [*H. antiquatiformis*] *(nom. nud.)* (fide LEE, 1939, p. 486). [MUIR-WOOD.]

Marginella GEMMELLARO, 1897, p. 113 *(nom. nud.)* [*non* LAMARCK, 1799, p. 70].

Martinella J. S. LEE, 1939 *(nom. nud.)* [*non* JOUSSEAUME, 1887, p. 173].

Megorima RAFINESQUE, 1818, p. 107 *(nom. nud.)*.

Neoproductella GRABAU & TIEN, ?date *(nom. nud.)* (fide LEE, 1939, p. 487).

Obovites RAFINESQUE, ?1820, p. 7 *(?nom. nud.)*.

Oxyrhynchus QUENSTEDT, 1868 [*non* LEACH, 1818, *nec* LAUGIER, 1822; *nec* SCHOENHERR, 1823, etc. *(nom. nud.)*].

Pachiloma RAFINESQUE, ?1820, p. 8 *(nom. nud.)* [=*Plachiloma* FERUSSAC, 1835, p. 23 *(nom. null.)*].

Pleurinia RAFINESQUE, ?1820, p. 8 *(?nom. nud.)* [=*Pleurinea* DALL, 1877, p. 56 *(nom. null.)*].

Praemagas FISCHER & OEHLERT, 1892, p. 751 *(nom. nud.)*.

Stropheria RAFINESQUE, 1820, p. 232 *(nom. nud.)*.

Strophesia RAFINESQUE, ?1820, p. 8 *(?nom. nud.)*.

Styriasis RAFINESQUE, ?1820, p. 8 *(?nom. nud.)*.

Tangkouella GRABAU, 1931 *(nom. nud.)*.

Tectorthis MAILLIEUX, 1940, p. 11 *(nom. nud.)*. [WILLIAMS.]

Trigorima RAFINESQUE, ?1820, p. 7 *(nom. nud.)*.

Trunculites RAFINESQUE, ?1820, p. 8 *(?nom. nud.)*.

Velciella HAVLÍČEK & ŠNAJDR, 1952, p. 258 [*V. pompechiana*; OD] *(nom. nud.)*. [WILLIAMS.]

REFERENCES

The composite list of references given below, which relates to the preceding systematic descriptions, has been prepared by the authors of this text.

Ager, D. V.

(1) 1956-62 (in progress), *A monograph of the British Liassic Rhynchonellidae:* Palaeontograph. Soc., London, pt. 1-3, p. 1-136, pl. 1-11.

(2) 1957, *The true Rhynchonella:* Palaeontology, v. 1, p. 1-15, pl. 1-2.

(3) 1959, *The classification of the Mesozoic Rhynchonelloidea:* Jour. Paleontology, v. 33, p. 324-332, pl. 49.

(4) 1960, *Nomenclatural problems in the Mesozoic Rhynchonelloidea:* Geol. Mag., v. 97, p. 157-162.

Aigner, Gustava, & Heritsch, Franz

(5) 1931, *Das Genus Isogramma im Carbon der Südalpen:* Akad. Wiss. Wien, Math.-Naturwiss. Kl., Denkschr., v. 102, p. 303-316, pl. 1-5.

Alekseeva, R. E.

(6) 1959, *Novyy rod semeystva Atrypidae Gill (Brachiopoda):* Akad. Nauk SSSR, Doklady, v. 126, no. 2, p. 389-391. [*A new genus of the family Atrypidae Gill (Brachiopoda).*]

(7) 1960, *O novom podrode Atrypa (Desquamatia) subgen. semeystva Atrypidae Gill (Brakhiopody):* Same, v. 131, no. 2, p. 421-424. [*A new subgenus Atrypa (Desquamatia) subgen. of the family Atrypidae Gill (Brachiopoda).*]

(8) 1962, *Devonskie Atripidy Kutznetskogo i Minusinskogo Basseynov i Vostochnogo Sklona Severnogo Urala:* Akad. Nauk SSSR, Sibir. Otdel., Inst. Geol. & Geofiz., 196 p., 12 pl., 80 text-fig. [*Devonian atrypids of Kutznetsk and Minusinsk basins and east slope of north Urals.*]

Alexander, F. E. S.

(9) 1947, *On Dayia navicula (J. de C. Sowerby) and Whitfieldella canalis (J. de C. Sowerby) from the English Silurian:* Geol. Mag., v. 84, no. 5, p. 304-316.

(10) 1948, *A revision of the genus Pentamerus (James Sowerby, 1813) and a description of the new species Gypidula bravonium from the Aymestry limestone of the main outcrop:* Geol. Soc. London, Quart. Jour., v. 103, pt. 3, p. 143-161, 1 pl.

(11) 1949, *A revision of the brachiopod species Anomia reticularis (Linnaeus), genolectotype of Atrypa (Dalman):* Same, v. 104, p. 207-220, pl. 10 & 11.

(12) 1951, *Proposed use of the plenary powers to prevent the confusion which would result, under a strict application of the "Règles", from the sinking of the name "Conchidium" as a synonym of "Pentamerus" Sowerby, 1813 (Class Brachiopoda) and the transfer of the latter name to the genus now known as "Conchidium":* Bull. Zool. Nomencl., v. 2 [Opinions and declarations rendered by ICZN, v. 8, pt. 14], p. 89-96.

Alichova, T. N.

(13) 1953, *Rukovodyashchaya fauna brachiopod Ordovitskikh otlozheniy severozapadnoi chasti Russkoy Platformy:* Vses. Nauchno-Issledov. Geol. Inst. (VSEGEI), Minist. Geol. i Okhrany Nedr., SSSR, Trudy, 162 p., 17 pl. [*Handbook of the brachiopod fauna of Ordovician deposits of the northwest part of the Russian platform.*]

(14) 1960, *Otryad Orthida:* Mshanki, Brakhiopody, SARYCHEVA, T. G. (ed.), in Osnovy Paleontologii, ORLOV, YU. A. (ed.), p. 183-197, pl. 7-19, text-fig. 85-97 (Moskva). [*Order Orthida.*]

Allan, R. S.

(15) 1931, *Descriptions of Tertiary Brachiopoda from New Zealand:* Same, v. 62, pl. 1-5, pl. 20-22.

(16) 1932, *Tertiary Brachiopoda from the Chatham Islands, New Zealand:* New Zealand Inst., Trans., v. 63, p. 11-23, pl. 4-6.

(17) 1933, *On the system and stage names applied to subdivisions of the Tertiary strata in New Zealand:* Same, v. 63, p. 81-108.

(18) 1935, *The fauna of the Reefton beds (Devonian) New Zealand:* New Zealand Geol. Survey, Paleont., Bull. 14, p. 1-72, pl. 1-5.

(19) 1937, *Type Brachiopoda in the Canterbury Museum:* Canterbury Museum, Records, v. 4, no. 3, p. 115-128, pl. 15-16.

(20) 1937, *Two new Hutchinsonian (Tertiary) brachiopods from New Zealand:* Same, v. 4, no. 3, p. 129-130, pl. 17.

(21) 1937, *Tertiary Brachiopoda from the Mount Brown Beds of Mount Donald, Weka Pass District, New Zealand:* Same, v. 4, no. 3, p. 131-137, pl. 18.

(22) 1937, *Tertiary Brachiopoda from the Forest Hill Limestone (Hutchinsonian) of Southland, New Zealand:* Same, v. 4, no. 3, p. 139-153, pl. 19-20.

(23) 1937, *On a neglected factor in Brachiopoda migration:* Same, v. 4, no. 3, p. 157-165.

(24) 1939, *Studies on the Recent and Tertiary Brachiopoda of Australia and New Zealand:* Same, v. 4, no. 5, pt. 1, p. 231-248, pl. 29-31.

(25) 1940, *A revision of the classification of the terebratelloid Brachiopoda:* Same, v. 4, no. 6, p. 267-275.

(26) 1940, *Studies on the Recent and Tertiary Brachiopoda of Australia and New Zealand:* Same, v. 4, no. 6, p. 277-297, pl. 35-37.

(27) 1947, *A revision of the Brachiopoda of the Lower Devonian strata of Reefton, New Zealand:* Jour. Paleontology, v. 21, no. 5, p. 436-452, pl. 61-63.

Amos, Arturo, & Boucot, A. J.

(27a) 1963, *A revision of the brachiopod family Leptocoeliidae:* Palaeontology, v. 6, pt. 3, p. 440-457, pl. 62-65.

Amsden, T. W.

(28) 1949, *Stratigraphy and paleontology of the Brownsport Formation (Silurian) of western Tennessee:* Peabody Museum Nat. History, Bull. 5, 134 p., 34 pl.

(29) 1949, *Two new genera of brachiopods from the Henryhouse Formation (Silurian) of Oklahoma:* Washington Acad. Sci., Jour., v. 39, no. 6, p. 202-203.

(30) 1951, *Brachiopods of the Henryhouse Formation (Silurian) of Oklahoma:* Jour. Paleontology, v. 25, no. 1, p. 69-96, pl. 15-20, 1 text-fig.

(31) 1952, *Request for a ruling that the alleged name "Antirhynchonella" Quenstedt, 1871, is a "nomen nudum" (Phylum Brachiopoda), class Articulata:* Bull. Zool. Nomencl., v. 6, no. 8, p. 242-244.

(32) 1953, *Some notes on the Pentameracea, including a description of one new genus and one new subfamily:* Washington Acad. Sci., Jour., v. 43, p. 137-147, 7 text-fig.

(33) 1958, *Haragan articulate brachiopods; Supplement to the Henryhouse brachiopods:* in AMSDEN, T. W. & BOUCOT, A. J., Stratigraphy and paleontology of the Hunton Group in the Arbuckle Mountain region, Oklahoma Geol. Survey, Bull. 78, pt. 2-3, p. 9-157, pl. 1-14, text-fig. 1-39.

(34) 1964, *Brachial plate structure in the brachiopod family Pentameridae:* Palaeontology, v. 7, p. 220-239, pl. 40-43.

————, **& Ventress, Wm. P. S.**

(35) 1963, *Early Devonian brachiopods of Oklahoma:* Oklahoma Geol. Survey, Bull. 94, 238 p., 21 pl.

Andreeva, O. N.

(36) 1955, *Polevoi atlas Ordovisskoi i Siluriiskoi fauny Sibirskoi platformy:* NIKIFOROVA, O. I. (ed.), Vses, Nauchno-Issledov. Geol. Inst. (VSEGEI), Minist. Geol. i Okhrany Nedr., 268 p., 62 pl. [*Field atlas of the Ordovician and Silurian faunas of the Siberian Platform.*]

(37) 1960, *Novye vidy drevnikh rasteniy i Bezprovonochnykh S.S.S.R.: Part I:* MARKOW-

SKII, B. P. (ed.): 288 p., 73 pl., 7 text-fig. (Moskva). [*New species of ancient plants and invertebrates of the USSR.*]

Andronov, S. M.

(38) 1961, *Nekotorye predstaviteli semeystva Pentameridae iz devonskikh otlozheniy okrestnostey severouralvska:* Akad. Nauk SSSR, Inst. Geol. Nauk, Trudy, no. 55, p. 38-108, 32 pl. [*Some representatives of the family Pentameridae from the Devonian deposits in the region of northern Urals.*]

Anthula, D. J.

(39) 1899, *Über die Kreidefossilien des Kaukasus:* Beiträge Paläont. & Geol. Osterreich-Ungarns Orients, v. 12, pt. 1, p. 55-159, pl. 2-14.

Archiac, E. J. A. D. d'

(40) 1847, *Rapport sur les fossiles du Tourtia:* Soc. Géol. France, Mém. 2, v. 2, no. 2, p. 291-351, pl. 13-25.

Armstrong, A. A.

(41) 1962, *Stratigraphy and paleontology of the Mississippian System in southwestern New Mexico and adjacent southeastern Arizona:* New Mexico State Bur. Mines & Min. Res., Mem. 8, 99 p., 22 pl., 4 text-fig.

Asatkin, B. P.

(41a) 1932, *Ecardines iz nizhnego silura Sibirskoi platformy:* Izvestiya Vses. Geol.-Razv. Ob'edineniya, SSSR, v. 51, p. 483-495, 2 pl. (Leningrad). [*Ecardines from the Lower Silurian of the Siberian Platform.*]

Atkins, D.

(42) 1958, *A new species and genus of Kraussinidae (Brachiopoda) with a note on feeding:* Zool. Soc. London, Proc., v. 131, pt. 4, p. 559-581, text-fig. 1-14, pl. 1.

(43) 1960, *A new species and genus of Brachiopoda from the western approaches, and the growth stages of the lophophore:* Marine Biol. Assoc., United Kingdom. Jour., v. 39, p. 71-89, fig. 1-14, pl. 1.

Backhaus, Egon

(44) 1959, *Monographie der cretacischen Thecideidae (Brach.):* Hamburg Geol. Staatsinst., Mitt., v. 28, p. 5-90, pl. 1-7.

Balashova, E. A.

(44a) 1955, *Produktidy Turneiskikh otlozheniy Ber-chogura (Mugodzhary):* Leningrad. Gosudar. Univ., Uch. Zap., Ser. Geol.-Pochv. Nauk, no. 189, v. 6, p. 124-156. [*Producti of the Tournaisian deposits of Ber-Chogura (Mugodzhary).*]

Bancroft, B. B.

(45) 1928, *On the notational representation of the rib-system in Orthacea:* Manchester Lit. & Philos. Soc., Mem. & Proc., v. 72, p. 53-90, pl. 1-3.

(46) 1928, *The Harknessellinae:* Same, v. 72, p. 173-196, pl. 1-2.

(47) 1929, *Some new genera and species of Strophomenacea from the Upper Ordovician of Shropshire:* Same, v. 73, pt. 4, p. 33-65, pl. 1-2.

(48) 1933, *Correlation tables of the stages Costonian-Onnian in England and Wales:* 4 p., 3 tables, The Author (Blakeney, England).

(49) 1945, *The brachiopod zonal indices of the stages Costonian-Onnian in Britain:* Jour. Paleontology, v. 19, p. 181-252, pl. 22-38.

(50) 1949, *Welsh Valentian brachiopods:* Quarry Managers' Jour., p. 2-10, pl. 1-2.

(51) 1949, *The Strophomena antiquata group of fossil brachiopods:* Same, p. 11-16, 1 pl.

Barrande, Joachim

(52) 1848, *Über die Brachiopoden der silurischen Schichten von Boehmen:* v. 2, no. 2, p. 153-256, pl. 15-23.

(53) 1879, *Systême Silurien du centre de la Bohême, Pt. I, Recherches paléontologiques, v. 5, classe des Mollusques. Ordre des Brachiopodes:* 226 p., 153 pl. (Praha).

Bassler, R. S.

(54) 1915, *Bibliographic index of American Ordovician and Silurian fossils:* Smithsonian Misc. Coll., Bull. 92, v. 1, p. 1-718; v. 2, p. 719-1521, 4 pl.

Bayle, C. E.

(55) 1878, *Explication de la carte géologique de la France:* v. 4, pt. 1, 158 pl. (Paris).

Beecher, C. E.

(56) 1890, *On Leptaenisca, a new genus of brachiopod from the Lower Helderberg group:* Am. Jour. Sci., ser. 3, v. 140, p. 238-240, pl. 1-9.

(56a) 1891, *Development of the Brachiopoda. Pt. I. Introduction:* Same, v. 41, p. 343-357, pl. 17.

(57) 1895, *Revision of the families of loop-bearing Brachiopoda:* Connecticut Acad. Arts & Sci., Trans., v. 9, pt. 2, p. 376-399, 3 pl.

(57a) 1897, *Morphology of the brachia,* in SCHUCHERT, CHARLES, A synopsis of American fossil Brachiopoda, including bibliography and synonymy: U.S. Geol. Survey, Bull. 87, 464 p.

Beets, C.

(58) 1943, *On Waisiuthyrina, a new articulate brachiopod genus from the Upper Oligocene of Buton (S.E. Celebes), Dutch East Indies:* Leidsche Geol. Med., v. 13, p. 341-347, pl. 33-34.

Bekker, Hendrik

(58a) 1921, *The Kuckers stage of the Ordovician rocks of N.E. Estonia:* Tartu Univ. (Dorpat) Acta & Commentationes, v. A2, 92 p., 12 pl., 15 text-fig.

(59) 1922, *A new brachiopod (Leptestia) from the Kuckers stage in Esthonia:* Geol. Mag., v. 59, p. 361-365, 4 text-fig.

(60) 1924, *The Devonian rocks of the Irboska district:* Naturk. Estlands Naturforsch. Gesell., v. 10, pt. 1, p. 1-55, pl. 1-6.

(61) 1924, *A new cemented brachiopod Irboskites and some other fossils of the Irboska stage:* Tartu Univ. (Dorpat) Acta & Commentationes, no. 2, p. 48-51, pl. 6.

Belanski, C. H.

(62) 1928, *Pentameracea of the Devonian of northern Iowa:* Iowa Univ. Studies, Nat. History, new ser., v. 12, no. 7, 34 p., 4 pl.

(63) 1928, *Terebratulacea of the Devonian of northern Iowa:* Same, v. 12, no. 8, 29 p., 4 pl., 15 text-fig.

Bell, W. A.

(64) 1929, *Horton-Windsor district, Nova Scotia:* Canada Geol. Survey, Dept. Min. & Res., Mem. 155, 268 p., 36 pl.

Bell, W. C.

(64a) 1938, *Prototreta, a new genus of brachiopod from the Middle Cambrian of Montana:* Michigan Acad. Sci., Arts & Letters, Paper 23, p. 403-408, 1 pl.

(65) 1941, *Cambrian Brachiopoda from Montana:* Jour. Paleontology, v. 15, no. 3, p. 193-255, pl. 28-37, text-fig. 1-20.

(66) 1944, *Early Upper Cambrian brachiopods,* in LOCHMAN, CHRISTINA, & DUNCAN, DONALD, Early Upper Cambrian faunas of central Montana: Geol. Soc. America, Spec. Paper 54, p. 144-155, pl. 18-19.

——, & Ellinwood, H. L.

(67) 1962, *Upper Franconian and Lower Trempealeauan Cambrian trilobites and brachiopods, Wilberns Formation, central Texas:* Jour. Paleontology, v. 36, p. 385-423, pl. 51-64.

Benediktova, R. N.

(68) 1956, *Spiriferidy Ostrogskoy svity Kuzbassa:* Voprosy Geol. Kuzb., p. 169-181, pl. 1-3. [*Spiriferids of the Ostrog Series of Kuzbassa.*]

Benson, W. N., & Dun, W. S.

(69) 1920, *The geology of the Great Serpentine belt of New South Wales:* Linnean Soc. New South Wales, Proc., v. 45, pt. 3, sec. B, p. 337-363, pl. 18-23. [Correct to: Dun & Benson.]

Berdan, J. M.

(70) 1960, *Revision of the ostracode family Beecherellidae and redescription of Ulrich's types of Beecherella:* Jour. Paleontology, v. 34, no. 3, p. 467-478, pl. 66.

Biernat, Gertrude

(71) 1959, *Middle Devonian Orthoidea of the Holy Cross Mountains and their ontogeny:* Palaeont. Polonica, v. 10, p. 1-78, pl. 1-12.

Billings, Elkanah

(71a) 1858, *Report for the year 1857 of E. Billings, Esq., Palaeontologist:* Canada Geol. Survey, Rept. of Progress for the year 1857, p. 147-192, 24 text-fig.

(72) 1859, *On some new genera and species of Brachiopoda from the Silurian and Devonian rocks of Canada:* Canadian Nat. & Geol., v. 4, p. 1-131, text-fig. 1-5.

(73) 1861-1865, *Palaeozoic fossils: containing descriptions and figures of new or little known species of organic remains from the Silurian rocks:* Canada Geol. Survey, v. 1, 426 p. [p. 1-24, Nov. 1861].

(74) 1866, *Catalogues of the Silurian fossils of the Island of Anticosti, with descriptions of some new genera and species:* Same, Spec. Publ., 93 p., 28 text-fig.

(74a) 1871, *On some new species of Palaeozoic fossils:* Canadian Nat. Quart. Jour. Sci., new ser., v. 6, p. 213-222, 7 text-fig.

Bittner, Alexander

(75) 1888, *Ueber das Auftreten von Terebrateln aus der Subfamilie der Centronellinen in der Alpinen Trias:* K.K. Geol. Reichsanst. for 1888 (5), p. 125-128.

(76) 1890, *Brachiopoden der Alpinen Trias:* Same, Abhandl., v. 14, p. 1-325, pl. 1-41.

(77) 1892, *Brachiopoden der Alpinen Trias. Nachtrag. 1:* Same, v. 17 (2), p. 1-40, pl. 1-4.

(78) 1893, *Neue Koninckiniden des Alpinen Lias:* Same, Jahrbuch, v. 43, no. 1, p. 133-144, pl. 4.

(79) 1896, *Eine neue Form der triadischen Terebratulidengruppe der Neocentronellinen oder Juvavellinen:* Same, Verhandl., p. 131-132.

(80) 1903, *Brachiopoden und Lamellibranchiaten aus der Trias von Bosnien, Dalmatien und Venetian:* Same, Jahrbuch, v. 52, no. 3 & 4, p. 495-843.

(81) 1912, *Brachiopoden aus der Trias. des Kokonyer Waldes:* Result. der Wissen. Erforschung des Balatonsees (1911), v. 1, pt. 1, p. 3-59, pl. 1-5 (Vienna).

Blainville, H. M. D. de

(82) 1825-27, *Manuel de malacologie et de conchyliologie:* Atlas, 109 pl. (Paris).

Blochmann, Friedrich

(83) 1906, *Neue Brachiopoden der Valdivia- und Gauss-Expedition:* Zool. Anzeiger, v. 30, no. 21-22, p. 690-702, 824, text-fig. 1-3.

(84) 1908, *Zur Systematik und geographischen Verbreitung der Brachiopoden:* Zeitschr. Wiss. Zool., v. 90, p. 596-644, pl. 36-40.

(85) 1912, *Die Brachiopoden der schwedischen Südpolar-Expedition:* Wiss. Ergebn. Schwed. Südpolar-Exped., 1901-1903, v. 6, no. 7, p. 1-12, pl. 1-3.

Bolkhovitinova, M. A., & Markov, P. N.

(86) 1926, *Les traits faunistiques des dépôts carbonifères dans la région des mines Jouravlinsky, gouvernement de Perm:* Inst. Priklad. Min. Tsvet. Met., Trudy, v. 20, p. 1-56, 5 pl.

Bond, Geoffrey

(87) 1941, *Species and variation in British and Belgian Carboniferous Schizophoriidae:* Geologists' Assoc., Proc., v. 52 (4), p. 285-303, pl. 21-22.

Booker, F. W.

(88) 1926, *The internal structure of some of the Pentameridae of New South Wales:* Royal Soc. New S. Wales, Jour., v. 60, p. 130-145, pl. 5-8.

(89) 1929, *Preliminary note on new subgenera of Productus and Strophalosia from Branxton district:* Same, v. 63, p. 24-32, pl. 1-3.

(90) 1930, *A review of some of the Permo-Carboniferous Productidae of New South Wales with a tentative reclassification:* Same, v. 64, p. 65-77, 3 pl.

Born, I. E.

(91) 1778, *Index rerum naturalium Musei Caesarei Vindobonensis, Pars. I Testacea. Verzeichniss der natürlichen Seltenheiten des K. K. Natüralien Cabinets zu Wien. Erster Theil, Schaltiere:* 458 p. (Vindobonae).

Böse, Emil

(92) 1894, *Monographie des Genus Rhynchonellina Gemm.:* Palaeontographica, v. 41, p. 49-80, pl. 6-7.

Bosquet, J.

(93) 1860, *Monographie des Brachiopodes Fossiles du terrain Crétacé Supérieur du Duché de Limbourg. 1e. Craniadae et Terebratulidae (Subfamilia Thecidiidae):* Mém. Serv. Descr. Géol. Néerlande, v. 3, p. 1-50, pl. 1-5.

(94) 1862, *Notice sur deux nouveaux brachiopodes trouvés dans le terrain tertiaire Oligocène du Limbourg néerlandais et du Limbourg belge:* K. Akad. Wetensch., Proc., v. 14, p. 345-350, 1 pl. (not numbered).

Bouchard-Chantereaux, N. R.

(95) 1849, *Mémoire sur un nouveau genre de brachiopode formant le passage des formes articulées à celles qui ne sont pas:* Ann. Sci. Nat., sér. 3, v. 12, p. 84-93, 1 pl.

Boucot, A. J.

(96) 1957, *A Devonian brachiopod, Cyrtinopsis, redescribed:* Senckenbergiana Lethaea, v. 35, no. 1/2, p. 37-48, pl. 1-2.

(97) 1957, *Revision of some Silurian and Early Devonian spiriferid genera and erection of Kozlowskiellinae, new subfamily:* Same, v. 38, no. 5/6, p. 311-334, pl. 1-3.

(98) 1958, *Kozlowskiellina, new name for Koz-lowskiella Boucot, 1957:* Jour. Paleontology, v. 32, no. 5, p. 1031.

(99) 1959, *Early Devonian Ambocoeliinae (Brachiopoda):* Same, v. 33, no. 1, p. 16-24, pl. 1-2.

(100) 1959, *A new family and genus of Silurian orthotetacid brachiopods:* Same, v. 33, no. 1, p. 25-28, pl. 3, text-fig. 1-2.

(101) 1959, *Brachiopods of the Lower Devonian rocks at Highland Mills, New York:* Same, v. 33, p. 727-769, 14 pl.

(102) 1960, *A new Lower Devonian stropheodontid brachiopod:* Same, v. 34, p. 483-485, 1 pl.

(103) 1960, *A Late Silurian fauna from the Sutherland River Formation, Devon Island Canadian Arctic Archipelago:* Canada Geol. Survey, Bull. 65, p. 1-51, pl. 1-10.

(104) 1963, *The Eospiriferidae:* Palaeontology, v. 5, pt. 4, p. 682-711, pl. 98-104.

————, & Amos, Arturo

(105) *See* 27a.

————, & Amsden, T. W.

(106) 1958, *New genera of brachiopods:* in AMSDEN, T. W., & BOUCOT, A. J., Stratigraphy and paleontology of the Hunton Group in the Arbuckle Mountain region, Oklahoma Geol. Survey, Bull. 78, pt. 4, p. 159-170, pl. 14, text-fig. 40-42.

(107) 1963, *Virgianidae, a new family of pentameracid brachiopods:* Jour. Paleontology, v. 37, p. 296.

————, Caster, K. E., Ives, David, & Talent, J. A.

(108) 1963, *Relationships of a new Lower Devonian terebratuloid (Brachiopoda) from Antarctica:* Bull. Am. Paleontology, v. 46, no. 207, p. 77-151, pl. 16-41.

————, & Ehlers, G. M.

(109) 1963, *Two new genera of stricklandid brachiopods:* Michigan Univ., Museum Paleont. Contrib., v. 18, p. 47-66, 5 pl.

(110) *See* 109.

————, & Gill, E. D.

(111) 1956, *Australocoelia, a new Lower Devonian brachiopod from South Africa, South America, and Australia:* Jour. Paleontology, v. 30, no. 5, p. 1173-1178, pl. 126.

————, & Johnson, J. G.

(112) 1963, Appendix in BOUCOT, et al., *Relationships of a new Lower Devonian terebratuloid (Brachiopoda) from Antarctica:* Bull. Am. Paleontology, v. 46, no. 207, p. 123.

————, ————, & Staton, R. D.

(113) 1964, *On some atrypoid, retzioid, and athyridoid Brachiopoda:* Jour. Paleontology, v. 38, p. 805-822, pl. 125-128, text-fig. 1-6.

————, & Pankiwskyj, Kost

(114) 1962, *Llandoverian to Gedinnian stratigraphy of Podolia and adjacent Moldavia:* Symposium on the Silurian-Devonian boundary, 11 p. (Stuttgart).

————, & Siehl, Agemar

(115) 1962, *Zdimir Barrande (Brachiopoda) redefined:* Hess. Landesamt Bodenf., Notizbl., v. 90, p. 117-131, pl. 15-20.

Branson, E. B.

(116) 1938, *Stratigraphy and paleontology of the Lower Mississippian of Missouri:* Missouri Univ. Studies, pt. 1, v. 13, no. 3, p. 1-208, pl. 1-20; pt. 2, v. 13, no. 4, p. 1-56, 179-189.

Breger, C. L.

(117) 1906, *On Eodevonaria, a new subgenus of Chonetes:* Am. Jour. Sci., ser. 4, v. 22, p. 534-536.

Brill, K. G.

(118) 1940, *Brachiopods of the Whitehorse Sandstone,* in NEWELL, N. D., The invertebrate fauna of the late Permian Whitehorse Sandstone: Geol. Soc. America, Bull., v. 51, p. 316-319.

Broderip, W. J.

(119) 1833, *Descriptions of some new species of Cuvier's family of Brachiopoda:* Zool. Soc. London, Proc., pt. 1, p. 124-125.

Broili, Ferdinand

(120) 1916, *Die permischen Brachiopoden von Timor:* Paläont. von Timor, no. 7, pt. 12, 104 p., pl. 115-127.

Bronn, H. G.

(121) 1862, *Die Klassen und Ordnungen der Weichthiere (Malacozoa):* v. 3, pt. 1, 518 p., 44 pl. (Leipzig and Heidelberg).

Brown, Ida A.

(122) 1953, *Martiniopsis Waagen from the Salt Range of India:* Royal Soc. New S. Wales, Jour. and Proc., v. 86, p. 100-107, pl. 9, 3 text-fig.

(123) 1953, *Permian spirifers from Tasmania:* Same, v. 86, p. 55-63, pl. 5-6.

Bruguière, J. G.

(124) 1797, *Tableau encyclopédique et méthodique des trois règnes de la nature:* v. 2, Vers, Coquilles, Mollusques et Polypiers: pl. 190-286 (Paris).

Bublichenko, N. L.

(125) 1927, *Die Brachiopoden des unteren Palaeozoicum aus der Umgegend des Dorfes Sara-Tschumysch aus dem Kohlenbassin von Kusnetsk:* Comité Géol., Bull., v. 46, no. 8, p. 979-1008, pl. 49-50.

(126) 1956, *Nekotorye novyi predstaviteli brakhiopod Devona i Karbona Rudnogo Altaya i Sary-arka:* Akad. Nauk Kazak. SSSR, Izvestiya, v. 23, p. 93-104, pl. 1, 4 text-fig. [*Some new representatives of Devonian and Carboniferous brachiopods from Rudno-Altai & Sary-Arka.*]

Buckman, S. S.
(127) 1895, *The Bajocian of the Mid-Cotteswolds. Pt. 3, Appendix, Notes on certain Brachiopoda:* Geol. Soc. London, Quart. Jour., v. 51, p. 445-462, pl. 14.
(128) 1906, *Brachiopod homoeomorphy: Pygope, Antinomia, Pygites:* Same, v. 62, p. 433-455, pl. 41.
(129) 1907, *Brachiopod morphology: Cincta, Eudesia, and the development of ribs:* Same, v. 63, p. 338-343.
(130) 1907, *Some species of the genus Cincta:* Cotteswold Naturalists Field Club, Proc., v. 16, pt. 1, p. 41-63, pl. 5-6.
(131) 1907, *Brachiopod nomenclature: the genotype of Terebratula:* Ann. & Mag. Nat. History, ser. 7, v. 19, p. 525-531, pl. 12.
(132) 1908, *Brachiopod nomenclature: the Terebratulae of the Crag:* Same, ser. 8, v. 1, p. 444-447.
(133) 1910, *Antarctic fossil Brachiopoda collected by the Swedish South Polar Expedition:* Wiss. Ergebn. Schwed. Südpolar-Exped., 1901-03, v. 3, no. 7, p. 1-43, pl. 1-3.
(134) 1914, *Genera of some Jurassic Brachiopoda:* 2 p., W. Wesley (London).
(135) 1915, *The Brachiopoda of the Namyau Beds of Burma:* India Geol. Survey, Rec., v. 45, pt. 1, p. 75-81.
(136) 1917[1918], *The Brachiopoda of the Namyau Beds, Northern Shan States, Burma:* Palaeont. Indica, new ser., v. 3, Mem. 2, 299 p., 21 pl.

———, & Walker, J. F.
(137) 1889, *On the spinose Rhynchonellae (Genus Acanthothyris, d'Orbigny), found in England:* Yorkshire Philos. Soc., Rept. (1888), p. 41-57.

Burri, Fritz
(138) 1956, *Die Rhynchonelliden der unteren Kreide (Valanginien-Barremien) im westschweizerischen Juragebirge:* Eclogae Geol. Helv., v. 49, p. 599-701, pl. 1-15.

Campbell, K. S. W.
(139) 1953, *The fauna of the Permo-Carboniferous Ingelara Beds of Queensland:* Queensland Univ., Dept. of Geol., Paper, v. 4, no. 3, p. 1-44, pl. 1-7, 4 text-fig.
(140) 1957, *A Lower Carboniferous brachiopod-coral fauna from New South Wales:* Jour. Paleontology, v. 31, p. 34-98, pl. 11-17, 27 text-fig.

(141) 1959, *The Martiniopsis-like spiriferids of the Queensland Permian:* Palaeontology, v. 1, p. 333-350, pl. 56-57.
(142) 1959, *The type species of three upper Palaeozoic punctate spiriferoids:* Same, v. 1, pt. 4, p. 351-363, pl. 58-60, text-fig. 1-6.
(143) 1961, *Carboniferous fossils from the Kuttung rocks of New South Wales:* Same, v. 4, pt. 3, p. 428-474, pl. 53-63.
(144) 1965, *Australian Permian terebratuloids:* Australia Bur. Mineral Res., Geol. & Geophys., Bull. 68 (in press).

Caneva, George
(145) 1907, *La fauna del Calcare a Bellerophon:* Soc. Geol. Italiana, Boll. 25, p. 427-452.

Carpenter, W. B.
(146) 1853, *On the intimate structure of the shells of Brachiopoda,* in DAVIDSON, THOMAS, British fossil Brachiopoda: Palaeontograph. Soc., v. 1, p. 23-40, pl. 4-5.

Castellaro, H. A.
(147) 1959, *Braquiopodos gotlandicos de la Precordillera de San Juan:* Rev. Asoc. Geol. Argentina, v. 13, p. 41-65, pl. 1-5.

Caster, K. E.
(148) 1939, *A Devonian fauna from Colombia:* Bull. Am. Paleontology, v. 24, no. 83, p. 1-218, pl. 1-14.
(149) 1945, *New names for two homonyms:* Jour. Paleontology, v. 19, p. 319.

Catullo, T. A.
(150) 1827, *Saggio di zoologia fossile:* 348 p., 8 pl. (Padova).
(151) 1851, *On the epiolitic rocks of the Venetian Alps:* Geol. Soc. London, Quart. Jour., v. 7, p. 66-76, 4 text-fig.

Chao, Y. T.
(152) 1927, *Productidae of China. Pt. 1. Producti:* China Geol. Survey, Palaeont. Sinica, ser. B, v. 5, pt. 2, 192 p., 16 pl.
(153) 1928, *Productidae of China. II. Chonetinae, Productinae and Richthofeniinae:* Same, v. 5, pt. 3, p. 1-103, pl. 1-6.
(154) 1929, *Carboniferous and Permian spiriferids of China:* Same, v. 11, pt. 1, p. 1-101, pl. 1-11, 20 text-fig.

Chapman, Frederick
(154a) 1935, *Primitive fossils, possibly atrematous and neotrematous Brachiopoda, from the Vindhyans of India:* India Geol. Survey, Rec., v. 69, p. 109-120, pl. 1-2.

Chernyshev [Tschernyschew], T. N.
(155) 1885, *Die Fauna des unteren Devon am West-Abhange des Urals:* Comité Géol., Mém., v. 3, no. 1, 107 p., 9 pl.
(156) 1887, *Die Fauna des mittleren und oberen*

Devon am Westabhange des Urals: Same, v. 3, no. 3, p. 1-208, 14 pl.

(157) 1893, *Die Fauna des unteren Devon am Ostabhange des Ural:* Same, v. 4, no. 3, p. 1-221, 14 pl.

(158) 1902, *Die obercarbonischen Brachiopoden des Ural und des Timan:* Same, v. 16, no. 2, p. 1-749, Atlas, 63 pl.

(159) 1914, *Die Fauna der oberpalaeozoischen Ablagerungen des Darvas. Lieferung I:* Same (new ser.), v. 104, p. 1-66, pl. 1-10.

————, & Stepanov, P. I.

(160) 1916, *Verkhnekamennougol'naya faunas Zemli Korolya Oskara i Zemli Geyberga:* Materialy dlya Geol. Rossii, v. 27, p. 1-105, pl. 1-12, 6 text-fig. [*Upper Carboniferous faunas from Zemli Korolya Oscara and Zemli Geyberga.*]

Chu, S.

(161) 1933, *Corals and brachiopods of the Kinling limestone:* Natl. Res. Inst. Geol., Acad. Sinica Mon., ser. A, v. 2, p. 1-73, pl. 1-5.

Clark, T. H.

(161a) 1917, *New blastoids and brachiopods from the Rocky Mountains:* Harvard Univ., Museum Comp. Zool., Bull. 61, no. 9, p. 361-380, 2 pl., 5 text-fig.

Clarke, J. M.

(162) 1907, *Some new Devonic fossils:* N.Y. State Museum, Bull. 107, p. 153-291.

(163) 1912, *El Devoniano de la Argentina occidental:* Ann. Ministerio Agric., Argentina, Sec. Geol., v. 8, no. 2, p. 1-19.

(164) 1913, *Ninth report of the director of the science division, including the 66th report of the State Museum, the 32nd report of the State geologist, and the report of the State paleontologist for 1912:* N.Y. State Museum, Bull. 164, p. 140-214.

(165) 1913, *Fosseis devonianos do Parana:* Serv. Geol. & Mineral. Brasil, Mon., v. 1, xx+353 p., 27 pl., 27 text-fig.

(166) 1921, *The heavenly twins:* Nautilus, v. 34, p. 138-139.

Cloud, P. E., Jr.

(167) 1942, *Terebratuloid Brachiopoda of the Silurian and Devonian:* Geol. Soc. America, Spec. Paper 38, xi+182 p., pl. 1-26.

(168) 1948, *Brachiopods from the Lower Ordovician of Texas:* Harvard Univ., Museum Comp. Zool., v. 100, no. 5, p. 468-470, pl. 2, text-fig. 32-45.

(169) 1948, *Dicaelosia versus Bilobites:* Jour. Paleontology, v. 22, p. 373-374.

Cobbold, E. S.

(169a) 1921, *The Cambrian horizons of Comley (Shropshire) and their Brachiopoda, Pteropoda, Gasteropoda, etc.:* Geol. Soc. London,

Quart. Jour., v. 76, p. 325-386, pl. 21-24, 4 text-fig.

Cockerell, T. D. A.

(169b) 1911, *The name Glossina:* Nautilus, v. 25, p. 96.

(170) 1929, *The brachiopod called Mimulus:* Same, v. 42, p. 105.

Comte, Pierre

(171) 1938, *Brachiopodes dévoniens des gisements de Ferrones (Asturies) et de Sabero (Léon):* Ann. Paléontologie, v. 27, p. 41-87, pl. 5-8.

Conrad, T. A.

(172) 1839, *Descriptions of new species of organic remains:* N.Y. State Geol. Survey, 3rd Ann. Rept., p. 57-66.

(173) 1842, *Observations on the Silurian and Devonian systems of the U.S. with descriptions of new organic remains:* Acad. Nat. Sci. Philadelphia, Jour., v. 8, p. 228-280, pl. 12-17.

(174) 1843, *Observations on the lead-bearing limestone of Wisconsin and descriptions of a new genus of trilobites and fifteen new Silurian fossils:* Same, v. 1, p. 329-335.

Cooper, G. A.

(175) 1930, *The brachiopod genus Pionodema and its homoeomorphs:* Jour. Paleontology, v. 4, no. 4, p. 369-382, pl. 35-37, 1 text-fig.

(176) 1936, *New Cambrian brachiopods from Alaska:* Same, v. 10, p. 210-214, pl. 26.

(177) 1942, *New genera of North American brachiopods:* Washington Acad. Sci., Jour., v. 32, no. 8, p. 228-235.

(178) 1944, *Phylum Brachiopoda,* in SHIMER, H. W., & SHROCK, R. R., Index Fossils of North America: p. 277-365, pl. 105-143, Wiley & Sons (New York).

(179) 1951, *New brachiopods from the Lower Cambrian of Virginia:* Washington Acad. Sci., Jour., v. 41, no. 1, p. 4-8, 1 pl.

(180) 1952, *Unusual specimens of the brachiopod family Isogrammidae:* Jour. Paleontology, v. 26, no. 1, p. 113-119, pl. 21-23.

(181) 1952, *New and unusual species of brachiopods from the Arbuckle group in Oklahoma:* Smithsonian Misc. Coll., v. 117, no. 14, p. 1-35, pl. 1-4.

(182) 1953, *Permian fauna et El Antimonio, western Sonora, Mexico:* Same, v. 119, no. 2, p. 21-77, pl. 4-23.

(183) 1954, *Unusual Devonian brachiopods:* Jour. Paleontology, v. 28, p. 325-332, pl. 36-37, 5 text-fig.

(184) 1954, *Recent brachiopods: Bikini and nearby Atolls, Marshall Islands:* U.S. Geol. Survey, Prof. Paper 260-G, p. 315-318, pl. 80-81.

(185) 1955, *New genera of Middle Paleozoic brachiopods:* Jour. Paleontology, v. 29, no. 1, p. 45-63, pl. 11-14.

(186) 1955, *New brachiopods from Cuba:* Same, v. 29, no. 1, p. 64-70, pl. 15.

(187) 1955, *New Cretaceous Brachiopoda from Arizona:* Smithsonian Misc. Coll., v. 131, no. 4, 18 p.

(188) 1956, *New Pennsylvanian brachiopods:* Jour. Paleontology, v. 30, no. 3, p. 521-530, pl. 61, 1 text-fig.

(189) 1956, *Chazyan and related brachiopods:* Smithsonian Misc. Coll., v. 127, pt. I, p. 1-1024, pt. II, p. 1025-1245, pl. 1-269.

(190) 1956, *Pustulatia, new name for the Devonian brachiopod Pustulina, preoccupied:* Jour. Paleontology, v. 30, no. 3, p. 769.

(191) 1957, *Loop development of the Pennsylvanian terebratuloid Cryptacanthia:* Smithsonian Misc. Coll., v. 134, no. 3, p. 1-18, 2 pl.

(192) 1957, *Permian brachiopods from central Oregon:* Same, v. 134, no. 12, 79 p., 12 pl., 2 text-fig.

(193) 1959, *Genera of Tertiary and Recent rhynchonelloid brachiopods:* Same, v. 139, p. 1-90, 22 pl.

(193a) 1960, *Correction of brachiopod names:* Jour. Paleontology, v. 34, p. 601.

———, & **Kindle, C. H.**

(194) 1936, *New brachiopods and trilobites from the Upper Ordovician of Percé, Quebec:* Jour. Paleontology, v. 10, p. 355-356, pl. 51, text-fig. 38, 42.

———, & **Muir-Wood, H. M.**

(195) 1951, *Brachiopod homonyms:* Washington Acad. Sci., Jour., v. 41, no. 6, p. 195-196.

———, & **Stehli, F. G.**

(196) 1955, *New genera of Permian brachiopods from West Texas:* Jour. Paleontology, v. 29, no. 3, p. 469-474, pl. 52-54.

———, & **Whitcomb, Lawrence**

(197) 1933, *Salonia, a new Ordovician brachiopod genus:* Washington Acad. Sci., Jour., v. 23, p. 496-503, 23 text-fig.

———, & **Williams, J. S.**

(198) 1935, *Tully Formation of New York:* Geol. Soc. America, Bull., v. 46, p. 781-868, pl. 54-60, 3 pl., 7 text-fig.

Cope, F. W.

(199) 1934, *Airtonia, a new brachiopod from the Lower Carboniferous of Yorkshire:* Ann. & Mag. Nat. History, ser. 10, v. 14, p. 273-289, pl. 6.

(200) 1935, *On Daviesiella carinata (Garwood):* Yorkshire Geol. Soc., Proc., v. 23, p. 79-90, pl. 3.

(201) 1940, *Daviesiella llangollensis Davidson and related forms: morphology, biology and distribution:* Manchester Geol. Assoc., Jour., v. 1, p. 199-231, 13 text-fig.

Cossmann, M.

(201a) 1898, *Errata et rectifications:* Rev. Crit. Paléozool., 2nd year, no. 2, p. 76-77.

Crickmay, C. H.

(202) 1950, *Some Devonian Spiriferidae from Alberta:* Jour. Paleontology, v. 24, p. 219-225, pl. 36-37.

(203) 1952, *Discrimination of late Upper Devonian:* Same, v. 26, no. 4, p. 585-609, pl. 70-78.

(204) 1952, *Nomenclature of certain Devonian brachiopods:* 2 p., The Author (Calgary).

(205) 1953, *Warrenella, a new genus of Devonian brachiopod:* Jour. Paleontology, v. 27, p. 596-600, text-fig. 25.

(206) 1953, *New Spiriferidae from the Devonian of western Canada:* 11 p., 6 pl., The Author (Calgary).

(207) 1954, *Paleontological correlation of Elk Point and equivalents, western Canada sedimentary basin: A symposium of the Alberta Society of Petroleum Geologists:* Am. Assoc. Petroleum Geologists, Ralph Leslie Rutherford Memorial Volume, p. 143-158.

(208) *See* 201a.

Cuvier, G. L. C. F. D.

(208a) 1800-05, *Leçons d'Anatomie comparée de G. Cuvier recueillies et publiées sous ses yeux par G. L. Duvernoy:* v. 1-5; v. 1 (1800), xxxi+521 p., 9 tab.; v. 2 (1800), xvi+696 p.; v. 3 (1805), xxviii+558 p.; v. 4 (1805), xii+539 p.; v. 5 (1805), vii+368 p., 52 pl., Baudouin (Paris).

Dagis, A. S.

(209) 1958, *Razvitie petli u nekotovykh Triasovykh Terebratulida:* Leit. TSR Mokslu, Akad. Geol. ir Geog. Inst. Mokslinai Dranesimai, SSR, Trudy, ser. B, v. 3, no. 15, p. 175-182, 5 text-fig. [*Loop development in some Triassic Terebratulida.*]

(210) 1959, *Novye Triasovye rody Terebratulida:* Same, v. 9, p. 23-41, 1 pl., 5 text-fig. [*New genera of Triassic Terebratulida.*]

(211) 1961, *Dva novykh roda Triasovykh rinkhonellid:* Paleont. Zhurnal, no. 4, p. 93-99, pl. 8 *(partim).* [*Two new genera of Triassic rhynchonellids.*]

(212) 1961, *Novyy rod Triasovykh spiriferid Triadispira, gen. n.:* Akad. Nauk SSSR, Doklady, v. 141, no. 2, p. 457-460, 2 text-fig. [*New genus of Triassic spiriferid, Triadispira, gen. nov.*]

(212a) 1963, *Verkhnetriasovye brakhiopody yuga S.S.S.R.:* Akad. Nauk SSSR, Sibir. Otdel., p. 5-248, 31 pl., 106 text-fig. [*Upper Triassic brachiopods of the southern U.S.S.R.*]

Dahmer, Georg

(212b) 1942, *Die Fauna der "Gedinne"-Schichten von Weismes in der Nordwest-Eifel:* Senckenbergiana, v. 25, no. 1/3, p. 111-156, 40 text-fig.

Dall, W. H.

(213) 1870, *A revision of the Terebratulidae and Lingulidae:* Am. Jour. Conchology, v. 6, p. 88-168, pl. 6-8.

(214) 1871, *Report on the Brachiopoda obtained by the United States Coast Survey Expedition in charge of L. F. de Pourtalès, with a revision of the Craniidae and Discinidae:* Harvard Univ., Museum Comp. Zool., Bull. 3, no. 1, p. 1-45, pl. 1-2.

(215) 1877, *Index to the names which have been applied to the subdivisions of the Class Brachiopoda:* U.S. Natl. Museum, Bull. 8, p. 1-88.

(216) 1889, *Preliminary report on the collection of Mollusca and Brachiopoda obtained in 1887-88:* Same, Proc., v. 12, p. 219-362, pl. 5-14.

(217) 1895, *Report on Mollusca and Brachiopoda dredged in deep water, chiefly near the Hawaiian Islands, with illustrations of hitherto unfigured species from Northwest America:* Same, v. 17, p. 675-733, pl. 23-32.

(218) 1903, *Contributions to the Tertiary fauna of Florida:* Wagner Free Inst. Sci. Trans., v. 3, pt. 6, p. 1219-1620, pl. 48-60.

(219) 1908, *Some new brachiopods:* Nautilus, v. 22, no. 3, p. 28-30.

(220) 1908, *Reports on the dredging operations off the west coast of Central America to the Galapagos, to the west coast of Mexico, and in the Gulf of California, in charge of Alexander Agassiz, carried on by the U.S. Fish Commission Steamer "Albatross," during 1891, Lieut. Commander Z. L. Tanner, U.S.N., commanding. 37. Reports on the scientific results of the expedition to the eastern tropical Pacific, in charge of Alexander Agassiz, by the U.S. Fish Commission Steamer "Albatross," from October, 1904, to March, 1905, Lieut. Commander L. M. Garrett, U.S.N., commanding. 14. The Mollusca and the Brachiopoda:* Harvard Univ., Museum Comp. Zool., Bull. 43, p. 205-487, 22 pl.

(221) 1920, *Annotated list of the Recent Brachiopoda in the collection of the United States National Museum, with descriptions of thirty-three new forms:* U.S. Natl. Museum, Proc., v. 57, no. 2314, p. 261-377.

Dalman, J. W.

(222) 1828, *Uppställning och Beskrifning af de i sverige funne Terebratuliter:* K. Svenska Vetenskapsakad. Handl. for 1827, p. 85-155, pl. 1-6.

Dana, J. D.

(223) 1847, *Descriptions of fossil shells of the collections of the exploring expedition under the command of Charles Wilkes, U.S.N., obtained in Australia, from the lower layers of the coal formation in Illawarra, and from a deposit of nearly the same age at Harper's Hill, Valley of the Hunter:* Am. Jour. Sci., v. 54, p. 151-160.

Davidson, Thomas

(224) 1847, *Descriptions of some species of Brachiopoda:* Ann. & Mag. Nat. History, v. 20, p. 250-257, pl. 18-19.

(225) 1848, *Sur les brachiopodes du système silurien supérieur de l'Angleterre:* Soc. Géol. France, Bull., ser. 2, v. 5, p. 309-338, 370-374.

(226) 1850, *Notes on an examination of Lamarck's species of fossil Terebratulae:* Ann. & Mag. Nat. History, ser. 2, v. 5, no. 30, p. 433-450, pl. 13-15.

(227) 1850, *On the internal structure of Terebratula deslongchampsi nov.:* Same, v. 5, no. 30, p. 449-450, pl. 15.

(228) 1851-52, *British fossil Brachiopoda: The Oolitic and Liasic Brachiopoda:* Paleontograph. Soc., v. 1, pt. 3, 64 p., 13 pl.

(229) 1851-1886, *A monograph of the British fossil Brachiopoda:* Palaeontograph. Soc., v. 1, Intro., 1853, p. 1-136, pl. 1-9, pt. 1, Tert., 1852, p. 1-23, pl. 1-2, pt. 2, Cret., 1852-55, p. 1-117, pl. 1-12, pt. 3, Oolit. & Lias., 1851-52, p. 1-100, pl. 1-18; v. 2, pt. 4, Perm., 1858, p. 1-51, pl. 1-4, pt. 5, Carb., 1858-63, p. 1-280, pl. 1-55; v. 3, pt. 6, Dev., 1864-65, p. 1-131, pl. 1-20, pt. 7, Sil., 1866-71, p. 1-397, pl. 1-50; v. 4, pt. 1, Cret.-Rec., Suppl., 1874, p. 1-72, pl. 1-8, pt. 2, Jur.-Trias., Suppl., 1876-78, p. 73-242, pl. 9-29, pt. 3, Carb.-Perm., Suppl., 1880, p. 243-316, pl. 30-37, pt. 4, S. & D. in Tr., 1881, p. 317-368, pl. 38-42, pt. 5, 1882, p. 369-383; v. 5, pt. 1, D. & S. Suppl., 1882, p. 1-134, pl. 1-7, pt. 2, Sil. Suppl., 1883, p. 135-242, pl. 8-17, pt. 3, Appendix, etc., 1884, p. 243-476, pl. 18-21; v. 6, Bibliography, 1886, p. 1-163 (by T. Davidson & W. H. Dalton).

(230) 1852, *Notes and descriptions of a few Brachiopoda; including a monograph of the French Liasic spirifers:* Ann. & Mag. Nat. History, ser. 2, v. 9, p. 249-267, pl. 13-15.

(230a) 1853, *On some fossil Brachiopoda of the Devonian age from China:* Geol. Soc. London, Quart. Jour., v. 9, p. 353-359, pl. 15.

(231) 1854, *Observations on the Chonetes comoides:* Same, v. 10, p. 202-207, pl. 8.

(232) 1862, *On British Carboniferous Brachiopoda:* Geologist, v. 4, p. 41-59.

(233) 1862, *On some Carboniferous Brachiopoda collected in India by A. Fleming, M.D., and W. Purdon, Esq. F. G. S.:* Geol. Soc. London, Quart. Jour., v. 18, p. 25-35, 2 pl.

(234) 1866, *Notes on the Carboniferous Brachiopoda collected by Captain Godwin-Austen in the Valley of Kashmere:* Same, v. 22, p. 39-45, pl. 2.

(235) 1869, in TAWNEY, E. B., *On the occurrence of Terebratula diphya in the Alps of the Canton de Vaud:* Same, v. 25, Proc. for May 26, 1869, p. 305-309, text-fig. 1.

(236) 1870, *On Italian Tertiary Brachiopoda:* Geol. Mag., v. 7, p. 359-370, 399-408, 460-466, pl. 17-21.

(237) 1874, *On the Tertiary Brachiopoda of Belgium:* Same, dec. 2, v. 1, p. 150-159, pl. 7-8.

(238) *See* 244.

(239) 1878, *On the Brachiopoda dredged by H.M.S. Challenger:* Royal Soc. London, Proc., v. 27, no. 188, p. 428-439.

(240) 1880, *Report on the Brachiopoda dredged by H.M.S. Challenger during years 1873-1876:* Rept. Sci. Res. Challenger (Zool.), v. 1, p. 1-67, pl. 1-4.

(241) 1881, *On genera and species of spiral-bearing Brachiopoda, from specimens developed by the Rev. Norman Glass:* Geol. Mag., new ser., dec. 2, v. 8, p. 1-13, 15 text-fig.

(242) 1881, *Description of new Upper Silurian Brachiopoda from Shropshire:* Same, v. 8, p. 145-156, pl. 5.

(243) 1881, *On the genera Merista, Suess, 1851, and Dayia, Davidson, 1881:* Same, v. 8, p. 289-293.

(244) 1886-88, *A monograph of Recent Brachiopoda:* Linnean Soc. London, Trans., ser. 2, v. 4 (Zool.), 248 p., 30 pl.

——, & King, William

(244a) 1872, *Remarks on the genera Trimerella, Dinobolus and Monomerella:* Geol. Mag., v. 9, p. 442-445.

(245) 1874, *On the Trimerellidae, a Palaeozoic family of the palliobranchs or Brachiopoda:* Geol. Soc. London, Quart. Jour., v. 30, p. 124-173, pl. 12-19, 3 text-fig.

Defrance, M. J. L.

(246) 1825-27, in BLAINVILLE, H. M. D., *Manuel de malacologie et de conchyliologie:* text (1825), viii+647 p., atlas (1827), p. 649-664, 109 pl., Levrault (Paris, Strasbourg).

(247) 1827, *Dictionnaire des sciences naturelles:* CUVIER, F. G. (ed.), v. 51, p. 152 (Paris, Strasbourg).

(248) [Koninck, L. G. de. *See* 484a.]

Derby, O. A.

(249) 1874, *On the Carboniferous Brachiopoda of Itaitúba, Rio Tapajos, Prov. of Pará, Brazil:* Cornell Univ., Sci., Bull., v. 1, ser. 2, p. 1-63, pl. 1-9.

(250) 1896, *Nota sobre a geologia e paleontologia de Matto Gross:* Rio de Janeiro, Museu Nac., Arch., v. 9, p. 81-84.

Deslongchamps, E. Eudes-

(251) 1853, *Mémoire sur les genres Leptaena et Thecidea des terrains jurassiques du Calvados:* Soc. Linnéenne Normandie, Mém. 9, p. 213-250, pl. 11-13.

(252) 1856, *Note sur deux nouvelles térébratules du Lias moyen de Précigné (Sarthe):* Same, v. 10, p. 1-4, pl. 17.

(253) 1862-85, *Paléontologie française ou description des animaux invertébrés fossiles de la France: Terrain jurassique I. Brachiopodes:* 448 p., 131 pl., Masson & Fils (Paris).

(254) 1863-87, *Études critiques sur des brachiopodes nouveaux ou peu connus:* Soc. Linnéenne Normandie, Bull., ser. 2, v. 7, p. 248-295, pl. 1-8; v. 8, p. 249-286, pl. 9-11; ser. 3, v. 8, p. 161-350, pl. 1-14; v. 10, p. 31-158, pl. 27-28.

(255) 1865, *Récherches sur l'organisation du manteau chez les brachiopodes articulés et principalement sur les spicules calcaires contenus dans son intérieur:* Same, Mém., v. 14, no. 2, 36 p., 3 pl.

(256) 1884, *Sur l'appareil brachial de diverses Térébratules du Lias et du système Oolithique inférieur:* Same, Bull., ser. 3, v. 8, p. 303-312.

Diener, Carl

(257) 1897, *The Permo-Carboniferous fauna of Chitichun No. 1: Himalayan fossils:* Palaeont. Indica, ser. 15, v. 1, pt. 3, 105 p., 13 pl.

(258) 1908, *Ladinic, Carnic and Noric faunae of Spiti:* Same, ser. 15, v. 5, pt. 3, p. 1-157, pl. 1-24.

(259) 1927, *Leitfossilien des marinen Perm:* Leitfossilien, v. 5, p. 19-42, pl. 3-9(Berlin).

Dittmar, Alphons von

(260) 1872, *Ueber ein neues Brachiopoden Geschlecht aus dem Bergkalk:* Verh. der Russische-Kaiser. Min. Gesell., ser. 2, v. 7, p. 1-14, pl. 1.

Douglas, J. A.

(261) 1940, *The genus Mentzelia Quenstedt, and its affinities to other members of the*

subfamily Martiniinae Waagen: Geol. Mag., v. 77, p. 330-333, text-fig. 1-3c.

———, & **Arkell, W. J.**

(262) 1932, *The stratigraphical distribution of the Cornbrash: II. The North-eastern Area*: Geol. Soc. London, Quart. Jour., v. 88, p. 112-170, pl. 10-12.

Douvillé, Henri

(263) 1879 [Sep. 1880], *Sur quelques genres de brachiopodes Terebratulidae et Waldheimiidae*: Soc. Géol. France, Bull., ser. 3, v. 7, p. 251-277, 19 text-fig.

(264) 1886, *Sur quelques brachiopodes du terrain jurassique*: Soc. Sci. Hist. Nat. de l'Yonne, Bull. 39, p. 43-102, pl. 1-4.

(265) 1916, *Les terrains secondaires dans le Massif du Moghara à l'est de l'isthme de Suez*: Acad. Sci. Paris, Mém., v. 54, p. 1-184, pl. 1-21.

Dovgal, V. N.

(265a) 1953, *Leiorhynchoides — novyy podrod plechenogikh iz srednego devona Gornogo Altaya*: Akad. Nauk SSSR, Gorno-geol. Inst., Trudy, no. 13, p. 139-141, 1 pl. [*Leiorhynchoides—a new subgenus of brachiopod from the Middle Devonian of Gorny Altay.*]

Dresser, Hugh

(266) 1954, *Notes on some brachiopods from the Itaituba Formation (Pennsylvanian) of the Tapajos River, Brazil*: Bull. Am. Paleontology, v. 35, p. 15-84, 8 pl.

Dubar, Gonzaque

(267) 1942, *Études paléontologiques sur le Lias du Maroc. Brachiopodes Térébratules et Zeilléries multiplissées*: Maroc Service Géol. Div. Mines & Géol., Notes & Mém., v. 57, p. 1-103, pl. 1-10.

Duméril, A. M. C.

(267a) 1806, *Zoologie analytique ou méthode naturelle de classification des animaux*: xxiv+344 p., Allais (Paris).

Dunbar, C. O.

(268) 1917, *Rensselaerina, a new genus of Lower Devonian brachiopods*: Am. Jour. Sci., ser. 4, v. 43, p. 466-470, pl. 2.

(269) 1955, *Permian brachiopod faunas of central east Greenland*: Meddel. Grønland, v. 110, no. 3, 169 p., 32 pl.

———, & **Condra, G. E.**

(270) 1932, *Brachiopoda of the Pennsylvanian system in Nebraska*: Nebraska Geol. Survey, ser. 2, Bull. 5, 377 p., 44 pl., 25 text-fig.

Ehlers, G. M., & Kline, V. H.

(271) 1934, *Revision of Alexander Winchell's types of Brachiopoda from the Middle Devonian Traverse Group of rocks of Michigan*: Michigan Univ., Museum Paleont., Contrib., v. 4, p. 143-176, pl. 1-4.

———, & **Wright, J. D.**

(272) 1955, *The type species of Spinocyrtia Fredericks and new species of this brachiopod genus from southwestern Ontario*: Michigan Univ., Museum Paleont., Contrib., v. 13, no. 1, p. 1-32, 11 pl.

Eichwald, Eduard von

(272a) 1829, *Zoologia specialis, quam expositis animalibus tum vivis, tum fossilibus potissium Rossiae in universum et Poloniae in specie, etc.*: v. 1, 314 p., 5 pl. (Vilnae).

Einor, O. L.

(273) 1939, *Brakhiopody Nizhney Permi Taymyra*: Arkt. Nauchno-Issledov. Inst., Trudy, v. 135, p. 1-150, pl. 1-15, 10 text-fig. [*Lower Permian brachiopods of Taimyr.*]

Elias, M. K.

(274) 1957, *Late Mississippian fauna from the Redoak Hollow Formation of southern Oklahoma*: Jour. Paleontology, v. 31, no. 3, p. 487-527, pl. 51-58, 26 text-fig.

Elliott, G. F.

(275) 1940, *Deux brachiopodes nouveaux de l'Auversien du bassin de Paris*: Soc. Géol. France, Bull., ser. 5, v. 9, p. 539-598, 3 text-fig.

(276) 1947, *The development of a British Aptian brachiopod*: Geologists' Assoc., Proc., v. 58, p. 144-159.

(277) 1948, *Palingenesis in Thecidea (Brachiopoda)*: Ann. & Mag. Nat. History, ser. 12, v. 1, p. 1-30, pl. 1-2.

(278) 1949, *The brachial development of Kraussina (Brachiopoda)*: Same, ser. 12, v. 2, p. 538-546, pl. 8-9.

(279) 1950, *The genus Hamptonina (Brachiopoda) and the relation of post-Palaeozoic brachiopods to coral-reefs*: Same, ser. 12, v. 3, p. 429-446, pl. 4.

(280) 1951, *On the geographical distribution of terebratelloid brachiopods*: Same, ser. 12, v. 4, p. 305-334.

(281) 1952, *The internal structure of some western Australian Cretaceous brachiopods*: Royal Soc. West. Australia, Jour., v. 36, p. 1-21, pl. 1-2.

(282) 1953, *The classification of the thecidean brachiopods*: Ann. & Mag. Nat. History, ser. 12, v. 6, p. 693-701, pl. 18.

(283) 1953, *Brachial development and evolution in terebratelloid brachiopods*: Biol. Reviews, v. 28, p. 261-279.

(284) 1954, *New Brachiopoda from the Eocene of England, France and Africa:* Ann. & Mag. Nat. History, ser. 12, v. 7, p. 721-728, pl. 15.

(285) 1955, *Shell-structure of thecidean brachiopods:* Nature, v. 175, p. 1124.

(286) 1957, *The origin of the Terebratellacea (Brachiopoda):* Ann. & Mag. Nat. History, ser. 12, v. 10, p. 259-264.

(287) 1958, *Classification of thecidean brachiopods:* Jour. Paleontology, v. 32, p. 373.

(288) 1959, *Six new genera of Mesozoic Brachiopoda:* Geol. Mag., v. 96, p. 146-148.

(289) 1960, *A new Mesozoic terebratellid brachiopod:* Geologists' Assoc., Proc., v. 71, p. 25-30, pl. 2.

Etheridge, Robert, Jr.
(290) 1876, *On an adherent form of Productus and a small Spiriferina from the Lower Carboniferous limestone group of the east of Scotland:* Geol. Soc. London, Quart. Jour., v. 32, p. 454-465, pl. 24-25.

(291) 1876, *On some species of Terebratulina, Waldheimia and Terebratella from the upper Tertiary deposits of Mount Gambier and the Murray-River Cliffs, South Australia:* Ann. & Mag. Nat. History, ser. 4, v. 17, p. 15-22, pl. 1-2.

(292) 1913, *Palaeontological contributions to the geology of western Australia:* West Australia Geol. Survey, Bull. 55, p. 1-34, pl. 1-4.

Fabiani, Ramiro, & Ruiz, Carmela
(293) 1933, *Sui giacimenti permiani del Sosio (Palermo) e sugli Strofomenidi in essi trovati:* Soc. Geol. Italiana, Mem., v. 1, no. 8, p. 1-22, pl. 1-2, text-fig. 1-4.

Fahrenkohl, Augustus
(294) 1856, *Flüchtiger Blick auf die Bergkalk- und Jura-Bildung in der Umgebung Moskwas:* Russ.-Kais. Min. Gesell. Verhandl. (1855-56), p. 219-236, pl. 3.

Fenton, C. L.
(295) 1931, *Studies of evolution in the genus Spirifer:* Wagner Free Inst. Sci., v. 2, 436 p., 50 pl., 204 text-fig.

———, & Fenton, M. A.
(296) 1924, *The stratigraphy and fauna of the Hackberry stage of the Upper Devonian:* Michigan Univ., Museum Geol., Contrib., v. 1, p. 1-260.

Finlay, H. J.
(297) 1927, *New specific names for austral Mollusca:* New Zealand Inst. Trans. & Proc., v. 57, p. 532-533.

Fischer, P., & Oehlert, D. P.
(298) 1890, *Diagnoses de nouveaux brachiopodes:* Jour. Conchyliologie, ser. 3, v. 38, no. 1, p. 70-74.

(299) 1891, *Brachiopodes:* Exped. Sci. Travailleur et du Talisman (1880-1883), 139 p., 8 pl. (Paris).

(300) 1892, *Brachiopodes de l'Atlantique Nord:* Résultats des campagnes scientifiques du Prince de Monaco, no. 3, 30 p., 2 pl.

Fischer de Waldheim, Gotthelf
(301) 1825, *Notice sur la Choristite:* Programme d'invitation à la Société Impériale des Naturalistes de Moscou, 12 p., 1 pl. (Moskva).

(302) 1829, *Quelques fossiles du gouvernement de Moscou:* Soc. Impér. Nat. Moscou, Bull., v. 1, no. 12, p. 375-376.

(303) 1830, *Oryctographie du gouvernement de Moscou:* 1st edit., 1830, ix+26 p., 60 pl.; 2nd edit., 1837, 202 p., 51 pl., A. Semen (Moskva).

(304) 1850, *Orthotetes genre de la famille des brachiopodes:* Soc. Impér. Nat. Moscou, v. 23(11), p. 491-494, pl. 10.

Foerste, A. F.
(305) 1909, *Fossils from the Silurian formations of Tennessee, Indiana and Kentucky:* Denison Univ. Sci. Lab., Bull., v. 14, p. 61-116, pl. 1-4.

(306) 1909, *Preliminary notes on Cincinnatian fossils:* Same, v. 14, p. 209-228, pl. 4.

(307) 1909, *Preliminary notes on Cincinnatian and Lexington fossils:* Same, v. 14, p. 289-324, pl. 7-11.

(308) 1912, *Strophomena and other fossils from the Cincinnatian and Mohawkian horizons, chiefly in Ohio, Indiana and Kentucky:* Same, v. 17, p. 17-173, pl. 1-8.

(309) 1914, *Notes on Lorraine faunas of New York and the province of Quebec:* Same, v. 17, p. 247-328, pl. 1-5.

(310) 1920, *The Kimmswick and Plattin limestones of northeastern Missouri:* Same, v. 19, p. 175-224, 3 pl.

(311) 1924, *Upper Ordovician faunas of Ontario and Quebec:* Canada, Geol. Survey, Mem., v. 138, p. 1-255, pl. 10-15.

Ford, S. W.
(311a) 1886, *Note on the recently proposed genus Billingsia:* Am. Jour. Sci., ser. 3, v. 32, p. 325.

Frech, Fritz
(311b) 1891, *Ueber das Devon der Ostalpen, II:* Deutschen geol. Gesell., Zeitschr., v. 43, p. 672-687, pl. 44-47.

(312) 1901, *Die Dyas:* Lethaea geognostica; 1, Theil Lethaea Palaeoz., v. 2, no. 3, p. 435-578 (Stuttgart).

(313) 1911, *Die Dyas,* in RICHTHOFEN, F. VON: China, v. 5, p. 103-202, pl. 19-28, D. Reimer (Berlin).

Frederiks [Fredericks], George

(314) 1916, *Uber einige oberpaläozoic Brachiopoden von Eurasien:* Comité Géol., Mém., v. 156, p. 1-87, 5 pl., 24 text-fig. [*Paleontological notes on some upper Paleozoic Brachiopoda of Eurasia.*]

(315) 1918, *Diagnoses generum et specierum novorum:* Ann. Soc. Paléont. Russie, v. 2, p. 142.

(316) 1918, *O primenenii podrazdeleniy Apikalvnago apparata k sistematike brakhiopod:* Russkoe Paleont. Obshch., v. 2 (1917), p. 85-91. [*Concerning the application of the subdivision of the apical apparatus to brachiopod classification.*]

(317) 1923, *New Lyttoniinae from the Up. Pal. of the Urals:* Rec. Geol. Com. Russian Far East No. 28, 52 p., 1 pl.

(318) 1924, *O Verkhne - Kamennougol'nykh spiriferidakh Urala:* Geologich. Komitet., Izvestiya, v. 38 (1919), no. 2, p. 295-324, 7 text-fig. [*On Upper Carboniferous spiriferids from the Urals.*]

(319) 1926, *Tablitsa dlya opredeleniya rodov semeystva Spiriferidae King:* Akad. Nauk SSSR, Izvestiya, ser. 6, v. 20, p. 393-423. [*Table for determination of the genera of the family Spiriferidae King.*]

(320) 1929, *Fauna Kynovskogo izvestnyaka na Urala:* Geologich. Komitet., Izvestiya, v. 48, no. 3, p. 369-413, pl. 20-21, 6 text-fig. [*Fauna of the Kyn Limestone of the Urals.*]

(321) 1931, *Faune paléozoique supérieure des monts Kharaoulakh:* Acad. Sci. URSS, Bull., ser. 7, no. 2, p. 199-221.

Fuchs, Alexander

(322) 1923, *Über die Beziehungen des sauerländischen Faciesgebietes zur belgischen Nord- und Südfacies und ihre Bedeutung für das Alter der Verseschichten:* K. Preuss. geol. Landesanst., Jahrbuch., v. 42, p. 839-859, 1 pl., 2 text-fig.

(323) 1929, *Beitrag zur Kenntnis der unteren Gedinnefauna:* Same, v. 50, p. 194-201, 3 pl.

Gabb, W. M.

(324) 1864, *Description of the Triassic fossils of California:* California Geol. Survey, Palaeont., v. 1, pt. 2, p. 17-35, pl. 3-6.

Garwood, E. J.

(325) 1916, *The faunal succession in the Lower Carboniferous rocks of Westmorland and N. Lancashire:* Geologists' Assoc., Proc., v. 27, p. 1-43, pl. 12-18.

Gatinaud, G.

(326) 1949, *Contributions à l'étude des brachiopodes Spiriferidae. 1. Exposé d'une nouvelle méthode d'étude de la morphologie externe des Spiriferidae à sinus plissé:* Muséum Histoire Nat. (France), Bull., ser. 2, v. 21, no. 1, p. 153-159; no. 2, p. 300-307; no. 3, p. 408-413; no. 4, p. 487-492.*

Geinitz, H. B.

(327) 1847, *Orthothrix Geinitz:* Soc. Impér. Nat. Moscou, Bull., v. 20, pt. 2, p. 84-86.

(328) 1866, *Carbonformation und Dyas in Nebraska:* Nova Acta, Leopoldina, v. 33, p. 1-91, pl. 1-5.

Gemmellaro, G. G.

(329) 1871[1876], *Studi paleontologici sulla fauna del calcare a Terebratula janitor del Nord di Sicilia:* Giornale Sci. Nat. & Econ. Palermo, v. 7, p. 73-108, pl. 1-5.

(330) 1894, *Sopra tre famiglie de Brachiopodi: (Spiriferidae, Rhynchonellidae e Terebratulidae) provenienti dei calcari con Fusulina della valle del fiume Sosio nella provincia di Palermo:* Soc. Sci. Nat. & Econ. Palermo, Bull., no. 1, p. 1-6.

(331) 1897, *Sopre due nuovi generi di brachiopodi provenienti dei calcari con Fusulina della provincia di Palermo:* Same, Giornale, v. 21, p. 8-10 [often cited as 1896].

(332) 1899, *La fauna dei calcari con Fusulina della Valle del Fiume Sosio nella provincia di Palermo:* Same, v. 22, p. 95-214, pl. 25-36, 46 text-fig.

George, T. N.

(333) 1927, *Studies in Avonian Brachiopoda: I. The genera Brachythyris and Martinia:* Geol. Mag., v. 64, no. 753, p. 106-119, 13 text-fig.

(334) 1931, *Ambocoelia Hall and certain similar British Spiriferidae:* Geol. Soc. London, Quart. Jour., v. 87, p. 30-61, pl. 3-5, 3 text-fig.

(335) 1932, *The British Carboniferous reticulate Spiriferidae:* Same, v. 88, p. 516-575, pl. 31-35, text-fig. 1-14.

Geyer, Georg

(336) 1889, *Über die liasischen Brachiopoden des Hierlatz bei Halstatt:* K. K. Geol. Reichsanst., Abhandl., v. 15, p. 1-88, pl. 1-9.

Gill, E. D.

(337) 1950, *The biological significance of exoskeletal structures in the Palaeozoic brachiopod genus Chonetes:* Royal Soc. Victoria, Proc., v. 60, p. 45-56.

(338) 1950, *Palaeontology and palaeoecology of Eldon Group:* Royal Soc. Tasmania, Paper 1949, p. 231-258, pl. 1.

(339) 1951, *Two new brachiopod genera from Devonian rocks in Victoria:* Natl. Museum Victoria, Mem. 17, p. 187-205, pl. 1.

(340) 1951, *Further studies in Chonetidae (Palaeozoic Brachiopoda) from Victoria:*

Royal Soc. Victoria, Proc., v. 63, p. 57-72, pl. 3.

Gill, Theodore
(341) 1871, *Arrangement of the families of molluscs prepared for the Smithsonian Institution:* Smithsonian Misc. Coll., no. 227, 49 p.

Girty, G. H.
(341a) 1898, *Description of a fauna found in the Devonian black shale of eastern Kentucky:* Am. Jour. Sci., ser. 4, v. 6, p. 384-395, 6 text-fig.
(342) 1903, *The Carboniferous formations and faunas of Colorado:* U.S. Geol. Survey, Prof. Paper 16, p. 1-546, pl. 1-10.
(343) 1904, *New molluscan genera from the Carboniferous:* U.S. Natl. Museum, Proc., v. 27, p. 721-736, pl. 16-18.
(344) 1908, *On some new and old species of Carboniferous fossils:* Same, v. 34, p. 281-303, pl. 14-21, text-fig. 6-15.
(345) 1908, *The Guadalupian fauna:* U.S. Geol. Survey, Prof. Paper 58, 651 p., 31 pl.
(346) 1910, *New genera and species of Carboniferous fossils from the Fayetteville Shale of Arkansas:* N.Y. Acad. Sci., Ann., v. 20, no. 3, pt. 2, p. 189-238.
(346a) 1911, *The fauna of the Moorefield Shale of Arkansas:* U.S. Geol. Survey, Bull. 439, 148 p., 15 pl.
(347) 1920, *Carboniferous and Triassic faunas,* Append. to BUTLER, B. S., & others, The ore deposits of Utah: U.S. Geol. Survey, Prof. Paper 111, p. 641-657.
(348) 1926, *Mississippian formations of San Saba County, Texas, III. The macrofauna of the limestone of Boone age:* Same, 146, p. 24-43, pl. 5-6.
(349) 1929, *New Carboniferous invertebrates. II:* Washington Acad. Sci., Jour., v. 19, p. 406-415, text-fig. 1-35.
(350) 1934, *New Carboniferous invertebrates:* Same, v. 24, p. 251.
(351) 1938, *Descriptions of a new genus and a new species of Carboniferous brachiopods:* Same, v. 28, p. 278-284, text-fig. 1-5.

Glenister, B. F.
(352) 1955, *Devonian and Carboniferous spiriferids from the North-West Basin, western Australia:* Royal Soc. West. Australia, Jour., v. 39, pt. 2, no. 6, p. 46-71, pl. 1-8, text-fig. 1-7.

Goldring, Roland
(353) 1955, *Some notes on the cardinal process in the Productidae:* Geol. Mag., v. 92, no. 5, p. 402-412.
(354) 1957, *The last toothed Productellinae in Europe (Brachiopoda, Upper Devonian):* Paläont. Zeitschr., v. 31, no. 3-4, p. 207-228.

Gortani, Michele, & Merla, G.
(355) 1934, *Fossili del Paleozoico: Spedizioni Italiana de Filippi nell' Himalaia etc.:* (1913-14), 323 p., 27 pl. (Bologna).

Goryansky, V. Yu.
(356) 1960, *Klass Inarticulata,* Mshanki, Brakhiopody, SARYCHEVA, T. G. (ed.) in Osnovy Paleontologii, ORLOV, YU. A. (ed.), p. 172-182, pl. 1-6, text-fig. 76-84 (Moskva). [*Class Inarticulata.*]

Grabau, A. W.
(357) 1923-4, *Stratigraphy of China: Pt. 1, Palaeozoic and older:* China Geol. Survey, p. 1-528, 306 text-fig., 6 pl.
(358) 1931, *Devonian Brachiopoda of China, I. Devonian Brachiopoda from Yunnan and other districts in South China:* Same, Palaeont. Sinica, ser. B, v. 3, pt. 3, 545 p., 54 pl., 6 text-fig.
(359) 1931, *Studies for students; Series I, Palaeontology; The Brachiopoda, pt. 2:* Peking, Natl. Univ., Sci. Quart., v. 2, p. 397-422, 21 text-fig.
(360) 1931, *The Permian of Mongolia:* Am. Museum Nat. History, Nat. History of Central Asia, v. 4, 665 p., 35 pl.
(361) 1932, *The significance of the sinal formula in Devonian and post-Devonian spirifers:* Geol. Soc. China, Bull. 11 (1931), no. 1, p. 93-96, pl. 1-2.
(361a) 1932, *Studies for students, studies of Brachiopoda III:* Peking, Natl. Univ., Sci. Quart., v. 3, no. 2, p. 75-112, fig. 22-46.
(362) 1934, *Early Permian fossils of China Pt. I, early Permian brachiopods, pelecypods, and gastropods of Kueichow:* China, Geol. Survey, Palaeont. Sinica, ser. B, v. 8, pt. 3, p. 1-168, pl. 1-11.
(362a) 1936, *Early Permian fossils of China Pt. II, fauna of the Maping Limestone of Kwangsi and Kweichow:* Same, v. 8, pt. 4, p. 1-441, pl. 1-31.

———, & Chao, Y. T.
(363) 1927, *Brachiopod fauna of the Chihsia Limestone:* Geol. Soc. China, Bull., v. 6, p. 83-120.

———, & Sherzer, W. H.
(364) 1910, *The Monroe formation of southern Michigan and adjoining regions:* Michigan Geol. & Biol. Survey, pub. 2, Geol. ser. 1, 248 p., 32 pl.

Grant, R. E.
(365) 1965, *The brachiopod superfamily Stenoscismatacea:* Smithsonian Misc. Coll., v. 148, no. 2, 185 p., 24 pl., 34 text-fig.

Gray, J. E.
(366) 1840, *Synopsis of the contents of the British Museum:* 42nd edit., 370 p. (London).

Greco, Benedetto

(367) 1938, *Revisione degli Strofomenidi permiani del Sosio conservati nel Museo di Geologia della R. Universita di Palermo:* Giornale Sci. Nat. & Econ. Palermo, v. 39, Mem. 11, p. 1-46, pl. 1-2.

Greene, F. C.

(368) 1908, *The development of a Carboniferous brachiopod (Chonetes granulifer) Owen:* Jour. Geology, v. 16, p. 654-663, pl. 1-4.

Greger, D. K.

(369) 1920, *Notes on certain brachiopod genera:* Nautilus, v. 34, p. 70.

Gregorio, A. de

(370) 1930, *Sul Permiano di Sicilia (Fossili del calcare con Fusulina di Palazzo Adriano non descritti del Prof. G. Gemmellaro conservati nel mio privato gabinetto):* Ann. Géologie & Paléontologie (Palermo), v. 52, p. 18-32, pl. 4-11.

Grubbs, D. M.

(371) 1939, *Fauna of the Niagaran nodules of the Chicago area:* Jour. Paleontology, v. 13, p. 543-560, fig. 1-2, pl. 61-62.

Gümbel, C. W.

(372) 1861, *Geognostische Beschreibung des bayerischen Alpengebirges:* xx+950 p., 42 pl. (Gotha).

Gürich, Georg

(373) 1896, *Das Palaeozoicum im Polnischen Mittelgebirge:* Russisch-Kaiserl. Min. Gesell. Verhandl., ser. 2, v. 32, p. 1-539, pl. 1-15.

(374) 1909, *Leitfossilien, Zweite Lieferung Devon:* p. 97-199, pl. 29-52, Gebrüder Borntraeger (Berlin).

Haas, H. J.

(375) 1885-91, *Étude monographique et critique des Brachiopodes Rhétiens et Jurassiques des Alpes Vaudoises et des contrées environnantes:* Schweiz. Paläont. Gesell., Abhandl., v. 11, p. 1-66, pl. 1-4; v. 14, p. 67-126, pl. 5-10; v. 18, p. 127-158, pl. 11.

(376) 1889-93, *Kritische Beiträge zur Kenntniss der jurassischen Brachiopodenfauna des schweizerischen Juragebirges und seiner angrenzenden Landestheile:* Same, v. 16, p. 1-35, pl. 1-2; v. 17, p. 36-102, pl. 3-10; v. 20, p. 103-147, pl. 11-23.

———, & Petri, C.

(377) 1882, *Die Brachiopoden der Juraformation von Elsass-Lothringen:* Geol. Spezialk. Elsass-Loth., Abhandl., v. 2, no. 2, p. 161-320, Atlas, 18 pl.

Hall, James

(378) 1843, *Natural History of New York:* Geology, pt. 4, ix+525 p., 19 pl. (Albany).

(378a) 1850, *On the Brachiopoda of the Silurian Period; particularly the Leptaenidae:* Am. Assoc. Adv. Science, v. 2, p. 347-351.

(379) 1852, *Containing descriptions of the organic remains of the lower middle division of the New York System (equivalent in part to the Middle Silurian rocks of Europe):* N.Y. State Geol. Survey, Palaeont. N.Y., 353 p., 85 pl.

(380) 1857, *Descriptions of Palaeozoic fossils:* N.Y. State Cab. Nat. History, 10th Annual Report, p. 41-186.

(381) 1858, *Description of new species of fossils from the Carboniferous limestones of Indiana and Illinois:* Albany Inst., Trans., v. 4, p. 1-36.

(382) 1858, in HALL, J., and WHITNEY, J. D., Report on the Geological Survey of the State of Iowa; embracing the results of investigations made during portions of the years 1855-1857: Paleont., v. 1, pt. 2, p. 473-724, 29 pl. (Albany, N.Y.).

(383) 1859, *Observations on genera of Brachiopoda:* 12th Ann. Rept. N.Y. State Cabinet, p. 8-110.

(384) 1859-61, *Containing descriptions and figures of the organic remains of the Lower Helderberg Group and the Oriskany Sandstone:* N.Y. State Geol. Survey, Palaeont. N.Y., v. 3, pt. 1 (1859), text, 532 p.; pt. 2 (1861), plates, 120 pl.

(385) 1860, *Descriptions of new species of fossils, from the Hamilton group of western New York, with notices of others from the same horizon in Iowa and Indiana:* N.Y. State Cab. Nat. History, Ann. Rept. 13, p. 76-94.

(386) 1860, *Contributions to palaeontology:* Same, 13, p. 55-125.

(387) 1861, *Observations upon some new and other species of fossils, from the rocks of Hudson-river group of Ohio and the western states; with descriptions:* Same, 14, Appendix C, p. 89-92.

(388) 1861, *Descriptions of new species of fossils from the Upper Helderberg, Hamilton, and Chemung groups:* Same, 14, p. 99-109.

(389) 1863, *Descriptions of new species of Brachiopoda from the Upper Helderberg, Hamilton, and Chemung groups:* Same, 16, p. 48, text-fig. 22-23.

(390) 1863, *Contributions to palaeontology:* 16th annual report of the Regents of the Univ. of the State of N.Y. on the condition of the State Cab. of Nat. History, p. 3-226, pl. 1-11 (Albany).

(391) 1867, *Notice of volume IV of the Paleontology of New York:* N.Y. State Cab. Nat. History, Ann. Rept. 20, p. 163.

(392) 1867, *Descriptions and figures of the fossil Brachiopoda of the Upper Helderberg,*

H880   *Brachiopoda*

Hamilton, Portage and Chemung groups:
N.Y. Geol. Survey, Palaeont. N.Y., v. 4,
pt. 1 (1862-66), 428 p.

(392a) 1868, *Note on the genus Eichwaldia:* 20th
Ann. Rept. of the Regents of the Univ. of
the State of N.Y. on the condition of the
State Cab. Nat. History, p. 274-278, 7 text-
fig. (Albany).

(392b) 1871, *Notes on some new or imperfectly
known forms among the Brachiopoda:* Preliminary Notice, 23rd Ann. Rept. on the
State Cab. Nat. History (Abstract), p. 1-5
(Albany).

(392c) 1871, *Descriptions of some new species of
fossils, from the shales of the Hudson
River Group, in the vicinity of Cincinnati,
Ohio:* Advance copy of the 24th Rept. on
the State Cab. Nat. History, p. 1-8, 4 pl.
(Albany).

(392d) 1872, *Notes on some new or imperfectly
known forms among the Brachiopoda:*
Advance copy, 23rd Ann. Rept. on the
State Cab. Nat. History, p. 244-247, pl. 13
(Albany).

(393) 1879, *The fauna of the Niagara Group:*
N.Y. State Museum Nat. History, 28th
Rept., p. 98-203, pl. 3-34.

(394) 1889, *8th Ann. Report N.Y. State Geol.
for the year 1888:* Same, 42nd Rept., p.
349-496.

(395) 1891, *Preliminary notice of Newberria, a
new genus of brachiopods, with remarks
on its relations to Rensselaeria and Amphigenia:* N.Y. State Geologist, Ann. Rept.,
10, p. 97-98.

———, & Clarke, J. M.

(395a) 1890, *Extract from Volume VIII, Palaeontology of New York:* p. 2 (120)-20(160),
pl. 4E-4F (Albany).

(396) 1892-95, *An introduction to the study of
the genera of Palaeozoic Brachiopoda:* N.Y.
Geol. Survey, v. 8, pt. 1, p. 1-367, pl. 1-20
(1892); pt. 2, p. 1-317 (1893), p. 319-
394, pl. 21-84 (1895).

(397) 1894, *An introduction to the study of the
Brachiopoda:* 13th Ann. Rept. N.Y. State
Geologist for the year 1893, Palaeont.,
pt. 2, p. 751-943, text-fig. 287-669, pl.
23-54.

———, & Whitfield, R. P.

(397a) 1875, *Descriptions of invertebrate fossils
mainly from the Silurian System: fossils
of the Hudson River Group:* Ohio Geol.
Survey, Rept., v. 2, Geol. & Palaeont., pt.
2, Palaeont., p. 67-110, 4 pl.

Harrington, H. J.

(398) 1955, *The Permian Eurydesma fauna of
eastern Argentina:* Jour. Paleontology, v.
29, no. 1, p. 112-128, pl. 23-26.

Hatai, K. M.

(399) 1940, *The Cenozoic Brachiopoda from

Japan: Tohoku Imper. Univ., Sci. Rept., ser.
2 (Geol.), v. 20, p. 1-413, 12 pl.

(400) 1941, *On some Brachiopoda from Kago-
sima-ken, Kyushu:* Geol. Soc. Japan, Jour.,
v. 48, no. 577, p. 491-495, pl. 13.

(401) 1948, *New Tertiary Brachiopoda from
Japan:* Jour. Paleontology, v. 22, p. 494-
499, pl. 78.

Havlíček, Vladimír

(402) 1949, *Orthoidea a Clitambonoidea z českého tremadoku:* Ústřed. Ústavu Geol.,
Sborník, v. 16, no. 1, p. 93-144, pl. 1-5
(English Summary, *Orthoidea and Clitambonoidea of the Bohemian Tremadoc,* p.
122-144).

(403) 1950, *Ramenonožci Českého Ordoviku:*
Same, Rozpr., v. 13, p. 1-72 (in Czech.),
p. 75-135 (in English), 13 pl., 17 text-
fig. [*The Ordovician Brachiopoda from
Bohemia.*]

(404) 1951, *A paleontological study of the Devonian of Čelechovice; brachiopods (Pentameracea, Rhynchonellacea, Spiriferacea):*
Same, Sborník v. 18 (1951), Pal., p. 1-20,
4 pl., 1952.

(405) 1952, *O ordovických zástupcích čeledi
Plectambonitidae (Brachiopoda):* Same,
Sborník, v. 19, p. 397-428 (English Sum-
mary, p. 423-428), pl. 1-3.

(406) 1953, *O několika nových ramenonožcích
českého a moravského středního devonu:*
Same, Věstník, v. 28, p. 4-9, pl. 1-2.

(407) See 408.

(408) 1956, *Ramenonožci vápenců branických a
hlubočepských z nejbližšího pražského
okolí:* Same, Sborník, v. 22, 1955, p. 535-
650, pl. 1-12 (English Summary, p. 651-
665). [*Brachiopods of the Baník and
Hlubočepy Limestones in the immediate
vicinity of Prague.*]

(409) 1957, *O nových rodech českých spiriferidie
(Brachiopoda):* Same, Věstník, v. 32, pt.
4, p. 245-248.

(410) 1957, *Další nove rody čelidi Spiriferidae v
Českém Siluru a Devonu:* Same, Ročnik,
v. 32, pt. 6, p. 436-440.

(411) 1959, *Spiriferidae v Českém Siluru a
Devonu:* Same, Rozpr., v. 25, p. 1-275, pl.
1-28, text-fig. 1-101.

(411a) 1961, *Rhynchonelloidea des böhmischen
mährischen Mitteldevon (Brachiopoda):*
Same, Rozpr., v. 27, p. 1-211, pl. 1-27, 87
text-fig.

(412) 1961, *Plectambonitacea im böhmischen
Palaözoikum (Brachiopoda):* Same, Věstník, v. 36, p. 447-451, pl. 1.

(413) 1962, *Oberfamilie Strophomenacea im
mährischen Mitteldevon (Brachiopoda):*
Same, v. 37, no. 6, p. 471-472, 1 text-fig.

(413a) 1963, *Zlichorhynchus hiatus n. g. et n. sp.,
neuer Brachiopode vom Unterdevon

Böhmens: Zvláštní Otisk Věstníku Ústřed. Ústavu Geol., Ročník 38, no. 8, p. 403-404, pl. 1.

————, & Šnajdr, Milan
(414) 1952, *Cambrian and Ordovician in the Brdské Hřebeny and in the Jince Area:* Ústřed. Ústavu Geol., Sborník, v. 18, p. 145-237 (English Summary, p. 258-276).

Hayasaka, Ichiro
(415) 1922, *On some Tertiary Brachiopoda from Japan:* Tohoku Imper. Univ., Sci. Rept., ser. 2 (Geol.), v. 6, no. 2, p. 139-163, pl. 7-8.
(416) 1946, *On fossil and Recent Brachiopoda of Formosa:* Taiwan Ocean. Inst. Bull., no. 1, p. 9-28, pl. 1.
(417) 1953, *Hamletella, a new Permian genus of Brachiopoda and a new species from the Kitakami Mountains, Japan:* Paleont. Soc. Japan, Trans. & Proc., new ser., no. 12, p. 89-95, pl. 9.

————, & Uozumi, Satoru
(417a) 1952, *On some Recent and fossil Brachiopoda from Hokkaido:* Hokkaido Univ., Faculty Sci., Jour., ser. 4 (Geol. and Min.), v. 8, no. 2, p. 86-96.

Hector, James
(418) 1879, *On the fossil Brachiopoda of New Zealand:* New Zealand Inst., Trans. & Proc., v. 11, p. 537-39.

Hedström, Herman
(419) 1916, *Ueber einige mit der Schale befestigte Strophomenidae aus dem Obersilur Gotlands:* Sver. Geol. Undersök., Arsb., ser. C, no. 276, p. 1-14, pl. 1-3.

Helmbrecht, W., & Wedekind, Rudolf
(420) 1923, *Versuch einer biostratigraphischen Gliederung der Siegener Schichten auf Grund von Rensselaerien und Spiriferen:* Glückauf, Berg.- und Hüttenmännische Zeitschrift, Jahrg. 59, no. 41, p. 949-953.

Helmcke, J. G.
(421) 1939, *Kraussina mercatori n. sp. und die Verbreitung der Gattung Kraussina:* Résultats Scientifiques des Croisières du Navire-École Belge "Mercator":* Musée Royal Histoire Nat. Belgique, Mém., ser. 2, v. 15, p. 135-139, 1 pl.
(422) 1939, *Die Muskeln der Brachiopoden:* Zool. Jahrb. (Systematik), v. 72, no. 1/2, p. 99-140.
(423) 1939, *Die Brachiopoden des zoologischen Museums zu Berlin:* Sitzungsbericht. Gesell. Naturf. Freunde, p. 221-268.
(424) 1939, *Waldheimiathyris nom. nov. für Brachiopoden Gattung Macandrevia King:* Zool. Anzeiger, v. 126, no. 11/12, p. 331-332.
(425) 1940, *Die Brachiopoden der deutschen Tiefsee-Expedition:* Wiss. Ergeb. Deutsch.

Tiefsee-Exped. Valdivia (1898-99), v. 24, no. 3, p. 217-316, 43 text-fig.

Helmersen, G. von
(426) 1847, *Aulosteges variabilis im Zechstein Russlands, ein neues Brachiopoden-Genus:* Neues Jahrb. Mincral., Geol. & Paläont., Mitt., p. 330.

Henningsmoen, Gunnar
(426a) 1948, *The Tretaspis Series of the Kullatorp Core,* in WAERN, BERTIL, THORSLUND, PER and HENNINGSMOEN, GUNNAR, Deep boring through Ordovician and Silurian strata at Kinnekulle, Vestergötland: Uppsala Univ., Geol. Inst., Bull., v. 32, p. 374-432, pl. 23-25.

Hertlein, L. G., & Grant, U. S.
(427) 1944, *The Cenozoic Brachiopoda of western North America:* Calif. Univ. Publ. Math. Phys. Sci., v. 3, p. 1-172, 21 pl.

Hessland, Ivar
(428) 1949, *Investigations of the Lower Ordovician of the Siljan District, Sweden. Notes on Swedish Ahtiella species:* Uppsala Univ., Geol. Inst., Bull., v. 33, p. 511-527, 2 pl., 4 tables.

Hoare, R. D.
(429) 1960, *New Pennsylvanian Brachiopoda from southwest Missouri:* Jour. Paleontology, v. 34, no. 2, p. 217-232, pl. 1-2.
(430) 1961, *Desmoinesian Brachiopoda and Mollusca from southwest Missouri:* Missouri, Univ. Studies, v. 36, 263 p., 23 pl.

Hoek, H. von
(430a) 1912, *Versteinerungen des Cambriums und Silurs,* in STEINMANN, G. and HOEK, H., Das Silur und Cambrium des Hochlandes von Bolivia und ihre Fauna: Neues Jahrb. Mineral., Geol., & Paläont., v. 34, p. 209-252, pl. 7-14.

Holtedahl, Olaf
(431) 1915, *Strophomenidae of the Kristiania Region:* K. Norske Vidensk. Selsk., Skrift., no. 12, p. 1-118, 16 pl.

Holzapfel, Eduard
(432) 1895, *Die Fauna der Schichten mit Maeneceras terebratum:* K. Preuss. Geol. Landesanst., Abhandl., no. 16, p. 234-237.
(433) 1912, *Beitrag zur Kenntnis der Brachiopodenfauna des rheinischen Stringocephalen-Kalkes:* Same, Jahrb., v. 29, pt. 2, p. 119-120.

Horný, Radvan
(433a) 1961, *New genera of Bohemian Monoplacophora and patellid Gastropoda:* Ústřed. Ústavu Geol., Věstník, v. 36, p. 299-302, 2 pl.
(433b) 1963, *Lower Paleozoic Monoplacophora and patellid Gastropoda (Mollusca) of Bohemia:* Same, Sborník, v. 28, p. 7-83, 18 pl., 19 text-fig.

Hou, Hun-fei
(433c) 1963, *Some new Middle Devonian brachiopods from the provinces of Guaunsi and Yun'nan':* Acta Palaeont. Sinica, v. 11, p. 421-432, pl. 1, 2.

Howell, B. F.
(434) 1947, *Spiriferid brachiopods new to the Silurian Cobleskill Formation of New York:* Wagner Free Inst. Sci., Bull., v. 22, no. 1, p. 1-10, pl. 1-3.

Huang, T. K.
(435) 1932, *Late Permian Brachiopoda of southwestern China. Pt. II:* China, Geol. Survey, Palaeont. Sinica, ser. B, v. 9, pt. 1, p. 1-138, pl. 1-9.

(436) 1933, *Late Permian Brachiopoda of southwestern China. Pt. II:* Same, ser. B, v. 9, pt. 2, p. 1-172, 11 pl.

Hudson, R. G. S., & Jefferies, R. P. S.
(437) 1961, *Upper Triassic brachiopods and lamellibranchs from the Oman Peninsula, Arabia:* Palaeontology, v. 4, p. 1-41, pl. 1-2.

——, **& Sudbury, Margaret**
(438) 1959, *Permian Brachiopoda from southeast Arabia:* Muséum Histoire Nat., Notes et Mémoires sur le Moyen-Orient, v. 7, p. 19-55, pl. 1-6, 12 text-fig.

Huene, Friedrich
(439) 1899, *Die silurischen Craniaden der Ostseeländer mit Ausschluss Gotlands:* Russiche-Kaiser. Min. Gesell. Verhandl., ser. 2, v. 36, p. 181-359, pl. 9-14, 18 text-fig.

Huxley, T. H.
(439a) 1869, *An introduction to the classification of animals:* 147 p., 47 text-fig., John Churchill & Sons (London).

Hyde, J. E.
(440) 1908, *Camarophorella, a Mississippian meristelloid brachiopod:* Boston Soc. Nat. History, Proc., v. 34, no. 3, p. 35-65, pl. 6-10.

(441) 1953, *Mississippian formations of central and southern Ohio:* Ohio Geol. Survey, Bull. 51, p. 1-355, pl. 1-54, 19 text-fig.

Hyman, L. H.
(441a) 1959, *The invertebrates: smaller coelomate groups:* v. 5, 783 p., McGraw-Hill (New York).

ICZN
(442) 1928, *Opinion 100:* Smithsonian Misc. Coll., v. 73, no. 5, p. 369-96.

(443) 1956, *Opinion 420, Addition to the "Official List of Specific Names in Zoology of the specific names for eleven species of the Class Brachiopoda and for 2 species of the Class Cephalopoda originally published by Martin (W.) in 1809 in the nomenclatorially invalid work entitled "Petrificata Derbiensia" and now available as from the* first subsequent date on which they were severally published in conditions satisfying the requirements of the "Regles": Opinions and Declarations, v. 14, pt. 4, p. 131-167.

(444) 1956, *Opinion 421, Designation under the Plenary Powers of a type species in harmony with accustomed usage for the genus Martinia McCoy, 1844 (Class Brachiopoda):* Opinions and Declarations, v. 14, pt. 5, p. 169-180.

Imbrie, John
(445) 1959, *Brachiopods of the Traverse Group (Devonian) of Michigan:* Am. Museum Nat. History, Bull. 116, art. 4, p. 349-409, pl. 48-67.

Ireland, H. A.
(445a) 1961, *New phosphatic brachiopods from the Silurian of Oklahoma:* Jour. Paleontology, v. 35, p. 1137-1142, pl. 137.

Ivanov, A. P.
(446) 1925, *Sur la systématique et la biologie du genre Spirifer et de quelques brachiopodes de C_{II} et C_{III} du Gouvernement de Moscou:* Soc. Impér. Nat. Moscou (Sect. Geol.), Bull., v. 33, p. 105-123.

——, **& Ivanova, E. A.**
(447) 1937, *Fauna brakhiopod srednego i verkhnego Karbona podmoskovnogo basseyna (Neospirifer, Choristites):* Akad. Nauk SSSR, Paleozool. Inst., Trudy, v. 6, pt. 2, p. 1-215, pl. 1-23, 55 text-fig. [*Brachiopod fauna of the Middle and Upper Carboniferous of the Submoscow Basin (Neospirifer, Choristites).*]

Ivanova, E. A.
(447a) 1959, *K systematike i evolyutsii spiriferid (Brachiopoda):* Paleont. Zhurnal 1959, no. 4, p. 47-63, pl. 2, text-fig. 1-9 [*To systematics and evolution of spiriferids (Brachiopoda).*]

(448) 1960, *Otryad Spiriferida:* Mshanki, Brachiopody, SARYCHEVA, T. G. (ed.), in Osnovy Paleontologii, ORLOV, YU. A. (ed.), p. 264-280, pl. 57-64, text-fig. 336-411 (Moskva). [*Order Spiriferida.*]

Jaanusson, Valdar
(449) 1962, *Two plectambonitacean brachiopods from the Dalby Limestone (Ord.) of Sweden:* Uppsala Univ., Palaeont. Inst., Publ. 40, p. 1-8, pl. 1.

——, **& Strachan, Isles**
(450) 1954, *Correlaton of the Scandinavian Middle Ordovician with the graptolite succession:* Geol. Fören. Stockholm, Förhandl., v. 76, no. 4, p. 684-696, 2 text-fig.

Jackson, W. J.
(451) 1918, *Brachiopoda. British Antarctic ("Terra Nova") Expedition (1910):* British Museum (Nat. History), Zool., v. 2, no. 8, p. 177-202, pl. 1.

(452) 1918, *The new brachiopod genus, Liothyrella, of Thomson*: Geol. Mag., new ser., decade 6, v. 5, p. 73-79.

Jacob, Claude, & Fallot, Paul
(453) 1913, *Etude sur les Rhynchonelles portlandiennes néocomiennes et mésocrétacées du sud-est de la France*: Soc. Paléont. Suisse Genève, Mém., v. 39, p. 1-82, pl. 1-11.

Jaekel, O. M. J.
(453a) 1902, *Ueber verschiedene Wege phylogenetischer Entwickelung*: 5th internationalen Zool.-Congresses zu Berlin, Verhandl., p. 1058-1117, 28 text-fig., Gustav Fischer (Jena).

Johnson, J. G., & Reso, Anthony
(454) 1964, *Probable Ludlovian brachiopods from the Sevy Dolomite of Nevada*: Jour. Paleontology, v. 38, p. 74-84, pl. 19-20, text-fig. 1-2.

Johnston, Joan
(454a) 1941, *Studies in Silurian Brachiopoda*: Linnean Soc. New S. Wales, Proc., v. 66, p. 160-168, pl. 7, 2 text-fig.

Jones, O. T.
(455) 1928, *Plectambonites and some allied genera*: Great Britain, Geol. Survey, Mem. Palaeont., v. 1, pt. 5, p. 367-527, 5 pl.

Joubin, Louis
(456) 1907, *Note sur les brachiopodes recueillis au cours des dernières croisières du Prince de Monaco*: Inst. Océan., Monaco, Bull. 103, p. 1-9.

Kashirtsev, A. S.
(457) 1959, *Novyy rod brakhiopod Jakutoproductus iz Nizhnepermskikh otlozheniy vostochnoy Sibiri*: Akad. Nauk SSSR, Paleont. Inst., Trudy, v. 3, p. 28-31. [*Jakutoproductus a new genus of brachiopod from the Lower Permian of East Siberia.*]

Kayser, Emanuel
(458) 1871, *Die Brachiopoden des Mittel- und Ober-Devon der Eifel*: Zeitschr. Deut. Geol. Gesell., v. 23, no. 3, p. 491-647, pl. 9-14.
(459) 1881, *Mittheilungen über die Fauna des chinesischen Kohlenkalks von Lo-Ping*: Same, v. 33, p. 351-352.
(460) 1882-3, *Ergebnisse eigener Reisen und darauf gegründeter Studien*: in RICHTHOFEN, F., China, v. 4, Paläont. Theil, 288 p., 54 pl. (Berlin).
(461) 1883, *Beschreibung einiger neuen Goniatiten und Brachiopoden aus dem rheinischen Devon*: Zeitschr. Reutsch. Geol. Gesell., v. 35, p. 306-317, 2 pl.

Kegel, Wilh.
(462) 1913, *Der Taunusquarzit von Katzenelnbogen*: K. Preuss. Geol. Landesanst., Abhandl., v. 76, p. 126.

Keyes, C. R.
(463) 1888, *On the fauna of the Lower Coal Measures of Central Iowa*: Acad. Nat. Sci. Philadelphia, Proc., v. 2, p. 222-246, pl. 12.

Khalfin, L. L.
(464) 1948, *Fauna i stratigrafiya Devonskikh otlozhenii Gornogo Altaya*: Zapad. Sibir. Geol. Uprav.-Tomsk. Politekh. Inst. Izvestiya, v. 65, pt. 1, p. 1-464, pl. 1-36, fig. 1-54. [*Fauna and stratigraphy of the Devonian deposits of the Gorny Altay.*]

Khalfina, V. K.
(465) 1955 (1956), *Atlas Rukovodyashchikh Form iskopaemykh fauny i flory zapadnoy Sibiri*: Same, Gosudar. Nauch.-Tekh. Izd. Lit. Geol. i Okhrany Nedr., v. 1, 502 p., 85 pl., 202 text-fig.; v. 2, 320 p. [*Atlas of leading fossil forms of fauna and flora of western Siberia.* L. L. KHALFIN, ed.]

Khodalevich, A. N.
(466) 1939, *Verkhne-siluriiskie brakhiopody vostochnogo sklona Urala*: Urals. Geol. Uprav. Trans., 135 p., 28 pl. [*Upper Silurian brachiopods of eastern slope of the Urals.*]
(467) 1951, *Nizhnedevonskie i Eifel'skie brakhiopody Ivdelakogoi i Serovskogo rayonov Sverdlovskoy oblast*: Sverd. Gornogo Inst., Trudy, v. 18, p. 1-107, pl. 1-30. [*Lower Devonian and Eifelian brachiopods from the Ivdel and Serov areas of the Sverdlovsk region.*]

————, & Breivel, M. G.
(468) 1959, *Brakhiopody i korally iz Eyfelskikh boksitonosnykh otlozheniy vostochnogo sklona srednego i severnogo Urala*: Urals Geol. Uprav., 282 p., 61 pl. [*Brachiopods and corals of the Eifelian bauxite deposits of the eastern slope of the middle and north Urals.*]

Kindle, E. M.
(469) 1909, *The Devonian fauna of the Ouray limestone*: U.S. Geol. Survey, Bull. 391, p. 1-60, 10 pl.

King, R. E.
(470) 1931, *The geology of the Glass Mountains, Texas, Part II, Faunal summary and correlation of the Permian formations with descriptions of Brachiopoda*: Texas Univ., Bull. 3042, p. 1-245, pl. 1-44, fig. 3-10.

————, Dunbar, C. O., Cloud, P. E., Jr., & Miller, A. K.
(470a) 1944, *Geology and paleontology of the Permian area of Las Delicias, southwestern Coahuila, Mexico. Pt. 3, Brachiopods*: Geol. Soc. America, Spec. Paper, v. 52, p. 49-69, pl. 17, 19.

King, R. H.
(471) 1938, *New Chonetidae and Productidae*

from Pennsylvanian and Permian strata of North-Central Texas: Jour. Paleontology, v. 12, p. 257-279, pl. 36-39.

King, William

(471a) 1846, *Remarks on certain genera belonging to the class Palliobranchiata:* Ann. & Mag. Nat. History, v. 18, p. 26-42.

(472) 1850, *A monograph of the Permian fossils of England:* Palaeontograph. Soc., Mon. 3, xxxvii+258 p., 29 pl.

(473) 1859, *On Gwynia, Dielasma, and Macandrevia, three new genera, etc.:* Dublin Univ., Zool. Bot. Assoc., Proc., v. 1, pt. 3, p. 256-262.

(474) 1871, *On Agulhasia davidsonii, a new palliobranchiate genus and species:* Ann. & Mag. Nat. History, ser. 4, v. 7, p. 109-112, pl. 11.

Kirchner, Heinrich

(475) 1933, *Die Fossilien der Würzburger Trias; Brachiopoda:* Neues Jahrb. Mineral., Geol. & Paläont., Beil. Bd. 71, p. 88-138, pl. 2, 11 text-fig.

Kirk, Edwin, & Amsden, T. W.

(476) 1952, *Upper Silurian brachiopods from southeastern Alaska:* U.S. Geol. Survey, Prof. Paper 233-C, p. 53-66, pl. 7-10, 7 text-fig.

Kitchin, F. L.

(477) 1897, *Zur Kenntnis der jurassischen Brachiopodenfauna von Kutch:* Inaug. Dissert. Doktorwürde phil. Fak. K. Ludwig-Max. Univ., München, p. 1-56.

(478) 1900, *Jurassic fauna of Cutch, pt. 1. The Brachiopoda:* Palaeont. Indica, ser. 9, v. 3, pt. 1, p. 1-87, pl. 1-15.

Kobayashi, Teiichi

(479) 1935, *The Cambro-Ordovician formations and faunas of South Chosen: pt. 3. Cambrian faunas of South Chosen:* Tokyo Imper. Univ. Fac. Sci. Jour., sec. 3, v. 4, pt. 2, p. 49-344, pl. 1-24.

(480) 1937, *A brief summary of the Cambro-Ordovician shelly faunas of South America:* Tokyo, Imper. Acad., Proc., v. 13, no. 1, p. 12-15, fig. 1-4.

Koch, C. H.

(481) 1843-48, in KÜSTER, H. C., Mollusca Brachiopoda, Terebratulacea: Conchylien Cabinet, v. 7, pt. 1, pt. 19-49, pl. 2, 2b-d, 3, 4.

Koken, E. F. R.

(481a) 1889, *Ueber die Entwickelung der Gastropoden vom Cambrium bis zur Trias:* Neues Jahrb. Mineral., Geol. & Paläont., v. 6, p. 305-484, pl. 10-14.

Koninck, L. G. de

(482) 1841-44, *Description des animaux fossiles qui se trouvent dans le terrain Carbonifère de Belgique:* iv+650 p., 55 pl. (Liège).

(483) 1847, *Recherches sur les animaux fossiles. Pt. 1. Monographie des genres Productus et Chonetes:* 246 p., 20 pl., H. Dessain (Liège).

(484) 1851, *Description des animaux fossiles qui se trouvent dans le terrain Carbonifére de Belgique:* Supplément, p. 651-716 (esp. pl. 56) (Liège).

(484a) 1887, *Fauna du calcaire carbonifère de la Belgique:* Musée Royal Histoire Nat. Belgique, v. 14, pt. 6, p. 30, pl. 8, text-fig. 23-43.

Kozlowski, Roman

(485) 1914, *Les brachiopodes du Carbonifère supérieur de Bolivie:* Ann. Paléontologie, v. 9, p. 1-100, pl. 1-11, 24 text-fig.

(486) 1927, *Sur certains Orthides ordoviciens des environs de St. Pétersburg:* Bibl. Univ. Lib. Palonae, 17 Wolna Wszechnica Polska, ser. A, v. 17, p. 3-21, 1 pl., 2 text-fig. (Warszawa).

(487) 1929, *Les brachiopodes gothlandiens de la Podolie Polonaise:* Palaeont. Polonica, v. 1, 254 p., 12 pl., 95 text-fig., 1 map.

(488) 1930, *Andobolus gen. nov. i kilka innych ramienionogów bezzawiasowych z ordowiku Boliwji:* Spraw. Polsk. Inst. Geol., v. 6, pt. 2, p. 293-313, pl. 3-4, 5 text-fig.

(489) 1946, *Alexander Kelus:* Soc. Géol. Pologne, Ann., v. 19, p. 63-64.

(490) 1946, *Howellella, a new name for Crispella Kozlowski, 1929:* Jour. Paleontology, v. 20, no. 3, p. 295.

Krotov, P.

(491) 1885, *Artinskische Étage——Geologisch-palaeontologische Monographie des Sandsteins von Artinsk:* Kazan Univ., Obshch. Estestv., Trudy, v. 13, no. 5, 314 p., 4 pl.

(492) 1888, *Geologische Forschungen am westlichen Ural-Abhange in den Gebieten von Tscherdyn und Ssolikamsk:* Comité Géol., Mém., v. 6, pt. 2, p. 297-563, pl. 1-2. [*Geological investigations in the western border of the Urals in the regions of Tscherdyn and Ssolikamsk.*]

(493) 1950, *O Sistematike spiriferov iz Verkhnepermiskikh otlozheniy Evropeyskoy chasti S.S.S.R.:* Vses. Nauchno-Issledov. Geol. Inst. (VSEGEI), Minist. Geol. i Okhrany Nedr., Trudy, v. 1, p. 3-7. [*On classification of spiriferids of the Upper Permian deposit of the European part of the USSR.*]

Kulkov, N. P.

(493a) 1963, *Brakhiopody Solovikhinskikh sloev Nizhnego Devona Gornogo Altaya:* Akad. Nauk SSSR, 131 p., 9 pl. [*Lower Devonian brachiopods of the Solovikhinskikh beds of the Altay Mountains.*]

Kutorga, S. S.

(494) 1844, *Zweiter Beitrag zur Paläontologie*

Russlands: Russisch-Kaiserl. Min. Gesell., Verhandl., p. 62-104, pl. 1-10.

(494a) 1848, *Über die Brachiopoden-Familie der Siphonotretaeae:* Same, p. 250-286, pl. 6-7.

Kyansep, N. P.

(495) 1959, *Zeillerina gen. nov.——novyy rod iz semeystva Zeilleriidae Rollier:* Leningrad. Univ., Vestnik, no. 18, ser. geol., v. 3, p. 118-123, 5 text-fig. [*Zeillerina, gen. nov.——new genus of the family Zeilleriidae Rollier.*]

(496) 1961, *Terebratulidy luzitanskogo yarusa i nizhnego Kimeridzha yugo-zapadnogo Kryma:* Akad. Nauk. SSSR, Trudy, v. 8, p. 1-101, pl. 1-8. [*Terebratulida of the lucite strata in Lower Kimmeridgian, southwestern Crimea.*]

Lacaze-Duthiers, F. J. H. de

(497) 1861, *Histoire naturelle des brachiopodes vivants de la Méditerranée: Première Monographie: Historie de la Thécidie (Thecidium mediterraneum):* Ann. Sci. Nat., ser. 4 (Zool.), v. 15, p. 259-330, pl. 1-5.

Laird, W. M.

(498) 1947, *An Upper Devonian brachiopod fauna from northwestern Montana:* Jour. Paleontology, v. 21, no. 5, p. 453-59, pl. 64.

Lamansky, V. V.

(499) 1904, *Die ältesten silurischen Schichten Russlands:* Comité Géol., Mém., new ser., v. 20, p. 1-203.

Lamarck, J. B. P. A. de M. de

(500) 1799, *Prodrome d'une nouvelle classification des Coquilles:* Soc. Histoire Nat., Mém., p. 63-91.

(500a) 1819, *Histoire naturelle des animaux sans vertèbres:* v. 6, pt. 1, 343 p., Lamarck (Paris).

Lamont, Archie

(501) 1935, *The Drummuck Group, Girvan; a stratigraphical revision, with descriptions of new fossils from the lower part of the group:* Geol. Soc. Glasgow, Trans., v. 19, p. 288-334, pl. 7-9, text-fig. 1-4.

——, & Gilbert, D. L. F.

(502) 1945, *Upper Llandovery Brachiopoda from Coneygore Coppice and Old Storridge Common, near Alfrick, Worcestershire:* Ann. & Mag. Nat. History, ser. 11, v. 12, p. 641-682, pl. 3-7, 7 text-fig.

Lane, N. G.

(502a) 1963, *A silicified Morrowan brachiopod faunule from the Bird Spring Formation, southern Nevada:* Jour. Paleontology, v. 37, p. 379-392, pl. 43-45, 6 text-fig.

Leidhold, Claus

(503) 1920, *Beitrag zur genaueren Kenntnis und Systematik einiger Rhynchonelliden des reichsländischen Jura:* Neues Jahrb. Mineral., Geol., & Paläont., Beil.-Bd. 44, p. 343-368, pl. 4-6.

(504) 1928, *Beitrag zur Kenntnis der Fauna des rheinischen Stringocephalenkalkes, insbesondere seiner Brachiopodenfauna:* K. Preuss. Geol. Landesanst., Abhandl., new ser., v. 109, p. 1-99, pl. 1-7, 43 text-fig.

Lesnikova, Aldona

(505) 1924, *Palaeontologische Charakteristik des Untersilurs., zwischen den Stationen Swanka und Nasja, längs der Nord Bahn:* Comité Géol., Bull. 42, pt. 5-9, p. 129-181, pl. 4.

Léveillé, Charles

(506) 1835, *Aperçu géologique de quelques localités tres riches en coquilles sur les frontières de France et de Belgique:* Soc. Géol. France, Mém., v. 2, p. 29-40, pl. 2.

Levy, Regina

(506a) 1961, *Sobre algunos Terebratellidae de Patagonia (Argentina):* Rev. Asoc. Paleont. Argentina, Ameghiniana, v. 2(5), p. 79-88, pl. 1-4.

Likharev [Licharew], B. K.

(507) 1925, *Über einen neuen Vertreter der Fam. Lyttoniidae aus dem Obercarbon des Ural:* Acad. Sci. URSS, Comptes Rendus, p. 1-7, 2 text-fig.

(508) 1925, *Un nouveau représentant des brachiopodes du Paléozoique supérieur de Caucase du Nord:* Comité Géol., Bull., v. 43, no. 6, p. 713-721, pl. 5.

(509) 1928, *Über einige seltene und neue Brachiopoden aus dem Unterperm des nördlichen Kaukasus:* Paläont. Zeitschr., v. 10, p. 258-289, pl. 3-4.

(510) 1931, *Über eine problematische Brachiopode aus dem unterpermischen Ablagerungen des nördlichen Kaukasus:* Ann. Soc. Paléont. Russie, v. 9, p. 157-161.

(511) 1932, *Fauna Permskikh otlozheniy Severniy Kavkaza. I Brachiopoda podsemeystvo Orthotetinae Waagen:* Vses. Geol.-Razv. Obed. SSSR, Trudy 215, p. 1-54. [*Fauna of the Permian deposits of northern Caucasus, 1. Brachiopod subfamily Orthotetinae (Waagen).*]

(512) 1934, *On some new genera of upper Palaeozoic Brachiopoda:* Acad. Sci. URSS, Comptes Rendus, new ser., v. 1, pt. 4, p. 210-213 (Translation in English).

(513) 1934, in ZITTEL, K. A. von, Grundzüge der Paläontologie Abt. I. Invertebrata: p. 458-552, fig. 707-843. (Leningrad-Moskva).

(514) 1935, *Bemerkungen über einige oberpaläozoische Brachiopoden:* Zentralbl. Mineral., Geol., & Paläont., abt. B, no. 9, p. 369-373, 2 text-fig.

(515) 1936, *Über einige palaeozoische Gattungen der Terebratulacea aus Eurasien:* Problem Paleont., v. 1, pt. 1, p. 263-271, 1 pl., 7 text-fig.

(516) 1947, *O novom podrode Muirwoodia roda Productus Sow., s. 1:* Acad. Sci. URSS, Comptes Rendus, v. 57, no. 2, p. 187-190. [*On a new subgenus Muirwoodia of the genus Productus Sow., s. 1.*]

(517) 1956, *Brachiopoda:* in KIPARISOVA, L. D., MARKOVSKY, B. P., & RADCHENKO, G. P., Materialy po paleontologii, novye semeystva i rody: Vses. Nauchno.-Issledov. Geol. Inst. Mater., new ser., v. 12, Paleont., 267 p. [*Materials for paleontology, new families and genera.*]

(517a) 1957, *O rode Goniophoria Yanisch. i drugikh blizkikh k nemu rodakh:* Vses. Paleont. Obshch., Ezhegodnik, v. 16, p. 134-141, pl. 1, text-fig. 1-4. [*On the genus Goniophoria Yanisch. and other related genera.*]

(518) *See* 694.

———, & Rzhonsnitskaya, M. A.

(518a) 1956, *Nadsemeistvo Rhynchonellacea Gray, 1848,* in Materialii dlya Paleontologii: Vses. Nauchno-Issledov. Geol. Inst. (VSEGEI), Trudy, new ser., v. 12, p. 53-61.

Lindström, Gustaf
(519) 1860, *Bidrag till Kännedomen om Gotlands Brachiopoder:* Öfvers. Vetenskapsakad. Förhand., v. 17, p. 337-382, pl. 12-13.

Lindström, Mauritz
(520) 1935, *On the Lower Chasmops beds in the Fågelsång District (Scania):* Geol. Fören. Stockholm, Förhandl., v. 75, no. 2, p. 125-148, pl. 1.

Link, H. F.
(521) 1830, *Handbuch der physikalischen Erdbeschreibung:* pt. 2, Abt. 1, 498 p. (Berlin).

Linnarsson, J. G. O.
(521a) 1876, *Brachiopoda of the Paradoxides beds of Sweden:* Bihang. Svensk. Vetenskakad. Handl., v. 3, no. 12, p. 1-34, 4 pl.

Linné, Carl von [Linnaeus, Carolus]
(522) 1758 and 1767, *Systema naturae:* 10th ed. (1758), 823 p.; 12th ed. (1767), 1154 p. (Stockholm).

Lundgren, Bernhard
(523) 1885, *Untersokningar ofver Brachiopoderna i Sveriges Krit-system:* Lunds Univ. Årsskrift., v. 20, p. 1-72, pl. 1-3.

Lyashenko, A. I.
(524) 1951, *Sopostavlenie Devonskikh otlozheniy Russkoy platformy i Urala:* Akad. Nauk SSSR, Doklady, v. 78, no. 1, p. 117-119. [*Correlation of Devonian deposits of the Russian Platform in the Urals.*]

(525) 1957, *Novyy rod Devonskikh brakhiopod Uchtospirifer:* Same, v. 117, no. 5, p. 885-888, 1 pl. [*New genus of Devonian brachiopod, Uchtospirifer.*]

Mabuti, Sei-iti
(526) 1937, *On a Permian brachiopod, Gemmellaroia (Gemmellaroiella) ozawai, subgen. et sp. nov. from Japan:* Tokyo Imper. Acad., Proc., v. 13, no. 1, p. 16-19, text-fig. 1-11.

Maillieux, Eugene
(527) 1912, *Apparition de deux formes siegeniennes dans les schistes de Mondrepuits:* Soc. Belge Géol., Paléont. d'Hydrol., Bull. 25, Proc.-verb. (1911), p. 176-180, pl. B.

(528) 1931, *La faune des grès et schistes de Solières:* Musée Royal Histoire Nat. Belgique, Mém. 51, p. 1-90, 2 pl.

(529) 1933, *Terrains, roches, et fossiles de la Belgique:* Same, p. 1-217, 252 text-fig.

(529a) 1935, *Contribution à la connaissance de quelques brachiopodes et pélécypodes dévoniens:* Same, Mém. 70, 42 p., 4 pl.

(530) 1939, *La faune des schistes de Barvaux-sur-Ourthe (Frasnien supérieur):* Same, Bull. 15, no. 53, p. 1-8, text-fig. 1-6.

(531) 1940, *Contribution à la connaissance du Frasnien moyen (assise de Frasnes) de la Belgique:* Same, v. 16, no. 14, p. 1-44.

(532) 1941, *Les brachiopodes de l'Emsien de l'Ardenne:* Same, Mém. 96, p. 1-74.

Makridin, V. P.
(532a) 1955, *Nekotorye Iurskie rinkhonellidy Evropeiskoi chasti S.S.S.R.:* Kharkov. Gosud. Univ., Izdatel., v. 12, p. 81-91. [*Some Jurassic rhynchonellids from the European part of the USSR.*]

Mansuy, H.
(532b) 1913, *Faunes des calcaires à Productus de l'Indochine, première série:* Serv. Géol. Indochine, Mém., v. 2, pt. 4, 133 p., 13 pl.

Martin, William
(533) 1793, *Figures and descriptions of petrifications collected in Derbyshire:* no. 1-4, 7 pl. (Wigan).

(534) 1809, *Petrificata derbiensia; or figures and descriptions of petrifactions collected in Derbyshire:* 28 p., 52 pl. (Wigan).

Martinsson, Anders, & Størmer, Leif
(535) 1960, *Report of the 21st session Norden; Part 7, Ordovician and Silurian stratigraphy and correlations:* Internat. Geol. Congress, 157 p. (Copenhagen).

Marwick, John
(536) 1953, *Divisions and faunas of the Hokonui System (Triassic and Jurassic):* New Zea-

land Geol. Survey, Palaeont., Bull. 21, p. 1-141, 17 pl., 3 text-fig.

Matthew, G. F.
(536a) 1891, *Illustrations of the fauna of the St. John Group No. 5:* Royal Soc. Canada, Proc. & Trans., ser. 1, v. 8, sec. 4, p. 123-166, pl. 11-16.
(536b) 1893, *Trematobolus. An articulate brachiopod of the inarticulate order:* Canadian Rec. Sci., v. 5, p. 276-279, 1 text-fig.
(536c) 1895, *Traces of the Ordovician System on the Atlantic Coast:* Royal Soc. Canada, Proc. & Trans., ser. 2, v. 1, sec. 4, p. 253-271, 2 pl.
(536d) 1899, *Preliminary notice of the Etcheminian fauna of Cape Breton:* New Brunswick Nat. History Soc., Bull. 4, p. 198-208, 4 pl.
(536e) 1901, *Acrothyra. A new genus of Etcheminian brachiopods:* Same, v. 4, p. 303-304, 6 text-fig.
(536f) 1903, *Report on the Cambrian rocks of Cape Breton:* Canada Geol. Survey, p. 5-246, 18 pl.

Maxwell, W. G. H.
(537) 1950, *An Upper Devonian brachiopod (Cyrtospirifer reidi sp. nov.) from the Mount Morgan District:* Queensland Univ., Dept. Geol., Paper, v. 3, no. 12, p. 1-8, 1 pl., 1 text-fig.
(538) 1951, *Upper Devonian and Middle Carboniferous brachiopods of Queensland:* Same, v. 3, no. 14, new ser., p. 1-27, pl. 1-4, text-fig. 1-4.
(539) 1954, *Upper Palaeozoic formations in the Mt. Morgan District-Faunas:* Same, v. 4, no. 5, p. 1-69, pl. 1-6, 3 text-fig.
(540) 1960, *Tournaisian brachiopods from Baywulla, Queensland:* Same, v. 5, no. 8, p. 1-11, 1 pl.
(541) 1961, *Lower Carboniferous brachiopod faunas from Old Cannindah, Queensland:* Jour. Paleontology, v. 35, no. 1, p. 82-103, pl. 19-20, 2 text-fig.

M'Coy, Frederick
(542) 1844, *A synopsis of the characters of the Carboniferous limestone fossils of Ireland:* 207 p., 29 pl., 34 text-fig. (Dublin).
(543) 1847, *On the fossil botany and zoology of the rocks associated with coal in Australia:* Ann. & Mag. Nat. History, v. 20, p. 145-157, 226-236, 298-331.
(544) 1851, *On some new Cambro-Silurian fossils:* Same, ser. 2, v. 8, p. 387-409.
(545) 1851, *A systematic description of the British Palaeozoic fossils in the Geological Museum of the University of Cambridge,* in SEDGWICK, A., A synopsis of the classification of the British Palaeozoic rocks, 1: 184 p., 11 pl. (London & Cambridge).

(546) 1855, *Systematic descriptions of the British Palaeozoic fossils in the Geological Museum of the University of Cambridge:* v. 3, p. 407-661 (London).

McEwan, E. D.
(547) 1939, *Convexity of articulate brachiopods as an aid in identification:* Jour. Paleontology, v. 13, p. 617-620.

McKee, E. D.
(548) 1938, *The environment and history of the Toroweap and Kaibab formations of northern Arizona and southern Utah:* Carnegie Inst. Washington, Publ., no. 492, 268 p., 48 pl.

McLaren, D. J.
(548a) 1961, *Three new genera of Givetian and Frasnian (Devonian) rhynchonelloid brachiopods:* Inst. Royal Sci. Nat. Belgique, Bull. 37, no. 23, 7 p., 2 pl.
(548b) 1962, *Middle and Early Upper Devonian rhynchonelloid brachiopods from western Canada:* Canada Geol. Survey, Bull 86, p. 1-122, pl. 1-18, 29 text-fig.

——, **Norris, A. W., & McGregor, D. C.**
(549) 1962, *Illustrations of Canadian fossils, Devonian of western Canada:* Canada Geol. Survey, Papers 62-64, 35 p., 16 pl.

McLearn, F. H.
(550) 1924, *Palaeontology of the Silurian rocks of Arisaig, Nova Scotia:* Canada Geol. Survey, Mem. 137, p. 1-179, pl. 1-30.

Meek, F. B.
(551) 1865, *Observations on the microscopic shell structure of Spirifer cuspidatus, Sowerby, and some similar American forms:* Acad. Nat. Sci. Philadelphia, Proc., v. 17, p. 275-277.
(552) 1872, *Descriptions of a few new species, and one new genus of Silurian fossils, from Ohio:* Am. Jour. Sci., ser. 3, v. 4, p. 274-281.
(552a) 1873, *Preliminary palaeontological report:* 6th Ann. Rept. of the U.S. Geol. Survey of Montana, Idaho, Wyoming and Utah; being a report of progress of the explorations for the year 1872, p. 429-518 (Washington).

——, **& Worthen, A. H.**
(553) 1866, *Palaeontology of Illinois, Sec. 2. Descriptions of invertebrates from the Carboniferous System:* Illinois State Geol. Survey, Rept., v. 2, p. 143-411, pl. 14-32.
(553a) 1870, *Descriptions of new species and genera of fossils from the Palaeozoic rocks of the Western States:* Acad. Nat. Sci. Philadelphia, Proc., p. 22-56.

Megerle von Mühlfeld, J. K.
(554) 1811, *Entwurf eines neuen System's der*

Schalthiergehäuse: Gesell. Naturforsch. Freunde Mag., v. 5, p. 38-72, pl. 3.

Mendes, J. C.

(555) 1959, *Chonetacea e Productacea Carboniferos da Amazónia:* São Paulo, Univ., Fac. Filoso. Ciências e Letras, Bull. 236 (Geol. 17), p. 1-83, pl. 1-7.

(555a) 1961, *Langella, Novo Gênero de Lingulídeo da Série Subarão:* Parana Univ., Bull. 5, p. 1-8, 2 pl., 5 text-fig.

Menke, C. T.

(555b) 1828, *Synopsis methodica molluscorum generum omnium et specierum earum quae in Museo Menkeano adservantur:* 91 p. (Pyrmonti).

Merla, Giovanni

(556) 1928, *Contributo alla conoscenza della fauna dei calcari a Schwagerina della Valle del Sosio:* Soc. Toscana Sci. Nat., Mem. 38, p. 70-87, 1 pl.

Meyer, C. J. A.

(557) 1864, *Notes on Brachiopoda from the Pebble-Bed of the Lower Greensand of Surrey . . . :* Geol. Mag., ser. 1, v. 1, no. 6, p. 249-257, pl. 11-12.

Mickwitz, August

(557a) 1896, *Über die Brachiopodengattung Obolus Eichwald:* Acad. Impér. Sci. St. Pétersbourg, Mém., ser. 8, v. 4, no. 2, 215 p., 3 pl., 7 text-fig.

(557b) 1909, *Vorläufige Mitteilung über das Genus Pseudolingula Mickwitz:* Same, ser. 6, v. 3, p. 765-772, 3 text-fig.

Middlemiss, F. A.

(558) 1959, *English Aptian Terebratulidae:* Palaeontology, v. 2, no. 1, p. 94-142, pl. 15-18.

Miloradovich, B. V.

(559) 1947, *O. dvukh novykh rodakh brakhiopod iz verkhnego Paleozoya Arktiki:* Soc. Impér. Nat. Moscou, Bull., new ser., v. 52, Otdel Geol., v. 22, pt. 3, p. 91-99. [*On two new brachiopod genera from the upper Paleozoic of the Arctic.*]

Minato, Masao

(560) 1951, *On the Lower Carboniferous fossils of the Kitakami Massif, northeast Honshu, Japan:* Hokkaido Univ., Fac. Sci. Jour., ser. 4 (Geol. & Min.), v. 7, no. 4, p. 355-382, pl. 1-5.

(561) 1952, *A further note on the Lower Carboniferous fossils of the Kitakami Mountainland, northeast Japan:* Same, v. 8, no. 2, p. 136-174, pl. 2-11.

(562) 1953, *On some reticulate Spiriferidae:* Palaeont. Soc. Japan, Trans. & Proc., new ser., no. 11, p. 65-73, 3 text-fig.

Mitchell, John

(563) 1921, *Some new brachiopods from the middle Palaeozoic rocks of New South Wales:* Linnean Soc., New S. Wales, v. 45, pt. 4 (no. 180), p. 543-551, pl. 31.

———, & Dun, W. S.

(564) 1920, *The Atrypidae of New South Wales, with references to those recorded from other states of Australia:* Linnean Soc., New S. Wales, v. 45, pt. 2 (no. 178), p. 266-276, pl. 14-16.

Moberg, J. C., & Segerberg, C. O.

(565) 1906, *Bidrag till Kännedomen om Ceratopygeregionen med särskild hänsyn till dess utveckling i Fogelsångstrakten:* Lund Univ. Årsskrift, new ser., v. 2, pt. 2, no. 7, p. 1-113, pl. 1-7.

Möller, V. I.

(565a) 1870, in *Obyknovennoe zasedanie, II Fevralya 1869 Goda:* Russisch-Kaiserl. Min. Gesell. Verhandl., ser. 2, v. 5, p. 408-413.

Moisseev, A. S.

(566) 1934, *Brakhiopody Iurskikh otlozhenii Kryma i Kavkasa:* Vses. Geol. Razv. Obed. SSSR, Trudy, v. 203, p. 1-213, pl. 1-19. [*Brachiopods of the Jurassic deposits of the Crimea and the Caucasus.*]

(567) 1936, *O novykh Triasovykh i Leiasovykh rodakh Rhynchonellidae:* Leningrad. Obshsch. Estestv., Otdel. Geol. & Mineral., Trudy, v. 65, p. 39-50, pl. 1. [*New Triassic and Liassic genera of the Rhynchonellidae.*]

Moore, Charles

(568) 1855, *On new Brachiopoda, from the Inferior Oolite of Dundry, etc.:* Somersetshire Arch. & Nat. History Soc., Proc., v. 5 for 1854, p. 107-128, pl. 1-3.

(569) 1860, *On new Brachiopoda, and on the development of the loop in Terebratella:* Geologist, v. 3 for 1860, p. 438-445.

Moore, R. C., Lalicker, C. G., & Fischer, A. G.

(570) 1952, *Invertebrate fossils:* 766 p., McGraw-Hill (New York).

Morris, J.

(571) 1845, *Descriptions of fossils;* in Strezelecki, P. E. de, Physical description of New South Wales and Van Diemen's Land, p. 270-291, pl. 10-19.

Morton, S. G.

(572) 1828, *Description of the fossil shells which characterize the Atlantic Secondary Formation of New Jersey and Delaware; including four new species:* Acad. Nat. Sci. Philadelphia, Jour., v. 6, p. 72-100, pl. 3-6.

Muir-Wood, Helen M.

(573) 1925, *Notes on the Silurian brachiopod*

genera *Delthyris, Uncinulina,* and *Meristina:* Ann. & Mag. Nat. History, ser. 9, v. 15, p. 83-95, 11 text-fig.

(574) 1928, *The British Carboniferous Producti. II. Productus (sensu stricto) semireticulatus and longispinus groups:* Great Britain, Geol. Survey (Palaeont.), Mem. 3, pt. 1, 217 p., 12 pl.

(575) 1930, *The classification of the British Carboniferous brachiopod subfamily Productinae:* Ann. & Mag. Nat. History, ser. 10, v. 5, no. 25, p. 100-108.

(576) 1934, *On the internal structure of some Mesozoic Brachiopoda:* Royal Soc. London, Philos. Trans., ser. B, v. 223, p. 511-567, pl. 62-63.

(577) 1935, *Jurassic Brachiopoda;* in MACFADYEN, W. A., and OTHERS, Geology & Palaeontology of British Somaliland: II. The Mesozoic Palaeontology of British Somaliland, p. 75-147, pl. 8-13, 33 text-fig. (London).

(578) 1936, *On the Liassic brachiopod genera Orthoidea and Orthotoma:* Ann. & Mag. Nat. History, ser. 10, v. 17, p. 221-242, 18 text-fig.

(579) 1936, *A monograph on the Brachiopoda of the British Great Oolite Series:* Palaeontograph. Soc. Mon., 144 p., 5 pl., 34 text-fig.

(580) 1938, *Notes on British Eocene and Pliocene Terebratulas:* Ann. & Mag. Nat. History, ser. 11, v. 2, p. 154-181, 13 text-fig.

(581) 1951, *The Brachiopoda of Martin's Petrificata Derbiensia:* Ann. & Mag. Nat. History, ser. 12, v. 4, p. 97-118, pl. 3-6.

(582) 1952, *Some Jurassic Brachiopoda from the Lincolnshire Limestone and Upper Estuarine Series of Rutland and Lincolnshire:* Geologists' Assoc., Proc., v. 63, pt. 2, p. 113-142, pl. 5-6.

(583) 1955, *A history of the classification of the phylum Brachiopoda:* 124 p., 12 text-fig., British Museum (Natural History) (London).

(584) 1959, *Report on the Brachiopoda of the John Murray Expedition:* John Murray Exped. 1933-34, Sci. Rept., v. 10, no. 6, p. 283-317, 5 pl., 4 text-fig., British Museum (Nat. History) (London).

(585) 1960, *Homoeomorphy in Recent Brachiopoda: Abyssothyris and Neorhynchia:* Ann. & Mag. Nat. History, ser. 13, v. 3, p. 521-525, 527, pl. 7.

(586) 1962, *On the morphology and classification of the brachiopod suborder Chonetoidea:* British Museum (Nat. History), Mon., viii+132 p., 16 pl., 24 text-fig.

——, & **Cooper, G. A.**
(587) 1960, *Morphology, classification & life habits of the Productoidea (Brachiopoda):*

Geol. Soc. America, Mem. 81, 447 p., 135 pl.

Müller, O. F.
(588) 1776, *Zoologiae Danicae Prodromus seu Animalium Daniae et Norvegiae indigenarum characteres, nomina, et synonyma imprimis popularium:* xxxii+282 p. (Havniae).

Murchison, R. I.
(589) 1840, *Description de quelques unes des coquilles fossiles les plus abondantes dans les couches devoniennes du Bas-Boulonnais:* Soc. Géol. France, Bull. 11, p. 250-256, pl. 2.

Nalivkin, D. V.
(590) 1925, *Gruppa Spirifer anossofi Vern. i Devon Evropeyskoy chasti S.S.S.R.:* Russische-Kaiserl., Min. Gesell. Zapiski, v. 54, no. 2, p. 267-358, pl. 4-5, 5 text-fig. [*The group of Spirifer anossofi Verneuil in the Devonian of the European part of the USSR.*]

(591) 1930, *Brakhiopod Verkhnego i Srednego Devona Turkestana:* Comité Géol., Mém., new ser., 180, 221 p., 10 pl. [*Brachiopods from the Upper and Middle Devonian of the Turkestan.*]

(592) 1937, *Brakhiopody Verkhnego i Srednego Devon i Nizhnego Karbona severovostochnogo Kazakhstana:* Tsentral. nauchnoissledov. Geol. Inst., Trudy, v. 99, p. 1-200, pl. 1-39. [*Brachiopoda of the Upper and Middle Devonian and Lower Carboniferous of northeastern Kazakhstan.*]

(593) 1937, *The Permian excursion southern part:* International Geol. Congress, 17th session, p. 1-131 (Moskva).

(594) 1941, *Brakhiopody glavnogo Devonskogo Polya,* in BATALINA, M. A., *et al.,* Fauna Glavnogo Devonskogo Polya: Akad. Nauk SSSR, Paleont. Inst., p. 139-226, pl. 1-8. [*Fauna of the main Devonian field.*]

(595) 1947, *Atlas Rukovodyashchikh form iskopaemykh faun S.S.S.R.: Devonskaya sistema Moskva-Leningrad:* Vses. Nauchno-Issledov. Geol. Inst., Minist. Geol. i Okhrany Nedr., v. 3, p. 1-245, pl. 1-56. [*Atlas of the guide forms of fossil faunas of the USSR: Devonian System Moscow-Leningrad.*]

Nebe, Baldwin
(596) 1911, *Die Culmfauna von Hagen i. Wien Beitrag zur Kenntnis des westfälischen Untercarbons:* Neues Jahrb. Mineral., Geol., & Paläont., Beil.-Bd. 31, p. 421-495, pl. 12-16.

Nechaev [Netschajew], A. W.
(597) 1894, *Die Fauna der permischen Ablagerungen des östlichen Theils des europäischen*

Russlands: Kazan Univ., Obshch. Estestv., Trudy, v. 27, no. 4, 503 p., 12 pl.

(598) 1911, *Fauna Permskikh otlozheniy vostoka i Kraynyago severa Evropeyskoy Rossii, Vyp. 1, Brachiopoda:* Geologich. Komitet., Trudy, new ser., v. 61, p. 1-164, pl. 1-15. [*Fauna of Perman deposits of eastern and extreme northern European Russia.*]

Nikiforova, O. I.

(599) 1937, *Brakhiopody Verkhnego Silura Sredneaziatskoy chasti SSSR:* Akad. Nauk SSSR, Paleont. Inst., Monographii po paleontologii SSSR, v. 35, pt. 1, 94 p., 14 pl. [*Upper Silurian Brachiopoda of the central Asiatic part of U.S.S.R.*]

(600) 1960, *Novyy rod Kulumbella iz semeystva Stricklandiidae:* Akad. Nauk SSSR, Paleont. Zhurnal, 1960, no. 3, p. 61-65, pl. 5. [*A new genus Kulumbella of the family Stricklandiidae.*]

(601) 1960, *Otryad Pentamerida:* Mshanki, Brakhiopody, SARYCHEVA, T. G. (ed.) in Osnovy Paleontologii, ORLOV, YU. A. (ed.), p. 197-205, pl. 20-25, text-fig. 98-135 (Moskva). [*Order Pentamerida.*]

———, & Andreeva, O. N.

(602) 1961, *Stratigrafiya Ordovika i Silura Sibirskoy Platformy i ee paleontologicheskoe obosnovanie (Brakhiopody):* Vses. Nauchno-Issledov. Geol. Inst. (VSEGEI), Trudy, v. 56, 412 p., 56 pl. [*Ordovician and Silurian stratigraphy of the Siberian Platform and its paleontological basis (Brachiopoda).*]

Nikitin, S. N.

(603) 1890, *Kamennougolnya otlozheniya podmoskovnago kraya i artezianskiya vody pod Moskvoyu:* Geologich. Komitet., Trudy, v. 5, no. 5, p. 1-182, pl. 1-3. [*Carboniferous deposits and artesian water in the Moscow area.*]

(604) 1900, *Zhmetka o geologicheskoy karte Zheleznykh Rudakh Saratovskoy Gub. Mestorozhdenie Margantsovoy Audys v Morshanskom Uezdy:* Same (1899), v. 18, no. 8, p. 383-410. [*Report on the geological map of the iron ores of Saratovsk Province and the bedded manganese ore in Morshansk District.*]

Noetling, Fritz

(605) 1905, *Untersuchungen über die Familie der Lyttoniidae Waagen emend. Noetling:* Palaeontographica, v. 51, p. 129-153.

Norford, B. S.

(606) 1960, *A well-preserved Dinobolus from the Sandpile Group (Middle Silurian) of Northern British Columbia:* Palaeontology, v. 3, p. 242-244, pl. 41.

North, F. J.

(607) 1920, *On Syringothyris Winchell and certain Carboniferous Brachiopoda referred to Spiriferina d'Orbigny:* Geol. Soc. London, Quart. Jour., v. 76, pt. 2, p. 162-227, pl. 11-13, 6 text-fig.

Northrop, S. A.

(608) 1939, *Paleontology and stratigraphy of the Silurian rocks of the Port Daniel-Black Cape region, Gaspé:* Geol. Soc. America, Spec. Paper 21, 302 p., 28 pl., 1 text-fig.

Norwood, J. G., & Pratten, H.

(609) 1855, *Notice of the genus Chonetes as found in the western states and territories, with descriptions of eleven new species:* Acad. Nat. Sci., Philadelphia, Jour., v. 3, p. 23-32, pl. 2.

Oehlert, D. P.

(610) 1886, *Étude sur quelques fossiles Dévoniens de l'ouest de la France:* Ann. Sci. Géologiques, v. 19, p. 1-80, pl. 1-5.

(611) 1887, in FISCHER, P. H., Manuel de conchyliologie et de paléontologie conchyliologique, ou Histoire naturelle des mollusques vivants et fossiles: pt. 11, p. 1189-1334, pl. 15, text-fig. 892-1138, F. Savy (Paris).

(612) 1890, *Note sur différents groupes établis dans le genre Orthis et en particulier sur Rhipidomella Oehlert (=Rhipidomys Oehlert, olim):* Jour. Conchyliologie, ser. 3, v. 30, p. 366-374.

(613) 1901, *Fossiles Dévoniens de Santa Lucia:* Soc. Géol. France, Bull., ser. 4, v. 1, p. 233-250, 12 text-fig.

Ohuye, Toshio

(614) 1936, *A note on the formed elements in the coelomic fluid of a brachiopod Terebratalia coreanica:* Tohoku Imper. Univ., Sci. Rept., ser. 4 (Biol.), v. 11, no. 2, p. 231-238, pl. 4.

(615) 1937, *Supplementary note on the formed elements in the coelomic fluid of some Brachiopoda:* Same, v. 12, no. 2, p. 241-253.

Oliveira, Euzebio de

(616) 1934, *Proposes the new name Oliveirella to replace the brachiopod generic name Brasilia (preoccupied):* Acad. Brasil. Sci., Ann., v. 6, no. 3, p. 167-168.

Öpik, A. A.

(617) 1930, *Brachiopoda Protremata der estländischen Ordovizischen Kukruse-Stufe:* Tartu Univ. (Dorpat), Acta & Commentationes, ser. A, v. 17, p. 1-262, pl. 1-22.

(618) 1932, *Über die Plectellinen:* Same, ser. A, v. 23, no. 3, p. 1-85, 12 pl.

(619) 1933, *Über Plectamboniten:* Same, ser. A, v. 24, no. 7, p. 1-79, 12 pl.

(620) 1933, *Über einige Dalmanellacea aus Estland:* Same, ser. A, v. 25, no. 1, p. 1-25, 6 pl.

(621) 1934, *Über Klitamboniten:* Same, ser. A, v. 26, no. 3, p. 1-239, 48 pl., 55 text-fig.

(622) 1939, *Brachiopoden und Ostrakoden aus dem Expansusschiefer Norwegens:* Norsk Geol. Tidsskrift, v. 19, p. 117-142, pl. 1-6, 2 text-fig.

(623) 1953, *Lower Silurian fossils from the Illaenus Band, Heathcote, Victoria:* Victoria Geol. Survey, Mem., no. 19, p. 1-42, pl. 1-13, 14 text-fig.

Orbigny, Alcide d'

(624) 1845, in MURCHISON, R. I., VERNEUIL, E. DE, KEYSERLING, A. DE, Géologie de la Russie d'Europe et des Montagnes de l'Oural; v. 2, Système Jurassique. Mollusques lamellibranches ou acéphales: p. 419-488, pl. 38-42 (London and Paris).

(625) 1847, *Considérations zoologiques et géologiques sur les brachiopodes ou palliobranches:* Acad. Sci. Paris, Comptes Rendus, v. 25, p. 193-195, 266-269.

(626) 1848-1851, *Paléontologie française, Terrains Crétacés, 4, Brachiopoda:* p. 1-390, pl. 490-599 (Paris).

(627) 1850, *Prodrôme de paléontologie stratigraphique universelle:* v. 1 (1849), 394 p. (Paris).

Orlov, Yu. A.

(627a) See 694.

Owen, E. F.

(628) 1961, *Palaeontology. Phylum Brachiopoda,* in CASEY, RAYMOND, The stratigraphical palaeontology of the Lower Greensand: Palaeontology, v. 3, pt. 4, p. 573-575, pl. 83, text-fig. 6a-c.

(629) 1962, *The brachiopod genus Cyclothyris:* British Museum (Nat. History), Bull., v. 7, no. 2, p. 39-63, pl. 4-5.

Ozaki, Kin-emon

(630) 1931, *Upper Carboniferous brachiopods from North China:* Shanghai Sci. Inst. Bull., v. 1, no. 6, p. 1-205, pl. 1-15.

Paeckelmann, Werner

(631) 1913, *Das Oberdevon des bergischen Landes:* K. Preuss. Geol. Landesanst., Abhandl., new ser., v. 70, p. 1-356, 7 pl.

(632) 1930, *Die Brachiopoden des deutschen Unterkarbons. I:* Same, new ser., v. 122, p. 1-326, pl. 9-24.

(633) 1931, *Die Fauna des deutschen Unterkarbons. II. Die Brachiopoden des deutschen Unterkarbons. Pt. 2. Die Productinae und Productus-ähnlichen Chonetinae:* Same, v. 136, 440 p., 41 pl.

(634) 1931, *Versuch einer zusammenfassenden Systematik der Spiriferidae King:* Neues Jahr. Mineral., Geol., & Paläont., Beil.-Bd. 67, Abt. B, p. 1-64.

Pahlen, A. von der

(635) 1877, *Monographie der baltisch-silurischen Arten der Brachiopodengattung Orthisina:* Acad. Impér. Sci., St. Pétersbourg. Mém., ser. 7, v. 24, no. 8, iv+52 p.

Palmer, A. R.

(635a) 1955, *The faunas of the Riley Formation in Central Texas:* Jour. Paleontology, v. 28, p. 709-786, pl. 86-92, 6 text-fig.

Pander, C. H.

(636) 1830, *Beiträge zur Geognosie des Russischen Reiches:* 165 p., 31 pl. (St. Petersburg).

(636a) 1861, in HELMERSEN, GREGOR VON, Die geologische Beschaffenheit des untern Narovathals und die Versandung der Narovamündung: Acad. Impér. Sci., St. Pétersbourg, Bull. 3, columns 46-49, pl. 2.

Paulus, Bruno

(637) 1957, *Rhynchospirifer n. gen. im Rheinischen Devon:* Senckenbergiana, v. 38, no. 1/2, p. 49-72, pl. 1-3.

————, **Struve, Wolfgang, & Wolfart, Reinhard**

(637a) 1963, *Chimaerothyris n.g. (Spiriferacea) aus dem Eifelium der Eifel:* Senckenbergiana, v. 44, p. 459-497, pl. 63-66, text-fig. 1-15.

Peetz, H. von

(638) 1901, *Beiträge zur Kenntniss der fauna aus den devonischen Schichten am Rande des Steinkohlenbassins von Kuznetz:* Travaux Sect. geol. cabinet de la Majesté St. Petersbourg, 394 p., 6 pl.

Perner, Jaroslav

(638a) 1903, *Bellerophontidae,* in BARRANDE, JOACHIM, Système silurien du centre de la Bohême: v. 4, p. 1-164, pl. 1-21 (Stockholm).

Pettitt, N. E.

(639) 1950-54, (in progress), *A monograph on the Rhynchonellidae of the British Chalk:* Palaeontograph. Soc., pt. 1-2, p. 1-52, pl. 1-3.

Phillips, John

(640) 1836, *Illustrations of the geology of Yorkshire: Pt. 2, the Mountain Limestone district:* 253 p., 25 pl., John Murray (London).

(641) 1841, *Figures and descriptions of the Palaeozoic fossils of Cornwall, Devon, and West Somerset:* xii+231 p., 60 pl., Longman & Co. (London).

Pictet, F. J.

(642) 1863-68, *Melanges paléontologiques:* v. 1, pt. 1-4, 308 p., 43 pl. (extr. Mém. soc. hist. nat. Genève, v. 17, pt. 1).

Pilsbry, H. A.
(643) 1892, *On some recent Japanese Brachiopoda, with a description of a species believed to be new:* Acad. Nat. Sci. Philadelphia, Proc. (1891), p. 165-171.

Posselt, H. J.
(644) 1894, *Brachiopoderne i den danske Kridtformation:* Danmarks Geol. Unders., v. 2Rk, no. 4, p. 1-59, 3 pl.

Poulsen, Christian
(645) 1943, *The fauna of the Offley Island formation, pt. 2, Brachiopoda:* Meddel. Grønland, v. 72, no. 3, 60 p., 6 pl.

Prendergast, K. L.
(646) 1935, *Some western Australian upper Palaeozoic fossils:* Royal Soc. West Australia, Jour., v. 21, p. 9-35, pl. 2-4.
(647) 1943, *Permian Productinae and Strophalosiinae of western Australia:* Same, v. 28 (1941-42), 73 p., 6 pl.

Prentice, J. E.
(648) 1950, *The genus Gigantella Sarycheva:* Geol. Mag., v. 87, no. 6, p. 436-438.

Prozorovskaya, E. L.
(649) 1962, *Some new brachiopods from the Upper Jurassic of western Turkmen:* Leningrad Univ., Vestník, no. 12 (Geol.-Geog.), pt. 2, p. 108-114, 5 text-fig.

Quenstedt, F. A.
(650) 1849-75, *Atlas zu den Cephalopoden, Brachiopoden und Echinodermen:* tab. 1-89, Ludwig Friedrich Fues (Tübingen).
(651) 1868-71, *Petrefactenkunde Deutschlands: v. 2, Brachiopoden:* 748 p., atlas, pl. 37-61 (Tübingen & Leipzig).

Quenstedt, Werner
(652) 1931, *Review of A. Öpik, Brachiopoda Protremata der estländischen ordovizischen Kukruse-Stufe:* Neues Jahrb. Mineral., Geol., & Paläont., v. 3, p. 477.

Rakusz, Gyula
(653) 1932, *Die oberkarbonischen Fossilien von Dobsina und Nagyvisnyó:* Geol. Hungarica, Palaeont., v. 8, p. 1-223, pl. 1-9, 28 text-fig.

Ramsbottom, W. H. C.
(654) 1952, *The fauna of the Cefn Coed Marine Band in the Coal Measures at Aberbaiden, near Tondu, Glamorgan:* Great Britain, Geol. Survey, Palaeont., Bull., v. 4, p. 8-32, pl. 2-3.

Rau, Karl
(655) 1905, *Die Brachiopoden des mittleren Lias Schwabens mit Ausschluss der Spiriferinen:* Geol. & Paläont., Abhandl., v. 10, p. 263-355, pl. 23-24.

Raymond, P. E.
(656) 1904, *The Tropidoleptus fauna at Canandaigua Lake, New York with the ontology of twenty species:* Carnegie Museum, Ann., 3, p. 79-177, pl. 1-8.
(657) 1911, *The Brachiopoda and Ostracoda of the Chazy:* Same, v. 7, no. 2, p. 215-259.
(658) 1923, *New fossils from the Chapman sandstone:* Boston Soc. Nat. History, Proc., v. 36, no. 7, p. 467-472.

Reed, F. R. C.
(659) 1906, *New fossils from the Bokkeveld Beds, South Africa:* Geol. Mag., decade 5, v. 3, p. 306-310.
(660) 1917, *The Ordovician and Silurian Brachiopoda of the Girvan District:* Royal Soc. Edinburgh, Trans., v. 51, pt. 4, p. 795-998, pl. 1-24.
(661) 1931, *New fossils from the Productus Limestones of the Salt Range, with notes on other species:* Palaeont. Indica, new ser., v. 17, p. 1-56, pl. 1-8.
(662) 1936, *The Lower Palaeozoic faunas of the Southern Shan States:* Same, v. 21, Mem. 3, p. 1-130, pl. 1-7.
(663) 1943, *Notes on certain Upper Devonian brachiopods figured by Whidborne, Pt. 1-2:* Geol. Mag., v. 80, no. 2-3, p. 69-78, 95-106.
(664) 1944, *Brachiopods and Mollusca from the Productus Limestones of the Salt Range:* Palaeont. Indica, new ser., v. 23, no. 2, 678 p., 65 pl.
(665) 1949, *Notes on some Carboniferous Spiriferidae from Fife:* Ann. & Mag. Nat. History, ser. 12, v. 1, no. 7, p. 449-487, pl. 7-12.

Renz, Carl
(666) 1932, *Brachiopoden des südschweizerischen und westgriechischen Lias:* Schweiz. Paläont. Gesell., Abhandl., v. 52, p. 1-61, pl. 1-3.

Retzius, A. J.
(666a) 1781, *Crania oder Todtenkopfs-Muschel:* Schrift. Berlin. Gesell. Naturforsch. Freunde, v. 2, p. 66-76, pl. 1.

Richards, J. R.
(666b) 1952, *The ciliary feeding mechanism of Neothyris lenticularis (Deshayes):* Jour. Morphology, v. 90, p. 65-91, 6 text-fig.

Richter, Rudolf, & Richter, Emma
(667) 1918, *Paläontologische Beobachtungen im rheinischen Devon:* Jahrb. Nassau. Vereins Naturk. Wiesbaden, v. 70 [1917], p. 143-161, 1 pl., 6 fig.

Rigaux, M. E.
(668) 1872, *Notes pour servir à la Géologie du Boulonnais, 1. Description de quelques Brachiopodes du terrain Dévonien de*

Ferques: Soc. Acad. Boulogne-sur-Mer, Mem., v. 5, p. 47-60, pl. 1.

Roemer, C. F.

(669) 1844, *Das Rheinische Uebergangsgebirge. Eine paläontologisch-geognostiche Darstellung:* 96 p., 6 pl. (Hanover).

Roemer, F. A.

(670) 1840-41, *Die Versteinerungen des norddeutschen Kreidegebirges:* pt. 1 (1840), p. 1-48, pl. 1-7; pt. 2 (1841), p. 49-145, pl. 8-16 (Hanover).

Roger, Jean

(671) 1952, *Classe des Brachiopodes,* in PIVETEAU, JEAN, Traité de Paléontologie, v. 2, p. 1-160, pl. 1-12 (Paris).

Rollier, H. L.

(672) 1915-19, *Synopsis des Spirobranches (Brachiopodes) Jurassiques Celto-souabes:* Soc. Paléont. Suisse, Mém., *(a)* v. 41 (1916), pt. 1 (Lingulidés, Spiriféridés), p. 1-69; *(b)* v. 42 (1917), pt. 2 (Rhynchonellidés), p. 71-184; *(c)* v. 43 (1918), pt. 3 (Terebratulidés), p. 187-275; *(d)* v. 44 (1919), pt. 4 (Zeilleridés, Repertoires), p. 279-422.

(673) *See 672(b).*

(674) *See 672(d).*

Rōōmusoks, A. K.

(675) 1956, *Luhaia, novyy rod strofomenid iz verkhnego ordovica Estonskoy S. S. R.:* Akad. Nauk SSSR, Doklady, v. 106, no. 6, p. 1091-1092, 1 pl., 1 text-fig. [*Luhaia, new strophomenid genus from the Upper Ordovician of the Estonian SSR.*]

(676) 1959, *Strophomenoidea Ordovika i Silura Estonii 1. Rod Sowerbyella, Jones:* Tartu Univ. (Dorpat), Acta & Commentationes, no. 75, p. 11-41, pl. 1-8. [*Strophomenoidea of the Ordovician and Silurian of Estonia 1. Genus Sowerbyella Jones.*]

Rosenstein, Elsa

(677) 1943, *Eine neue Gattung der Dalmanellacea aus dem Untersilur Estlands:* Tartu Univ. (Dorpat), Acta & Commentationes, no. 66, p. 471-478, pl. 1, text-fig. 1-3.

Ross, R. J.

(678) 1959, *Brachiopod fauna of Saturday Mountain Formation, Southern Lemhi Range, Idaho:* U.S. Geol. Survey, Prof. Paper, 294-L, p. 437-461, pl. 54-56.

Rotay [Rotai], A. P.

(678a) 1931, *Novye predstaviteli brachiopod iz nizhnego Karbona Donetskogo basseyna:* Glav. Geol.-Razv. Upr., Trudy, v. 73, p. 1-34. [*New types of brachiopods from the Lower Carboniferous of the Donetz Basin.*]

Rothpletz, August

(679) 1886, *Geologisch-palaeontologische Mono-*

graphie der Vilser Alpen: Palaeontographica, v. 33, p. 1-180, pl. 1-17.

Rowell, A. J.

(680) 1962, *The genera of the brachiopod superfamilies Obolellacea and Siphonotretacea:* Jour. Paleontology, v. 36, no. 1, p. 136-152, pl. 29-30.

(681) 1962, *The brachiopod genus Valdiviathyris Helmcke:* Palaeontology, v. 3, p. 542-545, pl. 68.

(681a) 1963, *Some nomenclatural problems in the inarticulate brachiopods:* Geol. Mag., v. 100, p. 33-43.

————, & **Bell, W. C.**

(682) 1961, *The inarticulate brachiopod Curticia Walcott:* Jour. Paleontology, v. 35, no. 5, p. 927-931, pl. 104.

Rowley, R. R., & Williams, J. S.

(683) 1933, *Unique coloration of two Mississippian brachiopods:* Washington Acad. Sci., Jour., v. 23, no. 1, p. 46-58, text-fig. 1-4.

Rozman, Kh. S.

(683a) 1962, *Stratigrafiya i brakhiopody Famenskogo Yarusa Mugodzhar i smezhnykh Rayonov:* Akad. Nauk SSSR, Inst. Geol., Trudy, v. 50, p. 1-196, pl. 1-31, 49 text-fig. [*Stratigraphy and brachiopods of the Famennian Stage of the Mugodzhar and adjacent areas.*]

Rübel, M. P.

(684) 1961, *Brakhiopody nadsemeystv Orthacea, Dalmanellacea i Syntrophiacea iz nizhnego Ordovika pribaltiki:* Akad. Nauk. Eston. SSSR, Inst. Geol., Trudy, v. 6, p. 141-226, pl. 1-27, 24 text-fig. [*Lower Ordovician brachiopods of the superfamilies Orthacea, Dalmanellacea, and Syntrophiacea of Eastern Baltic.*]

(684a) 1963, *O gonambonitakh (Clitambonitacea, Brach.) nizhnego Ordovika Pribaltiki:* Eesti. NSV Teaduste Akad. Geol. Inst., Uurimused, v. 13, p. 91-108, pl. 1-7. [*On Baltic Lower Ordovician gonambonitids (Clitambonitacea, Brach.).*]

Rudwick, M. J. S.

(685) 1959, *Growth and form of brachiopod shells:* Geol. Mag., v. 96, p. 1-24.

(686) 1962, *Filter-feeding mechanisms in some brachiopods from New Zealand:* Linnean Soc. London, Zool., Jour., v. 44, p. 592-615.

Rukavishnikova, T. B.

(687) 1956, in KELLER, B. M., *et al.,* Ordovik Kazakhstana, II: Stratigrafiya Chu-Iliiskikh gor.: Akad. Nauk SSSR, Inst. Geol. Nauk, Trudy, p. 105-168, 5 pl., 7 text-fig. [*Ordovician of Kazakhstan, II: Stratigraphy of the Iliiskikh Mountains.*]

Rusconi, Carlos

(688) 1956, *Oldhamias ordovicicas de Mendoza:* Rev. Museo Historia Nat. Mendoza, v. 9, p. 3-15, 1 pl.

Rzhonsnitskaya, M. A.

(689) 1952, *Spiriferidy Devonskikh otlozheniy okrain Kuznetskogo Basseyna:* Vses. Nauchno-Issledov. Geol. Inst.(VSEGEI), Minist. Geol. i Okhrany Nedr., Trudy, 232 p., 25 pl. [*Spiriferids from Devonian deposits at the edge of the Kuznetsk Basin.*]

(689a) 1956, *Semeystvo Pentameridae i sem. Camarotoechiidae:* p. 49-50, 53-55, in Khalfin, L. L. (*See* 465). [*Family Pentameridae and family Camarotoechiidae.*]

(690) 1960, *Otryad Atrypida,* Mshanki, Brakhiopody, Sarycheva, T. G. (ed.), in Osnovy Paleontologii, Orlov, Yu. A. (ed.), p. 257-264, pl. 53-56, text-fig. 310-335 (Moskva). [*Order Atrypida.*]

(691) 1960, in Markowski, B. P., Novye vidy Drevnikh Rasteniy i Bezpozvonochnykh SSSR: Vses. Nauchno-Issledov. Geol. Inst. (VSEGEI), Minist. Geol. i Okhrany Nedr., pt. 1, p. 1-612, pl. 1-93. [*New species of ancient plants and invertebrates of U.S.S.R.*]

(692) 1960, *K. systematike i filogeniya Pentameracea:* Akad. Nauk SSSR, Paleont. Zhurnal, no. 1, p. 38-49, 2 pl. [*On the systematics and phylogeny of the Pentameracea.*]

(693) Omit.

————, **Likharev, B. K., & Makridin, V. P.**

(694) 1960, *Otryad Rhynchonellida:* Mshanki, Brakhiopody, Sarycheva, T. G. (ed.), in Osnovy Paleontologii, Orlov, Yu. A. (ed.), p. 239-257, pl. 43-52, text-fig. 243-309 (Moskva). [*Order Rhynchonellida.*]

Sahni, M. R.

(695) 1925, *Morphology and zonal distribution of some chalk terebratulids:* Ann. & Mag. Nat. History, ser. 9, v. 16, p. 353-385, pl. 23-26.

(696) 1925, *Diagnostic value of hinge-characters and evolution of cardinal process in the terebratulid genus Carneithyris:* Same, ser. 9, v. 16, p. 497-502, pl. 25.

(697) 1929, *A monograph of the Terebratulidae of the British Chalk:* Palaeontograph. Soc. (1927), 62 p., 10 pl.

(698) 1955, *Recent researches in the palaeontologic division,* Geological Survey of India: Current. Sci. Bangalore, v. 24, no. 6, p. 187-188.

(699) 1960, *Revision of the Cretaceous Terebratulidae of southern India with descriptions of two species from the East Coast Gondwanas:* Palaeont. Indica, new ser., v. 35, no. 1, p. 1-34, 5 pl.

————, **& Bhatnager, N. C.**

(700) 1958, *New fossils from the Jurassic rocks of Jaisalmer, Rajasthan:* India, Geol. Survey, Rec., v. 87, pt. 2, p. 418-437, pl. 3-4.

————, **& Srivastava, J. P.**

(701) 1956, *Discovery of Eurydesma and Conularia in the eastern Himalaya and description of associated faunas:* Palaeont. Soc. India, Jour., v. 1, no. 1, p. 202-214, pl. 34-37.

St. Joseph, J. K. S.

(702) 1938, *The Pentameracea of the Oslo region, being a description of the Kiaer collections of pentamerids:* Norsk Geol. Tidsskrift, v. 17, p. 225-336, 8 pl.

(703) 1942, *A new pentamerid brachiopod genus from Yass, New South Wales:* Ann. & Mag. Nat. History, ser. 11, v. 9, p. 245-252.

Saito, Kazuo

(703a) 1936, *Older Cambrian Brachiopoda, Gastropoda, etc. from north-western Korea:* Tokyo Imper. Univ. Fac. Sci., Jour., sec. 2, v. 4, p. 345-367, 3 pl.

Salmon, E. S.

(704) 1942, *Mohawkian Rafinesquininae:* Jour. Paleontology, v. 16, p. 564-603, pl. 85-87, text-fig. 1-8.

Salter, J. W.

(704a) 1866, *Appendix. On the fossils of North Wales:* Great Britain Geol. Survey, Mem., v. 3, p. 240-381, pl. 1-26.

Sandberger, Guido, & Sandberger, Fridolin

(705) 1850-56, *Die Versteinerungen des rheinischen Schichtensystems in Nassau:* 564 p., Atlas, 41 pl., Kreidel & Niedner (Wiesbaden).

Sanders, J. E.

(705a) 1958, *Brachiopoda and Pelecypoda,* in Easton, W. H., et al., Mississippian Fauna in northwestern Sonora, Mexico: Smithsonian Misc. Coll., v. 119, no. 3, p. 41-72, pl. 3-7.

Sando, W. J.

(706) 1957, *Beekmantown Group (Lower Ordovician) of Maryland:* Geol. Soc. America, Mem. 68, p. 1-161, pl. 1-15, text-fig. 37-43.

Sapelnikov, V. P.

(707) 1960, *Novyy Nizhnevenlokskii rod Jolvia (Pentameracea) Srednego Urala:* Paleont. Zhurnal (1960), no. 4, p. 54-62, 2 pl. [*A new lower Wenlock genus Jolvia (Pentameracea) from the central Urals.*]

(708) 1963, *Novoe podsemeystvo i novye vidy Siluriyskikh pentamerid:* Same, no. 1, p. 63-69. [*A new subfamily and new species of Silurian pentamerids.*]

Sardeson, F. W.
(709) 1892, *The range and distribution of the Lower Silurian fauna of Minnesota with descriptions of some new species:* Minnesota Acad. Nat. Sci., Bull. 3, p. 326-343.

Sartenaer, Paul
(709a) 1961, *Late Upper Devonian (Famennian) rhynchonelloid brachiopods:* Inst. Royal Sci. Nat. Belgique, Bull., v. 37, no. 24, 10 p., 2 pl.

Sarycheva, T. G.
(710) 1960, [asst. ed.] *Osnovy paleontologii: Mshanki, brakhiopody:* 343 p., 75 pl. (Moskva). [*Principles of paleontology.*]
(710a) 1964, *Oldgaminoidnye brakhiopody iz permi Zakavkazya:* Paleont. Zhurnal, 1964, no. 3, p. 58-72, pl. 7-8, 2 text-fig. [*Oldhaminoid brachiopods from the Permian of Trans-Caucasia.*]

———, & Sokolskaya, A. N.
(711) 1952, *Opredelitel Paleozoiskikh brakhiopod Podmoskovnoy Kotloviny:* Akad. Nauk SSSR, Paleont. Inst., Trudy, v. 38, p. 1-307, pl. 1-71, 231 text-fig. [*Index of Paleozoic brachiopods of the Moscovian basin.*]

———, ———, Beznosova, G. A., & Maksimova, S. V.
(711a) 1963, *Brakhiopody i paleogeografiya Karbona Kuznetskoy Kotloviny:* Akad. Nauk SSSR, Paleont. Inst., Trudy, v. 95, 547 p., 64 pl., 151 fig. [*Carboniferous brachiopods and paleogeography of the Kuznetsk Basin.*]

Schellwien, E. T. T.
(712) 1898, *Die Auffindung einer permo-carbonischen Fauna in den Ostalpen:* K. K. Geol. Reichsanst., Verhandl., no. 16, p. 358-363, 1 text-fig.
(713) 1900, *Beiträge zur Systematik der Strophomeniden des oberen Paläozoicum:* Neues Jahrb. Mineral., Geol., & Paläont., v. 1, p. 1-15.

Schindewolf, O. H.
(714) 1955, *Über einige kambrische Gattungen inartikulater Brachiopoden:* Neues Jahrb. Mineral., Geol., & Paläont., v. 12, p. 538-557, 7 text-fig.

Schloenbach, G. J. C. V.
(715) 1866, *Beiträge zur Paläontologie der Jura- und Kreide-Formation im nordwestlichen Deutschland. 2:* Palaeontographica, v. 13, p. 267-339, pl. 38-40.

Schlotheim, E. F. von
(716) 1816, *Beiträge zur Naturgeschichte der Versteinerungen in geognostischer Hinsicht:* Akad. Wiss. München, math.-phys. Kl., Denkschrift., v. 6, p. 13-36.

(717) 1820, *Die Petrefactenkunde auf ihrem jetzigen Standpunkte durch die Beschreibung einer Sammlung versteinerter und fossiler Überreste der Tier- und Pflanzenreichs der Vorwelt erläutert:* v. 1, lxii+378 p. (Gotha).
(718) 1823, *Nachträgen zur Petrefactenkunde. Erklärung der Kupfertafeln:* 114 p., 37 pl., text-fig. 5 (Gotha).

Schmidt, Friedrich
(718a) 1888, *Über eine neuentdeckte untercambrische Fauna in Estland:* Acad. Impér. Sci. St. Pétersbourg, Mém., ser. 7, v. 36, no. 2, 27 p., 2 pl.

Schmidt, Herta
(718b) 1941, *Die mitteldevonischen Rhynchonelliden der Eifel:* Senckenberg. naturforsch. Gesell., Abhandl., v. 459, 78 p., 7 pl.
(718c) 1941, *Rhynchonellidae aus rechtsrheinischem Devon:* Senckenbergiana, v. 23, p. 277-290.
(719) 1943, *Die Terebratulidae des Wetteldorfer Richtschnittes:* Same, v. 27, no. 1/3, p. 67-75.
(719a) 1964, *Neue Gattungen paläozoischer Rhynchonellacea (Brachiopoda):* Same, v. 45, no. 6, p. 505-506.

Schnur, J.
(720) 1851, *Die Brachiopoden aus dem Uebergangsgebirge der Eifel:* Programm vereinigt. höhern Bürger- und Provinzial-Gewerbeschule Trier (1850-51), p. 2-16, Trier (Lintz).
(721) 1853, *Zusammenstellung und Beschreibung sämmtlicher im Uebergangsgebirge der Eifelvorkommenden Brachiopoden:* Palaeontographica, v. 3, p. 169-248, pl. 22-45.

Schuchert, Charles
(722) 1893, *A classification of the Brachiopoda:* Am. Geologist, v. 11, no. 3, p. 141-167.
(723) 1894, *A revised classification of the spire-bearing Brachiopoda:* Same, v. 13, p. 102-107.
(724) 1896, *Brachiopoda:* in ZITTEL, K. A. VON (transl. & ed. by EASTMAN, C. R.), Text-book of Palaeontology, v. 1, 1st ed., p. 291-343, text-fig. 489-587, Macmillan & Co., Ltd. (London).
(725) 1897, *A synopsis of American fossil Brachiopoda including bibliography and synonymy:* U.S. Geol. Survey, Bull. 87, p. 1-464.
(725a) 1911, *Paleogeographic and geologic significance of Recent Brachiopoda:* Geol. Soc. America, Bull. 22, p. 258-275.
(726) 1913, *Class 2. Brachiopoda:* in ZITTEL, K. A. VON (transl. & ed. by EASTMAN, C. R.), Text-book of Palaeontology, v. 1, 2nd edit., p. 355-420, text-fig. 526-636, Macmillan & Co., Ltd. (London).

(726a) 1913, *Class Brachiopoda:* Maryland Geol. Survey, p. 290-449, pl. 53-74.

——, & Cooper, G. A.
(727) 1930, *Upper Ordovician and Lower Devonian stratigraphy and paleontology of Percé, Quebec; Pt. 2, New species from the Upper Ordovician of Percé:* Am. Jour. Sci., ser. 5, v. 20, p. 265-288, pl. 1-3.
(728) 1931, *Synopsis of the brachiopod genera of the suborders Orthoidea and Pentameroidea, with notes on the Telotremata:* Same, ser. 5, v. 22, p. 241-251.
(729) 1932, *Brachiopod genera of the suborders Orthoidea and Pentameroidea:* Peabody Museum Nat. History, Mem., v. 4, pt. 1, p. 1-270, pl. A and 1-29.

——, & LeVene, C. M.
(730) 1929, *Brachiopoda (Generum et genotyporum index et bibliographia):* Fossilium Catalogus, 1, Animalia, Pars 42, 140 p., Junk (Berlin).
(731) 1929, *New names for brachiopod homonyms:* Am. Jour. Sci., v. 17, p. 119-122.

——, & Maynard, T. P.
(732) 1913, *Systematic paleontology of the Lower Devonian deposits of Maryland, Brachiopoda:* Maryland Geol. Survey, p. 290-449.

Schulz, E.
(733) 1914, *Über einige Leitfossilien der Stringocephalen Schichten der Eifel:* Verh. Naturh. Ver. Preuss. Rhein., Jahrg. 40, p. 335-383.

Scupin, Hans
(734) 1896, *Versuch einer Classification der Gattung Spirifer:* Neues Jahrb. Mineral., Geol., & Paläont., v. 2, p. 239-48.

Sdzuy, Klaus
(734a) 1955, *Die Fauna der Leinitz-Schiefer (Tremadoc):* Senckenberg. Naturforsch. Gesell., Abhandl., no. 492, p. 1-73, 8 pl.

Seidlitz, Wilfried von
(735) 1913, *Misólia, eine neue Brachiopoden-Gattung aus den Athyridenkalken von Buru und Misól:* Palaeontographica, suppl. 4, pt. 2, p. 163-193, pl. 12-14.

Seifert, Ilse
(735a) 1963, *Die Brachiopoden des oberen Dogger der schwäbischen Alb:* Palaeontographica, v. 121, A, p. 156-203, pl. 10-13.

Semikhatova, S. V.
(736) 1936, *Materialy k stratigrafii Nizhnego i Srednego Karbona Europeyskoy chasti SSSR:* Soc. Impér. Nat. Moscou, Bull., new ser., v. 44, sec. Géol., v. 14, p. 189-224, 3 pl. [*Contributions to the stratigraphy of the Middle and the Lower Carboniferous in the European part of the USSR.*]
(737) 1939, *Stratigraphic value of spirifers in Serpukhov beds of the Lower Carboniferous of the Moscow Basin:* Acad. Sci. URSS, Comptes Rendus, Doklady, v. 23, no. 3, p. 319-324.
(738) 1941, *Brakhiopody Bashkirskikh sloev SSSR, 1. Rod Choristites Fischer:* Akad. Nauk SSSR, Paleont. Inst., Trudy, v. 12, pt. 4, p. 1-152, pl. 1-13. [*Brachiopods of Bashkirian beds of USSR, 1. Genus Choristites Fischer.*]

Shaler, N. S.
(739) 1865, *List of the Brachiopoda from the island of Anticosti sent by the Museum of Comparative Zoology to different institutions in exchange for other specimens, with annotations:* Harvard Univ., Museum Comp. Zool., Bull., v. 1, p. 61-70.

Sharpe, Daniel
(739a) 1848, *On Trematis, a new genus belonging to the family of brachiopodous Mollusca:* Geol. Soc. London, Quart. Jour., v. 4, p. 66-69, 3 text-fig.

Shaw, A. B.
(739b) 1955, *Paleontology of northwestern Vermont. pt. 5. The Lower Cambrian fauna:* Jour. Paleontology, v. 29, p. 775-805, pl. 73-76.

Shimer, H. W., & Shrock, R. R.
(740) 1944, *Index fossils of North America:* 837 p., 303 pl., Mass. Inst. Tech. (Cambridge, Mass.).

Shrock, R. R., & Twenhofel, W. H.
(741) 1953, *Principles of invertebrate Paleontology:* 2nd edit., xx+816 p. (New York, Toronto, & London).

Sibly, T. F.
(742) 1908, *The faunal succession in the Carboniferous Limestone (Upper Avonian) of the Midland Area, (North Derbyshire and North Staffordshire):* Geol. Soc. London, Quart. Jour., v. 64, p. 34-80, pl. 1.

Sidyachenko, A. I.
(743) 1961, *Verkhnedevonskiy podrod tsirtospiriferid Dmitria:* Paleont. Zhurnal (1961), no. 2, p. 80-85, plate 11, 1 text-fig. [*The Upper Devonian cyrtospiriferid subgenus Dmitria.*]

Siehl, Agemar
(744) 1962, *Der Greifensteiner Kalk (Eiflium, Rheinisches Schiefergebirge) und seine Brachiopodenfauna. 1. Geologie; Atrypacea und Rostrospiracea:* Palaeontographica, pt. A, v. 119, p. 173-221, pl. 23-40.

Siemiradzki, J.
(745) 1906, *Monografia warstw paleozoicznych Podola:* Akad. Umiej. Krakow, Spraw. Kom. Fizj., Czesc 2, 39, p. 87-196. [*Monograph of the Paleozoic deposits of Podola.*]

Sinclair, G. W.
(745a) 1945, *Some Ordovician lingulid brachio-pods:* Royal Soc. Canada, Trans., ser. 3, v. 39, sec. 4, p. 55-82, 4 pl.
(746) 1946, *Bancroftina, a new brachiopod name:* Jour. Paleontology, v. 20, p. 295.

Slusareva, A. D.
(746a) 1958, *O Kazanskikh spiriferakh:* Akad. Nauk SSSR, Doklady, v. 118, p. 581-583. [*On Kazanian spiriferids.*]

Smirnova, T. N.
(747) 1960, *O novom podsemeystve Nizhnemelovykh dallinid:* Paleont. Zhurnal, no. 2, p. 114-120, pl. 11, 2 text-fig. [*A new sub-family of the Lower Cretaceous dallinids.*]
(748) 1962, *Novye dannye po Nizhnemelovym dallinidam (Brakhiopody):* Same, no. 2, p. 97-105. [*New data on Lower Cretaceous dallinids (Brachiopoda).*]

Smith, J. P.
(749) 1927, *Upper Triassic marine invertebrate faunas of North America:* U.S. Geol. Survey, Prof. Paper 141, p. 1-262, pl. 1-121.

Smith, Stanley
(750) 1925, *Notes upon the small species of Chonetes found in the Lower Carboniferous around Bristol:* Geol. Mag., v. 62, p. 85-88.

Smycka, František
(751) 1897, *Beitrag zur Kenntnis der Brachiopoden-Fauna im mährischen Devon bei Rittberg und Cekchovik:* Acad. Sci. de l'Empereur François Joseph I, Bull., p 32-44, 2 pl.

Sokolskaya, A. N.
(752) 1941, *Brakhiopody Osnovaniya Podmoskovnogo Karbona i Perekhodnykh Devonsko-Kamennougolnykh Otlozheniy (Chernyshinskie, Upinskie, i Malevko-muraevninskie sloi) chasti i Spiriferidae:* Akad. Nauk SSSR, Paleont. Inst., Trudy, v. 12, no. 2, p. 1-138, 12 pl., 39 text-fig. [*Lower Carboniferous and Devonian-Carboniferous brachiopods of the Moscow Basin (Tschernyschino, Upa and Malevka-Muraevnya beds.)*]
(753) 1948, *Evolyutsiya roda Productella Hall i smezhnykh s nim form v. Paleozoe podmoskovnoy kotloviny:* Same, v. 14, no. 3, 168 p., 10 pl. [*Evolution of the genus Productella Hall and related forms in the Paleozoic of the Moscow region.*]
(754) 1950, *Chonetidae Russkoy Platformy:* Same, v. 27, p. 1-108, pl. 1-13. [*Chonetidae of the Russian Platform.*]
(755) 1954, *Strofomenidy Russkoy Platformy:* Same, v. 51, p. 1-191, pl. 1-18. [*Strophomenids of the Russian Platform.*]
(756) 1959, *Osobennosti morfologii i rasprostraneniya spiriferid gruppy Spirifer dar-*

wini Morris: Paleont. Zhurnal, no. 1, p. 58-70, pl. 3. [*Peculiarity of morphology and distribution of the spiriferid group Spirifer darwini Morris.*]

——, **& Likharev, B. K.**
(757) 1960, *Otryad Productida:* Mshanki, Brakhiopody, SARYCHEVA, T. G. (ed.) in Osnovy Paleontologii, ORLOV, YU. A. (ed.), p. 221-238, pl. 33-42, text-fig. 194-242 (Moskva). [*Order Productida.*]

Solle, Gerhard
(758) 1938, *Sowerbyellinae im Unter- und Mitteldevon (Brachiopoda, Plectambonitidae):* Senckenberg. naturforsch. Gesell., v. 20, no. 3-4, p. 264-279, text-fig. 1-10.

Solomina, R. V., & Chernyak, G. E.
(759) 1961, *Orulgania - Novyy rod spiriferid iz verkhnego Paleozoya Arktiki:* Paleont. Zhurnal, no. 3, pl. 61-66, pl. 6. [*Orulgania, new spiriferid genus from the Upper Paleozoic of the Arctic.*]

Sowerby, G. B.
(760) 1846, *The Recent Brachiopoda:* Thesaurus Conchyliorum, or Monographs of genera of shells, v. 1, pt. 6-7, p. 337-371, pl. 67-73, Sowerby (London).

Sowerby, J., & Sowerby, J. de C.
(761) 1812-46, *The Mineral Conchology of Great Britain:* (a) v. 1(1812-15), p. i-vii, 1-234, pl. 1-102; (b) v. 2(1815-18), p. 1-235, pl. 103-203; (c) v. 3(1818-21), p. 1-184, pl. 204-306; (d) v. 4(1821-22), p. 1-114, pl, 307-383; (e) v. 4(1823), p. 115-160, pl. 384-406; (f) v. 5(1823-25), p. 1-168, pl. 407-503; (g) v. 6 (1826-29), p. 1-230, pl. 504-609; (h) v. 7 (1840-46), p. 1-80, pl. 610-648. [*a-d by J. Sowerby; e-h by J. de C. Sowerby.*]

Sowerby, J. de C.
(762) 1839, in MURCHISON, R. I., The Silurian System, xxxii+768 p., 36 pl. (London).
(763) 1835, *Mineral Conchology of Great Britain, systematical, stratigraphical and alphabetical indexes to the first six volumes:* p. 241-250 (London).

Spjeldnaes, Nils
(764) 1957, *The Middle Ordovician of the Oslo Region, Norway, 8, brachiopods of the suborder Strophomenida:* Norsk. Geol. Tidsskrift, v. 37, no. 1, p. 1-214, pl. 1-14.

Spriestersbach, Julius
(765) 1925, *Die Oberkoblenz-Schichten des bergischen Landes und Sauerlandes:* K. Preuss. Geol. Landesanst., Jahrb., v. 45, p. 367-450, pl. 15-17.

Stainbrook, M. A.
(766) 1943, *Strophomenacea of the Cedar Valley limestone of Iowa:* Jour. Paleontology, v. 17, p. 39-59, pl. 6-7.

(767) 1943, *Spiriferacea of the Cedar Valley Limestone of Iowa:* Jour. Paleontology, v. 17, no. 5, p. 417-450, pl. 65-70, text-fig. 1-14.

(768) 1945, *Brachiopoda of the Independence Shale of Iowa:* Geol. Soc. America, Mem. 14, p. 1-74, pl. 1-6, 2 text-fig.

(769) 1947, *Brachiopoda of the Percha Shale of New Mexico and Arizona:* Jour. Paleontology, v. 21, no. 4, p. 297-328, pl. 44-47.

(770) 1950, *Brachiopoda and stratigraphy of the Aplington Formation of northern Iowa:* Same, v. 24, p. 365-385, pl. 53-54.

Stefano, Giuseppe Di-

(771) 1887, *Sul Lias Inferiore di Taormina e de' suoi dintorni:* Giornale Sci. Nat. & Econ. Palermo, v. 18, p. 46-184, pl. 1-4.

(772) 1914, *Le Richthofenia dei calcari con Fusulina di Palazzo Adriano nelle valle del Fiume Socio:* Palaeont. Italica, v. 20, 29 p., 3 pl.

Stehli, F. G.

(773) 1954, *Lower Leonardian Brachiopoda of the Sierra Diablo:* Am. Museum Nat. History, Bull. 105, p. 257-358, pl. 17-27, text-fig. 1-55.

(774) 1955, *A new Devonian terebratuloid brachiopod with preserved color pattern:* Jour. Paleontology, v. 29, p. 868-870.

(775) 1956, *Evolution of the loop and lophophore in terebratuloid Brachiopoda:* Evolution, v. 10, p. 187-200, 9 text-fig.

(776) 1956, *A late Triassic terebratellacean from Peru:* Washington Acad. Sci. Jour., v. 46, p. 101-103.

(777) 1956, *Notes on oldhaminid brachiopods:* Jour. Paleontology, v. 30, no. 2, p. 305-313, pl. 41-42, 1 text-fig.

(778) 1961, *New terebratuloid genera from Australia:* Same, v. 35, p. 451-456.

(779) 1961, *New genera of upper Paleozoic terebratuloids:* Same, v. 35, p. 457-466.

(780) 1962, *Notes on some upper Paleozoic terebratuloid brachiopods:* Same, v. 36, p. 97-111.

(781) 1964, *New names for two homonyms:* Same, v. 38, p. 610.

Steinich, G.

(782) 1963, *Drei neue Brachiopodengattungen der Subfamilie Cancellothyrinae Thomson:* Geologie, Berlin, v. 12, pt. 6, p. 732-740, 8 text-fig.

Steininger, J.

(783) 1853, *Geognostische Beschreibung der Eifel:* 144 p., 9 pl., Trier (Lintz).

Stoliczka, Ferdinand

(784) 1872, *Cretaceous fauna of southern India:* Palaeontologia Indica, v. 4, no. 1, p. 1-31, 7 pl.

Stoyanow, A. A.

(785) 1915, *On some Permian Brachiopoda of Armenia:* Comité Géol., Mém., new ser., pt. 111, 95 p., 6 pl.

Struve, Wolfgang

(786) 1955, *Grünewaldtia aus dem Schönecker Richtschnitt (Brachiopoda, Mittel-Devon der Eifel):* Senckenbergiana Lethaea, v. 36, no. 3/4, p. 205-234, 4 pl.

(787) 1956, *Spinatrypa kelusiana n. sp., eine Zeitmarke im rheinischen Mittel-Devon (Brachiopoda):* Same, v. 37, no. 3/4, p. 383-409, 3 pl.

(788) 1961, *Zur Stratigraphie der südlichen Eifler Kalkmulden (Devon: Emsium, Eifelium, Givetium):* Same, v. 42, no. 3/4, p. 291-345, 3 pl.

(788a) 1964, *Über Alatiformia-Arten und andere, äusserlich ähnlich Spiriferacea:* Same, v. 45, p. 325-346, pl. 31, text-fig. 1-21.

Stubblefield, C. J.

(789) 1939, *Some Devonian and supposed Ordovician fossils from south-west Cornwall:* Great Britain, Geol. Survey, Bull., no. 2, p. 1-30, pl. 1-4.

Stuckenberg, A. A.

(790) 1905, *Die Fauna der obercarbonischen Suite des Wolgadurchbruches bei Samara:* Comité Géol., Mém., new ser., v. 23, 144 p., 13 pl.

Suess, Eduard

(791) 1854, *Über die Brachiopoden der Kössener Schichten:* Akad. Wiss. Wien, Math-naturwiss. Kl., Denkschrift, v. 7, p. 29-65, pl. 1-4.

(792) 1855, *Über die Brachiopoden der Hallstätter Schichten:* Same, v. 9, p. 23-32, pl. 1-2.

(793) 1855, *Über Meganteris, eine neue Gattung von Terebratuliden:* Same, Sitzungsber., v. 18, p. 51-64, 3 pl.

(794) 1858, *Die Brachiopoden der Stramberger Schichten:* Beiträge Paläont. & Geol. Österreich-Ungarns Orients, v. 1, p. 15-58, 6 pl.

Sutton, A. H.

(795) 1938, *Taxonomy of Mississippian Productidae:* Jour. Paleontology, v. 12, p. 537-569, pl. 62-66.

Talent, J. A.

(796) 1956, *Devonian brachiopods and pelecypods of the Buchan Caves Limestone, Victoria:* Royal Soc. Victoria, Proc., new ser., v. 68, p. 1-56, pl. 1-5, text-fig. 1-15.

Tate, Ralph

(797) 1896, in TATE, R., & DENNANT, J., *Correlation of marine Tertiaries of Australia:* Royal Soc. S. Australia, Trans., v. 20, p. 130 (footnote).

Teichert, Curt

(798) 1937, *Ordovician and Silurian faunas from Arctic Canada:* Rept. of the 5th Thule Exped., 1921-24; Danish Exped. to Arctic N. Amer., in charge of Knud Rasmussen, v. 1, no. 5 (1937), 169 p., 24 pl., Gyldendalske Borghandel Nordisk Forlag (Copenhagen).

Termier, Henri

(799) 1936, *Études géologiques sur le Maroc Central et le Moyen Atlas Septentrional, v. 3, Paléontologie:* Maroc. Service Géol. Div. Mines & Géol., Notes & Mém., no. 33, p. 1087-1421, pl. 1-23.

———, **& Termier, Geneviève**

(800) 1949, *Essai sur l'évolution des Spiriféridés:* Maroc. Service Géol., Div. Mines & Géol., Notes & Mém., no. 74, p. 85-112, 12 text-fig.

(801) 1960, *Les Oldhaminidés du Cambodge:* Soc. Géol. France, Bull., ser. 7, v. 1, p. 233-244, text-fig. 1-2.

Thomas, G. A.

(802) 1958, *The Permian Orthotetacea of Western Australia:* Australia Bur. Mineral Res., Geol., & Geophys., Bull. 39, p. 1-158, pl. 1-22.

Thomas, H. D.

(803) 1937, *Plicatoderbya, a new Permian brachiopod subgenus:* Jour. Paleontology, v. 11, p. 13-18, pl. 3.

Thomas, Ivor

(804) 1910, *British Carboniferous Orthotetinae:* Great Britain, Geol. Survey, Mem., v. 1, pt. 2, p. 83-134, pl. 13.

(805) 1914, *The British Carboniferous Producti. I. Genera Pustula and Overtonia:* Same, Mem., pt. 4, p. 197-366, pl. 17-20.

Thomson, J. A.

(806) 1913, *Material for the palaeontology of New Zealand:* New Zealand Geol. Survey, Palaeont. Bull., no. 1, p. 1-104, 6 pl.

(807) 1916, *Additions to the knowledge of the Recent and Tertiary Brachiopoda of New Zealand and Australia:* New Zealand Inst. Trans., v. 48 (1915), p. 41-47, pl. 1.

(808) 1919, *Brachiopod nomenclature: Clavigera, Hectoria, Rastelligera, Psioidea:* Geol. Mag., new ser., v. 6, p. 411-413.

(809) 1926, *A revision of the subfamilies of the Terebratulidae (Brachiopoda):* Ann. & Mag. Nat. History, ser. 9, v. 18, p. 523-530.

(810) 1927, *Brachiopod morphology and genera (Recent and Tertiary):* New Zealand Board Sci. & Art, Manual, no. 7, 338 p., 2 pl., 103 text-fig.

Tien, C. C.

(811) 1938, *Devonian Brachiopoda of Hunan:* Palaeont. Sinica, new ser., B, no. 4, whole ser. 113, 192 p., 22 pl.

Tokuyama, Akira

(812) 1957, *On some Upper Triassic spiriferinoids from the Sakawa Basin in Prov. Tosa, Japan:* Palaeont. Soc., Japan, Trans. & Proc., new ser., no. 27, p. 99-106, pl. 17.

(813) 1958, *On some terebratuloids from the Middle Jurassic Naradani Formation in Shikoku, Japan:* Japanese Jour., Geol. & Geog., v. 29, no. 1-3, p. 1-10, pl. 1.

(814) 1958, *On some terebratuloids from the late Jurassic Torinosu series in Shikoku, Japan:* Same, v. 29, no. 1-3, p. 119-131, pl. 9.

Tolmachev, I. P.

(814a) 1924, *Nizhnekamennougolnaya fauna kuznetsogo uglenoshogo Basseyna:* Geologich. Komitet., Materialy po obschestvo i prikladnoy geologii, v. 25, no. 1, p. 1-320 (pt. 2, p. 321-663, 1931). [*Lower Carboniferous fauna of the Kuznetsk coal basin.*]

Torley, K.

(815) 1934, *Die Brachiopoden des Massenkalkes der oberen Givet-Stufe von Bilveringsen bei Iserlohn:* Senckenberg. Naturforsch. Gesell. Abhandl., v. 43, p. 69-148, 9 pl., 82 text-fig.

Trechmann, C. T.

(816) 1918, *The Trias of New Zealand:* Geol. Soc. London, Quart. Jour., v. 73, pt. 3, no. 291, p. 165-246, pl. 17-25, 5 text-fig.

Twenhofel, W. H.

(817) 1914, *The Anticosti Island faunas:* Canada Geol. Survey, Dept. Min. & Res., Bull. 3, Geol. Ser. 19, p. 1-39, pl. 1.

(818) 1927, *Geology of Anticosti Island:* Same, Mem. 154, 481 p., 60 pl.

(819) 1954, *Correlation of the Ordovician formations of North America:* Geol. Soc. America, Bull. 65, p. 247-298, 1 pl., 2 text-fig.

Ulrich, E. O.

(819a) 1886, *Descriptions of new Silurian and Devonian fossils:* Am. Palaeont. Contrib., v. 1, p. 3-35, 3 pl., 2 text-fig.

(820) 1889, *On Lingulasma, a new genus, and eight new species of Lingula and Trematis:* Am. Geologist, v. 3, p. 377-391, 9 text-fig.

(821) 1926, *in* Butts, Charles, The Palaeozoic rocks in the geology of Alabama: Alabama Geol. Survey, Spec. Rept. 14, p. 41-230, pl. 3-76.

(822) 1927, *in* Poulsen, Christian, The Cambrian, Ozarkian & Canadian faunas of N.W. Greenland: Meddel. Grønland, v. 70, p. 233-343, pl. 14-21.

———, **& Cooper, G. A.**

(823) 1936, *New Silurian brachiopods of the*

family Triplesiidae: Jour. Paleontology, v. 10, no. 5, p. 331-347, pl. 48-50.

(824)   1936, *New genera and species of Ozarkian and Canadian brachiopods:* Same, v. 10, no. 7, p. 616-631.

(825)   1938, *Ozarkian and Canadian Brachiopoda:* Geol. Soc. America, Spec. Paper 13, 323 p., 58 pl., 14 text-fig.

(826)   1942, *New genera of Ordovician brachiopods:* Jour. Paleontology, v. 16, p. 620-626, pl. 90.

(827)   *See* 417a.

Vandercammen, Antoine

(828)   1955, *Septosyringothyris demaneti, nov. gen., nov. sp., un syringothyride nouveau du Dinantien de la Belgique:* Inst. Royal Sci. Nat. Belgique, Bull., v. 31, no. 30, p. 1-6, pl. 1.

(829)   1957, *Revision des Reticulariinae du Devonien de la Belgique:* Same, Bull., v. 33, no. 14, p. 1-19, pl. 1-3, 9 text-fig.

(830)   1957, *Revision de Spirifer euryglossus Schnur 1851, =Minatothyris nov. gen. euryglossa (Schnur):* Senckenbergiana, v. 38, no. 3/4, p. 177-193, pl. 1-3.

(831)   1957, *Revision du genre Gürichella Paeckelmann, 1913:* Inst. Royal Sci. Nat. Belgique, Mém., no. 138, p. 1-50, 2 pl., 47 text-fig.

(832)   1957, *Revision des Reticulariinae du Devonien de la Belgique. 2. Genre Plectospirifer A. Grabau, 1931:* Same, Bull., v. 33, no. 24, p. 1-23, pl. 1-2, 10 text-fig.

(833)   1958, *Revision des Reticulariinae du Devonien de la Belgique. III. Genre Tingella, A. Grabau, 1931:* Same, Bull., v. 34, no. 12, p. 1-19, pl. 1-2, 10 text-fig.

(834)   1959, *Essai d'étude statistique des Cyrtospirifer du Frasnien de la Belgique:* Same, Mém. 145, p. 1-175, pl. 1-5, 119 text-fig.

Vanuxem, Lardner

(835)   1842, *Geology of New York, pt. 3, comprising the survey of the Third Geological Districts:* Nat. History of N.Y., 306 p., 80 text-fig.

Vaughan, Arthur

(836)   1905, *The palaeontological sequence in the Carboniferous Limestone of the Bristol Area:* Geol. Soc. London, Quart. Jour., v. 61, p. 181-307, pl. 22-29.

Veevers, J. J.

(837)   1959, *The type species of Productella, Emanuella, Crurithyris, and Ambocoelia (Brachiopoda):* Jour. Paleontology, v. 33, no. 5, p. 902-908, 7 text-fig.

(838)   1959, *Devonian brachiopods from the Fitzroy Basin, Western Australia:* Australia Bur. Mineral Res., Geol. & Geophys., Bull. 45, p. 1-220, pl. 1-18.

(839)   1959, *Devonian and Carboniferous brach-iopods from north-western Australia:* Same, Bull. 55, p. 1-43, 4 pl.

Verco, J. C.

(840)   1910, *The brachiopods of South Australia:* Royal Soc. S. Australia, Trans., v. 34, p. 89-99, pl. 27-28.

Verneuil, Edouard de

(841)   1845, *Paléontologie, mollusques, brachiopodes:* in MURCHISON, R. I., VERNEUIL, E. DE, & KEYSERLING, A. DE, Géologie de la Russie d'Europe et des Montagnes de l'Oural: v. 2, pt. 3, Paléontologie, p. 17-395, 43 pl., John Murray (London) & Bertrand (Paris).

(842)   1848, *Note sur quelques brachiopodes de L'ile de Gothland:* Soc. Géol. France, Bull., ser. 2, v. 5, p. 339-347, pl. 4.

(843)   1850, *Notes sur les fossiles devoniens du district de Sabero (Leon):* Same, ser. 2, v. 7, p. 175-176.

Vincent, Emile Gérard

(844)   1893, *Contribution à la paléontologie des terrains tertiaires de la Belgique: Brachiopodes:* Soc. Malacol. Belgique, Ann., v. 28, p. 38-64, pl. 3-4.

Waagen, W. H.

(845)   1882-85, *Salt Range fossils, Part 4 (2) Brachiopoda:* Palaeont. Indica, Mem., ser. 13, v. 1, p. 329-770, pl. 25-86 [fasc. 1, p. 329-390, pl. 25-28, Dec., 1882; fasc. 2, p. 391-546, pl. 29-49, Aug., 1883; fasc. 3, p. 547-610, pl. 50-57, May, 1884; fasc. 4, p. 611-728, pl. 58-81, Dec., 1884; fasc. 5, p. 729-770, pl. 82-86, July, 1885].

Walcott, C. D.

(845a)   1884, *Paleontology of the Eureka district, Nevada:* U.S. Geol. Survey, Mon., v. 8, 298 p., 24 pl.

(845b)   1885, *Palaeontologic notes:* Am. Jour. Sci., ser. 3, v. 29, p. 114-117, 8 text-fig.

(845c)   1889, *Description of a new genus and species of inarticulate brachiopod from the Trenton Limestone:* Advance copy, U.S. Natl. Museum, Proc., v. 12, p. 365-366, 4 text-fig.

(845d)   1897, *Cambrian Brachiopoda: Genera Iphidea and Yorkia with descriptions of new species of each and of the genus Acrothele:* Same, Proc., v. 19, p. 707-718.

(845e)   1901, *Cambrian Brachiopoda: Obolella, subgenus Glyptias; Bicia; Obolus, subgenus Westonia; with descriptions of new species:* Same, v. 23, p. 669-695.

(845f)   1902, *Cambrian Brachiopoda: Acrotreta; Linnarssonella; Obolus; with descriptions of new species:* Same, v. 25, p. 577-612.

(846)   1905, *Cambrian Brachiopoda with descriptions of new genera and species:* Same, no. 1395, v. 28, p. 227-337.

(847) 1908, *Cambrian geology and Palaeontology. 3. Cambrian Brachiopoda, descriptions of new genera and species: 4. Classification and terminology of the Cambrian Brachiopoda*: Smithsonian Misc. Coll., v. 53, p. 53-165, pl. 7-12.

(848) 1912, *Cambrian Brachiopoda*: U.S. Geol. Survey, Mon. 51, pt. 1, 872 p., 76 text-fig.; pt. 2, 363 p., 104 pl.

Walker, J. F.

(849) 1868, *On the species of Brachiopoda, which occur in the Lower Greensand at Upware*: Geol. Mag., v. 5, no. 9, p. 399-407, pl. 18-19.

(850) [Huang, T. K.] (*see* 435).

Wang, Yü

(851) 1949, *Maquoketa Brachiopoda of Iowa*: Geol. Soc. America, Mem. 42, 55 p., 12 pl.

(852) 1955, *New genera of brachiopods*: Acad. Sinica, Scientia Sinica, v. 4, no. 2, p. 327-357, pl. 1-6, 2 text-fig.

Wanner, Joh., & Sieverts, Hertha

(853) 1935, *Zur Kenntnis der permischen Brachiopoden von Timor. 1, Lyttoniidae und ihre biologische und stammes-geschichtliche Bedeutung*: Neues Jahrb. Mineral., Geol., & Paläont., Beil.-Bd. 74, Abt. B, p. 201-281, 4 pl., 25 text-fig.

Watson, D. M. S.

(854) 1917, *Poikilosakos, a remarkable new genus of brachiopods from the Upper Coal-measures of Texas*: Geol. Mag., new ser., v. 4, p. 212-219, pl. 14.

Wedekind, Rudolf

(855) 1926, *Die Devonische Formation*: in SALOMON, W. H., Grundzüge der Geologie: v. 2, Erdgeschichte, p. 194-226, pl. 1-6, text-fig. 1-5, E. Schweizerbart (Stuttgart).

Weller, Stuart

(856) 1906, *Kinderhook faunal studies, IV, The fauna of the Glen Park limestone*: Acad. Sci. St. Louis, Trans., v. 16, p. 435-471, pl. 6, 7, text-fig. 13-16.

(856a) 1910, *Internal characters of some Mississippian rhynchonelliform shells*: Geol. Soc. America, Bull., v. 21, p. 497-516, 18 text-fig.

(857) 1911, *Genera of Mississippian loop-bearing Brachiopoda*: Jour. Geology, v. 19, p. 445-446.

(858) 1914, *The Mississippian Brachiopoda of the Mississippi Valley Basin*: Illinois State Geol. Survey, Mon. 1, p. 1-508, pl. 1-83.

White, C. A.

(859) 1862, *Description of new species of fossils from the Devonian and Carboniferous rocks of the Mississippi Valley*: Boston Soc. Nat. History, Proc., v. 9 (1862), p. 8-33, 5 text-fig.

————, & St. John, Orestes

(860) 1867, *Descriptions of new Subcarboniferous Coal-Measure fossils, collected upon the Geological Survey of Iowa; together with a notice of new generic characters involved in two species of Brachiopoda*: Chicago Acad. Sci., Trans., v. 1, p. 115-127.

Whitehouse, F. W.

(861) 1928, *Notes on upper Palaeozoic marine horizons in eastern and western Australia*: Australian & New Zealand Assoc. Adv. Sci. Rept. (1926), Trans., sec. C, v. 18, p. 281-283.

Whitfield, R. P.

(862) 1882, *Descriptions of some new species of fossils from Ohio*: N.Y. Acad. Sci., Ann., v. 11, p. 193-244.

(863) 1885, *Brachiopoda and Lamellibranchiata of the Raritan Clays and Greensand Marls of New Jersey*: U.S. Geol. Survey, Mon., v. 9, 268 p., 35 pl.

(863a) 1890, *Description of a new genus of inarticulate brachiopodous shell*: Am. Museum Nat. History, Bull., v. 3, p. 121-122, 8 text-fig.

(864) 1891, in WENDT, A. F., Notes on some fossils from Bolivia collected by Mr. A. F. Wendt and description of a remarkable new genus and species of brachiopod: Am. Inst. Min., Met. & Petrol. Eng. Trans., v. 19, p. 104-107.

(865) 1908, *Notes and observations on Carboniferous fossils and semifossil shells brought home by members of the Peary Expedition of 1905-1906*: Am. Museum Nat. History, Bull. 24, p. 51-58, pl. 1-4.

Whittington, H. B.

(866) 1938, *New Caradocian brachiopods from the Berwyn Hills*: Ann. & Mag. Nat. History, ser. 11, v. 2, p. 241-259, pl. 10-11.

Willard, Bradford

(867) 1928, *The brachiopods of the Ottosee and Holston Formations of Tennessee and Virginia*: Harvard Univ., Museum Comp. Zool., Bull., v. 68, no. 6, p. 255-292, 3 pl.

Williams, Alwyn

(868) 1949, *New Lower Ordovician brachiopods from the Llandeilo-Llangadock District*: Geol. Mag., v. 86, p. 161-174, p. 226-238, pl. 8-9.

(869) 1950, *New stropheodontid brachiopods*: Washington Acad. Sci., Jour., v. 40, no. 9, p. 277-282, 1 pl.

(870) 1951, *Llandovery brachiopods from Wales with special reference to the Llandovery District*: Geol. Soc. London, Quart. Jour., v. 107, pt. 1, p. 85-136, pl. 3-8.

(871) 1953, *North American and European stropheodontids: their morphology and*

systematics: Geol. Soc. America, Mem. 56, p. 1-67, pl. 1-13.

(872) 1953, *The classification of the strophomen-oid brachiopods:* Washington Acad. Sci., Jour., v. 43, no. 1, p. 1-13, text-fig. 1-13.

(873) 1953, *The morphology and classification of the oldhaminid brachiopods:* Same, v. 43, no. 9, p. 279-287, pl. 1-3.

(874) 1955, *Systematic description of Brachio-poda,* in WHITTINGTON, H. B., & WILLIAMS, A., The fauna of the Derfel Limestone of the Arenig District, North Wales: Royal Soc. London, Phil. Trans., ser. B, v. 238, p. 397-430, pl. 38-40.

(875) 1956, *The calcareous shell of the Brachio-poda and its importance to their classifica-tion:* Biol. Reviews, v. 31, p. 243-287, 7 text-fig.

(876) 1956, *Productorthis in Ireland:* Royal Irish Acad. Proc., v. 57, sec. B, no. 13, p. 179-183, pl. 9.

(877) 1962, *The Barr and Lower Ardmillan Series (Caradoc) of the Girvan District, South-west Ayrshire, with descriptions of the Brachiopoda:* Geol. Soc. London, Mem. 3, 267 p., 25 pl., 13 text-fig.

(878) 1963, *The Caradocian brachiopod faunas of the Bala District, Merionethshire:* British Museum (Nat. History), Geol., Bull., v. 8, no. 7, p. 327-471, pl. 1-16, text-fig. 1-13.

———, & Wright, A. D.

(879) 1961, *The origin of the loop in articu-late brachiopods:* Palaeontology, v. 4, p. 149-176, 13 text-fig.

(880) 1963, *The classification of the "Orthis testu-dinaria Dalman" group of brachiopods:* Jour. Paleontology, v. 37, no. 1, p. 1-32, pl. 1-2.

Williams, H. S.

(881) 1900, *The Paleozoic faunas of Maine:* U.S. Geol. Survey, Bull. 165, p. 15-92.

(882) 1908, *Dalmanellas of the Chemung forma-tion and a closely related new brachiopod genus Thiemella:* U.S. Natl. Museum, Proc., v. 34, p. 35-64, pl. 2-4.

———, & Breger, C. L.

(883) 1916, *The fauna of the Chapman Sand-stone of Maine including descriptions of some related species from the Moose River Sandstone:* U.S. Geol. Survey, Prof. Paper 89, p. 1-347, pl. 1-27.

Williams, J. S.

(883a) 1943, *Stratigraphy and fauna of the Louisi-ana Limestone of Missouri:* U.S. Geol. Sur-vey, Prof. Paper 203, 133 p., 9 pl.

Wilson, A. E.

(884) 1913, *A new brachiopod from the base of the Utica:* Canada Geol. Survey, Dept. Min. & Res., Bull. 1, p. 81-86, pl. 8, text-fig. 1-2.

(885) 1926, *An Upper Ordovician fauna from the Rocky Mountains, British Columbia:* Same, 44, p. 1-34, pl. 1-5.

(886) 1932, *Ordovician fossils from the region of Cornwall, Ontario:* Royal Soc. Canada, Trans., ser. 3, v. 26, sec. 4, p. 373-404, 6 pl.

(887) 1944, *Rafinesquina and its homeomorphs Öpikina and Öpikinella, from the Ottawa Limestone of the St. Lawrence Lowlands:* Same, ser. 3, v. 38, pt. 4, p. 145-203, fig. 1-10, pl. 1-2.

(888) 1945, *Strophomena and its homomorphs Trigrammaria and Microtrypa, from the Ottawa Limestone of the Ottawa-St. Law-rence Lowlands:* Same, v. 39, no. 4, p. 121-150, pl. 1-2.

Wiman, Carl

(889) 1914, *Über die Karbonbrachiopoden Spitz-bergens und Beeren-Eilands:* Nova Acta Regiae, ser. 4, v. 3, no. 8, p. 1-91, pl. 1-19.

Winchell, Alexander

(890) 1862, *Descriptions of fossils from the Mar-shall and Huron groups of Michigan:* Acad. Nat. Sci. Philadelphia, Proc., v. 14, p. 405-430.

(891) 1866, *The Grand Traverse region. A report on the geological and industrial resources in the Lower Peninsula of Michigan:* 97 p. (Ann Arbor).

Wirth, Eberhard

(892) 1936, *Uber "Clitambonites" giraldii Mar-telli und Yangtzeella poloi Martelli aus dem Ordoviz Chinas:* Paläont. Zeitschr., v. 18, p. 292-302, pl. 20.

Wiśniewska, Marja

(893) 1932, *Les Rhynchonellidés Jurassique sup. de Pologne:* Palaeont. Polonica, v. 2, no. 1, p. 1-71, pl. 1-6.

Worthen, A. H.

(894) 1884, *Descriptions of two new species of Crustacea, fifty-one species of Mollusca, and three species of crinoids from the Car-boniferous formation of Illinois and ad-jacent states:* Illinois State Museum Nat. History, Bull. 2, 27 p.

Wright, A. D.

(895) 1963, *The morphology of the brachiopod superfamily Triplesiacea:* Palaeontology, v. 5, no. 4, p. 740-64, pl. 109-110.

(895a) 1963, *The fauna of the Portrane Limestone, I. The inarticulate brachiopods:* British Mu-seum (Nat. History), Bull. 8, p. 221-254, 4 pl., 5 text-fig.

(895b) 1964, *The fauna of the Portrane Limestone, II:* Same, Bull. 9, no. 6, p. 157-256, pl. 1-11.

Yabe, Hisakatsu

(896) 1932, *Brachiopods of the genus Pictothyris*

Thomson, 1927: Tohoku Imper. Univ., Sci. Rept., ser. 2 (Geol.), v. 15, no. 3, p. 193-197, pl. 13.

———, & Hatai, K. M.

(897) 1934, *The Recent brachiopod fauna of Japan (1). New genera and subgenera:* Japan Imper. Acad., Proc., v. 10, no. 9, p. 586-589, 4 text-fig.

Yan, Shi-pu

(898) 1959, *Feng Xian Tong Shi Yen Xin Shu Grandispirifer:* Acta Palaeont. Sinica, v. 7, no. 2, p. 111-120, pl. 1-2, 4 text-fig. [Russian abst., *Novyy vizeyskiy rod spiriferid —Grandispirifer, gen. nov.—New Visean spiriferid genus—Grandispirifer, gen. nov.*]

Zeiler, Friedrich

(899) 1857, *Versteinerungen der älteren Rheinischen Grauwacke:* Verh. Naturh. Ver. Preuss. Rhein, v. 14, p. 45-51, pl. 3-4.

Zittel, K. A. von

(900) 1870, *Ueber den Brachial-Apparat bei einigen jurassischen Terebratuliden und über eine neue Brachiopodengattung Dimerella:* Palaeontographica, v. 17, 211-222, pl. 41.

(901) 1880, *Handbuch der Palaeontologie:* v. 1, no. 4, p. 641-722, text-fig. 473-558, R. Oldenbourg (München & Leipzig).

(902) 1913, *Text-book of Paleontology:* transl. C. R. EASTMAN, 2nd edit., v. 1, 839 p., 1594 text-fig., Macmillan & Co., Ltd. (London).

Zugmayer, Heinrich

(903) 1880, *Ueber rhätische Brachiopoden:* K. K. Geol. Reichsanst., Jahrb., v. 30, p. 149-156.

(904) 1880, *Untersuchungen über rhätische Brachiopoden:* Beiträge Paläont. & Geol. Österreich-Ungarns Orients, v. 1, p. 1-42, pl. 1-4.

ADDITIONAL SOURCES OF ILLUSTRATIONS

(905) Ager, D. V., a, 1954; b, 1958
(906) Amos, Arturo, 1958
(907) Biernat, Gertruda, 1957
(907a) Boucot, A. J., 1960
(907b) Brown, Ida, 1952
(908) Buckman, S. S., 1901
(909) Chatwin, C. P., 1948
(910) Chernyshev, Theodore, 1937
(911) Chiplonker, G. W., 1938
(912) Cloud, P. E., 1944
(913) ———, & Cooper, G. A., 1948
(914) Cooper, G. A., 1949
(915) Crickmay, C. H., a, 1933, b, 1963
(916) Dumortier, E., 1869
(917) Friele, Herman, 1877
(918) Gregorio, A. de, 1886
(919) Hall, James, 1861
(920) Hisinger, Wilhelm, 1837
(921) Ivanov, P. P., 1925
(922) Kozlowski, Roman, 1923
(923) Lamplugh, G. W., & Walker, J. F., 1903
(924) Laube, G. C., 1866
(925) Moisseev, A. C., 1956
(926) Muir-Wood, H. M., & Cooper, G. A., 1951
(927) Oehlert, D. P., 1884
(928) Oppel, A., 1861
(929) Rouillièr, C., 1846
(930) Sartenaer, Paul, a, 1955; b, 1956; c, 1961
(931) Schmidt, Herta, a, 1937; b, 1954; c, 1955; d, 1965
(932) Sowerby, J. de C., 1840
(932A) Struve, Wolfgang, 1964
(933) Thomson, J. A., 1915
(934) Trümpy, R., 1956
(935) Upton, Charles, 1904
(936) Warren, P. S., 1937
(937) Zieten, C. H. von, 1860
(938) Zittel, K. A., 1869

ADDENDUM

Alatiformia STRUVE, 1963, p. 499 [**Spirifer alatiformis* DREVERMANN, 1907, p. 126; OD]. Extremely transverse; fold with distinct round- or flat-bottomed median depression; micro-ornament consisting of delicate capillae and distinct growth lamellae; otherwise similar to *Spinocyrtia*. *L.Dev.(Ems.)-M.Dev.(Couvin.),* Eu. (Spinocyrtiidae, p. *H*691.) [PITRAT.]

Balakhonia SARYCHEVA, 1963, p. 231 [**B. ostrogensis;* OD]. Shell thin, medium-sized to large, concavo-convex, ears broad, flattened; both valves costellate, with numerous fine growth lines interrupting costellae, rugae on ears and umbonal slopes, spines in row along hinge, rare elsewhere, absent from brachial valve; cardinal process small, bilobate, with 2 separate lobes, septum posteriorly broad, becoming narrow ridge 1/3 length of valve; lateral ridges extending along outer margin of

longitudinally ribbed adductor scars, latter divided by longitudinal ridge, each forming 2 scars. *L. Carb.(Visean)-U. Carb.(L. Namur.),* Eu. (USSR)-Asia(Sib., Kuznetsk Basin); *?L.Perm.,* Eu. (Linoproductidae, Linoproductinae, p. *H*500.) [MUIR-WOOD.]

Callipentamerus BOUCOT, 1964, p. 887 [*Pentamerus corrugatus* WELLER & DAVIDSON, 1896, p. 173, pl. 7, figs. 1-4; OD]. Shell exterior with crisscross ornamentation produced by intersecting sets of concentric rugae; internally like *Pentameroides. L. Sil. (Llandover.),* USA (Iowa). (Pentameridae, Pentamerinae, p. *H*547.) [AMSDEN.]

Cancrinelloides USTRITSKY, 1963, p. 85 [**Productus (Productus) obrutschewi* LICHAREV, 1934, p. 24; OD]. Shell large, moderately concavo-convex, no median sulcus; valves capillate, spines numerous, scattered, fine rugae; cardinal process bilobate, supported by median septum, bifurcating anteriorly, and by lateral ridges, adductor scars dendritic in both valves. *U.Perm.*, Arctic Regions. (Linoproductidae, Linoproductinae, p. *H501*.) [MUIR-WOOD.]

Chimaerothyris PAULUS, STRUVE & WOLFART, 1963, p. 463 [**C. hotzi*; OD]. Pedicle valve interior lacking delthyrial plate, but delthyrium largely closed by callus deposits growing inward from sides of dental plates and upward from floor of valve; micro-ornament consisting of capillae and growth lamellae, former predominant; otherwise similar to *Spinocyrtia*. *M.Dev.(Couvin.)*, W.Eu. (Spinocyrtiidae, p. *H691*.) [PITRAT.]

Chonopectoides CRICKMAY, 1963, p. 23 [**C. catamorphus*; OD]. Near *Devonoproductus* but smaller, valves plano-convex, pedicle valve capillate, spines along hinge margin only and laterally directed, brachial valve with concentric lamellae; cardinal process bilobate, deeply cleft, breviseptum low, half valve length. *Up.M.Dev.*, W.Can. (Leioproductidae, Devonoproductinae, p. *H471*.) [MUIR-WOOD.]

Chonostrophiella BOUCOT & AMSDEN, 1964, p. 881 [**Chonetes complanata* HALL, 1857, p. 56; OD]. Ornamentation of fine radial costellae; notothyrium partly closed by chilidial plates; brachial valve with long median septum and 2 short lateral septa. *L.Dev.*, E. N. Am.-?S. Am. (Colombia). (Chonostrophiidae, p. *H434*.) [MUIR-WOOD.]

Clitambonites AGASSIZ, 1846 [*=Klitambonites* PANDER, 1830, p. 70 (obj.) *(nom. oblit.)*; *Prionites* FISCHER DE WALDHEIM, 1834, p. 228 (obj.) *(nom. van.)*]. (Clitambonitidae, Clitambonitinae, p. *H349*.) [WILLIAMS.]

Diabolirhynchia DROT, 1964, p. 111 [**D. hollardi*; OD] [Notes Service Géol. Maroc, v. 23, no. 172, p. 111-116, 1 pl., 1 text-fig., 1964]. *U.Sil.(Ludlov.)*, N. Afr. (Morocco) - N. Am. (Ind.). (Rhynchonellacea, fam. Uncertain, p. *H592*.) [SCHMIDT.]

Eccentricosta BERDAN, 1963, p. 254 [**Chonetes jerseyensis* WELLER, 1900, p. 8 footnote; OD]. Small, concavo-convex, costellae radiate from hinge instead of from umbo, and rarely bifurcate, spines nearly perpendicular, developed from hinge margin; cardinal process sessile, bilobed, supported by callus platform and 2 divergent septa, socket ridges massive, pedicle valve with triangular callus platform and low median septum. *U.Sil.*, N.Am.-*L.Dev.* Eu.(Ger.-Eng.). (Chonetidae, Subfamily Uncertain, p. *H433*.) [MUIR-WOOD.]

Eomarginifera MUIR-WOOD, 1930 [*=Lissomarginifera* LANE, 1962, p. 901 (type, *L. nuda*)]. (Marginiferidae, Marginiferinae, p. *H477*.) [MUIR-WOOD.]

Koninckina SUESS in DAVIDSON, 1853 [*=Koninckia* SUESS in WOODWARD, 1854, p. 231 *(nom. null.)*]. (Koninckinidae, p. *H666*.)

Magharithyris FARAG & GATINAUD, 1960, p. 77 [**M. triplicata*; OD]. Said to be near *Parathyridina* but shell more elongate, folding uniplicate, triplicate or quadriplicate to multiplicate. [Genus based on one imperfectly preserved specimen and internal characters unknown. *M.Jur.(Bathon.)*, Egypt. (Terebratulacea, Family Uncertain, p. *H816*.) [MUIR-WOOD.]

Monticulifera MUIR-WOOD & COOPER, 1960 [*=Sinoproductus* CHAN & LI, 1962, p. 477 (type, *Productus intermedius* var. *sinensis* FRECH, 1911, p. 176)]. (Linoproductidae, Monticuliferinae, p. *H505*.) [MUIR-WOOD.]

Nix EASTON, 1962, p. 46 [**N. angulata*; OD]. Small, concavo-convex, with shallow ventral median sulcus and dorsal fold, finely capillate and faint concentric growth lines, 6 to 8 spines along hinge margin, extending at angle of 45 degrees; brachial valve interior with septa reduced or absent, pit (alveolus) at anterior base of cardinal process, pedicle valve with median septum about 0.5 or 0.7 of valve length. *Miss.*, USA(Heath F., Mont., or Brazer of Rocky Mts.). (Chonetidae, Subfamily Uncertain, p. *H433*.) [MUIR-WOOD.]

Rensselandia HALL, 1867 [*=Macroplectane* COSSMAN, 1909, p. 215 *(nom. subst. pro Denckmannia* HOLZAPFEL, 1912, *non* BUCKMAN, 1898)]. (Stringocephalidae, Rensselandiinae, p. *H746*.)

Rugoclostus EASTON, 1962, p. 59 [**R. nivalis*; OD]. Medium-sized, quadrate outline, both valves geniculated, pedicle valve medially sulcate, narrow interarea (ginglymus) and delthyrium; ornament of concentric rugae only posteriorly, then costate, rugose, and moderately reticulate, costae coarse or obsolete anteriorly, spines fine, scattered over shell, row along hinge margin and group on ears; cardinal process trilobed? and dorsally recurved at angle of 90 degrees, breviseptum present, dorsal adductors smooth on club-shaped ridges. *Miss.* or *Penn.* (Cameron Creek F.), USA(Mont.). (?Dictyoclostidae, Subfamily Uncertain, p. *H500*.) [MUIR-WOOD.]

Scutepustula SARYCHEVA, 1963, p. 165 [**Productus (Waagenoconcha) scutelatus* BALASHOVA, 1955, p. 146; OD]. Shell thin, medium-sized, outline rounded, plano-convex; valves ornamented by prominent rugae bearing single row of very fine prostrate spines, appearing like capillation; cardinal process small, trilobate, geniculated, and curving dorsally, median septum 2/3 length of valve, supporting cardinal process, lateral ridges long, straight, brachial ridges not observed, adductor scars obscurely dendritic. *L.Carb.(Tournais.-L. Visean)*, W. Eu. (USSR) - Asia (Kuznetsk Basin); *Miss.*, N.Am. (Overtoniidae, Overtoniinae, p. *H474*.) [MUIR-WOOD.]

Stelckia CRICKMAY, 1963, p. 21 [**S. galerius*; OD]. Medium-sized, concavo-convex, adult shell trigonal in outline, greatest width along hinge, interareas narrow; exterior rather smooth, ill-defined costellae, growth lines, and spines mainly on pedicle valve along hinge and on ears, on median ridge, rare elsewhere; interior as in *Productella* with larger cardinal process and stronger breviseptum.

M.Dev., W.Can. (Productellidae, Productellinae, p. *H*465.) [MUIR-WOOD.]

Tomilia SARYCHEVA, 1963, p. 220 [**T. khalfini;* OD]. Shell massive, medium-sized, elongate, both valves rounded-geniculate, ears small, medium sulcus in some specimens, flanks almost parallel; both valves irregularly costellate with bifurcations and intercalations, obscure rugae posteriorly, more numerous on brachial valve, spines fine, scattered, 2 rows along hinge, grouped on ears, less numerous on brachial valve; cardinal process massive, trilobed, with median lobe dorsally directed, swollen, bifurcating in some, base of median septum supporting cardinal process, then contracting to low ridge 2/3 length of valve, lateral ridges short tapering, brachial ridges given off almost horizontally, adductor scars dendritic. *L.Carb.(Visean),* Asia(Sib., Kuznetsk Basin). (Buxtoniidae, Buxtoniinae, p. *H*492.) [MUIR-WOOD.]

?**Tomiproductus** SARYCHEVA, 1963, p. 201 [**Productus elegantulus* TOLMACHEV, 1924, p. 244; OD]. Shell thin, small, elongate, both valves rounded-geniculate; capillate or costellate, rugae on ears of pedicle valve, well developed on brachial valve, spines curving, scattered, and in row along hinge and ears, less numerous on brachial valve; cardinal process small, bilobate, 2 parallel vertical buttress plates extend from cardinal process base, breviseptum thin, inserted between plates and extending half valve length; breviseptum and buttress plates may fuse as in *Buxtonia,* lateral ridges diverging slightly from margin. *L.Carb. (Tournais.),* Asia (Kuznetsk Basin-?Kazakhstan-?Taimyr Penin.)-?W.Eu.; *?Miss.,* USA. (Buxtoniidae, Buxtoniinae, p. *H*492.) [MUIR-WOOD.]

Triadithyris DAGIS, 1963, p. 187 [**Terebratula gregariaeformis* ZUGMAYER, 1882, p. 13; OD]. Small, pentagonal, valves moderately biconvex, strongly biplicate on anterior half, anterior commissure sulciplicate; umbo suberect, foramen rounded, permesothyridid, pedicle collar present; cardinal process rather prominent, bilobed, medially depressed, hinge plates short, nearly horizontal, well demarcated from massive inner socket ridges, loop about 0.5 valve length, ventrally curved transverse band, adductors pear-shaped, median septum and dental plates absent. *U.Trias. (Rhaet.),* Eu.(Alps-Carpathians-Crimea)-Asia(Caucasus-Pamirs). (Terebratulidae, Terebratulinae, p. *H*789.) [MUIR-WOOD.]

Tulcumbella CAMPBELL, 1963, p. 68 [**T. microstriata;* OD]. Small, convexo-concave, or pedicle valve plane, interareas very low, chilidium and pseudodeltidium developed; valves capillate, capillae of uniform width, increasing by bifurcations and intercalations, vertical spine row along hinge; pedicle valve with very short median septum, brachial valve with trilobate cardinal process, having high medially cleft median lobe and 2 lower lateral lobes, median septum low, socket ridges thin, parallel to hinge. *L.Carb.(Tournais.),* Australia (New S. Wales). (Chonostrophiidae, p. *H*434.) [MUIR-WOOD.]

REFERENCES

Berdan, J. M.
(1) 1963, *Eccentricosta, a new Upper Silurian brachiopod genus:* Jour. Paleontology, v. 37, p. 254-256, 1 text-fig.

Campbell, K. S. W., & Engel, B. A.
(2) 1963, *The faunas of the Tournaisian Tulcumba Sandstone and its members in the Werrie and Belvue Synclines, New South Wales:* Geol. Soc. Australia, Jour., v. 10, pt. 1, p. 55-122, pl. 1-9, 11 text-fig.

Chan, L. P., & Li, Li
(3) 1962, *Chin Lin clong duan zao er dion shi Mao Kou su wan zu lei hua shi:* Acta Palaeont. Sinica, v. 10, p. 472-501, 4 pl. [*Lower Permian brachiopods from the Mao-Kou suite of the eastern part of Chin Lin.*]

Crickmay, C. H.
(4) 1963, *Significant new Devonian brachiopods from Western Canada:* 63 p., 16 pl., Evelyn de Mille Books (Calgary).

Easton, W. H.
(5) 1962, *Carboniferous formations and faunas of central Montana:* U.S. Geol. Survey, Prof. Paper 348, 126 p., 13 pl., strat. secs., 1 text-fig.

Farag, I. A. M., & Gatinaud, W.
(6) 1960, *Un nouveau genre de Térébratulidés dans le Bathonien d'Egypte:* Jour. Geology United Arab Republic, v. 4, no. 1, p. 77-79, pl. 1.

Lane, B. O.
(7) 1962, *The fauna of the Ely Group in the Illipah area of Nevada:* Jour. Paleontology, v. 36, p. 888-911, 4 pl.

Likharev, B. K.
(8) 1934, *Die Fauna der permischen Ablagerungen des Kolyma-Gebietes:* Akad. Nauk SSSR, Trudy, Soveta po lzucheniyu Proizvoditelnykh Sil, Yakutskaya Ser., no. 14, 148 p., 11 pl. (In Russian, German, p. 98-136) (Kolyma-Gebiet Geol. Exped. 1929-30, v. 1, pt. 2).

Ustritsky, V. I.
(9) 1963, in USTRITSKY, V. I. & CHERNIAK, G. E., *Biostratigrafiia i Brakhiopody Verkhnogo Paleozoiia Taimyra:* Nachno-issledovatelskii institut geologii Arktiki, Trudy, v. 134, p. 139. [*Biostratigraphy and brachiopods of the Upper Paleozoic Taimyr.*]

Weller, Stuart
(10) 1900, *A preliminary report on the stratigraphic paleontology of Walpack Ridge, in Sussex Co., New Jersey:* New Jersey Geol. Survey, Ann. Rept. for 1899, p. 1-46.

INDEX

Italicized names in the following index are considered to be invalid; those printed in roman type, including morphological terms, are accepted as valid. Suprafamilial names are distinguished by the use of full capitals and authors' names are set in small capitals with an initial large capital. Page references having chief importance are in boldface type (as **H332**).

Typeset by the University of Kansas Printing Service, Lawrence, Kansas
Printed in U.S.A. by Malloy Lithographing, Inc., Ann Arbor, Michigan